Proofs

A Long-Form Mathematics Textbook

JAY CUMMINGS

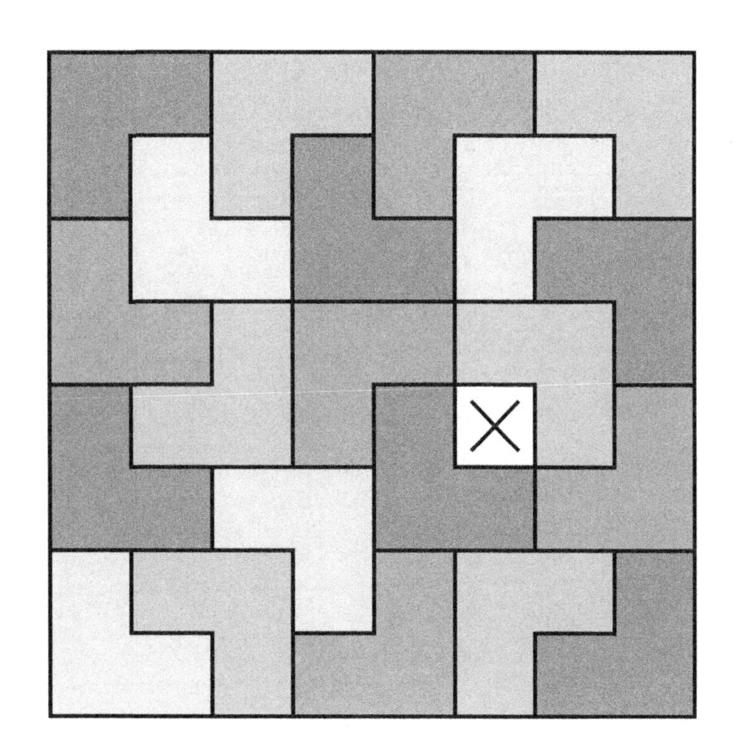

To my loving wife
who read this entire book
except the math parts.

Editorial Board

Contents

1 Intuitive Proofs **1**

1.1 Chessboard Problems 2
1.2 Naming Results 10
1.3 The Pigeonhole Principle 13
1.4 Bonus Examples 23
Exercises 31

Introduction to Ramsey Theory 41

2 Direct Proofs **47**

2.1 Working From Definitions 48
2.2 Proofs by Cases 53
2.3 Divisibility 54
2.4 Greatest Common Divisors 59
2.5 Modular Arithmetic 62
2.6 Bonus Examples 74
Exercises 81

Introduction to Number Theory 89

3 Sets **97**

3.1 Definitions 97
3.2 Proving $A \subseteq B$ 101
3.3 Proving $A = B$ 105
3.4 Set Operations 106
3.5 Bonus Examples 119
Exercises 125

Introduction to Topology 137

4 Induction **147**

4.1 Dominoes, Ladders and Chips 147
4.2 Examples 149
4.3 Strong Induction 164
4.4 Non-Examples 173

4.5 Bonus Examples . . . 175
Exercises . . . 188

Introduction to Sequences . . . 199

5 Logic **207**
5.1 Statements . . . 207
5.2 Truth Tables . . . 214
5.3 Quantifiers and Negations . . . 219
5.4 Proving Quantified Statements . . . 227
5.5 Paradoxes . . . 228
5.6 Bonus Examples . . . 232
Exercises . . . 242

Introduction to Real Analysis . . . 253

6 The Contrapositive **261**
6.1 Finding the Contrapositive of a Statement . . . 263
6.2 Proofs Using the Contrapositive . . . 264
6.3 Counterexamples . . . 269
6.4 Bonus Examples . . . 273
Exercises . . . 278

Introduction to Big Data . . . 285

7 Contradiction **293**
7.1 Two Warm-Up Examples . . . 295
7.2 Examples . . . 298
7.3 The Most Famous Proof in History . . . 300
7.4 The Pythagoreans . . . 304
7.5 Bonus Examples . . . 311
Exercises . . . 320

Introduction to Game Theory . . . 325

8 Functions **331**
8.1 Approaching Functions . . . 331
8.2 Injections, Surjections and Bijections . . . 335
8.3 The Composition . . . 347
8.4 Invertibility . . . 352
8.5 Bonus Examples . . . 356
Exercises . . . 362

Introduction to Cardinality . . . 371

9 Relations **379**
9.1 Equivalence Relations 379
9.2 Abstraction and Generalization 391
9.3 Bonus Examples 395
Exercises 402

Introduction to Group Theory 413

Appendices **421**

A Other Proof Methods **423**
A.1 Probabilistic Method 424
A.2 Linear Algebra Method 429
A.3 Combinatorial Method 434
A.4 Computer-Assisted Proofs 440
A.5 Proofs by Picture 448

B Proofs From The Book **453**
B.1 Merry Madness from March 454
B.2 Significant Sets of Shifting Shapes 455
B.3 A Flow of Factors From Fermat 460
B.4 A Pinpointed Proof Pausing Prussian Parades 462
B.5 Cleverly Cutting the Cruising Coins 464
B.6 An Antisocial Ant Avalanche 468
B.7 A Pack of Pretty (Book) Proofs by Picture 472
B.8 An Image's Insightful Illusion 476
B.9 Monotone Marches through Muddled Marks 479
B.10 Zigging Zeniths and Zagging Zones 484

C Writing Advice **487**
C.1 Writing Proofs 487
C.2 Writing in LaTeX 493

Index **497**

Chapter 1: Intuitive Proofs

> The History of every major Galactic Civilization tends to pass through three distinct and recognizable phases, those of Survival, Inquiry and Sophistication, otherwise known as the How, Why, and Where phases. For instance, the first phase is characterized by the question "How can we eat?" the second by the question "Why do we eat?"'and the third by the question "Where shall we have lunch?"
>
> – Douglas Adams, *The Hitchhiker's Guide to the Galaxy*

Your mathematics education will also pass through these phases. The first phase is characterized by the question, "How can I solve this integral?" The second phase by the question, "Why does my solution work?" And the third phase by the question, "Where should we explore next?"

Indeed, your previous "Phase 1" math classes were probably focused on computation, like how to use the fundamental theorem of calculus to solve a problem. In your future "Phase 2" math classes, you will seek to understand *why* the fundamental theorem of calculus is true.[1] In your earlier "Phase 1" math courses, you were taught to use the quadratic formula to solve second-degree polynomials. In your future "Phase 2" math courses, you will learn that there are similar formulas for third- and fourth-degree polynomials... but not for fifth-degree polynomials — and you will see precisely why that is the case. Your future courses will also introduce you to lots of topics that did not appear at all in your previous courses. In fact, I think the most interesting math topics are saved entirely for second-phase courses — so you have a lot to look forward to!

This book is the gateway to Phase 2. It will show you the techniques mathematicians use to understand our math (which we call *proof techniques*), and it will introduce you to new math topics that you will explore in detail in your future courses. So buckle up, because math is about to get a lot more interesting.

[1]When your curiosities guide you to seek out new math and pursue your own original ideas — perhaps by engaging in mathematical research — you will enter Phase 3. Much more on this later!

Our first topic is one that mathematicians genuinely care about, but which right now might feel more like a puzzle than a math problem to you. In doing so, it highlights the idea that math is no longer just about numbers and computations, but about *ideas*. Let's talk about chessboard problems.

1.1 Chessboard Problems

Suppose you have a chessboard (8×8 grid of squares) and a bunch of dominoes (2×1 block of squares), so each domino can perfectly cover two squares of the chessboard.[2]

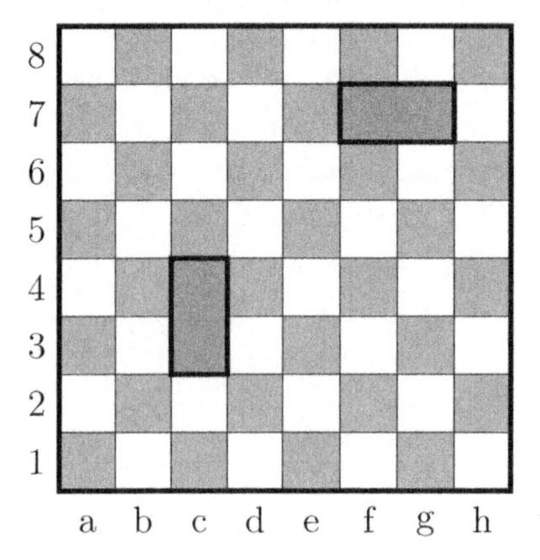

Note that with 32 dominoes you can cover all 64 squares of the chessboard. There are many different ways you can place the dominoes to do this, but one way is to cover the first column by 4 dominoes end-to-end, cover the second column by 4 dominoes, and so on.

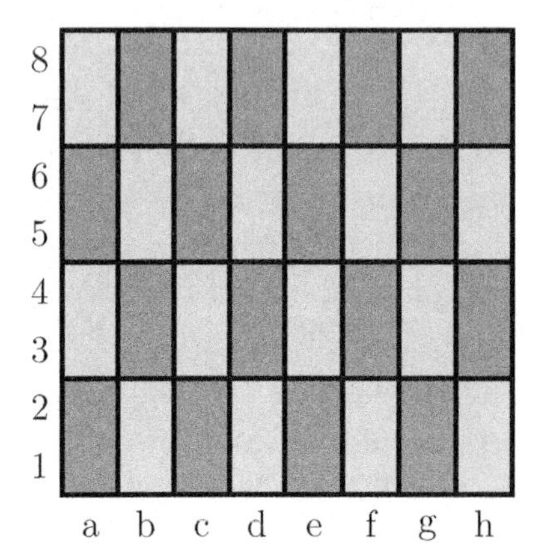

[2]Note: Along the left and bottom edges of the chessboard are numbers and letters. They are there simply to label the rows and the columns.

Of course, that's not the only way. Here's a nifty way to cover all the squares:

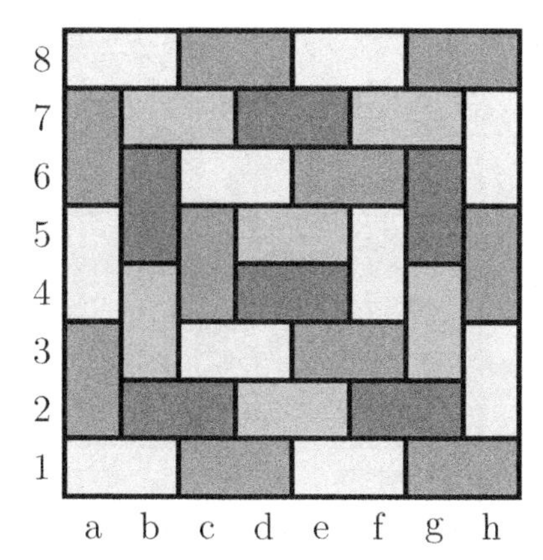

Math runs on definitions, so let's give a name to this idea of covering all the squares. Moreover, let's define it not just for 8×8 boards — let's allow the definition to apply to boards of other dimensions.

Definition.

Definition 1.1. A *perfect cover* of an $m \times n$ board with 2×1 dominoes is an arrangement of those dominoes on the chessboard with no squares left uncovered, and no dominoes stacked or left hanging off the end.

As the last two pictures explicitly showed, there do exist perfect covers of the 8×8 chessboard. This is a book about *proofs*, so let's write down our discovery as a proposition (something which is true and requires proof) and then let's write out a formal proof of this fact.

Proposition.

Proposition 1.2. There exists a perfect cover of an 8×8 chessboard.

Before most proofs, we will discuss some of the proof's key ingredients or ideas.

Proof Idea. This proposition is asserting that "there exists" a perfect cover. To say "there exists" something means that there is at least one example of it. Therefore, any proposition like this can be proven by simply presenting an example which satisfies the statement.

Proof. Observe that the following is a perfect cover.

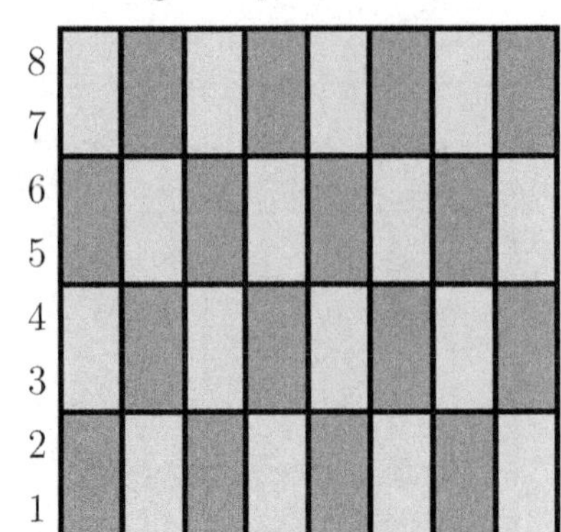

We have shown by example that a perfect cover exists, completing the proof. □

We typically put a small box at the end of a proof, indicating that we have completed our argument. Thus, this paragraph is not part of the proof. The practice of drawing a box to conclude a proof was brought into mathematics by Paul Halmos, and it is sometimes called the *Halmos tombstone.*[3]

We have seen two different perfect covers of the (8×8) chessboard. How many are there in total? This is a very hard question, but mathematicians have found the surprisingly large answer: there are exactly 12,988,816 perfect covers. This was discovered in 1961, long before modern computers could by brute force the answer.[4]

Getting back to perfect coverings, we proved that a standard 8×8 chessboard can be perfectly covered by dominoes. What if I cross out the bottom-left and top-left squares? Can we still perfectly cover the 62 remaining squares?

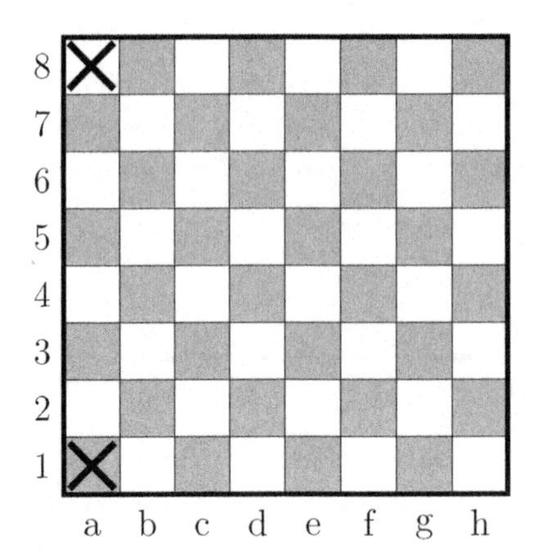

[3]One apocryphal story: Halmos regarded mathematical claims as "living" until they were proven. Once proved, they were defeated — killed. So he wrote a little tombstone to conclude his proofs.

[4]In fact, in that 1961 paper by Temperley & Fisher (and independently by Kasteleyn), they showed that the answer for a general $m \times n$ board is this crazy thing:

$$\prod_{j=1}^{\lceil \frac{m}{2} \rceil} \prod_{k=1}^{\lceil \frac{n}{2} \rceil} \left(4\cos^2\left(\frac{\pi j}{m+1}\right) + 4\cos^2\left(\frac{\pi k}{n+1}\right) \right).$$

Perhaps you can see that the answer is yes. For example, the first column can now be covered by 3 dominoes and the other columns can be covered by 4 dominoes each.

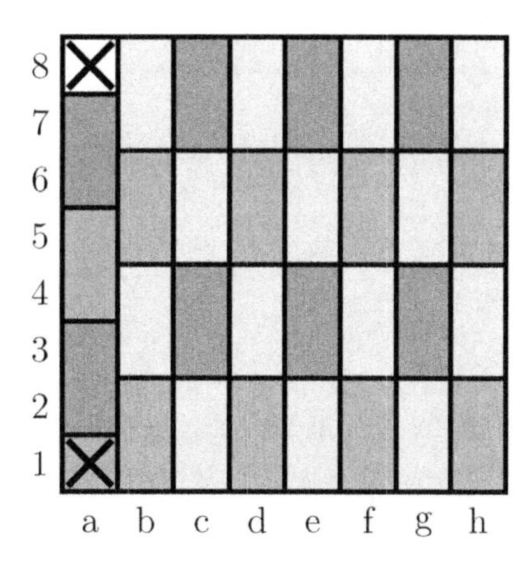

What if I cross out just one square, like the top-left square?

Can this new board be perfectly covered? This is a good opportunity to mention how important it is to reason through explanations at your own pace, and to try to solve things on your own before reading the explanations here. Doing so will deepen your understanding immensely. So, on that note, take a moment and come up with an answer before reading on.

⋮

...Ok, hopefully you did so! The answer is no...Do you see why? Hint: Think about *parity* — meaning, evenness vs. oddness. Try to convince yourself of the answer before moving on...

Let's again write this out formally as a proposition, and then include a proof of it.

Proposition.

Proposition 1.3. If one crosses out the top-left square of an 8×8 chessboard, the remaining squares cannot be perfectly covered by dominoes.

Once again, we begin with a "Proof Idea" section in which we discuss the central ideas in a more casual way. This is for intuition.

Proof Idea. The idea behind this proof is that one domino, wherever it is placed, covers two squares. And two dominoes must cover four squares. And three dominoes cover six squares. In general, the number of squares covered — 2, 4, 6, 8, 10, etc. — is always an *even* number. This insight is the key, because the number of squares left on this chessboard is 63 — an odd number. Ok, now here is the proof.[5]

Proof. Since each domino covers 2 squares and the dominoes are non-overlapping, if one places our k dominoes on the board, then they will cover $2k$ squares, which is always an even number.[6] Therefore, a perfect cover can only cover an *even* number of squares. Notice, though, that the board has 63 remaining squares, which is an *odd* number. Thus, it cannot be perfectly covered. □

Makes sense? One can never cover an odd number of squares, because any collection of dominoes can only cover an even number of squares. This reasoning is what prevents the existence of a perfect cover. Neat!

What if I take an 8×8 chessboard and cross out the top-left and the bottom-right squares? Then can it be covered by dominoes?

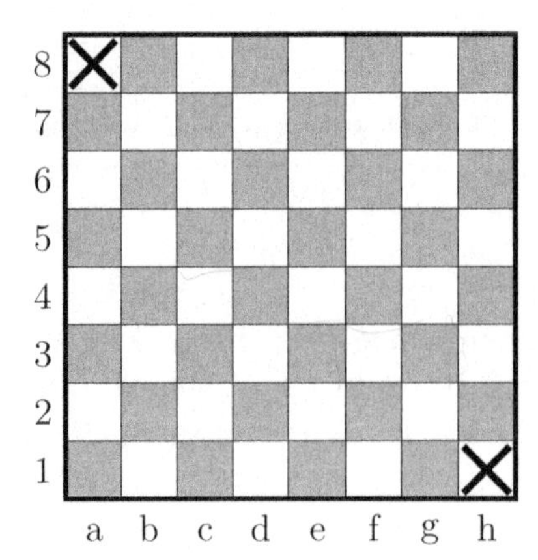

[5]While a "proof idea" may include the essence of why a proposition is true, a *proof* is more formal and thorough. Notions of formality and thoroughness are subjective, and it will take time to understand what level of rigor is required. We will discuss this much more throughout this book.

[6]Note: Since k is the number of dominoes, k must be a positive integer. We will be more formal about this beginning in Chapter 2, but the *integers* are these numbers: $\dots, -3, -2, -1, 0, 1, 2, 3, \dots$; that is, we include positive numbers, negative numbers and zero, but not numbers like 2.4. The *positive integers* are thus these numbers: $1, 2, 3, 4, \dots$.

We are back to an even number of squares, so there's no problem there. I encourage you to draw the board on some scratch paper and give it a try. See if you can find a perfect cover or discover a reason why one does not exist.

If you get stuck, another way to approach a problem like this is to try a smaller example;[7] this board has 62 squares and so would require 31 dominoes, which is quite a lot. Oftentimes a problem is too big to tackle as it is, but a smaller case will help get your brain cells firing. To this end, maybe you could try a 6×6 board with the top-left and bottom-right squares crossed out? Or, maybe even a 4×4 board? Give it a shot before moving on! Remember, learning math is an active endeavor!

In fact, in case it helps, here is a 4×4 board for you to work on:

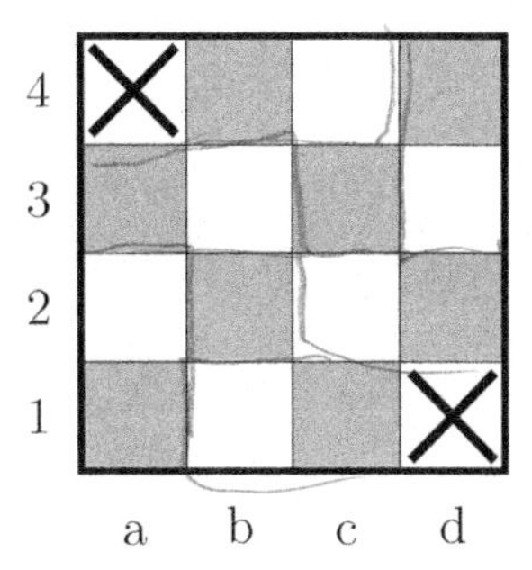

...I hope by now you have tried it on your own. If so, you probably got stuck. Indeed, no perfect cover exists. Did the small cases give you any intuition for why your attempts failed?[9] There is a really slick way to see it, which is contained in the proof below.

Proposition.

Proposition 1.4. If one crosses out the top-left and bottom-right squares of an 8×8 chessboard, the remaining squares cannot be perfectly covered by dominoes.

Proof. Observe that the chessboard has 62 remaining squares, and since every domino covers two squares, if a perfect cover did exist it would require

$$\frac{62}{2} = 31 \text{ dominoes.}$$

Also observe that every domino on the chessboard covers exactly one white square

[7] Something I learned in grad school: Even the best mathematicians do this. Because it works.[8]

[8] P.S. This works in life, too. Problem solving skills you learn in math class can have real applications beyond your coursework.

[9] The great Henri Poincaré said, "It is by logic we prove. It is by intuition we discover." An important aspect of learning math is fine-tuning your intuition. Proofs run on logic, but you will *discover* how to prove many things by following your intuition.

and exactly one black square. Two examples are shown here:

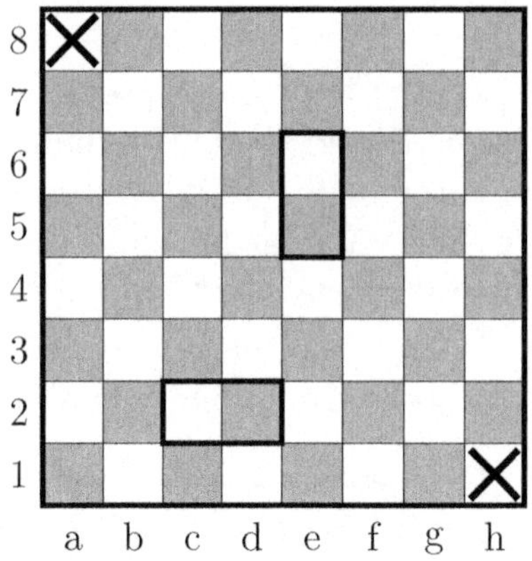

Thus, whenever you place 31 non-overlapping dominoes on a chessboard, they will collectively cover 31 white squares and 31 black squares.

Next, observe that since both of the crossed-out squares are white squares, the remaining squares consist of 30 white squares and 32 black squares. Therefore, it is impossible to have 31 dominoes cover these 62 squares. □

Did the proof make sense? We showed that any perfect cover using 31 dominoes must cover 31 white squares and 31 black squares. And since our chessboard has 30 white squares and 32 black squares, no perfect cover is possible.[10]

We also used a picture within our proof. Pictures can help the reader, but you must also be careful that your picture is not too simplistic and misses special cases. A good rule of thumb is that you want your proof to be 100% complete without the picture; the picture illustrates your words, but should not replace your words.

For many of you, your earlier math courses proceeded like this: You were introduced to a new type of problem, you learned The Way to solve those problems, you did a dozen similar problems on homework, and then if a similar problem was on your exam, you repeated The Way one more time.

Beginning now, this paradigm will begin to shift. This shift will not be abrupt, because there are many new skills which will require practice, but you will notice a change. In calculus, if two students submitted full-credit solutions, then it is likely their work looks very similar. For proofs, this is less likely.

Furthermore, when learning new ideas, it helps to think about them from multiple angles. For example, below is a slightly different method to prove Proposition 1.4.

- Assume you do have a perfect cover and think about placing dominoes on the board one at a time.
- At the start there are 62 squares — 32 black squares and 30 white squares.

[10] A common mistake after reading Proposition 1.3 is to assume that the *only* way to prevent perfect covers is by having an odd number of squares, and that as long as you have an even number there must be perfect covers. Proposition 1.4 shows that this is not the case. Perfect covers could be excluded for other reasons, too.

- After placing the first domino, no matter where it's placed, there will be 31 black squares and 29 white squares left.
- After placing the second domino, no matter where it's placed, there will be 30 black squares and 28 white squares left.
- After placing the third domino, no matter where it's placed, there will be 29 black squares and 27 white squares left.

$\vdots$

- After placing the 30th domino, no matter where it's placed, there will be 2 black squares and 0 white squares left.
- But since every domino must cover up 1 black square and 1 white square, and there are only 2 black squares to go, the final domino cannot possibly be placed.

The central idea is the same as in our earlier proof, but their presentations are different. In other cases, two different proofs will rely on two different central ideas.

– Asking Questions –

Earlier we asked whether removing the top-left and bottom-right squares of a chessboard prevents a perfect covering, and the answer was interesting. While good mathematicians can answer interesting questions, great mathematicians can *ask* interesting questions. Take a moment to look back at the propositions we have proven thus far, and see if you can come up with other interesting questions which one might ask. And only after doing so, take a look at a few which I included below.

Question 1: If I remove two squares of different colors from an 8×8 chessboard, must the result have a perfect cover?

Question 2: If I remove four squares—two black, two white—from an 8×8 chessboard, must the result have a perfect cover?

Question 3: For every pair of positive integers, m and n, does there exist a perfect cover of the $m \times n$ chessboard by 2×1 dominoes? If not, for which m and n is there a perfect cover?

These questions are asked of you in the Chapter 1 exercises. Other, more challenging questions include: How many ways can one cover the $m \times n$ chessboard with 2×1 dominoes? What if I change the domino to be another shape, and then ask all these same questions again? Can we generalize these questions to higher dimensions?[11] What does the image on this book's cover have to do with all this?

[11] A good math problem is like whatever a fun version of the Hydra monster is. You solve one problem, and three more appear in its place! The number of unsolved math problems is steadily increasing because of this. Indeed, pick up a math research paper and you will likely find more questions asked than answered. This provides wonderful job security for us academics.

1.2 Naming Results

So far, all of our results have been called "propositions." Here's the run-down on the naming of results:

- A *theorem* is an important result[12] that has been proved.
- A *proposition* is a result that is less important than a theorem. It has also been proved.
- A *lemma* is typically a small result that is proved before a proposition or a theorem, and is used to prove the following proposition or theorem.[13]
- A *corollary* is a result that is proved after a proposition or a theorem, and which follows quickly from the proposition or theorem. It is often a special case of the proposition or theorem.

All of the above are results that have been proved — a *conjecture*, though, has not.

- A *conjecture* is a statement that someone guesses to be true, although they are not yet able to prove it or disprove it.

As an example of a conjecture, suppose you were investigating how many regions are formed if one places n dots randomly on a circle and then connects them with lines.

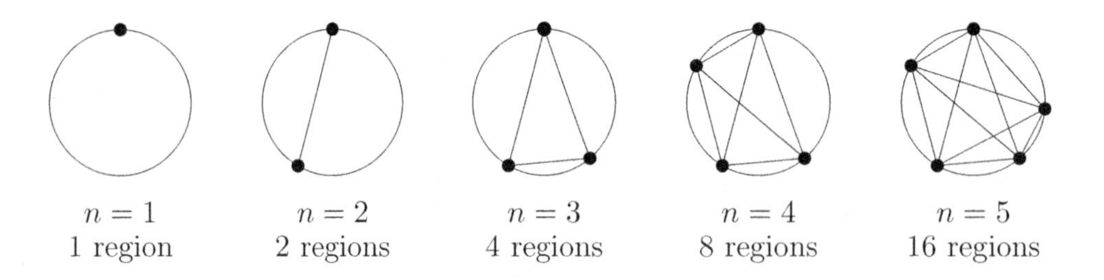

At this point, if you were to conjecture how many regions there will be for the $n = 6$ case, your guess would probably be 32 regions — the number of regions certainly seems to be doubling at every step. In fact, if it kept doubling, then with a little more thought you might even conjecture a general answer: that n randomly placed dots form 2^{n-1} regions; for example, the $n = 4$ case did indeed produce $2^{4-1} = 2^3 = 8$ regions.

If I saw such a conjecture, I know I'd be tempted to believe it! Yet, surprisingly,

[12] By "result" we mean a sentence or mathematical expression that is true. We will discuss this in much more detail in Chapter 5.

[13] It's like it's saying *"Yo, lemma help you prove that theorem."*

this conjecture would be incorrect. One way to disprove a conjecture is to find a *counterexample* to it. And as it turns out, the $n = 6$ case is such a counterexample:

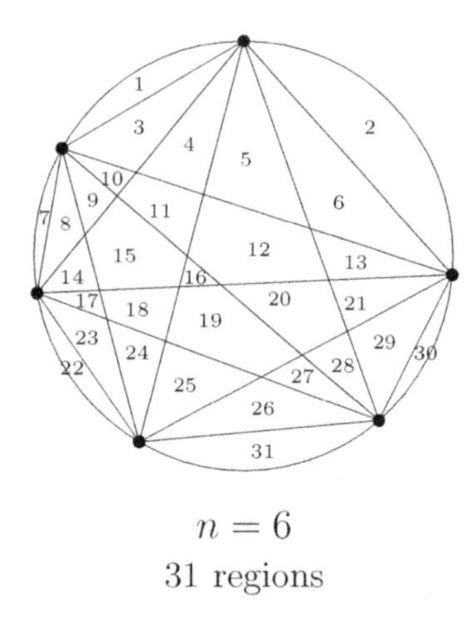

$n = 6$
31 regions

This counterexample also underscores the reason why we prove things in math. Sometimes math is surprising.[14] We need proofs to ensure that we aren't just guessing at what seems reasonable. Proofs ensure we are always on solid ground.[15]

Furthermore, proofs help us understand *why* something is true — and that understanding is what makes math so fun. When I showed you the chessboard with the upper-left and bottom-right squares removed, if I immediately told you that it is impossible to perfectly cover it with 31 dominoes, then you might not have found the result very interesting (especially if I said the reason why is because a computer just ran through all the cases and none worked). But when you understood precisely *why* such a tiling was impossible by counting white and black squares, I hope you found it much more interesting and insightful.

Lastly, we study proofs because they are what mathematicians do, and one goal of this book is to teach you how to think and act like a mathematician.[16] What else does this book aim to teach you? I'm glad you asked:

Textbook Goal. Develop the skills to read and analyze mathematical statements, learn techniques to prove or disprove such statements, and improve one's ability to communicate mathematics clearly. It also aims to give you a taste of the different areas of math, and show what it is like to be a mathematician by learning some of our discipline's practices, culture, history and quirks.

There is another set of goals that has to come from you. To go beyond rote learning — to really understand mathematics — requires you to struggle with the material. As you are introduced to a proof, I hope you do not just passively read it without challenging yourself to figure out portions on your own. I encourage you to

[14] It looked like 2^{n-1} was going to be the fomula. Actual formula: $\frac{1}{24}(n^4 - 6n^3 + 23n^2 - 18n + 24)$.

[15] Conjecture: All positive integers are smaller than a trillion. Computer: I've tested the first billion cases, and they all check out. Looks true to me, mate!

[16] And if you are using this book in a course, then there's one final reason: *It's on the test!*

work through plenty of exercises, to read extra proofs on your own, and to organize study groups to discuss the material with others. Challenge yourself and you will grow faster. These are the soft skills that only you can instill, and I hope you put in the work to do so.

Why do we prove things?

Now that you know the goal of this book, and have seen the first examples of proofs, it is reasonable to ask *why* we prove things. When you took algebra and geometry, you learned rules about solving equations and investigating geometric shapes. When you took linear algebra and differential equations, you learned deep results about systems of linear and differential equations. These results are really big deals, but they are not handed down to us from on high — mathematicians had to discover them.[17] Meanwhile, mathematicians also have discovered things that seemed true but are not. How do we distinguish truth from lies? We use proofs.

In some sense, a proof is like a computer program. The program has to compile. It has to run. It must be logically sound in order for it to work. But a proof is also much more than that. It has to convince. We write proofs to communicate with each other, to justify our thoughts, and to know that we and others are correct. Thus, we must write proofs which are comprehensible and persuasive. Your proofs should convince your readers that you are correct.[18]

It is a wonderful thing about mathematics that we can *know* things to be true. A physicist or psychologist relies on theories that they can test, but without an axiomatic basis like in math, they can't know whether they have reached truth.

Proofs also help us not just know math, but *understand* math. They help show *why* something is true. This is the "Phase 2" mathematics we discussed on this book's first page. In calculus and pre-calculus, you learned mathematical rules and procedures. Each of those has a proof which can be verified and understood. A few of them will be proven in this book (like the Pythagorean theorem), while many more will be shown in your later courses.

You may have been taught that there are infinitely many primes, and that every positive integer (larger than 1) can be broken down as a product of prime numbers. How would you *know* that there are infinitely many primes without a proof? No experiment could verify such a thing. You may have heard that irrational numbers exist, and every integer can be written in binary. Perhaps you have heard that there is a deep result from number theory that is used to securely encrypt our digital transactions. Everything mentioned in this paragraph will be explained and proved later in this book.

There is also so much more! We do not want to just verify that the boring calculations from pre-calc were correct — we want to explore new math, and prove that our discoveries are true. This is not just the beauty of math, it is the fun of math. We continue this endeavor with our second topic: the pigeonhole principle.

[17] Possible exception: Ramanujan.

[18] Also, computer programs can have bugs. Especially long programs. We insist on rigor in our proofs to prevent bugs in our proofs. Especially our long proofs.

1.3 The Pigeonhole Principle

Let's warm up with the following fact from the real world.

Proposition. There are 3 non-balding people in Sacramento, CA, who have *exactly* the same number of hairs on their head.

We will prove this using what is called *the pigeonhole principle*. This principle is fascinating because while it is clearly true, it has some remarkable consequences. Its name comes from a simple, real-world observation: If 6 pigeons live in just 5 pigeonholes, then at least one pigeonhole must have at least two pigeons living in it.

Likewise, if 11 or more pigeons are living in these 5 pigeonholes, then at least one pigeonhole has at least 3 pigeons living in it.[19]

Said in complete generality: If at least $kn + 1$ pigeons live in n pigeonholes, then at least one pigeonhole has at least $k + 1$ pigeons living in it.

The true power of the pigeonhole principle, is that it works for more than just pigeons![20] Let's now talk about the hairs on the heads of Sacramentans. Our proof will rely on a few real-world facts and a definition.

- The average person has between 100,000 and 150,000 hairs on their head, and essentially everyone has under 200,000 hairs. So we will focus on Sacramentans with at most 199,999 hairs.[21]

- For the sake of this problem we'll define "non-balding" to mean they have at least 50,000 strands of hair (what we choose doesn't change things much).

- There are 480,000 people in Sacramento. A quick search online shows that certainly less than 100,000 Sacramentans are balding. Therefore, a conservative estimate gives at least 380,000 non-balding Sacramentans.

***Proof*.** By the above facts, there are at least 380,000 non-balding Sacramentans. These are our "pigeons." What are our pigeonholes?

For each number between 50,000 and 199,999, imagine a box with that number written on it.

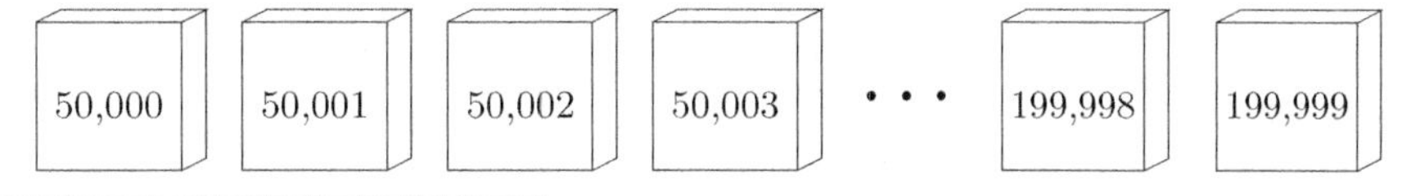

[19](Foot)Note: The most "balanced" case is if you have two pigeons living in each of the pigeonholes, except one pigeonhole has three pigeons living in it. But we do not require this! Perhaps they are all living in just one pigeonhole, or the breakdown is 3-4-0-2-2. In all these situations, at least one of the pigeonholes does indeed have at least three pigeons living in it.

[20]When my undergraduate combinatorics professor, Christine Kelley, told our class about the pigeonhole principle, she made this joke. I thought it was hilarious and have not forgotten it.

[21]Note that "at most 199,999 hairs" means "199,999 or fewer hairs." Or, said more math-y: "if n is the number of hairs, then $n \leq 199,999$." Likewise, "at least 5" means "5 or more." The phrases "at most" and "at least" can be confusing when you first hear them, but they are used a lot in math.

There are 150,000 boxes, and each of these becomes a "pigeonhole." If Sophie Germain has 122,537 hairs on her head, then write her name on a piece of paper and drop it into the box with the number 122,537 written on it.

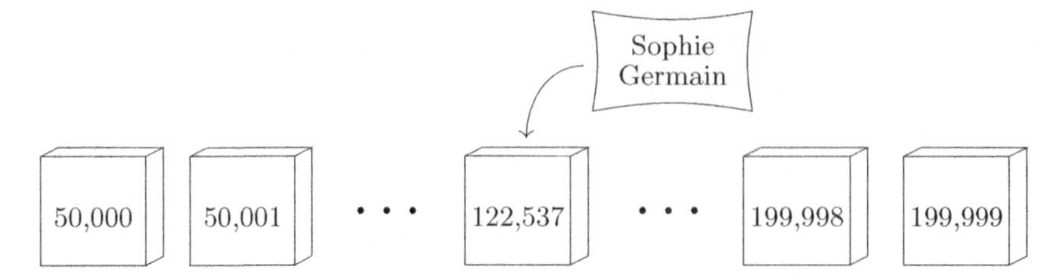

If Chris Webber has 101,230 hairs on his head, then place his name into the box with 101,230 written on it. Do this for every one of the 380,000 non-balding Sacramentans.

In the end, we have 380,000 names to put in just 150,000 boxes. So certainly (or by the pigeonhole principle) there must be at least two names in one of the boxes. These two people — being in the same box — must have the same number of hairs.

Moreover, if there were exactly two names in each box, that would be 300,000 names. But we have 380,000 names! With these extra 80,000 names to place, there must be a box with at least three names in it. Indeed, the pigeonhole principle tells us that if there are more than twice as many names as boxes, then there must be a box with at least three names in it. This proves the proposition. □

Note that with 300,000 people, it is extremely likely that three people have exactly the same number hairs on their heads — it would be remarkable for every number of hairs to have exactly two people with that number. But it is not until 300,001 people that it is *guaranteed* that three people have the same number.

Now, it is rare for a proof in math to rely on real-world data like the number of hairs on a human's head, but I think this is a fun example to introduce this important mathematical principle which is the focus of the rest of this chapter. Let's first restate the pigeonhole principle using the more common objects/boxes phrasing than the antiquated pigeon/pigeonhole phrasing.

Principle.

Principle 1.5 (*The pigeonhole principle*). The principle has a simple form and a general form. Assume k and n are positive integers.[22]

Simple form: If $n+1$ objects are placed into n boxes, then at least one box has at least two objects in it.

General form: If $kn+1$ objects are placed into n boxes, then at least one box has at least $k+1$ objects in it.

[22] Reminder: The positive integers are these numbers: $1, 2, 3, 4, \ldots$.

This principle makes use of variables. When possible, I find it helpful to plug in some specific values for those variables to better understand what it is saying. For example, you could plug in $k = 1$ and $n = 4$, or $k = 3$ and $n = 2$. After some specific cases make sense, you can begin to make sense of the general case.

Next, let's take a look at some basic examples of the pigeonhole principle, beginning with the one we already proved.

Example 1.6.

- There are 3 non-balding Sacramentans who have exactly the same number of hairs on their head.
- Among any 5 playing cards, there are at least two cards of the same suit.
- Among any 37 people, at least 4 must have the same birthmonth.

Notice that these are asserting how many are needed to *guarantee* that the property holds. With just four people, it is *possible* they all have the same birthmonth, but it is not until the 37th person that we are *guaranteed* such a quadruple.

Likewise, it takes 367 people to guarantee that two of them have the same birthday. But just as a quick fun fact, how many people do you think you need to have a 50% chance that two have the same birthday? Maybe $\frac{367}{2}$? The answer is remarkably few... you only need 23 people! With 23 random people, the odds that two have the same birthday is 51%. And the reason why is purely mathematical;[23] look up *the birthday problem* for the deets.

Let's discuss some more examples of the pigeonhole principle.

Example 1.7. You just washed n pairs of socks ($2n$ individuals), and suppose each pair is a different color than the other pairs. If you pull the socks out of your dryer one-at-a-time, how many must you pull out to be guaranteed to have a matching pair?

Solution. Imagine we have one box for each pair of socks, and each sock is considered an object; thus, we have n boxes.

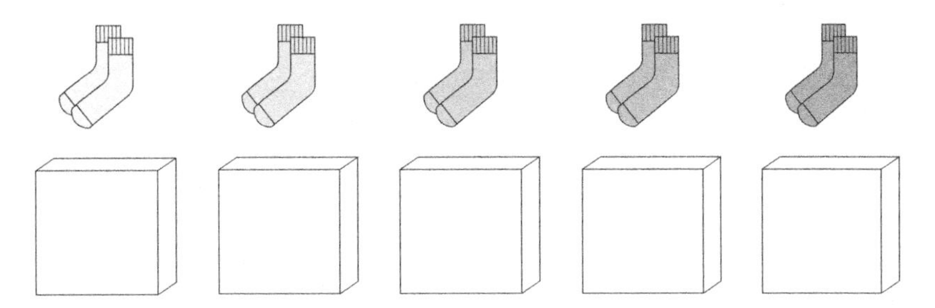

When you pull out a sock, put it in its box. As soon as a box has two socks in it, we have a pair. By the pigeonhole principle, once we have pulled out and placed

[23] Warning: In the real world, when people say "what are the odds" they usually mean it rhetorically and do NOT want a detailed mathematical analysis of the answer.

$n+1$ socks into the n boxes, we are guaranteed to have a box with two in it. So $n+1$ guarantees the property holds.[24] Could fewer also guarantee it?

In fact, $n+1$ is the smallest number that can guarantee a match, because it is possible that the first n socks were all from separate pairs — for example, if each pair has a left-foot sock and a right-foot sock, then it is possible that you pulled out the n left-foot socks.

Since $n+1$ socks guarantees the property but n does not, $n+1$ is the number of socks that you must pull out to be guaranteed a pair. □

Example 1.8. As of this writing, the population of the United States is about 330 million people. How many U.S. residents are guaranteed to have the same birthday, according to the pigeonhole principle?

Most solutions are discovered through scratch work, in which you try out ideas and test your hypotheses. At times I will include scratch work to help show how you could have discovered the main idea on your own.

Scratch Work. To determine this, let's see what would happen if each date of the year had exactly the same number of people born on it.[25] This is straightforward, just divide the 330 million people into 366 days:

$$\frac{330,000,000}{366} = 901,639.344\ldots$$

Since 901,639.344 people are born on an *average* day of the year, we should be able round up and say that at least one day of the year has had at least 901,640 people born on it. That is, with the pigeonhole principle we should be able to prove that there are at least 901,640 people in the USA with the same birthday.

Solution. Imagine you have one box for each of the 366 dates of the (leap) year, and each person in the U.S. is considered an object (sorry[26]). Put each person in the box corresponding to their birthday. By the general form of the pigeonhole principle (with $n = 366$ and $k = 901,639$ and thus $k+1 = 901,640$), any group of

$$(901,639)(366) + 1$$

people is guaranteed to contain 901,640 people which have the same birthday. And because

$$330,000,000 > (901,639)(366) + 1,$$

[24]Said differently, $n+1$ is an *upper bound* on the answer to the question "what is the minimum number of socks that you must pull out to be guaranteed a matching pair?" We now know that the minimum number is no bigger than $n+1$, since $n+1$ does guarantee a pair.

[25]No surges 9 months after New Year's Eve or Valentine's Day, or lulls on (the 75% absent) Leap Day of February 29th or on major holidays (when few induced labors are scheduled).

[26]*How to Objectify People with Math* was among the rejected titles for Chapter 1 of this book. Others: *The Hairs Within the Pigeon Holes*, and *Castles Going Mental.*[27]

[27]This last one will make sense at the end of the chapter.

there are enough people in the U.S. to guarantee that 901,640 people all have the same birthday.

Moreover, we cannot do any better. That is, the pigeonhole principle does not guarantee that 901,641 people all have the same birthday. In order to guarantee that, we would need $(901,640)(366)+1$ people, but there are not this many people in the U.S., because

$$330,000,000 < (901,640)(366)+1.$$

□

Mathematical Examples

One of the challenges of applying the pigeonhole principle is identifying what you should make your "boxes" and what you should make your "objects." The following examples highlight this difficulty, and since these are becoming a little more mathy and serious, we will begin to call them propositions.

The following example also refers to a *set*. In this case, the set is simply used to refer to a collection of eight numbers. We will study sets in detail in Chapter 3.

Proposition.

Proposition 1.9. Given any five numbers from the set $\{1,2,3,4,5,6,7,8\}$, two of the chosen numbers will add up to 9.

Let's again begin with some scratch work. When you work on homework, scratch work is the space to try out ideas and test hypotheses. It also makes you more efficient: By writing down ideas and trying out examples, you will likely discover a proof faster.[28]

Scratch Work. For propositions like this, it is a good idea to begin by testing it on your own. For example, when writing this I randomly chose these five numbers from the set: 1, 3, 5, 6 and 7. In this case, 3 and 6 are the two numbers which add up to 9. Or perhaps my five numbers were 2, 3, 4, 7 and 8. In this case, 2 and 7 are the two numbers which add to 9. Pick five more on your own and check that it works.

It seems to check out, but how do we prove it? Since we are trying to have two numbers add up to 9 from the set $\{1,2,3,4,5,6,7,8\}$, it would be natural to start

[28] It tends to be much faster than the stereotypical practice of just sitting back in an overstuffed armchair, sipping Scotch until the idea pops fully-formed into your head. This was actually a mistake of mine when I first learned proofs. Not the armchair or Scotch part, but I was hesitant to start writing until I knew where I was going. I've learned from my mistake, though, and I now jump right in to scratch work, and am a more efficient mathematician as a result.

writing down which pairs of numbers do that.

$$1 + 8 = 9$$
$$2 + 7 = 9$$
$$3 + 6 = 9$$
$$4 + 5 = 9.$$

And, of course, also $8 + 1$ and $7 + 2$ and so on. Writing these down, four sums appear! And we are told that we are to pick *five* integers from the set. This is highly suggestive of the pigeonhole principle: If each of the four sums is a box, and each number is an object, then we are placing five objects into four boxes — the simple form of the pigeonhole principle is perfectly set up for just that! Let's do it.

Proof. Let one box correspond to the numbers 1 and 8, a second box correspond to 2 and 7, another to 3 and 6, and a final box to 4 and 5. Notice that each of these pairs adds up to 9.

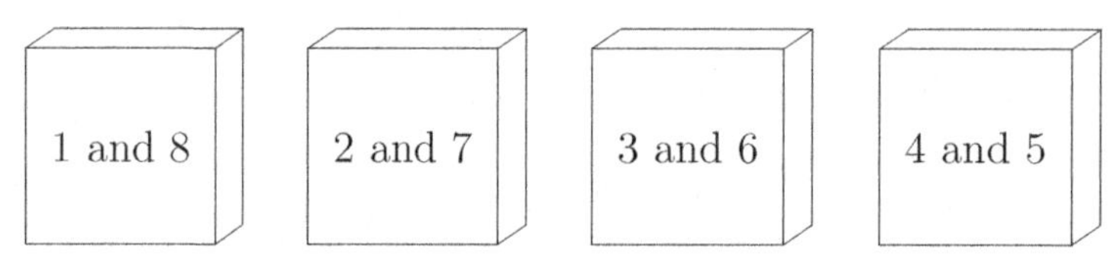

Given any five numbers from $\{1, 2, 3, 4, 5, 6, 7, 8\}$, place each of these five numbers in the box to which it corresponds; for example, if your first number is a 6, then place it in the box labeled "3 and 6." We just placed five numbers into four boxes, so by the simple form of the pigeonhole principle (Principle 1.5), there must be some box[29] which contains two numbers in it. These two numbers add up to 9, as desired. □

Let's do another!

Proposition.

Proposition 1.10. Given any collection of 10 points from inside the following square (of side-length 3), there must be at least two of these points which are of distance[30] at most $\sqrt{2}$ from each other.

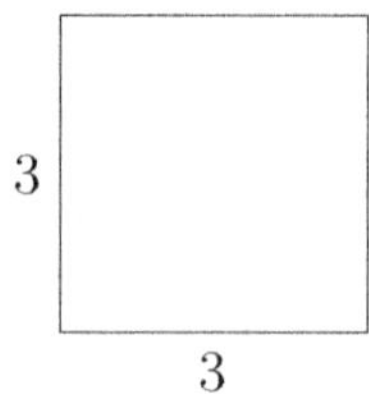

[29]Note: The word "some" can be confusing to new mathematicians. The phrase "some box" means "at least one box." It does not mean "exactly one box."

[30]Reminder: "At least two points" means "two or more points". Likewise, "of distance at most $\sqrt{2}$" means "of distance less than or equal to $\sqrt{2}$".

Scratch Work. We have 10 points. How can we use the pigeonhole principle? Since we are trying to show that two points have some property, and since the conclusion of the *simple* form of the pigeonhole principle regards two objects, it's probably the simple form of the principle that we will use... Can you see a way to get 9 (or fewer) "boxes" to put our points in? The 3×3 square has area 9... perhaps that's a sign of what to do...

Here's one idea: Divide up the 3×3 square into 9 "boxes," each 1×1:

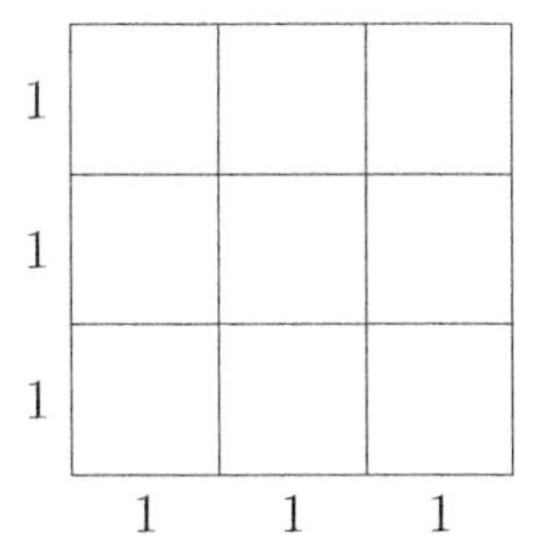

Then if you pick any 10 points from the 3×3 square, they will fall neatly into these boxes! For example:

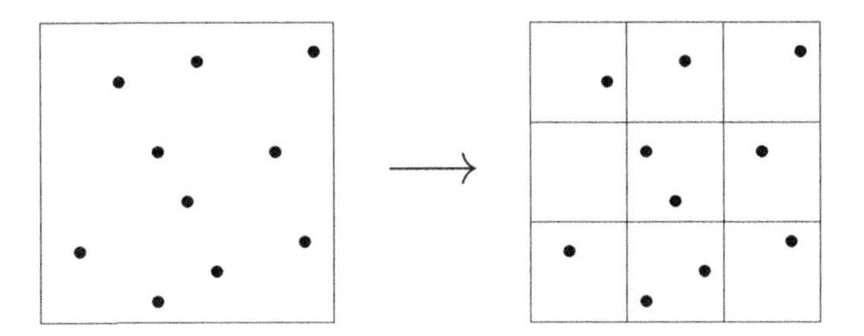

Now, it is possible that a point will fall exactly on the line between two boxes, so we will have to make up a rule for how to break a tie, but otherwise this does at least place 10 points into 9 boxes. And so by the pigeonhole principle we will get two points in the same box. But does that give us what we want?

If there are 2 points in the same 1×1 box, how far apart can two points be? As you an think about that, let's start the proof.

Proof. Take the 3×3 square and divide it into 9 boxes as follows:

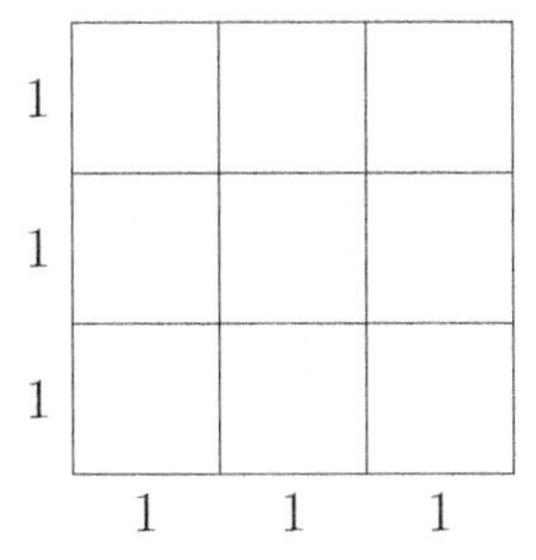

As for the points on the lines between squares, consider them part of the square above and/or to the right. Doing this, each of the points in the 3×3 square is assigned to one of the nine boxes. By the pigeonhole principle (Principle 1.5), by placing 10 points into these 9 boxes, at least one box must have at least two points in it; let's call these points x and y.

We now determine how far apart two points can be if they are in the same box. Indeed, observe that the maximum such distance occurs when the two points are on opposite corners, which by the Pythagorean theorem is of distance $\sqrt{2}$.

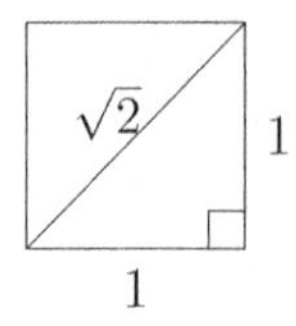

The distance between x and y must be at most this maximum distance of $\sqrt{2}$, which completes the proof. □

Paul Erdős[31] is one of the great mathematicians of the 20th century,[32] and is as unique and fascinating a person as you can imagine. I encourage you all to read *The Man Who Loved Only Numbers* for a really interesting look at an extraordinary genius. If you are interested in my own mathematical upbringing, then the book also tells of my Ph.D. advisor, Ron Graham, who struck up a lifelong friendship with Erdős. But beyond that, the book is an excellent collection of mathematics, mathematicians, and anecdotes that I think each of you will enjoy.

Erdős was famous for being a problem solver. More so than building theory, he largely spent his time solving problems—a staple of combinatorics. He also liked to share math with others, and liked giving problems to young and promising kids who were aspiring mathematicians. The following was his favorite problem to give, and it is this problem with which we end the main content of this chapter.

This problem will be a challenge and will introduce a few ideas which we have not yet discussed, but I want to share it with you anyways. Beginning in Chapter 2, we will methodically build new material from stuff we have already done, but our goal for this chapter is to get our feet wet and to have fun proving some interesting things.

Proposition.

Proposition 1.11. Given any 101 integers from $\{1, 2, 3, \ldots, 200\}$, at least one of these numbers will divide another.

Scratch Work. We will study divisibility in detail in Chapter 2, but for now you simply have to recall that, say, 3 divides 15 because $\frac{15}{3}$ is an integer. Likewise, 6 divides 30. However, 12 does not divide 30 because $\frac{30}{12} = 2.5$, which is not an integer.

[31] Pro-Tip: "Erdős" is pronounced "air-dish." It's Hungarian.

[32] He pioneered an area of math called *combinatorics*, which includes techniques like the pigeonhole principle and areas like graph theory (which we will discuss on page 24) and Ramsey theory (which we will discuss on page 41).

This is again set up perfectly for the simple form of the pigeonhole principle. If we can set up 100 boxes somehow, and we create some rule that tells us how to place the 101 numbers into these 100 boxes, then the pigeonhole principle guarantees that two of these numbers will land in the same box. So we just need it to be the case that once two numbers land in the same box, then one will divide the other...

Another proof strategy is to look at related problems and see how we solved them. Maybe a similar approach will work here. For example, for Proposition 1.9 the rule was that 1 and 8 go in the first box; 2 and 7 go in the second box; and so on. We need another rule like this, but instead of the two numbers adding to 9, one must be a multiple of the other...

This is tough! I would encourage you to go out to dinner tonight with your most boring friends, and when the conversation drifts you can spend the time pondering this problem. So feel free to stop reading now and go do that.

⋮

Ok, welcome back! Hope your friends didn't mind. Anyways, here are my stream-of-consciousness thoughts in my scratch work:[33]

- We need to choose 100 boxes. What could they be?

- There are 100 numbers between 1 and 100. Maybe we should make a box for each of those numbers. And maybe in Box n we can put n and $2n$? Like Box 3 will be where 3 and 6 go? But wait...should 6 go in Box 3 or Box 6? And where would a number like 135 go? Maybe Box 3 is for 3 and another number from $\{101, 102, \dots, 200\}$ which is divisible by 3? Like Box 5 could be for 5 and 135? Box 15 for 15 and 165? But what about prime numbers in $\{101, 102, \dots, 200\}$...

- Ok, new plan. The prime numbers[34] larger than 100, like 101, do not divide anything in $\{1, 2, 3, \dots, 200\}$ besides themselves, and nothing in $\{1, 2, 3, \dots, 200\}$ divides them (except 1, but 1 divides everything and can only go in one box, so let's ignore 1 for now). So these big primes have to be in their own box. Otherwise, if we got 101 and some other number in the same box, then once the pigeonhole principle gives us "two in the same box" we would not be guaranteed that one divides the other. Ok, so we start off with a box for each of them. And a random dude on `Quora.com`[35] says there are 20 primes between 101 and 200, so I'll trust him. So 80 boxes to go...Hmmm...Well we can now start doing what we had thought of before. We could pick a number less than 100 and pair it with a non-prime larger than 100. But then 20 of these numbers

[33]I will only do this once, but for Chapter 1, let alone the hardest problem in Chapter 1, I think it's worth it to emphasize the trial-and-error mental process when trying to prove something hard.

[34]Recall: A positive integer is *prime* if it is at least 2 and the only numbers which divide it are 1 and itself. The primes are $2, 3, 5, 7, 11, 13, 17, \dots$.

[35]The day that this reference stops making sense is the day I will start thinking about writing a second edition of this book.

can't have their own box... Ok this is getting too complicated. No way Erdős' favorite problem would have a solution this complicated...

- Ok, new plan. There are also 100 even numbers and 100 odd numbers in the set $\{1, 2, 3, \ldots, 200\}$. Maybe we can have a box for each even number? But now if you double it or multiply it by anything else to find what to pair it with, you keep getting even numbers! Ok, let's try odd numbers. If you have a box for every odd number... well, Box 3 could be where 3 and 6 go! And Box 5 can be where 5 and 10 go! IT'S WORKING!!! Oh wait. But what about, like, 12? Where does it go? I suppose it could go in Box 3 with the 3 and the 6... Because if any two of 3, 6 or 12 wind up together, then the smaller one still divides the larger one... Oh, and in that case, you might as well put 3, 6, 12, 24, 48, 96 and 192 all in the same box. Each is just 3 times a bunch of 2s, and so the smaller will always divide the larger! Likewise, in Box 5 we will put 5, 10, 20, 40, 80 and 160. In Box 7 we will put 7, 14, 28, 56 and 112. And so on.

- Ok, now that's feeling right. And the primes above 101 are all odd, and doubling them is larger than 200, so they are ending up in their own box, which earlier we said they would have to. So that's a good sanity check. Let's do one more sanity check. We have said before that it is often beneficial to test ideas on smaller cases. What would this look like if we instead chose 51 numbers from $\{1, 2, 3, \ldots, 100\}$? We are still choosing 1 more than half. Or 16 numbers from $\{1, 2, 3, \ldots, 30\}$? These are still too big to do by hand. Let's do 7 numbers from $\{1, 2, 3, \ldots, 12\}$. Following the strategy we just discovered, let's create a box for every odd number in this set:

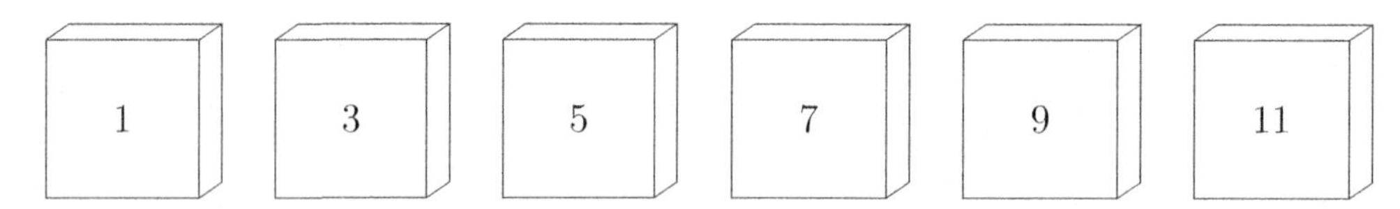

And in box m we will put any number of the form $2^k \cdot m$. Thus, these are the numbers that will go in each box:

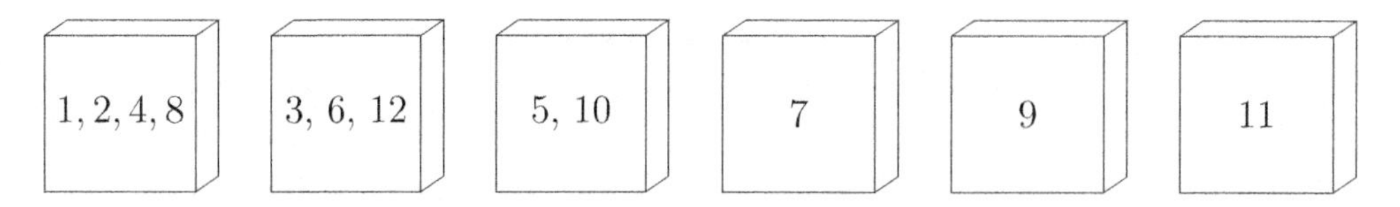

This seems right. Pick any 7 of these 12 numbers, and place each in the appropriate box. With 7 numbers but 6 boxes, by the pigeonhole principle at least two numbers will end up in the same box. If it is, say, 2 and 8, then yes, one divides the other. Or 3 and 12, or 5 and 10. Being in the same box means the smaller number divides the bigger one.

- Ok, yeah, this is feeling right. Sanity has been checked! And the bigger case should work in the same way. Now for the writeup!

***Proof*.** For each number n from the set $\{1, 2, 3, \ldots, 200\}$, factor out as many 2's as possible, and then write it as $n = 2^k \cdot m$, where m is an odd number. So, for example, $56 = 2^3 \cdot 7$, and $25 = 2^0 \cdot 25$. Now, create a box for each odd number from 1 to 199; there are 100 such boxes.

Remember that we are given 101 integers and we want to find a pair for which one divides the other. Place each of these 101 integers into boxes based on this rule:

If the integer is n, then place it in Box m if $n = 2^k \cdot m$ for some k.

For example, $72 = 2^3 \cdot 9$ would go into Box 9, because that's the largest odd number inside it.

Since 101 integers are placed in 100 boxes, by the pigeonhole principle (Principle 1.5) some box must have at least 2 integers placed into it; suppose it is Box m. And suppose these two numbers are $n_1 = 2^k \cdot m$ and $n_2 = 2^\ell \cdot m$, and let's assume the second one is the larger one, meaning $\ell > k$. Then, we have now found two integers where one divides the other; in particular n_1 divides n_2, because $\frac{n_2}{n_1}$ is an integer:

$$\frac{n_2}{n_1} = \frac{2^\ell \cdot m}{2^k \cdot m} = 2^{\ell - k}.$$

This completes the proof. □

This procedure might not seem optimal since some of the boxes have many numbers in them (the first box contains $\{1, 2, 4, 8, 16, 32, 64, 128\}$) while each of the fifty odd numbers larger than 100 is in a box all to itself. Moreover, many of these are divisible by other numbers. For instance, if 125 and 25 were among the 101 numbers chosen, then these two numbers would be placed in separate boxes and our procedure would fail to detect that one is divisible by the other. Thus, although our proof does still go through and in some other box we are still guaranteed to find a divisible pair, we will never catch a pair like 25 and 125.

It makes you wonder if 101 numbers are really needed. If we risk missing lots of pairs, maybe only 80 numbers guarantee that one divides another. Or maybe 51 numbers do so.

Alas, and perhaps surprisingly, our procedure is indeed optimal. Even if you chose just 1 fewer number — 100 — you would not be guaranteed that one divides another. You really do need 101. In Exercise 1.16, you will be asked to find 100 numbers from $\{1, 2, 3, \ldots, 200\}$ for which none divides another.

1.4 Bonus Examples

It is important that students take time to read and digest additional proofs beyond the ones covered in class. To support this, the last section of each chapter contains a few bonus examples which professors may safely omit from their lectures if they wish, but which I recommend students read anyway. Also, occasionally I will use these examples to introduce some new topics, ideas or theorems.

Your first bonus example comes from the field of graph theory. A *graph*[36] can be thought of as a collection of points on a piece of paper, called *vertices*, with a collection of line segments, called *edges*, each of which connects two vertices. Also, there is no rule saying a graph has to be in one piece, and there is no rule saying that a vertex has to have an edge touching it (if a vertex touches no edges, it is called a *lone vertex*). Here's an example of a single graph:

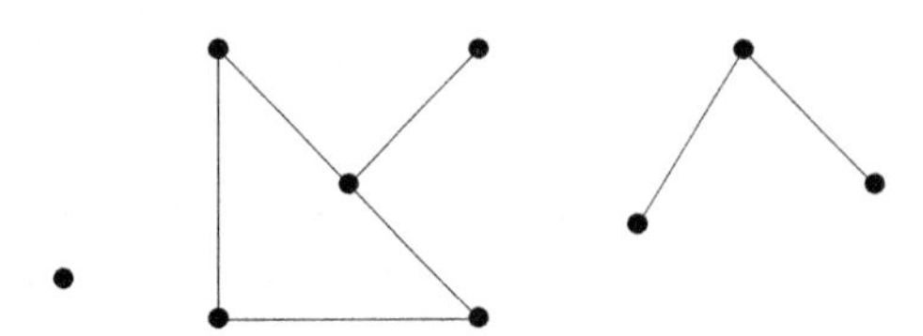

For graphs, we also do not care about how it's drawn, only how many vertices there are and which vertices are connected by an edge. For instance, here is the exact same graph as above, just drawn differently:

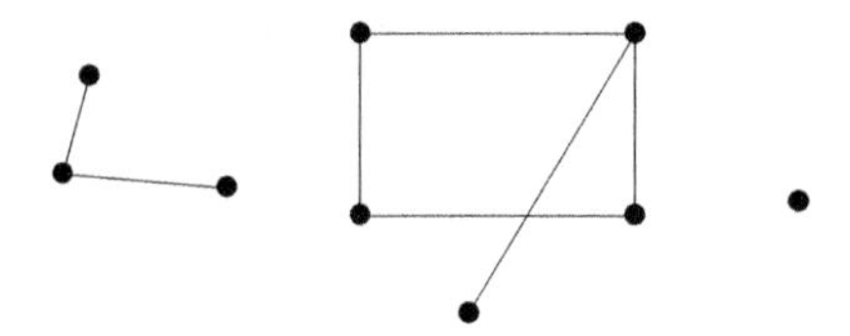

Notice that there is a spot at the bottom of the rectangle where two edges appear to intersect. This does not count as a vertex, though; only the solid dots count as vertices.

The question we want to ask is in regards to the *degree* of a vertex, which is defined to be the number of edges touching that vertex. For instance, here's the same graph, drawn the first way, with degrees labeled:

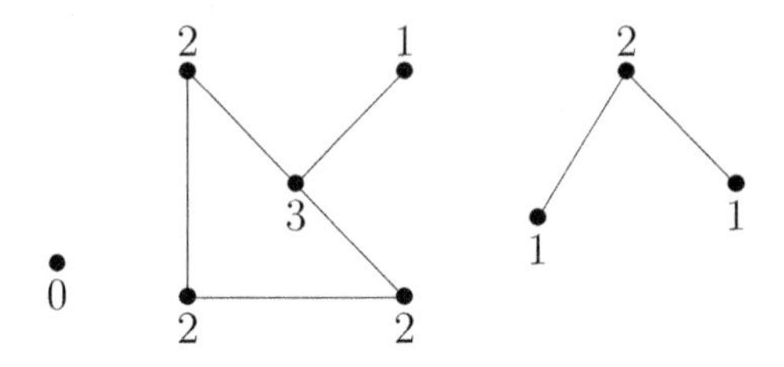

What we wish to prove is that in *any* graph (with at least two vertices), there must be two vertices which have the same degree.[37] In the above there are many

[36]Note that a graph in this context is nothing like the xy-plane graphs that you have used in every math class up to this point. Upper division math is very different than lower division math in many ways. At times, this even includes the vocabulary.

[37]To be clear, in math when we say "there must be two vertices which have the same degree," what we really mean is "there must be *at least* two vertices which have the same degree." If vertices v_1, v_2 and v_3 all have the same degree, then that still counts, because there *are* two vertices, v_1 and v_2, which have the same degree. Sure, there's v_3 out there too, but that doesn't take away from the fact that v_1 and v_2 are two vertices and they do have the same degree. This is a subtle point that's worth spending time wrapping your head around.

such pairs. Let's do one more sanity check: Among the graphs with "at least two vertices," the simplest examples are those with exactly two vertices. Let's quickly check whether those work. There are two such graphs: the first is just two lone vertices, while the second has an edge between its two vertices:

Yup! Both of these graphs contain a pair of vertices with the same degree. In fact, that's all they have! Next, on your own, draw out and check all of the graphs with three or four vertices. Then, when you're ready, let's state and prove the result.

Proposition.

Proposition 1.12. Suppose G is a graph with $n \geq 2$ vertices. Then, G contains two vertices which have the same degree.

Proof Idea. How many options are there for the degree of a vertex? The smallest possible degree is 0. What is the largest possible degree? Well, since G has n vertices, a vertex can be connected to a maximum of $n-1$ other vertices. And if it were connected to *all* other vertices, then its degree would be $n-1$. So a vertex's minimum degree is 0, and maximum degree is $n-1$. Therefore, the degree possibilities are:

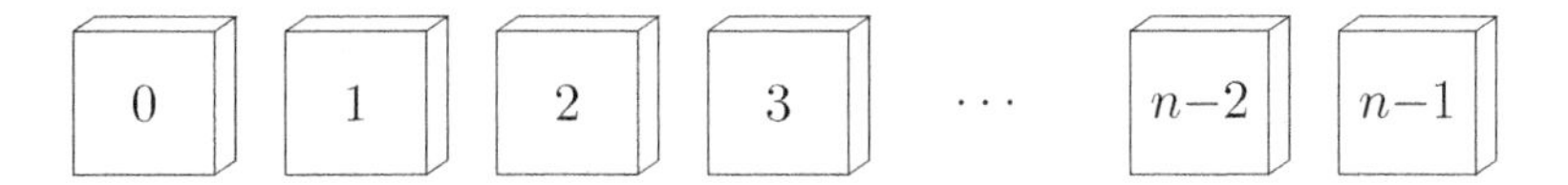

This is beginning to look like a pigeonhole principle problem, where the "objects" are the vertices, the "boxes" are the possible degrees, and you place a vertex into the box corresponding to its degree. But there are n vertices and n boxes! The pigeonhole principle cannot be applied in such a scenario. If there weren't a Box 0 we would be in business, but there is... And we certainly can't ignore that box, since we have already seen examples where it is needed. Take a moment and see if you can figure out how to get out of this pickle. And if you need a hint, check out the footnote[38] before reading the proof on the next page.

[38] Hint: Imagine you placed each vertex into its corresponding box. Is it possible that both of these outer boxes have a vertex in them?

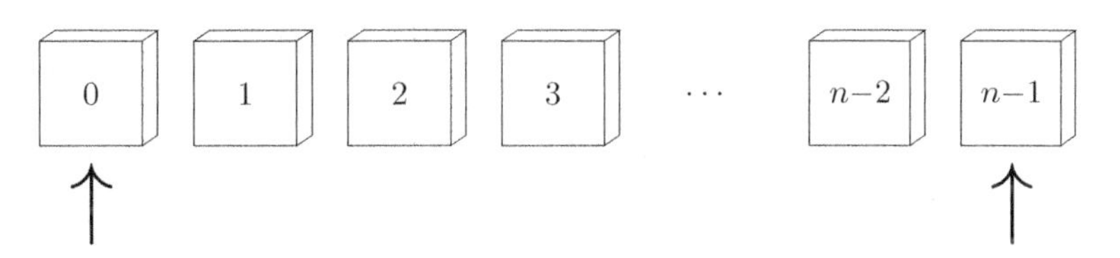

Proof. Let G be a graph with n vertices. Since each vertex may be connected to as few as zero other vertices, or as many as all $n-1$ other vertices, the possible degrees of a vertex are 0, 1, 2, ..., $(n-1)$. Next, note that G either has a lone vertex or it does not. Consider these two cases separately.[39]

Case 1: G does not have a lone vertex. Since G does not have a lone vertex, every vertex has degree at least 1. Therefore, there are n vertices and $n-1$ possible vertex degrees:

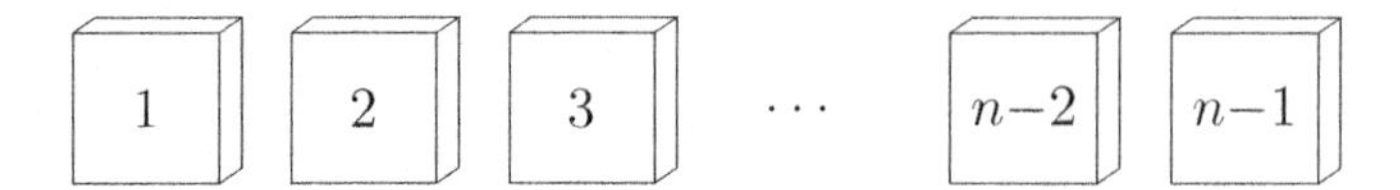

Since we have n vertices being placed into $n-1$ boxes, by the simple form of the pigeonhole principle (Principle 1.5) two vertices must be placed into the same box, which means they have the same degree.

Case 2: G has a lone vertex. Let v_0 be a lone vertex in G. Then, v_0 has degree zero. Moreover, if v_1 is any other vertex in G, we know that v_1 is not connected to v_0, implying that v_1 has only $n-2$ other vertices which it may be connected to. That is, the maximum possible degree of v_1 is $n-2$. Since v_1 was arbitrary, the maximum possible degree of any vertex in G is $n-2$. Therefore, there are n vertices and $n-1$ possible vertex degrees:

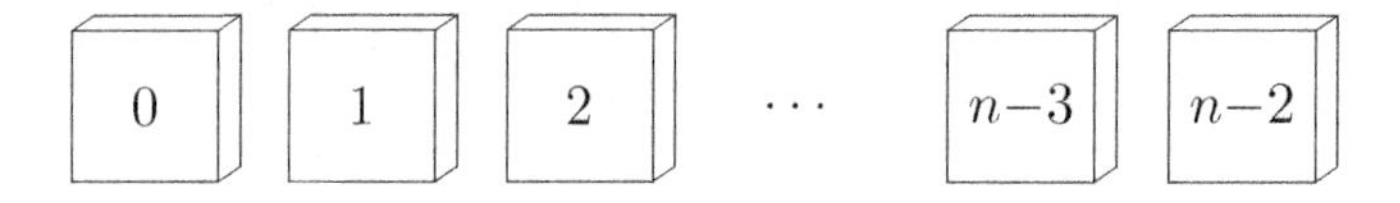

Since we have n vertices being placed into $n-1$ boxes, by the simple form of the pigeonhole principle (Principle 1.5) two vertices must be placed into the same box, which means they have the same degree.

In both of the two possible cases, we proved that G has two vertices of the same degree. Therefore, this is true in general, establishing the result. $\square$

The final proposition is a personal favorite. And while it could be phrased slightly more rigorously using spheres and circles... I believe it is best phrased in terms of fruit.

[39]We will discuss *proof by cases* much more in Chapter 2.

Proposition.

Proposition 1.13. If you draw five points on the surface of an orange in marker, then there is always a way to cut the orange in half so that four points (or some part of each of those points) all lie on one of the halves.

Proof Sketch. This should be surprising! When you cut an orange in half, you in essence create two boxes for these five points; but shouldn't the pigeonhole principle only guarantee us 3 points on one of the halves? How the heck do we get *four* points on one half?!

There are two subtle parts of the statement. First, it asserts that "there is always a way to cut the orange in half so that...." It doesn't assert that *any* such cut has this property; just that among all of the infinitely many angles your knife can take, at least one has this property.

Second, it is important that we say "or some part of each of those points." Here is how we will use that to our advantage: When you use a marker to make the points, the points are big enough that when you slice through any point, part of the point appears on *both* halves.

Perhaps this gives you some ideas. But, I confess, this is a sneaky problem because its solution also relies on a theorem that I haven't told you. It is in fact a classic theorem from geometry. It deals with so-called *great circles.* Given a sphere, there are infinitely many ways to cut it in half, and each of these paths of the knife is called a *great circle* (like Earth's equator or any of Earth's lines of longitude). Below are three examples, followed by the classic theorem from geometry.

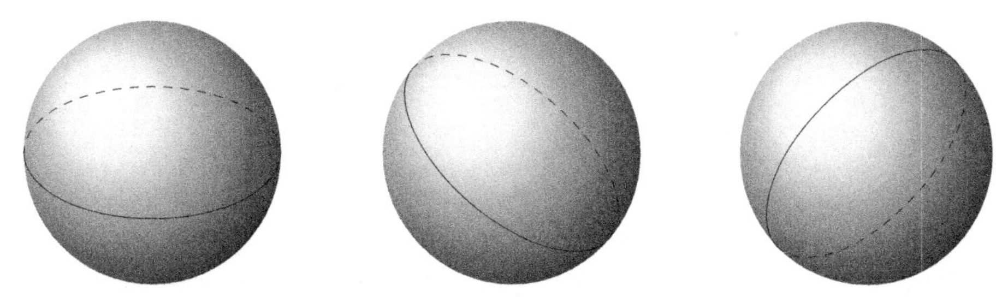

Classic Geometry Theorem. Given any two points on the sphere, there is a great circle that passes through those two points. For example:

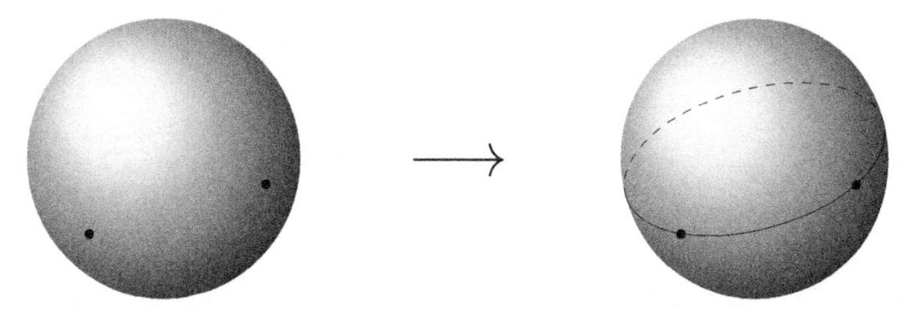

Ok, now you have all the tools and caveats you need to prove this result. See if you can piece it together before reading the proof on the next page.

***Proof*.** Consider an orange with five points drawn on it. Pick any two of these points, and call them p and q. By the Classic Geometry Theorem, there exists a great circle passing through these points; angle your knife to cut along this great circle. Because the points are drawn in marker, they are wide enough so that part of these two points appear on both halves.

Now consider the remaining three points and the two halves that you just cut the orange into. Viewing these three points as objects and the halves as boxes; by the simple form of the pigeonhole principle (Principle 1.5), at least two of these three points are on the same orange half. These two, as well as a portion of p and of q, give four points or partial points on the same half, as desired.[40] □

[40]If you feel slightly cheated by the fact that two points are in both halves, I will point out that in practice this is rarely needed. Here's how: (1) Pick any two points, and call them p and q; (2) Angle your knife so that you would cut through them and the orange in half; (3) Identify which side contains two of the remaining three points; (4) Shift and re-angle your knife just slightly so that you get all of p and q on the same half as these other two, giving a half with four complete dots. It is rare that doing this will cause you to lose the other two points, but if so just pick another two points and try again. Unless all the points lie on a single great circle, you will soon find the angle giving four complete points.

— Chapter 1 Pro-Tips —

Next up are some "Pro-Tips," with which I will end each chapter. These are short thoughts on things I wish I had known when I took my intro-to-proofs class. They are quite varied, and include finer comments on the material, study tips, historical notes, comments on mathematical culture, and more. I hope you find them beneficial.

- To master mathematical content, one must struggle with it. In order to not just learn the material but deeply understand it, you need to test it against your own knowledge and intuition. It can't be a passive enterprise; mathematics is a contact sport. There is a metaphor for this, that I worked on with its inventor Abigail Higgins, to explain the laborious-yet-exciting work to construct your mental conception of mathematics.

  Think about math as a giant, beautiful castle. No teacher can download this castle into your brain. We use definitions and theorems and proofs and examples and non-examples and conjectures to introduce you to a new room or alcove of the castle, or to help you make connections between different wings. But in the end, you must build your own mental castle.

  It takes effort, but I can assure you there is no greater satisfaction than standing back after completing a new course, reading a new research paper, completing a new project, or reflecting on a conversation, and realizing that there is a connection between two ballrooms you hadn't discovered before, or that a room has some amazing artwork in it that you had never noticed. These are the mental rewards when you're willing to fight through a mathematical difficulty rather than just looking up how the book does it. Furthermore, no advancement in mathematics research has been won without a personal struggle in which mistakes were made and small steps were taken. To be the first to discover a new feature of the castle is a reward reserved only for tenacious learners.

  These soft skills are not instilled easily. You must practice fighting through difficulties in order to become good at fighting through difficulties. You must practice solving a lot of problems to become good at solving a lot of problems. I encourage you to carry this attitude forward with you as you enter into the heart of the mathematics castle.

- It is strongly advised that you form a study group to practice and discuss the material with others. The best math is done collaboratively, and the best learning occurs from discussions with your peers. Also, I find that math is most fun in collaboration.

  Also, remember that while math is intrinsic, proofs are human. Math is a search for objective truths, while proofs are the search for subjective agreement. The goal of a proof is to communicate your ideas and convince others that you are correct, and so it is important to discuss your ideas and share your thoughts with others. So talk things out with your study group and read over each others' work. This is the field research of proof writing, and it is important.

- When writing out their homework solutions, students are far more likely to write *too little* than they are to write *too much*. As the author of long-form textbooks, it may not be surprising that I am against terse proofs in homework solutions, but I can assure you that this is not a personal quirk — a survey of my colleagues agrees that more is better, especially for a class like this.

  It is like that episode of The Office where Kevin tries to talk as simply as possible. He justified this by saying "Me think, why waste time say lot word when few word do trick." But it just causes mass confusion and wastes time. Don't be like Kevin. Say a little more to make sure your ideas are clear, and the readers of your proofs will thank you.

- When you start taking upper-division math classes, how you approach the material will make a big difference to how you do. Research suggests the importance of active learning, deliberate practice, metacognition, and having a growth mindset. These are more than just buzzwords, and I encourage you to check out some articles and videos that I collected at `longformmath.com`. If you plan to teach math at any level some day, they will be particularly helpful.

- We did not prove the pigeonhole principle. It was also not called a lemma, proposition, theorem or corollary, which do require proofs. Are there proofs of the pigeonhole principle? The answer is yes, and you are welcome to search the Internet for them — you will quickly find several. The problem is that the principle is such a basic idea that the proofs often rely on something that seems even less obvious than the principle itself, or they are written in terms that will likely be very confusing to you at the moment.

- When concluding proofs, it is common to include the □ proof symbol. But you will discover that there are many variants of this symbol which are also used. Some use a filled in square, like ■. Others make it skinnier and taller, like ▯ or ▮. And others use entirely different symbols altogether. I have had students use everything from smiley faces to cat drawings to spatulas. The late Paul Sally, in his book *Tools of the Trade*, ended each of his proofs with a self portrait — which is pretty bad ass because Sally wore an eye patch and smoked a pipe, so his end-of-proof symbol looked like this: .

  One could also end each proof with a short phrase. The ancient Greeks, including Euclid and Archimedes, ended their proofs with the Latin phrase "quod erat demonstrandum," which means "what was to be shown." This phrase, or its initialization of Q.E.D., was a popular way to conclude proofs for a couple thousand years, and is still used occasionally today.[41] You could also adopt your own phrase, if you wish. A few suggestions: "Bada bing bada boom!" or "Oh happy day!" or *"Do you believe me now?!"* or, for 90s music fans, "Proof, there it is!"

[41] Fun fact: The United States' most watched public TV station, and most listened to public radio station, is a PBS- and NPR-member station called KQED, the "K" means that its headquarters are west of the Mississippi, and the "QED" is from quod erat demonstrandum! The founder's wife, Beverly Day, apparently came up with it.

— Exercises —

Exercise 1.1. Read *The Secret to Raising Smart Kids* by Carol Dweck[42] and write a few paragraphs about what you learned and how it may help you be successful in a proof-based math class.

Exercise 1.2. Explain the error in the following "proof" that $2 = 1$.

Let $x = y$. Then,

$$\begin{aligned} x^2 &= xy \\ x^2 - y^2 &= xy - y^2 \\ (x+y)(x-y) &= y(x-y) \\ x + y &= y \\ 2y &= y \\ 2 &= 1. \end{aligned}$$

Exercise 1.3. Suppose that m and n are positive odd integers.

(a) Does there exist a perfect cover of the $m \times n$ chessboard?

(b) If I remove 1 square from the $m \times n$ chessboard, will it have a perfect cover?

Exercise 1.4. The game *Tetris* is played with five different shapes — the five shapes that can be obtained by piecing together four unit squares:

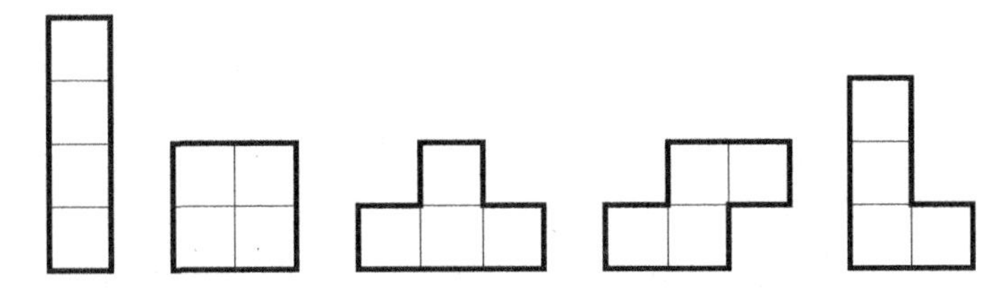

For the questions below, we also allow these pieces to be "flipped over." For example, and are both allowed.

(a) Is it possible to perfectly cover a 4×5 chessboard using each of these shapes exactly once? Prove that it is impossible, or show by example that it is possible.

(b) Is it possible to perfectly cover an 8×5 chessboard using each of these shapes exactly twice? Prove that it is impossible, or show by example that it is possible.

Exercise 1.5. If I remove two squares of different colors from an 8×8 chessboard, must the result have a perfect cover?

---

[42] `https://www.scientificamerican.com/article/the-secret-to-raising-smart-kids1/`

Exercise 1.6. If I remove four squares — two black, two white — from an 8×8 chessboard, must the result have a perfect cover?

→ If you believe a perfect cover must exist, justify why.

→ If you believe a perfect cover does not need to exist, give an example of four squares that you could remove for which the result does not have a perfect cover.

Exercise 1.7. In chess, a *knight* is a piece that can move two squares vertically and one square horizontally, or two squares horizontally and one square vertically. Below are two examples of knights, and the squares to which they could move.

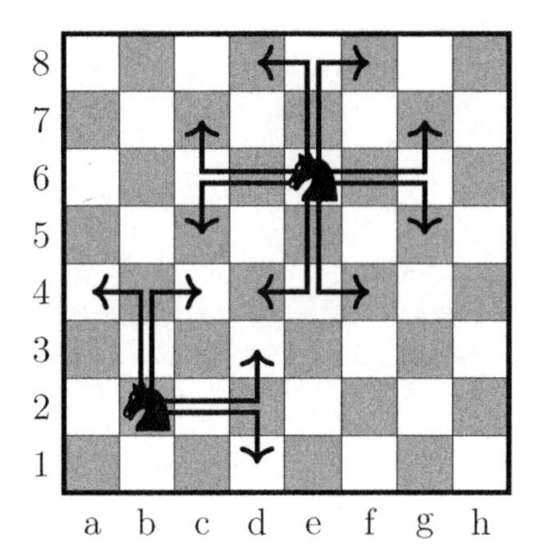

A knight can legally move to any one of these squares, provided there is not another piece on that same square.

(a) Suppose there is a knight on every square of a 7×7 chessboard. Is it possible for every one of these knights to simultaneously make a legal move?

(b) Suppose there is a knight on every square of a 8×8 chessboard. Is it possible for every one of these knights to simultaneously make a legal move?

Exercise 1.8. Prove that if one chooses $n + 1$ numbers from $\{1, 2, 3, \ldots, 2n\}$, it is guaranteed that two of the numbers they chose are consecutive. Also, before your proof, write down an example of 4 numbers from $\{1, 2, 3, 4, 5, 6\}$ and locate two of them which are consecutive. Then, repeat for 5 numbers from $\{1, 2, \ldots, 8\}$ and 6 numbers from $\{1, 2, \ldots, 10\}$.

Exercise 1.9. Assume that n is a positive integer. Prove that if one selects any $n + 1$ numbers from the set $\{1, 2, 3, \ldots, 2n\}$, then two of the selected numbers will sum to $2n + 1$. Also, before your proof, write down 4 numbers from $\{1, 2, 3, 4, 5, 6\}$ and locate two of them which sum to 7. Then, repeat for 5 numbers from $\{1, 2, \ldots, 8\}$ and 6 numbers from $\{1, 2, \ldots, 10\}$.

Exercise 1.10. Explain in your own words what the general pigeonhole principle says.

Exercise 1.11. Prove that there are at least two U.S. residents that have the same weight when rounded to the nearest *millionth* of a pound. Hint: Do a Google search for how many U.S. residents weigh over 300 pounds.[43]

Exercise 1.12. Determine whether or not the pigeonhole principle guarantees that two students at your school have the exact same 3-letter initials. (Include first, middle and last name in the initials. For instance, Natalie Laura Hobson = NLH).

Exercise 1.13. Find your own real-world example of the pigeonhole principle.

> **Definition.** Two integers m and n are said to be *relatively prime*[44] if there is no integer larger than 1 which divides both m and n. For example, 6 and 25 are relatively prime, because the only such divisors of 6 are 2, 3 and 6, and none of these divide 25.

This definition will be used in the following exercise.

Exercise 1.14. Prove that if one chooses 31 numbers from the set $\{1, 2, 3, \dots, 60\}$, that two of the numbers must be relatively prime.

Exercise 1.15. Assume that n is a positive integer. Prove that if one chooses any $n+1$ distinct odd integers from $\{1, 2, 3, \dots, 3n\}$, then at least one of these numbers will divide another. Also, before your proof, check all possible selections of 4 odd numbers from $\{1, 2, 3, \dots, 9\}$, and for each selection locate two of the numbers for which one divides the other.

Exercise 1.16. Give an example of 100 numbers from $\{1, 2, 3, \dots, 200\}$ such that none of your selected numbers divides any of the others. By doing so, this proves that Proposition 1.11 is optimal.

Exercise 1.17. Prove that any set of seven integers contains a pair whose sum or difference is divisible by 10. Also, before your proof, write down three different sets of seven integers, and for each set locate a pair whose sum or difference is divisible by 10. Have your sets contain a diverse collection of integers—some bigger, some smaller, some positive, some negative.

---

[43]Note: If you cut one finger nail, your weight changes by about 200 millionths of a pound. I believe I verified that a single eyelash is not quite enough to change your weight by over a millionth of a pound. I think you would need a hair on your head to fall out.

[44]Other terms which mean the same thing are *mutually prime* and *coprime*. Euclid used the phrase "numbers prime to one another," although that is literally ancient history.

Exercise 1.18. Prove that if one chooses any 19 points from the interior of a 6×4 rectangle and no three them form a straight line, then there must exist four of these points which form a quadrilateral of area at most 4.

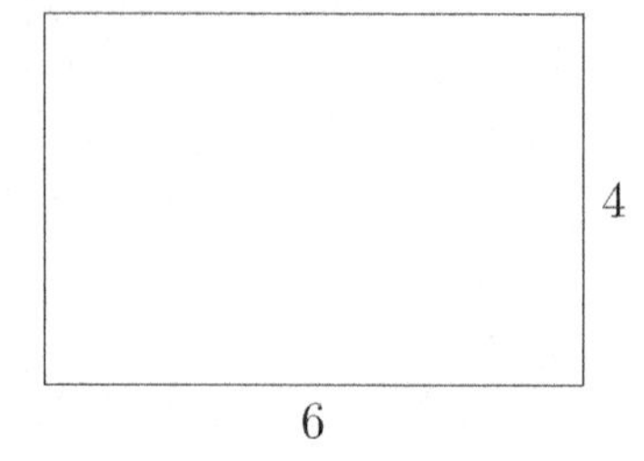

Note: a quadrilateral is a four-sided shape, and the condition that no three form a straight line is simply to guarantee that any four points form a quadrilateral.

Exercise 1.19. Assume that 9 points are chosen from the right triangle below and that no three of them form a straight line. Prove that there exist three of these points which form a triangle whose area is less than $1/2$.

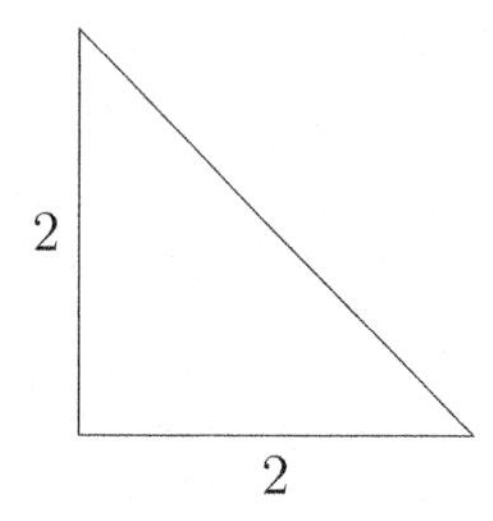

Note: the condition that no three form a straight line is simply to guarantee that any three of them can form a triangle.

Exercise 1.20. At a party, each person is *acquainted* with a certain number of others at the party (and is a stranger to everyone else). For example, Jessica may be acquainted with six people at the party while Fara is acquainted with eight. Suppose that there are $n \geq 2$ people at a party. Prove that at least two people at this party have the same number of acquaintances at the party.

You may also assume the following two things: (1) Being acquaintances is symmetric (if John is acquainted with Heidi, then Heidi is also acquainted with John — no stalkers are allowed at this party) and (2) Every person is acquainted with at least one person at the party (no party crashers are allowed).

Exercise 1.21.

(a) Determine the population of your hometown and how many non-balding people in your hometown, if any, are guaranteed to have the same number of hairs on their head, according to the pigeonhole principle.

(b) Determine, as best you can, the number of students who attended your high school while you were a senior. Then, determine how many of them, if any, are guaranteed to have the same birthday according to the pigeonhole principle.

Exercise 1.22. The following conjectures are all false. Prove that they are false by finding a counterexample to each.

(a) Conjecture 1: If x and y are real numbers, then $|x + y| = |x| + |y|$.

(b) Conjecture 2: If x is a real number, then $x^2 < x^4$.

(c) Conjecture 3: Suppose x and y are real numbers. If $|x+y| = |x-y|$, then $y = 0$.

Exercise 1.23. Suppose you deal a pile of cards, face down, from a shuffled deck of cards (this is a standard 52-card deck, where each card is one of 4 suits and one of 13 ranks). How many must you deal out until you are guaranteed...

(a) five of the same suit?

(b) two of the same rank?

(c) three of the same rank?

(d) four of the same rank?

(e) two of one rank and three of another?

In the terminology of the game Poker, part (a) asks how many cards must be dealt to guarantee a flush, part (b) asks this for a two-of-a-kind, part (c) for a three-of-a-kind, part (d) for a four-of-a-kind, and part (e) for a full house.

Exercise 1.24. Determine the U.S. population at the time that you are reading this.

(a) Does the pigeonhole principle guarantee that 1 million U.S. residents all have the same birthday?

(b) If the principle does not guarantee this, how many people are needed until that milestone is reached? If the USA grows by 2 million people per year, in what year will this occur?

Exercise 1.25. Imagine a friend gives you a deck of cards (a standard 52-card deck) and lets you shuffle it a few times. They then ask you to slowly deal out the cards, one at a time, into a new pile on the table. The entire time the cards are face-down, so they have no idea which cards you are dealing.

At a certain point in this procedure, they ask you to stop, and declare with confidence that the two stacks—the one still in your hand, and the one on the table—are in perfect balance. They say that the number of red cards in the stack in your hand is equal to the number of black cards in the stack on the table. They let you count, and sure enough, they were correct!

There were now gimmicks in this procedure—no trick cards or hidden cameras or outside help. How did your friend do it?

Exercise 1.26. An alien creature has three legs, and on each of his three alien feet he wears an alien sock. Suppose he just washed n triplets of alien socks ($3n$ individuals), and each triplet is a different color. If this alien pulls his alien socks out of his alien dryer one-at-a-time, how many must he pull out to be guaranteed to have a matching triplet?

Exercise 1.27. A *magic square* is an $n \times n$ matrix where the sum of the entries in each row, column and diagonal equal the same value. For example,

8	1	6
3	5	7
4	9	2

is a 3×3 matrix whose three rows, three columns, and two diagonals each sum to 15. Thus, this is a magic square.

An *antimagic square* is an $n \times n$ matrix where each row, column and diagonal sums to a distinct value. For example,

9	4	5
10	3	-2
6	9	7

is a 3×3 matrix whose rows sum to 18, 11 and 22, columns sum to 25, 16 and 10, and diagonals sum to 19 and 14. Notice that all eight of these numbers is different than the rest, showing that this is an antimagic square.

Prove that, for every n, there does not exist an $n \times n$ antimagic square where each entry is -1, 0 or 1.

Exercise 1.28. Read the *Introduction to Ramsey Theory* following this chapter. Then, let $r(n, m)$ be the smallest value N for which every red/blue coloring of K_N contains either a red K_n or a blue K_m. Prove that $r(n, 2) = n$.

— Open Question[45] —

In this chapter, we discussed perfect coverings of the chessboard with dominoes. What else can we cover? What other tile shapes can we use? Could we perfectly cover the entire xy-plane?[46] For example, perhaps you can see how the pattern below could continue forever (horizontally and vertically) to perfectly cover the entire plane.

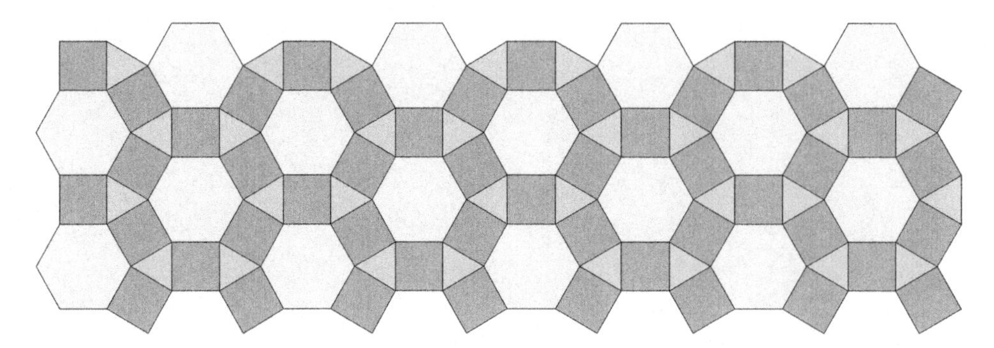

Imagine this pattern were perfectly covering the plane. This pattern is repeating in the exact same way forever, and therefore is called *periodic*. Here is the technical condition: It is possible to slide this pattern to the left and have it land right back on top of itself, so that every hexagon lands exactly on a hexagon, every square on a square, and every triangle on a triangle. Assuming you can slide it (in some direction), and the result looks exactly as it did before, it is periodic.

If a perfect cover of the plane is not periodic, then it is *aperiodic*. For example, if you took a simple perfect cover but "misplaced" a couple tiles, you can create an aperiodic cover. As an example:

Two of the tiles are rotated $90°$, and this makes it aperiodic. It would be impossible to shift this picture and have it land on itself, because those two rotated tiles appear nowhere else in the perfect cover, so any shift would cause them to land on shapes of the other orientation.

[45] An "open question" is a question that is unsolved. No one knows the answer. If you are my student and you solve one of them, I will give you one point of extra credit.

[46] Such a perfect covering is also called a *tessellation*.

This shows that having an aperiodic covering is nothing special. But here is where things get interesting. Roger Penrose discovered that there are many ways to perfectly cover the plane using tiles of shape and . Here is the start of one way to tile the plane with these shapes:

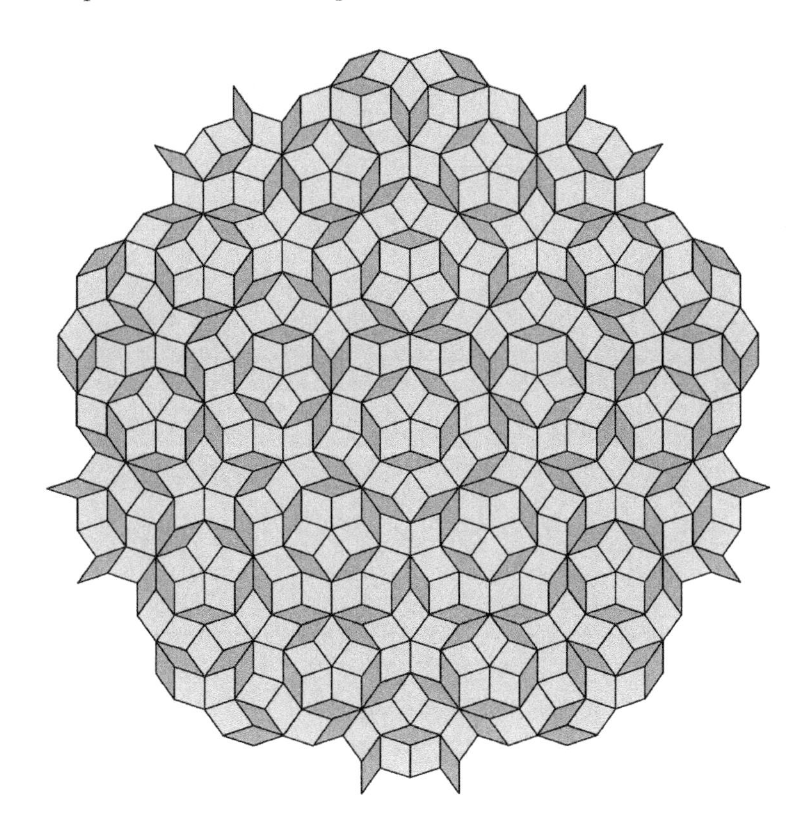

This is a particularly famous perfect cover, named after its founder, Roger Penrose; it is called a *Penrose tiling.* The above patten is aperiodic. At first glance, it might seem that each shape is the same as many of the others, and that you should be able to slide one on top of another and everything else will match up, too. This is not the case, though. In fact, not only is this arrangement of the and shapes aperiodic, but, amazingly, *every single possible perfect cover using these two shapes is guaranteed to be aperiodic!* Personally, I find this both strange and remarkable.

In 1964, Robert Berger found 20,426 shapes which, when used to perfectly cover the plane, were guaranteed to be aperiodic. The fact that Roger Penrose found just *two* shapes that do this is incredible. Moreover, they are not strange shapes. They are quite normal and natural shapes; both quadrilaterals with standard angles.

Now that we know there exist two shapes for which every perfect covering of the plane using them results in an aperiodic covering, the next question is this: Can a single shape do this? Nobody knows. This is Chapter 1's open question.

Open Question. Does there exist a single tile that can be used to perfectly cover the plane, such that every perfect covering using that tile is aperiodic?

Introduction to Ramsey Theory

If you observe a group of 20 children playing outside, you may notice that you can always find a group of four of them, each of which is friends with the other three, or a group of four of them, none of which is friends with the other three — there is always a group of four friends or four strangers. This may seem like an interesting sociological phenomenon, but the reason for this does not lie in the cerebral cortex or in Darwin's *On the Evolution of Species* — it is a necessary consequence from *Ramsey theory.* Ramsey theory not only has contextual formulations, like the one above, but is also an active research area in combinatorics, with over 100 papers published on the theory each year. (In fact, I research in this area!)

Ramsey theory is named after the very impressive work that Frank Ramsey accomplished before his untimely death at the age of 26. Ramsey theory is one of the few areas of math that has a nice philosophical question behind it. The underlying question is: "Does there exist complete randomness?" A little more concretely: "Can something have absolutely no structure to it?" Maybe more to the point: "If a set of random stuff is big enough, can we always find some sort of structure within it?" This is the sort of question that Ramsey theory attempts to model and answer, and many solutions to such problems rely on the pigeonhole principle. Let's begin with a question which is a simpler version of the playground example above.

Question. Suppose you are at a party with a large group of people. Among any 6 people there, must there be 3 people who mutually know each other, or 3 people who mutually do not know each other?

By "mutually know (or not know) each other," what I mean is that for every pair of people, that pair either knows each other or does not. We assume there are no stalkers who know someone while the other person does not know them, for instance.

Before we jump in, please review the short introduction to graphs on pages 24 and 25. One thing to add to that discussion is the definition of the *complete graph* on n vertices. This graph is denoted K_n and is the graph on n vertices that has all possible edges. Here are the first 5 complete graphs:

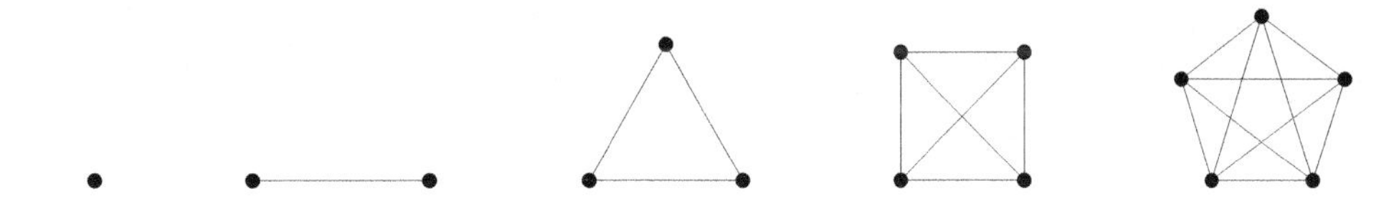

Graphs are particularly useful at modeling things. How does graph theory help us model our party question? Think about each person as being a vertex in a graph, and let's go ahead and add all the edges in. This gives the complete graph on 6 vertices:

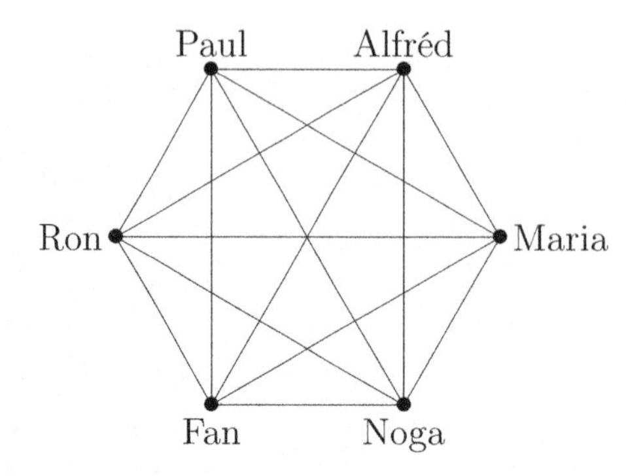

It is common in Ramsey theory to color the edges of a graph. If a pair of people know each other, the edge between them is colored blue; if they do not know each other, the edge between them is colored red.[47] But since one goal for this book is to have it be as inexpensive as possible, I have avoided using color.[48] Therefore for our purposes, red will be a thin line, and blue will be a bold line. So, for example, the 6 people could have this set of relationships:

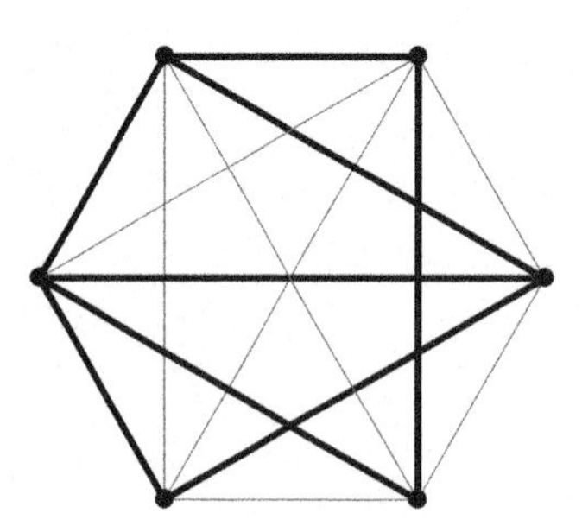

Or maybe they all know each other, in which case this is their graph:

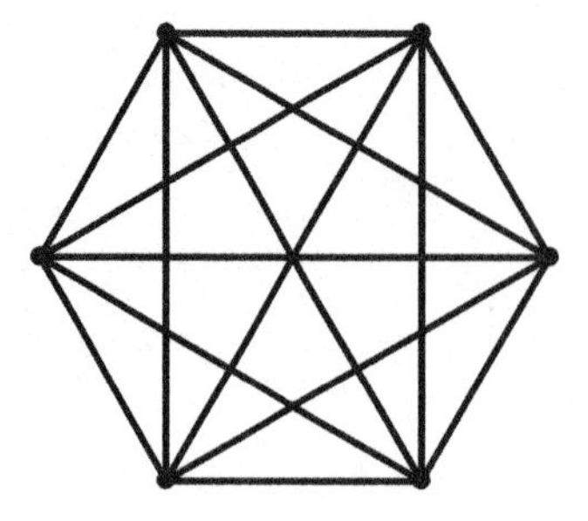

[47] I was going to include a colorful graph theory joke here, but it was just a bit too edgy.
[48] Pro-Tip: This is also good practice to keep your work accessible to color-blind folks.

So what do we want to know? We want to know if we must have either three people who all know each other (which in the graph would be three vertices with all the edges between them being bold blue) or three people none of which know each other (which in the graph would be three vertices with all the edges between them being thin red). The previous page had two colored K_6 graphs. The second colored K_6 has bold blue triangles everywhere (it has no thin red triangles but that's ok, we only require one or the other). Meanwhile, the first colored K_6 has a thin red triangle and two bold blue triangles. Must this property always hold no matter how the graph is colored (i.e., no matter which party we attend)? Said differently:

Same question, phrased in terms of graphs: Does every red/blue coloring of the edges of K_6 contain either a thin red triangle or a bold blue triangle?

The word *monochromatic* means "all the same color." So this is usually worded:

Same question, phrased in terms of graphs: Does every red/blue coloring of the edges of K_6 contain a monochromatic triangle?

It's an interesting question, and it's not obvious how to proceed... The answer is actually yes, and there is a really nice proof of the result.

Proposition.

Proposition 1.14. Every red/blue coloring of the edges of K_6 contains a monochromatic triangle.

Proof. First, pick any vertex, like the bottom left one. There are five edges coming out of this vertex.

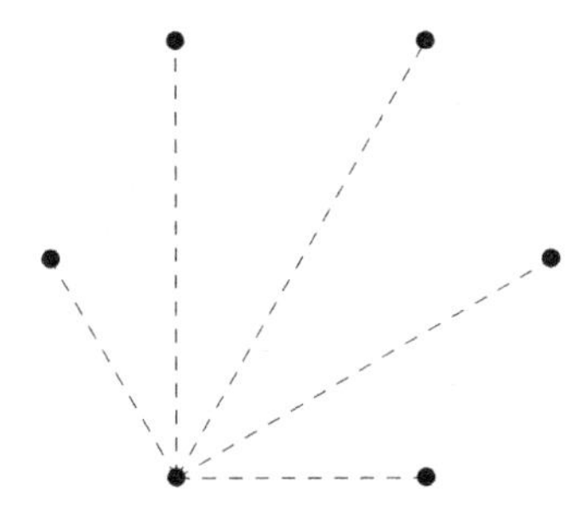

Now, remember that these five edges are actually colored. Each of the five edges received one of two colors, and so by the pigeonhole principle at least three of them received the same color.[49] Perhaps it looks like this:

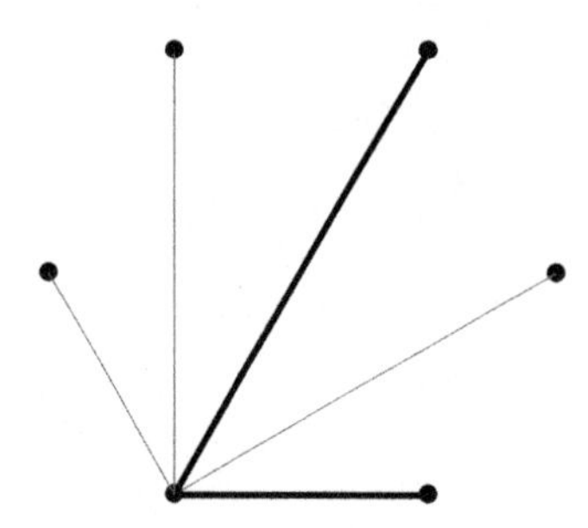

As you'll see, no matter which three (or four or five) are the same color, the proof proceeds in the same way. Indeed, focus now on the three thin red edges, and think about the three vertices on the other ends of these edges. These three vertices form the dashed triangle below.

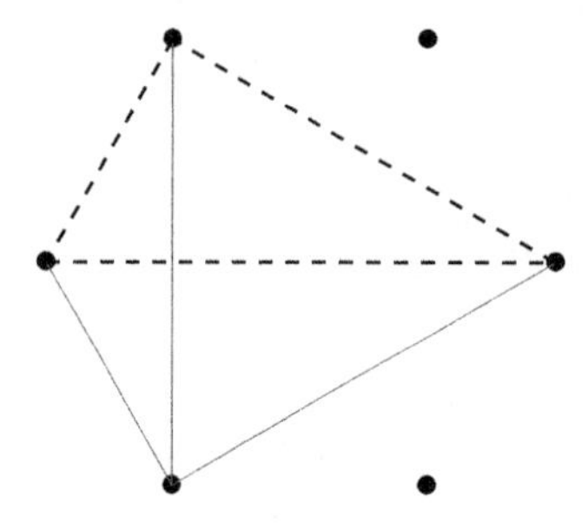

What color are these edges? There are two cases: Either at least one of these three edges is thin red, or they are all bold blue.

If at least one of them is thin red, then that thin red edge forms a red triangle. For example:

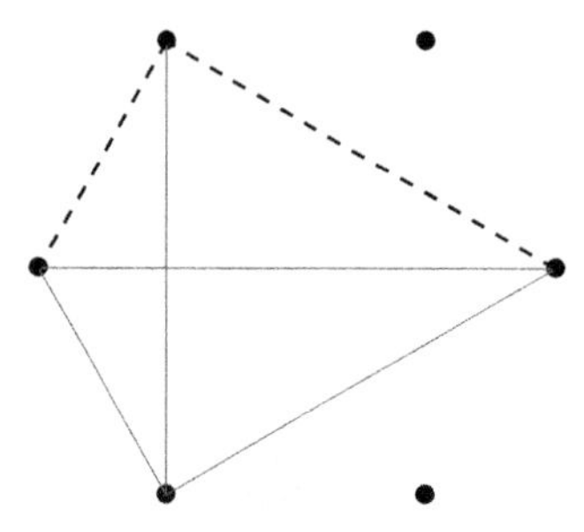

And since any of these edges being thin red makes a red triangle, this case is guaranteed to end in a monochromatic triangle.

[49]Yes, this is a very basic form of the pigeonhole principle, but there are later theorems in Ramsey theory which rely crucially on the principle, including a classic proof of van der Waerden's theorem, which is one of the most important theorems in the field. (Look it up!)

The second case is if none of these edges is thin red, but that also gives a monochromatic triangle, a bold blue one:

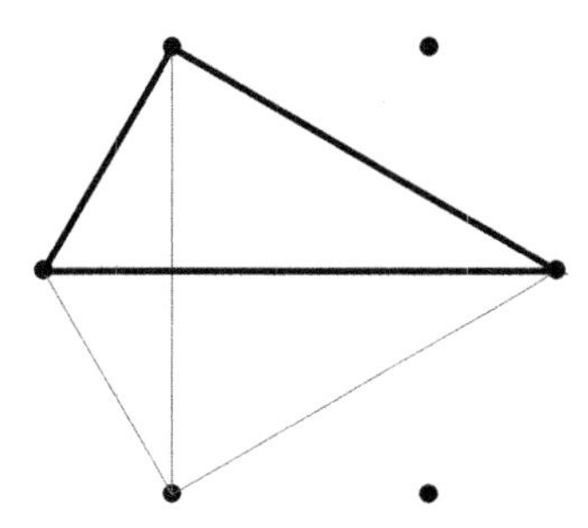

So either way we get a monochromatic triangle, completing the proof. □

So every red/blue coloring of K_6 gives a monochromatic triangle. Since K_6 is contained inside K_n for all $n > 6$, this also holds for larger complete graphs (convince yourself of this). Does it hold for smaller complete graphs? Well, certainly not *all* smaller ones. For instance, not all red/blue colorings of K_3 contain a monochromatic triangle, as here's one coloring that doesn't:

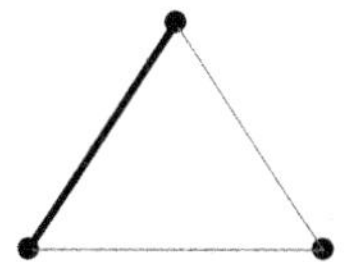

Of course, the all-red coloring or the all-blue coloring both contain a monochromatic triangle (the whole thing), but what we care about is whether *every* coloring contains a monochromatic triangle, and the above coloring shows that not every coloring does.

Likewise, there are many colorings of K_4 which contain a monochromatic triangle, but not *every* coloring does, since here's a coloring that does not:

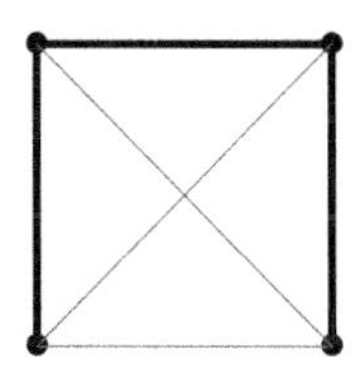

What about K_5? Once again, not every red/blue coloring has a monochromatic triangle. Indeed, here is such a coloring:

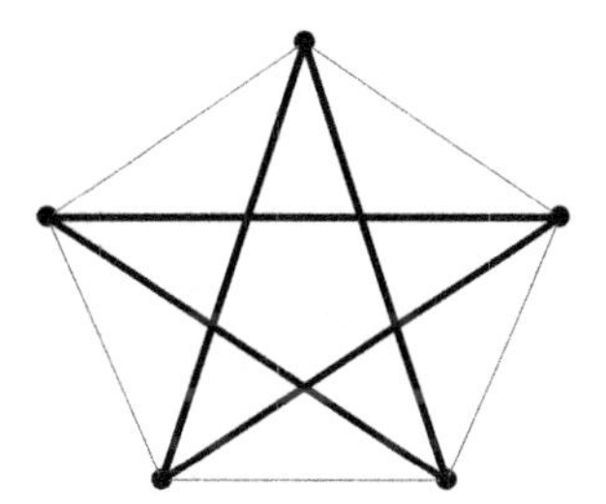

We have shown that 6 is the smallest n for which every 2-coloring of K_n contains a monochromatic K_3 (since a K_3 is a triangle). Thus, 6 is said to be the (symmetric) *Ramsey number* for the K_3 problem.[50]

Definition 1.15. The *Ramsey number* $R(t)$ is the smallest n for which every red/blue coloring of the edges of K_n contains a monochromatic K_t.

For example, above we showed that $R(3) = 6$. It is known that for any t, such an n must exist. Yet finding each n is a *really* hard problem. I've heard one top combinatorist assert with complete confidence that no human will ever find a general formula for $R(t)$. He claims that this problem is simply too hard. Humans are not smart enough.[51]

Here are the known (symmetric) Ramsey numbers:

- $R(1) = 1$ (trivial)
- $R(2) = 2$ (trivial)
- $R(3) = 6$ (We solved above.)
- $R(4) = 18$ (Takes a few pages of clever math and an algebraic construction.)
- $R(5) =$ ¯_(ツ)_/¯ (All we know is that it's between 43 and 48.)[52]
- $R(6) =$ ¯_(ツ)_/¯ (Humans may never know. It's between 102 and 165.)[53]

In Appendix A, we will prove a lower bound on $R(t)$, but this makes little progress in finding exact Ramsey numbers. Just look at the above list—see how quickly the problem moves from trivial to difficult to impossible? In fact, a witty Erdős quote sums this up best:

> "Suppose aliens invade the earth and threaten to obliterate it in a year's time unless human beings can find the Ramsey number for red five and blue five. We could marshal the world's best minds and fastest computers, and within a year we could probably calculate the value. If the aliens demanded the Ramsey number for red six and blue six, however, we would have no choice but to launch a preemptive attack."[54]
>
> – Paul Erdős

[50]The asymmetric cases are when you want to know whether every coloring contains, for example, a red K_3 or a blue K_5. Note that if you were trying to avoid both a red K_3 and a blue K_5, you would use many more blue edges than red edges.

[51]I am unable to comment on his general sense of optimism/pessimism.

[52]For what it's worth, some experts strongly believe it is 43.

[53]For what it's worth, some experts have absolutely no idea.

[54]Or, at least, we would have to get schwifty.

Chapter 2: Direct Proofs

If your professor asked you to prove that "every perfect number is even," then you would probably ask them what the heck a perfect number is. Definitions are really important in math — they give us precision. They are also subjective, human choices. The math is deep and intrinsic; definitions are our inventions to make it easier to discuss the math.

Deciding on a definition can be difficult, too. It can be a challenge to precisely write down what something is, and do so in a way that excludes the things that it is not, in such a way that makes it easy to work with and apply. As a fun example of this difficulty, imagine that you were writing a dictionary and were trying to define a *sandwich*. Right now, try to come up your own definition of a sandwich.

Got one? Good. Does your definition require bread? Meat? Cheese? Vegetables? Do you require a certain quantity of these things? Does it attempt to classify them abstractly? If you demand meat between bread, you rule out vegetarian sandwiches and grilled cheese, while counting hot dogs. Are you ok with that? And what counts as "bread" anyways? Any carb? Is a quesadilla a sandwich? You would have to carefully define that term if you plan on using it.

Must the bread be on top and bottom? Do you want to include open-faced sandwiches? Probably some, but probably not pizza or an "open-faced PB&J," let alone some buttered toast? Leniency with your bread is important, but if you are too lenient you may accidentally include burritos or veggie wraps. And if you don't want those (but maybe you do?), then demanding two slices of bread would exclude a submarine sandwich. You might say a sub is ok because it leaves one side open, but so does a taco and a bread bowl of clam chowder.

A club sandwich is definitely a sandwich, but it includes bread in the middle. But if that is ok, what about a Big Mac or a slice of lasagna? What about a mushroom burger? Can a sandwich be sweet? Which definition would allow this without including a Pop-Tart? Is a cookie an open-faced sandwich? As you can tell, it is sometimes tough to get a definition right.[1]

Indeed, when considering the statement "every perfect number is even," it is important that you know the definition of a perfect number; in fact, it would also be a good idea to ask for the precise definition of an *even number*. You intuitively know that $2, 4, 6, 8, \dots$ are the (positive) even numbers, but there are potentially multiple ways to define such a number, and we should all be on the same page as to

[1] Search the hashtag #HoagieHomies on Twitter. You... may be surprised.

which definition we are working with. In Chapter 1 we were more relaxed because we wanted to jump into making mathematical arguments, but from here on out we will be precise and deliberate. Indeed, in a moment we will define even and odd numbers. But first, recall that the set of *integers* is $\{\ldots, -3, -2, -1, 0, 1, 2, 3, \ldots\}$ and the following basic fact.[2]

Fact.

Fact 2.1. The sum of integers is an integer, the difference of integers is an integer, and the product of integers is an integer. Also, every integer is either even or odd.

We are calling these facts because, while they are true and one could prove them, we will not be proving them here. That would go beyond the scope of this text. We will use these facts and you are allowed to use them, too. They can be thought of as the *axioms* for this book—things which are assumed to be true and on which you build a theory.

2.1 Working From Definitions

Definition.

Definition 2.2.

- An integer n is *even* if $n = 2k$ for some integer k;
- An integer n is *odd* if $n = 2k + 1$ for some integer k.

Mathematical definitions are precise. If you change something small about a sandwich, you might still count it as a sandwich. The same is not true in math. If k is an integer and $n = 2k + 1$, then n is odd. However, if k is an integer and $n = 2k + 1.000001$, then n is no longer odd. The lines are sharp in mathematics.[3] Below are some examples of even and odd integers, where we justify each claim showing how it satisfies the definition.

[2]We have to start from somewhere, and we will begin with the assumption that you know what the integers are as well as their very basic properties as laid out in Fact 2.1. We will also use the standard arithmetic facts. Some examples: If a and b are real numbers, then $a + b = b + a$ and $ab = ba$. And if c is also a real number, then $(a + b) + c = a + (b + c)$ and $(ab)c = a(bc)$ and $a(b + c) = ab + ac$ and $a^b a^c = a^{b+c}$. But as you'll see, we will be assuming very little else—the rest we will prove ourselves.

[3]"When you are writing laws you are testing words to find their utmost power. Like spells, they have to make things happen in the real world, and like spells, they only work if people believe in them." –Hilary Mantel in her novel *Wolf Hall*.

Example 2.3.

- 6 is even because $6 = 2 \cdot 3$, and 3 is an integer;
- 9 is odd because $9 = 2 \cdot 4 + 1$, and 4 is an integer;
- 0 is even because $0 = 2 \cdot 0$, and 0 is an integer; and
- -15 is odd because $-15 = 2 \cdot (-8) + 1$, and -8 is an integer.

Without a definition, someone might wonder whether zero or negative integers should count as even or odd. The great thing about definitions is that there are no ambiguities. You either satisfy the definition or you don't.[4] Since zero and negative integers satisfy the definition, they count as even/odd.

Let's now prove some results by using this definition.

Proposition.

Proposition 2.4. The sum of two even integers is even.

Proof Idea. First, make sure it is clear to you what we are assuming and what you are trying to prove. The above is equivalent to saying

"If two integers are both even, then their sum is also even."

Or:

"If n and m are even integers, then $n + m$ is an even integer."

The proposition doesn't use the "if. . . , then. . . " format, and doesn't use variable names to refer to the numbers, but these are all equivalent. Indeed, the third statement, in which the numbers are named using variables, n and m, is going to be useful for the proof; the proof will begin by doing just that.

As for the mechanics of the proof, the big picture is this:

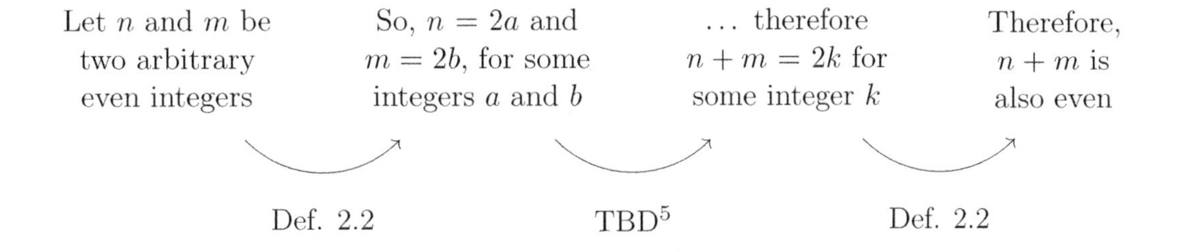

[4] *"No one is above the law!"* . . . (Also, no seven is above the law. And no zero or π either. No number is above the law, is what I think that quote is getting at.)

[5] This means "to be determined." This step depends on the problem. More on this later.

That is, we use the definition of even integers to translate the problem to one that is just about integers, then we solve the integer problem (that's the middle "to be determined" step), then we translate what we found back to a conclusion about even integers. The algebra will need to be worked out in our proof, but that is the overview. Ok, let's prove it.

Proof. Assume that n and m are even integers. By Definition 2.2, this means that $n = 2a$ and $m = 2b$, for some integers a and b. Then,

$$n + m = 2a + 2b = 2(a + b).$$

And since, by Fact 2.1, $a + b$ is an integer too, we have shown that $n + m = 2k$, where $k = a + b$ is an integer. Therefore, by Definition 2.2, this means that $n + m$ is even. □

That was fun. Let's do more!

Proposition.

Proposition 2.5. The sum of two odd integers is even.

Proof Idea. As with Proposition 2.4, this proposition is not phrased in the "if..., then..." form, but it is equivalent to saying "If n and m are odd integers, then $n+m$ is an even integer." The overview of this proof is very similar to the last one:

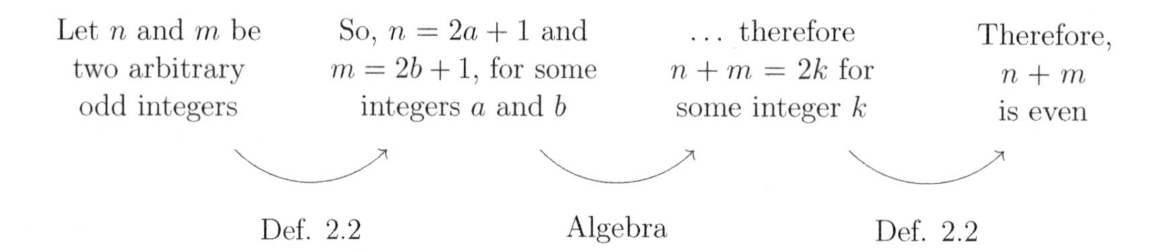

Let's do it!

Proof. Assume that n and m are odd integers. By Definition 2.2, this means that $n = 2a + 1$ and $m = 2b + 1$, for some integers a and b. Then,

$$n + m = (2a + 1) + (2b + 1) = 2a + 2b + 2 = 2(a + b + 1).$$

And since, by Fact 2.1, $a + b + 1$ is an integer too, we have shown that $n + m = 2k$, where $k = a+b+1$ is an integer. Therefore, by Definition 2.2, this means that $n+m$ is even. □

Let's do one more like this.

Proposition.

Proposition 2.6. If n is an odd integer, then n^2 is an odd integer.

This proof will be similar to the last two, and so this is an especially good proposition to try to prove on your own before reading on.

Proof. Assume that n is an odd integer. By Definition 2.2, this means that $n = 2a+1$ for some integer a. Then,

$$n^2 = (2a+1)^2 = 4a^2 + 4a + 1 = 2(2a^2 + 2a) + 1.$$

And since, by Fact 2.1, $2a^2 + 2a = 2 \cdot a \cdot a + 2 \cdot a$ is an integer too,[6] we have shown that $n^2 = 2k + 1$, where $k = 2a^2 + 2a$ is an integer. Therefore, by Definition 2.2, this means that n^2 is odd. □

For practice,[7] try to prove the following on your own:

- The sum of an even integer and an odd integer is odd;
- The product of two even integers is even;
- The product of two odd integers is odd;
- The product of an even integer and an odd integer is even;
- An even integer squared is an even integer.

— A Few Comments on "if…, then" statements —

You'll notice that — perhaps with a little rewriting, like with Propositions 2.4 and 2.5 — most of our results in this chapter take on this standard form:

If ≪statement≫ is true, then ≪other statement≫ is also true.

For example, "If you live in Los Angeles, then you live in California."[8] Or:

"If m and n are even, then $m + n$ is also even."

Another way to summarize such statements is this:

≪some statement is true≫ implies ≪some other statement is true≫.

[6] In case this is a little confusing, we are technically using Fact 2.1 many times: a and a are integers, so $a \cdot a = a^2$ is too. So $2a^2$ is too. Likewise, $2a$ is too. So $2a^2 + 2a$ is too.

[7] Or because your professor made you, see Exercise 2.3.

[8] But, again, perhaps it was not rewritten yet in this "if…, then…" form. Perhaps this implication was written as "You live in California if you live in Los Angeles" or "Every LA resident is a Californian."

For example, "Living in Los Angeles implies living in California." Or:

"m and n being even implies $m + n$ is even."

Since so many of our results can be broken down like this, mathematicians have given the word "implies" a special symbol: "$\Longrightarrow$". So the shorthand for the above is this:

"m and n being even $\Longrightarrow m + n$ is even."

And so, a general "if, then" statement is of the form "$P \Longrightarrow Q$," where P and Q are each statements.[9]

Symbols like this are commonly used in scratch work when you're still figuring out how to prove your homework problems, or when you are writing up solutions on an exam and are crunched for time. This is a good thing — proofs are usually obtained only after a lot of scratch work, and writing stuff down is a good way to generate ideas. However, when writing formally, like for the final draft of your homework, these symbols are rarely used. You should write out solutions with words, complete sentences, and proper grammar. Pick up any of your math textbooks, or look online at math research articles, and you will find that such practices are standard.[10]

— The Structure of Direct Proofs —

The proofs we did on evenness/oddness are called *direct proofs*. A direct proof is a way to prove a "$P \Rightarrow Q$" proposition by starting with P and working your way to Q. The "working your way to Q" stage often involves applying definitions, previous results, algebra, logic and techniques. Later on we will learn other proof methods. Here is the general structure of a direct proof:

Proposition. $P \Rightarrow Q$.

Proof. Assume P.

≪ An explanation of what P means ≫ ⟵ Apply definitions and/or other results.

⋮ apply algebra, logic, techniques

≪ Hey look, that's what Q means ≫

Therefore Q. □

Take a look at the proofs of Propositions 2.4, 2.5 and 2.6 and see if you can identify this general structure in each one.

[9]We will discuss all this in much finer detail in Chapter 5.

[10]For more advice on proof writing, check out Appendix C at the end of this book!

2.2 Proofs by Cases

A related proof strategy is *proof by cases*. This is a "divide and conquer" strategy where one breaks up their work into two or more cases. If you read the Bonus Examples section of Chapter 1, then you saw your first example of a proof by cases in the proof of Proposition 1.12.

The example below is a proof by cases which at the same time will give us more practice with direct proofs involving definitions. Indeed, when you break up a problem in two parts, those two parts still need to be proven, and a direct proof is often the way to tackle each of those parts.

Proposition.

Proposition 2.7. If n is an integer, then $n^2 + n + 6$ is even.

Proof Idea. At this point, if I asked you to prove "if n is even, then $n^2 + n + 6$ is even" or if I asked you to prove "if n is odd, then $n^2 + n + 6$ is even," then you would know what to do: You would prove it directly, just like what we did in Propositions 2.4, 2.5 and 2.6.

For example, to show that "if n is odd, then $n^2 + n + 6$ is even" you would write n as $2a + 1$ and you'd plug it in: This would turn $n^2 + n + 6$ into $(2a + 1)^2 + (2a + 1) + 6$. Finally, you would do some algebra to try to write this as $2k$ for some integer k. If you can do this, then you have successfully proved that if n is odd, then $n^2 + n + 6$ is even.

In this problem, you're asked to prove that every integer has this property. Do you see what to do? ... Since every integer is either even or odd, if we prove the proposition for even n, and we prove the proposition for odd n, then combined we have proven it for all integers! This is what a *proof by cases* is all about.

Proof. Assume that n is an integer. Then n is either even or odd.

Case 1: n is even. Assume n is even. Then $n = 2a$ for some integer a. Thus,

$$\begin{aligned} n^2 + n + 6 &= (2a)^2 + (2a) + 6 \\ &= 4a^2 + 2a + 6 \\ &= 2(2a^2 + a + 3). \end{aligned}$$

And since a is an integer, $2a^2 + a + 3$ is also an integer by Fact 2.1. Therefore, $n^2 + n + 6 = 2k$ where $k = 2a^2 + a + 3$ is an integer, which by the definition of an even integer (Definition 2.2) means that $n^2 + n + 6$ is even.

Case 2: n is odd. Assume n is odd. Then $n = 2a + 1$ for some integer a. Thus,

$$\begin{aligned} n^2 + n + 6 &= (2a+1)^2 + (2a+1) + 6 \\ &= (4a^2 + 4a + 1) + (2a + 1) + 6 \\ &= 4a^2 + 6a + 8 \\ &= 2(2a^2 + 3a + 4). \end{aligned}$$

And since a is an integer, $2a^2 + 3a + 4$ is also an integer by Fact 2.1. Therefore, $n^2 + n + 6 = 2k$ where $k = 2a^2 + 3a + 4$ is an integer, which by the definition of an even integer (Definition 2.2) means that $n^2 + n + 6$ is even.

We have shown that $n^2 + n + 6$ is even whether n is even or odd. Combined, this shows that $n^2 + n + 6$ is even for all integers n, completing the proof.[11] □

Here are four examples of cases that you might see in the future (and if you are wondering what a "mod" is, we will discuss that very soon!):

Case 1: n is prime
Case 2: n is composite

Case 1: f is continuous
Case 2: f is not continuous

Case 1: $\sum_{k=1}^{\infty} a_k$ converges
Case 2: $\sum_{k=1}^{\infty} a_k$ diverges

Case 1: $n \equiv 0 \pmod 3$
Case 2: $n \equiv 1 \pmod 3$
Case 3: $n \equiv 2 \pmod 3$

A proof by cases divides up the possibilities into more manageable chunks. If the theorem refers to a collection of elements and your proof is simply checking each element individually, then it is called a *proof by exhaustion* or a *brute force proof.*

2.3 Divisibility

In this section, we will use direct proofs to prove some propositions about divisibility. To begin, we must define what it means to say that one integer *divides* another.

But first, what *should* the definition be? We say that "2 divides 8" because $\frac{8}{2} = 4$, and 4 is an *integer*. Likewise, we say "3 divides 18" because $\frac{18}{3} = 6$, and 6 is an integer. On the other hand, we say "4 does not divide 10" because $\frac{10}{4} = 2.5$, and 2.5 is *not* an integer.

[11]Likewise, if I showed you the equation $4n^2 + 7$ and you said to yourself, "huh, that's an odd equation," then you'd be exactly right! For any integer n, $4n^2 + 7$ is odd. And the proof is by cases, similar to that of Proposition 2.7. If $n = 2a$ for an integer a, then $4n^2 + 7 = 16a^2 + 7 = 2(8a^2 + 3) + 1$, which is odd. And if $n = 2a + 1$, then $4n^2 + 7 = 16a^2 + 16a + 11 = 2(8a^2 + 8a + 5) + 1$, which is again odd. See if you can fill in the details.

So one definition could be

$$\text{“}a\text{ divides }b\text{” if }\tfrac{b}{a}\text{ is an integer.}$$

That's a perfectly good definition, but it will be easier to apply the definition if it includes what the integer actually is (as you will see in the proof of our next proposition). So another option would be

$$\text{“}a\text{ divides }b\text{” if }\tfrac{b}{a}=k\text{ where }k\text{ is an integer.}$$

But, as it turns out, we can do even better.[12] By multiplying over the 'a' we obtain

$$\text{“}a\text{ divides }b\text{” if }b=ak\text{ where }k\text{ is an integer.}$$

Although the definitions are all the same, this is the one that will be the easiest to work with. And although this may have seemed like a boring, pointless discussion, there is something significant underlying it: Definitions do not fall out of the sky — they are carefully chosen by mathematicians to do the work we seek. This will become an important theme as we move forward.

Definition.

Definition 2.8. A nonzero integer a is said to *divide* an integer b if $b = ak$ for some integer k. When a does divide b, we write "$a \mid b$" and when a does not divide b we write "$a \nmid b$."

Reminder: Saying that a divides b if "$b = ak$ for some integer k" means that there exists an integer k for which $b = ak$. For example, 3 divides 12 because $12 = 3 \cdot 4$ and 4 is an integer. So there does exist "some" integer k; in particular, $k = 4$ works. Below are more examples.

Example 2.9.

- $2 \mid 14$ because $14 = 2 \cdot 7$ and 7 is an integer.
- $-4 \mid 20$ because $20 = -4 \cdot (-5)$ and -5 is an integer.
- $12 \mid -48$ because $-48 = 12 \cdot (-4)$ and -4 is an integer.
- $-7 \mid -7$ because $-7 = -7 \cdot 1$ and 1 is an integer.
- $6 \nmid 9$ because $9 \neq 6k$ for any integer k. (We can write $9 = 6 \cdot 1.5$, but 1.5 is not an integer.)
- The $b = 0$ case: $a \mid 0$ for every nonzero integer a, because $0 = a \cdot 0$ for every such a, and 0 is an integer.

[12] And odd better.

Note: A common mistake is to see something like "$2 \mid 8$" and think that this equals 4. The expression "$a \mid b$" is either true or false, it never equals a number; $2 \mid 8$ is true, while $3 \mid 8$ is false. This mistake is understandable because $2 \mid 8$ looks a lot like $2/8$ or $8/2$, and while these are all related, they are also all different.

Armed with Definition 2.8, let's use a direct proof to prove our first result on divisibility—the *transitive* property of divisibility.

Proposition.

Proposition 2.10. Let a, b and c be integers. If $a \mid b$ and $b \mid c$, then $a \mid c$.

Note: By assuming that "$a \mid b$," this inherently means that a is nonzero, because according to Definition 2.8, the only a-values for which this is true are ones in which a is nonzero. Likewise, by assuming that "$b \mid c$," it must also be the case that b is nonzero. Let's now begin our scratch work to prove this proposition.

Scratch Work. When possible, it's a good idea to do a couple of examples to convince yourself that what you're asked to prove is actually true. Oftentimes, this will also help you prove the result, as working through some examples can help show you why it is true. Let's test it.

- If we choose $a = 3$, $b = 12$ and $c = 24$, then it is indeed true that $3 \mid 12$ (because $12 = 3 \cdot 4$ and 4 is an integer) and $12 \mid 24$ (because $24 = 12 \cdot 2$ and 2 is an integer). According to this proposition, it must then be true that $3 \mid 24$, and indeed that is true (because $24 = 3 \cdot 8$ and 8 is an integer).[13]

$$a \mid b \text{ and } b \mid c \Longrightarrow a \mid c\ ?$$
$$3 \mid 12 \text{ and } 12 \mid 24 \Longrightarrow 3 \mid 24\ \checkmark$$

- On your own, do another example. And when you do so, remember that this proposition is only referring to a, b and c for which $a \mid b$ and $b \mid c$. So if $a = 3$ and $b = 4$, then we already know $a \nmid b$ and so the proposition does not apply. But if $a \mid b$ and $b \mid c$, then the proposition guarantees that $a \mid c$.

Using the intuition we gained from these examples, let's now discuss how to prove the proposition.

[13] And you might notice that the "k" integer (from Definition 2.8) in the $a \mid b$ is 4, in the $b \mid c$ is 2, and in the $a \mid c$ is 8. And these three numbers seem connected: $8 = 4 \cdot 2$. How interesting! Perhaps that holds in general, and perhaps that will be useful for the proof! Test this again when you do another example on your own.

Remember the general structure of a direct proof:

> ***Proof.*** Assume P.
>
> $\langle\!\langle$ An explanation of what P means $\rangle\!\rangle$ $\longleftarrow$ Apply definitions and/or other results.
>
> ⋮ apply algebra, logic, techniques
>
> $\langle\!\langle$ Hey look, that's what Q means $\rangle\!\rangle$
>
> Therefore Q. $\square$

Here, P is our assumption: $a \mid b$ and $b \mid c$. And Q is what we are trying to prove: $a \mid c$. An explanation of what P means is simply applying the definition of divisibility: $a \mid b$ and $b \mid c$ mean $b = as$ for some integer s, and $c = bt$ for some integer t. What we are asked to show is that $a \mid c$, which by definition means that we need to show that $c = ak$ for some integer k. Updating the above outline gives us this:

> ***Proof.*** Assume that a, b and c are integers, and $a \mid b$ and $b \mid c$.
>
> Then by the definition of divisibility (Definition 2.8), $b = as$ for some integer s, and $c = bt$ for some integer t.
>
> ⋮ apply algebra, logic, techniques
>
> Therefore $c = ak$ for an integer k.
>
> Therefore $a \mid c$. $\square$

There's just a little work to go to bridge the gap, but it turns out that some algebra does the trick. Now, finally, here's the formal proof.

Proof. Assume that a, b and c are integers, $a \mid b$ and $b \mid c$. Then, by the definition of divisibility (Definition 2.8), $b = as$ for some integer s, and $c = bt$ for some integer t. Thus,

$$\begin{aligned} c &= bt \\ &= (as)t \\ &= a(st). \end{aligned}$$

We have shown that $c = a(st)$, and since s and t are integers, so is st by Fact 2.1. So it is indeed true that $c = ak$ for the integer $k = st$, which by the definition of divisibility (Definition 2.8) means $a \mid c$. $\square$

The Division Algorithm

Now, it's often the case that two integers do not divide each other. For example, $3 \nmid 7$. A common way to think about this is that if you tried to divide 7 by 3, you would get a remainder of 1:

$$7 = 3 \cdot 2 + 1.$$

In the above, the '2' is called the *quotient* and the '1' is called the *remainder*. The fact that we can relate numbers in such a way is called *the division algorithm*[14] and is important enough to be granted the stature of...[15]...our first theorem!

Theorem.

Theorem 2.11 (*The division algorithm*). For all integers a and m with $m > 0$, there exist unique integers q and r such that

$$a = mq + r$$

where $0 \leq r < m$.

Note that if $m = 2$, then the two options are $a = 2q + 0$ and $a = 2q + 1$; these are the definitions of even and odd numbers! And if $r = 0$, then this produces $a = mq$, which is the divisibility definition (in this case, for $m \mid a$)! Here are a few more examples of expressing numbers in the form of the division algorithm:

- If $a = 18$ and $m = 7$, then $18 = 7 \cdot 2 + 4$. Note that $0 \leq 4 < 7$.
- If $a = 13$ and $m = 3$, then $13 = 3 \cdot 4 + 1$. Note that $0 \leq 1 < 3$.
- If $a = 35$ and $m = 5$, then $35 = 5 \cdot 7 + 0$. Note that $0 \leq 0 < 5$.
- If $a = -18$ and $m = 7$, then $-18 = 7 \cdot (-3) + 3$. Note that $0 \leq 3 < 7$.
- If $a = 3$ and $m = 13$, then $3 = 13 \cdot 0 + 3$. Note that $0 \leq 3 < 13$.
- If $a = -3$ and $m = 5$, then $-3 = 5 \cdot (-1) + 2$. Note that $0 \leq 2 < 5$.

We will prove the division algorithm in the Bonus Examples section of Chapter 7.

[14] The fact that this theorem is called an *algorithm* is a misnomer. It got this name because there is a related algorithm. Kind of like how koala bears aren't actually bears, strawberries aren't actually berries but an avocado actually is, and this is called a footnote but is really just an excuse for me to say that a banana is a berry but a raspberry is not and the world needs to know!

[15] *...Drum roll please...*

2.4 Greatest Common Divisors

Definition.

Definition 2.12. Let a and b be integers. If $c \mid a$ and $c \mid b$, then c is said to be a *common divisor* of a and b.

The *greatest common divisor* of a and b is the largest integer d such that $d \mid a$ and $d \mid b$. This number is denoted $\gcd(a, b)$.

Suppose for a moment that a and b are nonzero. Note that since $1 \mid a$ and $1 \mid b$, the number 1 is guaranteed to be a common divisor of a and b, and hence the *greatest* common divisor will always be at least 1. It is also the case that the greatest common divisor is never larger than $|a|$ or $|b|$; this means that $1 \leq \gcd(a, b) \leq |a|$ and $1 \leq \gcd(a, b) \leq |b|$, which in particular means that $\gcd(a, b)$ always exists. Below are some examples of gcds.

- $\gcd(6, 8) = 2$
- $\gcd(6, -8) = 2$
- $\gcd(-5, -20) = 5$
- $\gcd(12, 8) = 4$
- $\gcd(7, 15) = 1$
- $\gcd(9, 9) = 9$

However, there is one pair of integers that does not have a greatest common divisor; if $a = 0$ and $b = 0$, then *every* positive integer d is a common divisor of a and b. This means that no divisor is the *greatest* divisor, since you can always find a bigger one. Thus, in this one case, $\gcd(a, b)$ does not exist.

Next up is a pretty neat theorem about greatest common divisors, which has a challenging but interesting proof.

Theorem.

Theorem 2.13 (*Bézout's identity*). If a and b are positive integers, then there exist integers k and ℓ such that

$$\gcd(a, b) = ak + b\ell.$$

Proof Idea. To make sure we understand this, let's jot down an example. Maybe $a = 12$ and $b = 20$, making $\gcd(12, 20) = 4$. The claim is that there are integers k and ℓ such that

$$\gcd(12, 20) = 12k + 20\ell.$$

Indeed, $\gcd(12, 20) = 4$, and by testing a few numbers one can find that

$$4 = (12)(2) + (20)(-1).$$

Or maybe you found that

$$4 = (12)(-3) + (20)(2).$$

Indeed, there are (infinitely) many solutions! Nevertheless, this theorem simply says that at least one solution must exist. Pretty cool! But how do we prove it? This will be our general structure:

1. Assume a and b are positive integers. These should be thought of as fixed numbers (like $a = 12$ and $b = 20$), which means that $\gcd(a, b)$ is also a fixed number (like $\gcd(a, b) = 4$). But despite being fixed, we don't know what they are. What we get to control is the k and ℓ—we want to choose those in such a way that $\gcd(a, b) = ak + b\ell$.

2. As it turns out, there is a clever way to choose the correct k and ℓ. But once we pick the correct k and ℓ, we still have to prove that they work. Once we choose them, the sum $ak + b\ell$ will be equal to something (we will call this sum d; that is, $d = ak + b\ell$), and then the goal turns to proving that d is what we want: We need to show $d = \gcd(a, b)$.

3. To show that d is in fact $\gcd(a, b)$, we will use the definition of the greatest common divisor. Once we accomplish this, then $d = ak + b\ell$ from the previous step will turn into $\gcd(a, b) = ak + b\ell$, completing the proof.

As you will see, the most difficult step will be step 3. It will be solved in two parts:

Part 1. Prove that d is a common divisor of a and b.

Part 2. Prove that d is greater than any other common divisor of a and b.

That's the idea behind the proof of Bézout's identity. Let's now formally prove it.

Proof. Assume that a and b are fixed positive integers. Notice that, for integers x and y, the expression $ax + by$ can take many different values, including positive values, negative values and (if $x = y = 0$) can even be zero. Let d be the *smallest positive* value that $ax + by$ can equal.[16] We now let k and ℓ be the x and y values that give this minimum value of d. That is, for these integers k and ℓ,

$$d = ak + b\ell. \qquad (☕)^{17}$$

Our goal in this proof is to find some k and ℓ such that $\gcd(a, b) = ak + b\ell$. As it turns out, $d = ak + b\ell$ is the exact equation we are looking for—we just need to prove that $d = \gcd(a, b)$. We defined d to be the smallest positive value that $ax + by$ can take; to prove that this same d is the $\gcd(a, b)$, we must prove that d is a common divisor of a and b, and then that it is the *greatest* common divisor. We will prove these two parts separately.

[16]That is, if there exist x and y such that $ax + by = 1$, then $d = 1$. But if no such x and y exist, but there exist x and y such that $ax + by = 2$, then $d = 2$. And so on. For example, if $a = 4$ and $b = 10$, then $d = 2$ because there are no x and y for which $ax + by = 1$ (try to convince yourself of this by thinking about even integers), but $4 \cdot (-2) + 10 \cdot 1 = 2$ does work, showing that $d = 2$ is the smallest positive value.

[17]Later in this proof, I will refer to this equation by writing "(by ☕)." Most books use a small star each time, but just for fun I will use a variety of little symbols, like this little cup of tea.

Part 1: d is a common divisor of a and b. By Definition 2.12, d is a common divisor of a and b if $d \mid a$ and $d \mid b$. To see that $d \mid a$, note that by the division algorithm there exist integers q and r such that

$$a = dq + r$$

with $0 \leq r < d$. By rewriting this,

$$\begin{aligned} r &= a - dq \\ &= a - (ak + b\ell)q && \text{(by ☕)} \\ &= a - akq - b\ell q \\ &= a(1 - kq) + b(-\ell q). \end{aligned}$$

And since $(1 - kq)$ and $(-\ell q)$ are both integers by Fact 2.1, we have found another expression of the form $ax + by$. But remember, d was chosen to be the *smallest positive* number that can be written like this. So, since r can be written like this too, and $0 \leq r < d$ (and remember, 0 is not considered positive), it must be that $r = 0$.

This is what we wanted to show, since $r = 0$ means that $a = dq + r$ is simply $a = dq$, which by the definition of divisibility (Definition 2.8) means that $d \mid a$, as desired.

In the same exact way, one can also show that $d \mid b$. Collectively, these prove that d is a common divisor of a and b.

Part 2: d is the *greatest* common divisor of a and b. Suppose that d' is some other common divisor of a and b. In order to conclude that d is the *greatest* among all common divisors, we must show that $d' \leq d$. To do this, observe that since d' is a common divisor, $d' \mid a$ and $d' \mid b$, which by the definition of divisibility (Definition 2.8) means that

$$a = d'm \qquad \text{and} \qquad b = d'n,$$

for some integers m and n. Then, applying the above to Equation (☕),

$$\begin{aligned} d &= ak + b\ell \\ &= d'mk + d'n\ell \\ &= d'(mk + n\ell). \end{aligned}$$

We have shown that $d = d'(mk + n\ell)$ where $(mk + n\ell)$ is an integer (by Fact 2.1). With d being positive and $d' = \frac{d}{mk+n\ell}$ where the denominator is an integer, this implies $d' \leq d$. We have shown that d is larger than any other divisor of a and b, which means that d is in fact the greatest common divisor of a and b.

Thus,

$$\gcd(a, b) = d = ak + b\ell,$$

which completes the proof. $\square$

That was a tough one! Spending most of our time on shorter proofs makes sense, as those are the ones which allow us to focus on the proof mechanics without getting too distracted by complicated ideas. However, it is also beneficial to go over some complicated ones, to see where you are headed.

A proof like this seems difficult now, and it is perfectly fine if you did not understand it completely, but once you have a few proof-based classes under your belt, proofs like this will seem much more manageable. Throughout this book, I will throw in a challenging proof from time to time for this very purpose — including another one before the end of this chapter.

We just proved Bézout's[18] identity, which is surprisingly useful to prove further results. For example, one can prove that for positive integers a, b and m,

$$\gcd(ma, mb) = m \cdot \gcd(a, b).$$

The proof roughly goes as follows:

$$\begin{aligned}\gcd(ma, mb) &= \text{the smallest positive value of } max + mby \\ &\qquad \text{among all possible choices of } x \text{ and } y \\ &= m \cdot (\text{the smallest positive value of } ax + by) \\ &\qquad \text{among all possible choices of } x \text{ and } y \\ &= m \cdot \gcd(a, b).\end{aligned}$$

2.5 Modular Arithmetic

The division algorithm (Theorem 2.11) told us that if an integer a is being divided by an integer $m > 0$, then there are unique integers q and r such that

$$a = mq + r,$$

where $0 \le r < m$. The way to think about this is that we are dividing a by m, and doing so produces a remainder of r.

It turns out that the relationship between a and its remainder r is surprisingly important. In fact, in such a case we say that a is *congruent* to r.

Definition.

Definition 2.14. For integers a, r and m, we say that *a is congruent to r modulo m*, and we write $a \equiv r \pmod{m}$, if $m \mid (a - r)$.

[18]Pro-Tip: Bézout is a French name and is pronounced bay-zoo.

Repeating the division algorithm examples from page 58:

- $18 \equiv 4 \pmod{7}$
- $35 \equiv 0 \pmod{5}$
- $3 \equiv 3 \pmod{13}$
- $13 \equiv 1 \pmod{3}$
- $-18 \equiv 3 \pmod{7}$
- $-3 \equiv 2 \pmod{5}$

because 18 divided by 7 leaves a remainder of 4, and 13 divided by 3 leaves a remainder of 1, and so on. Here are those same six examples showing these remainders:

- $18 = 7 \cdot 2 + 4$
- $35 = 5 \cdot 7 + 0$
- $3 = 13 \cdot 0 + 3$
- $13 = 3 \cdot 4 + 1$
- $-18 = 7 \cdot (-3) + 3$
- $-3 = 5 \cdot (-1) + 2$

Or, we can see that those six mod examples are correct by using Definition 2.14:

- $7 \mid (18 - 4)$ ✓
- $5 \mid (35 - 0)$ ✓
- $13 \mid (3 - 3)$ ✓
- $3 \mid (13 - 1)$ ✓
- $7 \mid (-18 - 3)$ ✓
- $5 \mid (-3 - 2)$ ✓

As our examples show, if a divided by m leaves a remainder of r, then $a \equiv r \pmod{m}$. However, this is not the only way to have $a \equiv r \pmod{m}$—it is not required that r be the remainder of a divided by m, all that is required is that a and r have the *same* remainder when divided by m. For example,

- $18 \equiv 11 \pmod{7}$
- $2 \equiv 2 \pmod{13}$
- $-3 \equiv 7 \pmod{5}$
- $1 \equiv 13 \pmod{3}$
- $32 \equiv 42 \pmod{5}$
- $-15 \equiv -13 \pmod{2}$

Consider the first example above: 18 divided by 7 gives a remainder of 4, and 11 divided by 7 gives a remainder of 4. Since 18 and 11 have the same remainder when divided by 7, we have $18 \equiv 11 \pmod{7}$. And if you don't believe me, just check the definition. Definition 2.14 says that $18 \equiv 11 \pmod{7}$ provided that $7 \mid (18 - 11)$. And this is true! This is just saying that 7 divides 7, which is true, since $\frac{7}{7} = 1$, which is an integer.

– The Boxes Metaphor –

One way to think about modular congruence is with boxes. Let's think about a specific case: the integers modulo 6. Suppose you have a box with balls in it, and you are allowed to remove 6 at a time. If you start with 14 balls, then you can remove six to give you 8, and six again to give you 2. You can no longer remove six at a time, thus we are done. The 14 balls turned into 2, thus $14 \equiv 2 \pmod{6}$. A number, modulo 6, is congruent to whatever the *remainder* is after removing 6 at a time until you can't remove any more.

Moreover, $m \equiv n \pmod{6}$ if a box with m things in it, and a box with n things in it, will end up with the same number after removing 6 balls at a time. And the same applies for other "mods." For example, $9 \equiv 13 \pmod{4}$, because a box with 9 balls in it, after removing 4 balls at a time, will leave you with 1 ball. And a box with 13 balls in it, after removing 4 balls at a time, will also leave you with 1 ball. Because they leave you with the same number when removing 4 at a time, they are congruent modulo 4.

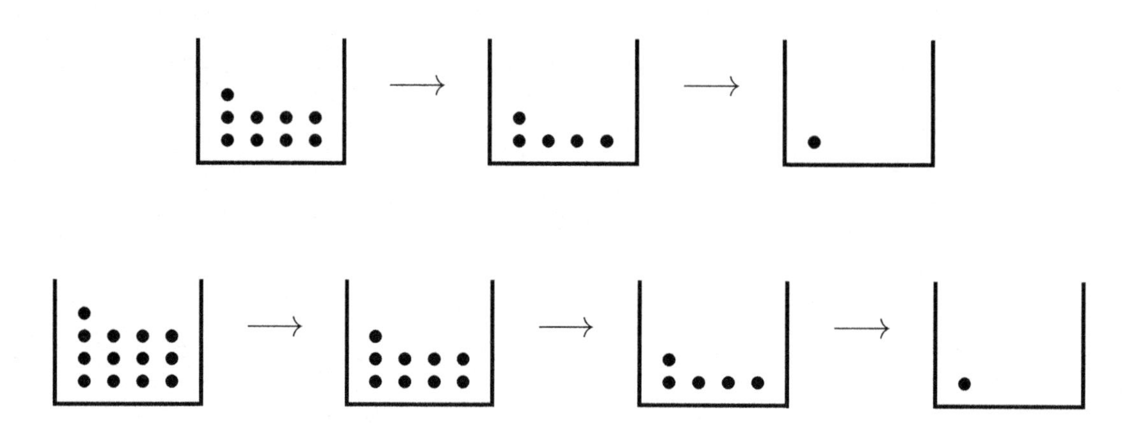

– The Clock Metaphor –

Another way to think about numbers modulo 12 is by thinking of a clock. If it is 10 o'clock, in four hours it won't be 14 o'clock,[19] it will be 2 o'clock, right? That's because $14 \equiv 2 \pmod{12}$. And from 2 o'clock, in 27 hours it won't be 29 o'clock, it will be 5 o'clock, because $29 \equiv 5 \pmod{12}$.

Likewise, you can think about congruence modulo 5 as being done on a clock with only 5 hours. If it is 2 o'clock on this special clock, then in six hours it won't be 8 o'clock, it will be 3 o'clock, which shows why $8 \equiv 3 \pmod{5}$. And from 3 o'clock, in nine hours it won't be 12 o'clock, it will be 2 o'clock. Thus, $12 \equiv 2 \pmod{5}$.

And if you can accept that a box can have a negative number of balls, and if you can ask questions like "what time was it five hours ago?", then these metaphors also show why, say, $-2 \equiv 3 \pmod{5}$.

These metaphors deal with adding numbers under a mod, and modular congruence does indeed have some nice arithmetic properties. The first three are detailed in the following proposition.

[19]Unless you're taking this class in West Point, Annapolis, or in most countries outside of North America.

Proposition.

Proposition 2.15 (*Properties of Modular Arithmetic*). Assume that a, b, c, d and m are integers, $a \equiv b \pmod{m}$ and $c \equiv d \pmod{m}$. Then,

(i) $a + c \equiv b + d \pmod{m}$;

(ii) $a - c \equiv b - d \pmod{m}$; and

(iii) $a \cdot c \equiv b \cdot d \pmod{m}$.

Scratch Work. A quick reminder: I recommend not being a passive learner, but an active one. Try to prove (i) on your own before moving on. I know, it would be so much easier to just read on. It's like exercising — the fact that it's strenuous is how you know it's working.[20] Challenge yourself! Be a mathlete, not a mathemachicken!

Good job! As for my scratch work, let's begin by seeing how far our general strategy for direct proofs gets us for part (i).

> ***Proof.*** Assume $a \equiv b \pmod{m}$ and $c \equiv d \pmod{m}$.
>
> ≪An explanation of what those mean≫ ⟵ Apply definitions and/or other results.
>
> ⋮ apply algebra, logic, techniques
>
> ≪Hey look, that's what $a + c \equiv b + d \pmod{m}$ means≫
>
> Therefore $a + c \equiv b + d \pmod{m}$. □

What does each modular congruence mean? Definition 2.14 tells us!

> ***Proof.*** Assume $a \equiv b \pmod{m}$ and $c \equiv d \pmod{m}$. Then,
>
> $$m \mid (a - b) \quad \text{and} \quad m \mid (c - d).$$ ⟵ Apply definitions and/or results.
>
> ⋮ apply algebra, logic, techniques
>
> Then, $m \mid \big[(a + c) - (b + d)\big]$
>
> Therefore $a + c \equiv b + d \pmod{m}$. □

[20] And you *definitely* don't get fit by watching someone else exercise!

How do we bridge the gap? Well, what does it mean to say one integer divides another? Definition 2.8 tells us!

> ***Proof.*** Assume $a \equiv b \pmod{m}$ and $c \equiv d \pmod{m}$. Then,
>
> $m \mid (a-b)$ and $m \mid (c-d)$. Then,
>
> $a - b = mk$ and $c - d = m\ell$ for some integers k, ℓ.
>
> ⋮ apply algebra,
> ⋮ logic, techniques
>
> $(a+c)-(b+d) = mt$ for some integer t.
>
> Then, $m \mid \big[(a+c)-(b+d)\big]$.
>
> Therefore $a + c \equiv b + d \pmod{m}$. □

And with that, I think we can bridge the gap. Now here's the proof.

Proof. Part (i). Assume that $a \equiv b \pmod{m}$ and $c \equiv d \pmod{m}$. By the definition of modular congruence (Definition 2.14),

$$m \mid (a-b) \qquad \text{and} \qquad m \mid (c-d).$$

Then, by the definition of divisibility (Definition 2.8),

$$a - b = mk \qquad \text{and} \qquad c - d = m\ell$$

for some integers k and ℓ. Adding these two equations together,

$$(a-b)+(c-d) = mk + m\ell.$$

Regrouping,

$$(a+c)-(b+d) = m(k+\ell).$$

Since $k + \ell$ is an integer, by the definition of divisibility (Definition 2.8),

$$m \mid \big[(a+c)-(b+d)\big],$$

which then by the definition of modular congruence (Definition 2.14) means that

$$a + c \equiv b + d \pmod{m},$$

completing the proof of part (i).

Parts (ii) and (iii). These are left to you as exercises (See Exercise 2.21). For part (iii) you may use the fact that if $a \equiv b \pmod{m}$, then a and b have the same remainder when divided by m. □

Modular arithmetic has nice properties for addition, subtraction and multiplication. What about division? Well, not always. Try to think of an example on your own where $ak \equiv bk \pmod{m}$, but $a \not\equiv b \pmod{m}$. Really, try it on your own! .... Now, if you found one, there's a good chance that it is an example where $k \equiv 0 \pmod{m}$; this is similar to saying "$2 \cdot 0 = 3 \cdot 0$, even though $2 \neq 3$." So here's your next challenge: Can you think of an example where $k \not\equiv 0 \pmod{m}$? Give it a shot on your own! Then, once you have, you can check out one answer in the footnote.[21]

See if you can convince yourself that if k and m have a common divisor larger than 1, then this phenomenon can occur. And then see if you can convince yourself that if $\gcd(k, m) = 1$, then the cancellation property will hold. As an example of this latter claim, note that $21 \equiv 6 \pmod{5}$, which means that $7 \cdot 3 \equiv 2 \cdot 3 \pmod{5}$. And since $\gcd(3, 5) = 1$, the cancellation property says that we can cancel the 3. And this does check out: $7 \equiv 2 \pmod{5}$.

This is indeed the next proposition. The proof of this proposition will make use of a lemma (our first lemma!). And this lemma requires that we know what a prime number is. So let's formally define a prime number, then state and prove the lemma, and then use that to prove the proposition.

Definition.

Definition 2.16. An integer $p \geq 2$ is *prime* if its only positive divisors are 1 and p. An integer $n \geq 2$ is *composite* if it is not prime. Equivalently, n is composite if it can be written as $n = st$, where s and t are integers and $1 < s, t < n$.

(To be clear, "$1 < s, t < n$" means that both s and t are between 1 and n. It means $1 < s < n$ and $1 < t < n$ both hold.)

This definition shows that every $n \geq 2$ is either prime or composite.[22] In Exercise 2.41 you are asked to justify the "equivalently" part of the definition.

Let's now state and prove the lemma, which is a three-parter, grouped together because they are all related. This lemma's proof will require a bit of work, but as momma always said, when life gives you lemmas, make lemmanade.

[21] Notice that $20 \equiv 8 \pmod{6}$, but yet when dividing both by 4, we get $5 \not\equiv 2 \pmod{6}$.

[22] We do not consider 1 to be prime, and we do not consider negative numbers like -7 to be prime. Definitions are human choices, and mathematicians decided that having those be considered prime would be detrimental. Here's one reason: One of the most fundamental properties of an integer $n \geq 2$ is that it can be written uniquely as a product of primes; for example, $12 = 2 \cdot 2 \cdot 3$. But if 1 and negative numbers could be prime, then also $12 = 1 \cdot 2 \cdot 2 \cdot 3$ and $12 = (-2) \cdot 2 \cdot (-3)$ would be ways to write 12 as a product of primes. We exclude 1 and negative numbers for many reasons, but having this unique factorization property is one such reason.

Lemma.

Lemma 2.17. Let a, b and c be integers, and let p be a prime.

(i) If $p \nmid a$, then $\gcd(p, a) = 1$.

(ii) If $a \mid bc$ and $\gcd(a, b) = 1$, then $a \mid c$.

(iii) If $p \mid bc$, then $p \mid b$ or $p \mid c$ (or both[23]).

Proof Idea for (i). Here is the main idea for part (i). Since p is a prime number, its only divisors are p, 1, -1 and $-p$. Therefore, these four numbers are the only possible *common divisors* between p and a. The question is: Which of these also divides a? And among the ones which do (i.e., the common divisors), which is the largest? Let's investigate, keeping in mind that the lemma assumed that $p \nmid a$.

$-p$	-1	1	p
is not a common divisor with a	is a common divisor with a	is a common divisor with a	is not a common divisor with a

Among the two which *are* common divisors, we can see that 1 is the greatest.

Proof Idea for (ii). In this part we need a tool. A tool that connects integers to their gcd... and Bézout's identity (plus a little algebra) will do just that!

Proof Idea for (iii). Part (iii) turns out to be just a mixture of parts (i) and (ii). So once those two are proven, we will be able to unite their powers to give us (iii).

Proof. Assume that a, b and c are integers, and p is a prime.

Proof of (i). Assume that p does not divide a. To be a common divisor of a and p means that you must divide both of them, but since we are assuming that p does not divide a, this also means that p is not a common divisor of a and p. And since p is not a common divisor of a and p, it is certainly not the greatest common divisor of a and p.

Since p is prime, the two largest divisors of p are 1 and p—and we just showed that p is not the $\gcd(p, a)$. Therefore, since 1 is a common divisor of a and p (since

[23]Note: In math, 'or' is always an *inclusive or*, as compared to an *exclusive or*. An 'inclusive or' allows the possibility that both are true, while an 'exclusive or' demands that only one is true. Notice that Lemma 2.17 part (iii) would be false if math used an 'exclusive or'. For example, if $p = 5$, $a = 10$ and $b = 15$, then $p \mid ab$, but it's not true that p only divides one of the two—it divides both!

$1 \mid a$ and $1 \mid p$), it must be the greatest common divisor of these numbers.

Proof of (ii). Assume $a \mid bc$ and $\gcd(a, b) = 1$. By Bézout's identity (Theorem 2.13), there exist integers k and ℓ such that

$$\gcd(a, b) = ak + b\ell.$$

And since $\gcd(a, b) = 1$, this means

$$1 = ak + b\ell.$$

By multiplying both sides by c, this gives

$$c = ack + bc\ell.$$

Now, we assumed that $a \mid bc$, which by the definition of divisibility (Definition 2.8) means $bc = am$ for some integer m. Plugging this in,

$$\begin{aligned} c &= ack + am\ell \\ &= a(ck + m\ell). \end{aligned}$$

And since c, k, m and ℓ are all integers, so is $ck + m\ell$. Since $c = at$ where $t = ck + m\ell$ is an integer, by the definition of divisibility (Definition 2.8) we have $a \mid c$, as required.

Proof of (iii). Now that we have proven that (i) and (ii) are true, we may use them to prove that (iii) is also true. To prove (iii), we begin by assuming that $p \mid bc$. We will use a proof by cases here. The two cases are: $p \mid b$ or $p \nmid b$.

Case 1. Assume that $p \mid b$. Our goal in part (iii) is to prove $p \mid b$ or $p \mid c$. So we are immediately done: Our assumption is what we wanted to prove, so no more work is needed![24]

Case 2. Assume that $p \nmid b$. Then, by part (i), $\gcd(p, b) = 1$. But at this point, we simply apply (ii): we know that $p \mid bc$ and $\gcd(p, b) = 1$, therefore $p \mid c$.

In either case we have deduced that $p \mid b$ or $p \mid c$, which shows that (iii) must be true. □

Euclid wrote down proofs of these results nearly 2500 years ago in his book *Elements*, making them among the first recorded and rigorously proven results in number theory.[25] And since we called them a lemma, you already know that we're about to use them to prove another result.

[24]The trickiest part is to not overthink this. In math, often the easiest proofs are the hardest to think about, because so little happens. The proof of Case 1 is basically this: "Assume Joe's last name is Smith. Prove that Joe's last name is either Smith or Anderson. We are done, because we already assumed at the start that it was Smith."

[25]Conspiracy theorist somewhere: "Proven? Whatever, man. Number theory is just a theory. Who knows if number is true?"

Proposition.

Proposition 2.18 (*Modular cancellation law*)**.** Let a, b, k and m be integers, with $k \neq 0$. If $ak \equiv bk \pmod{m}$ and $\gcd(k, m) = 1$, then $a \equiv b \pmod{m}$.

Proof Idea. The idea behind this proof is very similar to that of Proposition 2.15, in that both our assumption and conclusion may be expressed in terms of divisibility, which can in turn be expressed in terms of a product. This will again leave a gap that we will need to cross, but this time we will need the help of Lemma 2.17 to do so.[26] See if you can do it on your own before looking at the proof below!

Proof. Let a, b, k, and m be integers, and assume $ak \equiv bk \pmod{m}$ and $\gcd(k, m) = 1$. By the definition of modular congruence (Definition 2.14),

$$m \mid (ak - bk).$$

And by the definition of divisibility (Definition 2.8), this means that $ak - bk = m\ell$, for some integer ℓ. That is,

$$k(a - b) = m\ell. \qquad (✌)$$

By the same definition, and because $(a - b)$ must be an integer, the above also implies that

$$k \mid m\ell.$$

And since, by assumption, $\gcd(k, m) = 1$, by Lemma 2.17 part (ii) we must have $k \mid \ell$; by the definition of divisibility (Definition 2.8) this means that $\ell = kt$, for some integer t. This allows us to rewrite Equation (✌):

$$\begin{aligned} k(a - b) &= m\ell \\ k(a - b) &= mkt \\ a - b &= mt, \end{aligned}$$

where in the last line we used that $k \neq 0$. By the definition of modular congruence (Definition 2.14), this means that $m \mid (a - b)$. That is, $a \equiv b \pmod{m}$. □

Another way to think about this proposition is this: If $a \not\equiv b \pmod{m}$, under what conditions is it possible that, by multiplying a and b by some k, you can get $ak \equiv bk \pmod{m}$?

For example, consider what happens when you multiply $0, 1, 2, 3, 4$ and 5 by 2, and write the answers modulo 6:

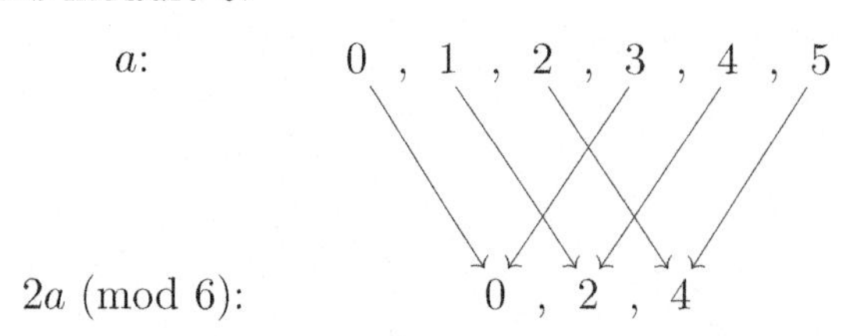

[26] *"Yo, lemma help you prove that proposition."*

You can see that $2 \cdot 0$ and $2 \cdot 3$ are the same, modulo 6. And $2 \cdot 1$ and $2 \cdot 4$ are the same, modulo 6. And $2 \cdot 2$ and $2 \cdot 5$ are the same, modulo 6. This allows us to have, say, $1 \cdot 2 \equiv 4 \cdot 2 \pmod 6$, and yet we are unable to divide out those 2s from each side, since doing so would produce $1 \equiv 4 \pmod 6$, which is false.

It's worth pausing for a moment to say that, despite this being just the second chapter, I expect many readers will find this section on mods to be the most challenging of the entire text. I am throwing some tough stuff at you right now! Hang in there and do your best. We will soon be changing topics, so keep working hard but do not feel discouraged if you feel confused at the moment. Most readers will, and you should not view it as a bad sign for the chapters to come. Here's a cute puppy for motivation.

We end this chapter on a challenging proof.[27] Let's use Lemma 2.17 to prove an important theorem from number theory. Now, if the following theorem had been proven by a mathematician like me, then it would be known as Cummings's Super Duper Important Theorem; but for the likes of Pierre de Fermat,[28] it is simply known as *Fermat's little theorem.* Which is a pretty nice tribute to the guy.

Theorem.

Theorem 2.19 (*Fermat's little theorem*). If a is an integer and p is a prime which does not divide a, then

$$a^{p-1} \equiv 1 \pmod p.$$

Proof Idea. The all-important observation is the following, which we explain through the example $a = 4$ and $p = 7$. Consider the two sets:[29]

$$\{a, 2a, 3a, 4a, 5a, 6a\} \qquad \text{and} \qquad \{1, 2, 3, 4, 5, 6\}.$$

In this example, since $a = 4$, this is the same as

$$\{4, 8, 12, 16, 20, 24\} \qquad \text{and} \qquad \{1, 2, 3, 4, 5, 6\}.$$

These look like completely different sets. But look what happens when you consider each of the numbers modulo p; the second set stays the same (e.g., $3 \equiv 3 \pmod 7$),

[27]To reiterate: Challenging is good! Struggle is good! Modular arithmetic is a tough topic, but nobody can download the Math Castle into your brain but yourself.

[28]Pro-Tip: "Fermat" is a French name and is pronounced Fer-mah.

[29]The next chapter is focused on sets, so if you have never learned much about them, then fear not, help is coming. For now, all you need to know is that a set is a collection of elements, and that the order in which the elements are listed does not matter. For example, $\{1, 2, 3\}$ and $\{3, 1, 2\}$ are considered the exact same set.

but the numbers in the first set do change (e.g., $12 \equiv 5 \pmod 7$). Indeed, here are the sets now:

$$\{4, 1, 5, 2, 6, 3\} \qquad \text{and} \qquad \{1, 2, 3, 4, 5, 6\}.$$

Notice anything interesting? These are the same set! Sure, the numbers in the first set are written in a different order, but since the exact same numbers are there, they are considered identical sets. In particular, since order does not matter with multiplication (e.g., $1 \cdot 2 \cdot 3 \cdot 4 = 3 \cdot 2 \cdot 4 \cdot 1$), this means that

$$a \cdot 2a \cdot 3a \cdot 4a \cdot 5a \cdot 6a \equiv 1 \cdot 2 \cdot 3 \cdot 4 \cdot 5 \cdot 6 \pmod 7.$$

Why? Because once the numbers are reduced mod 7, we are multiplying the same six numbers together—just perhaps in a different order. This in fact holds for any a and p, and is the key to prove Fermat's little theorem.

***Proof*.** Assume that a is an integer and p is a prime which does not divide a. We begin by proving that when taken modulo p,

$$\{a, 2a, 3a, \dots, (p-1)a\} \equiv \{1, 2, 3, \dots, p-1\}.$$

To do this, observe that the set on the right has every modulo except 0 (that is, has every remainder except 0), and each such modulo appears exactly once. Therefore, since both sets have $p - 1$ elements listed, in order to prove that the left set is the same as the right set, it suffices to prove this:

1. No element in the left set is congruent to 0, and
2. Each element in the left set appears exactly once.

In doing so, we will twice use the modular cancellation law (Proposition 2.18) to cancel out an a, and so we note at the start that by Lemma 2.17 part (i) we have $\gcd(p, a) = 1$.

Step 1. First we show that none of the terms in $\{a, 2a, 3a, \dots, (p-1)a\}$, when considered modulo p, are congruent to 0. To do this, we will consider an arbitrary term ia, where i is anything in $\{1, 2, 3, \dots, p-1\}$. Indeed, if we did have some

$$ia \equiv 0 \pmod p,$$

which is equivalent to

$$ia \equiv 0a \pmod p,$$

then by the modular cancellation law (Proposition 2.18) we would have

$$i \equiv 0 \pmod p.$$

That is, in order to have $ia \equiv 0 \pmod p$, that i would have to have $i \equiv 0 \pmod p$. Therefore we are done with Step 1, since no i from $\{1, 2, 3, \dots, p-1\}$ is congruent to 0 modulo p.

Step 2. Next we show that every term in $\{a, 2a, 3a, \ldots, (p-1)a\}$, when considered modulo p, does not appear more than once in that set. Indeed, if we did have

$$ia \equiv ja \pmod{p},$$

for i and j from $\{1, 2, 3, \ldots, p-1\}$, then by the modular cancellation law (Proposition 2.18) we have

$$i \equiv j \pmod{p}.$$

And since i and j are both from the set $\{1, 2, 3, \ldots, p-1\}$, this means that $i = j$. In other words, each term in $\{a, 2a, 3a, \ldots, (p-1)a\}$ is not congruent to any other term from that set — it is only congruent to itself. This completes Step 2.

We have succeeded in proving that when taken modulo p,

$$\{a, 2a, 3a, \ldots, (p-1)a\} \equiv \{1, 2, 3, \ldots, p-1\},$$

even though the numbers in these sets may be in a different order. But since the order does not matter when multiplying numbers, we see that

$$a \cdot 2a \cdot 3a \cdot 4a \cdot \ldots \cdot (p-1)a \equiv 1 \cdot 2 \cdot 3 \cdot 4 \cdot \ldots \cdot (p-1) \pmod{p}.$$

Then, since $\gcd(2, p) = 1$ by Lemma 2.17 part (i), by the modular cancellation law (Proposition 2.18) we may cancel a 2 from both sides:

$$a \cdot a \cdot 3a \cdot 4a \cdot \ldots \cdot (p-1)a \equiv 1 \cdot 3 \cdot 4 \cdot \ldots \cdot (p-1) \pmod{p}.$$

Then, since $\gcd(3, p) = 1$ by Lemma 2.17 part (i), by the modular cancellation law (Proposition 2.18) we may cancel a 3 from both sides:

$$a \cdot a \cdot a \cdot 4a \cdot \ldots \cdot (p-1)a \equiv 1 \cdot 4 \cdot \ldots \cdot (p-1) \pmod{p}.$$

Continuing to do this for the $4, 5, \ldots, (p-1)$ on each side (each of which has a greatest common divisor of 1 with p, by Lemma 2.17 part (i)), by the modular cancellation law (Proposition 2.18) we obtain

$$\underbrace{a \cdot a \cdot a \cdot a \cdot \ldots \cdot a}_{p-1 \text{ copies}} \equiv 1 \pmod{p},$$

which is equivalent to what we sought to prove:

$$a^{p-1} \equiv 1 \pmod{p}.$$

□

Fermat's little theorem is not only an important result in mathematics, but a crucial tool in cybersecurity. This connection is discussed in the *Introduction to Number Theory* following this chapter.

2.6 Bonus Examples

For this chapter's bonus examples, let's do some examples of direct proofs where the "apply algebra, logic, techniques" step is a little trickier, and let's also branch out a bit from the divisibility and modularity topics that we have focused on. Let's prove some things about inequalities instead! Below are two such examples.

Proposition.

Proposition 2.20. Assume that x and y are positive numbers. If $x \geq y$, then $\sqrt{x} \geq \sqrt{y}$.

Proof Sketch. Following our general direct proof strategy doesn't get us very far:

Proof. Assume $x \geq y$.

≪An explanation of what $x \geq y$ means≫ ⟵ Apply definitions and/or other results.

⋮ apply algebra, logic, techniques

≪Hey look, that's what $\sqrt{x} \geq \sqrt{y}$ means≫

Therefore $\sqrt{x} \geq \sqrt{y}$. □

It doesn't seem like we have any definitions and/or other results to apply. As it turns out, getting from $x \geq y$ to $\sqrt{x} \geq \sqrt{y}$ is just algebra. There are certain strategies you will pick up along your mathematical journey, and one is that it is often helpful to have 0 on one side of an equality or inequality, since that allows you to factor.

Proof. Assume $x \geq y$.

This is the same as $x - y \geq 0$.

⋮ apply algebra, logic, techniques

Which implies $\sqrt{x} - \sqrt{y} \geq 0$.

Therefore $\sqrt{x} \geq \sqrt{y}$. □

How do we bridge this gap? Well, we mentioned that when one side equals zero, it's a good idea to try to factor. If it were $a^2 - b^2$ you would probably notice a difference of squares and think $a^2 - b^2 = (a-b)(a+b)$. In fact, $x - y$ can also be viewed as a difference of squares: $x - y = \sqrt{x}^2 - \sqrt{y}^2$. And from this perspective and a little more algebra, the bridge can be formed. Below is this argument.

Proof. Assume that $x \geq y$, and that x and y are positive numbers. Since $x \geq y$,

$$x - y \geq 0.$$

Moreover, since x and y are positive, note that $x = \sqrt{x}^2$ and $y = \sqrt{y}^2$. This allows us to again rewrite our expression as

$$\sqrt{x}^2 - \sqrt{y}^2 \geq 0.$$

The left-hand side is a difference of squares, and hence can be factored:

$$(\sqrt{x} - \sqrt{y})(\sqrt{x} + \sqrt{y}) \geq 0.$$

Next observe that since x and y are positive, so is $\sqrt{x} + \sqrt{y}$, which allows us to divide both sides of the inequality by $(\sqrt{x} + \sqrt{y})$, which simply gives

$$(\sqrt{x} - \sqrt{y}) \geq 0.$$

Finally, by moving $\sqrt{y}$ to the right, we get what we sought:

$$\sqrt{x} \geq \sqrt{y}.$$ □

Deep results in math are typically built on other results. Indeed, let's now use the result we just proved to prove another, much less intuitive result. This is also a very important result in the world of inequalities, called the AM-GM inequality.[30]

Theorem.

Theorem 2.21 (*AM-GM inequality*)**.** If x and y are positive real numbers, then $\sqrt{xy} \leq \dfrac{x+y}{2}$.

[30] The 'AM' refers to the arithmetic mean of two numbers, and the 'GM' refers to their geometric mean. Given two numbers x and y, their arithmetic mean is simply their average: $\frac{x+y}{2}$. Their geometric mean is the multiplication version of this: instead of adding, you multiply; instead of dividing by 2, you take the second root. Thus, their GM is $\sqrt{xy}$. In fact, there's even a version with four inequalities, called the QM-AM-GM-HM inequalities. There is also a neat way to view all four as parts of a circle, as pictured on the right ⟶

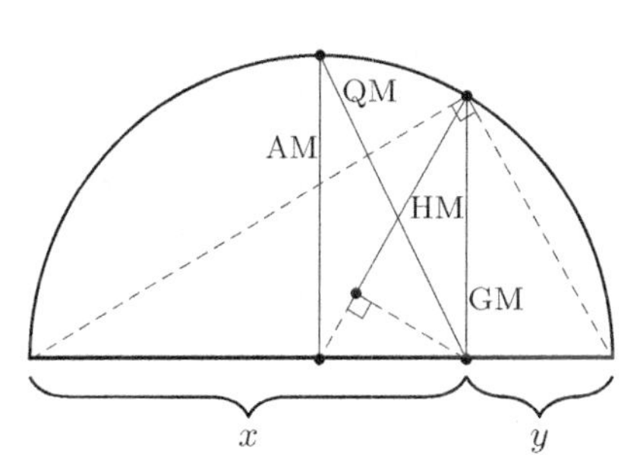

Scratch Work. Since not much is being assumed, let's jump straight to the conclusion and then do some algebra to see if we can reach something that we know to be true; this might seems strange, but bear with me for now. Starting at

$$\sqrt{xy} \leq \frac{x+y}{2},$$

let's multiply over the 2 (since denominators are annoying) which gives

$$2\sqrt{xy} \leq x + y.$$

Next, let's square both sides (since square roots are annoying), which gives

$$4xy \leq x^2 + 2xy + y^2.$$

Just like in the last proposition, having things equal to zero is commonly a wise step since it allows you to factor, so let's move the $4xy$ to the right:

$$0 \leq x^2 - 2xy + y^2.$$

Be grateful, my friends, for The Factoring Gods have smiled upon us. This is the same as

$$0 \leq (x - y)^2.$$

And this we know to be true, since squaring a real number always gives a nonnegative result.

Now, it might seem weird that we *started* scratch work at our conclusion and *ended* our scratch work at something we know to be true... Typically a direct proof is the exact opposite. But this is actually ok, because all of our steps can also be done in reverse! (The one questionable step might be when we squared both sides, but this can be done in reverse by Proposition 2.20!) Indeed, if we now start at the *bottom* of our scratch work and move upwards, we will have a proof.

Proof. Let x and y be positive real numbers. Observe that $0 \leq (x - y)^2$, because the square of a real number is always nonnegative. Rewriting this,

$$0 \leq x^2 - 2xy + y^2.$$

Adding $4xy$ to both sides then gives

$$4xy \leq x^2 + 2xy + y^2,$$

which allows us to factor the right-hand side:

$$4xy \leq (x + y)^2.$$

Finally, since everything is positive we may apply Proposition 2.20 and take the square root of both sides to get

$$2\sqrt{xy} \leq x + y,$$

and dividing over the 2 proves our result:

$$\sqrt{xy} \leq \frac{x+y}{2}.$$

□

Pretty neat, don't cha think? If you saw this proof without the scratch work it would seem like only a genius could have realized where to begin and foreseen the required algebra. But with a little reverse scratch work, it becomes clear.

One final note: A common mistake is to do the scratch work as we did, but to not reverse it for your actual proof. Remember, our scratch work started at our desired *conclusion*, and then worked its way back to something we knew to be true. But at the end of the day, our desired conclusion has to be at the end of our proof, not at the start.

If the theorem says "P implies Q," and what you prove is "Q implies P," well that's a very interesting result but it is not what you had to prove. Do you see how, if you had proved "Q implies P" instead, what you are actually doing is starting with the conclusion of "P implies Q" and working your way back to the assumptions?

Furthermore, if P implies Q is true, this does *not* mean that Q implies P is true. It is true that "living in California implies living in the United States," but it is false that "living in the United States implies living in California." In Chapter 5 we will spend a lot of time studying the many subtleties of this.

— Chapter 2 Pro-Tips —

- When reading a definition, get in the habit of asking "why is it called that?" Often the word that mathematicians chose provides some intuition for what's being defined, or suggests a connection to something else. Graph theory is especially rich in this way, given how tangible a graph is, but there are illustrative definitions throughout math.

  At times, a definition can even tell story. *Calculus*, for example, used to mean a "small stone," like what is used in an abacus, and *geometry* used to mean "land measurement," which was an early use of shapes, areas and perimeters (e.g., for tax purposes and governmental land surveys). As writer Jorge Luis Borges said, "every word is a dead metaphor."

- Definitions not only offer us precision in our language, but they present to us objects worthy of study. It had always been the case that 6 and 28 were the sum of their proper[31] divisors ($6 = 3 + 2 + 1$ and $28 = 14 + 7 + 4 + 2 + 1$), but when such a property was given a name — that of a *perfect number* — then it became a concrete, tangible thing. It allowed mathematicians to focus their attention and sharpen their dialogue.

  Do you believe math is deep and intrinsic, a consequence of logic and nature, existing beyond humans? Or is it created only in our minds, and would not exist without us? The mathematician Leopold Kronecker was far on one side of this argument, saying "God created the integers, the rest is the work of man."

  This topic makes for a fun debate, but in the end most people at least agree that what we choose to define, and what definitions we use, is a distinctly human decision.[32]

- In every class, some things are more important than others. In this book, some things we study here are mostly used as instruments to practice new proof techniques. You would be forgiven, for example, if you walked away from this book believing that even and odd integers are going to play a massive role in your later courses. They won't, but there is a good reason for why we will keep discussing them: When first learning a new proof method or technique, if the practice problems are phrased in simpler terms, then your focus can remain on the method.

  Modular arithmetic, though, is certainly not some arbitrary topic, and it is worth your time to learn it well. And while it is a challenging topic, putting in the time now will make your abstract algebra class go smoother, and you will have a big leg-up in your future studies of number theory, cryptography, and more. In Chapter 9, some of these connections are introduced.

- Suppose you are proving something by cases, and it turns out that two cases are exactly the same just with variable names switched. For example, suppose

[31] A *proper divisor* of n is a number d such that $1 \leq d < n$, and $d \mid n$.

[32] Indeed, there is a Kronecker-like belief in Judiasm and Christianity: According to those religions, God created the animals, but it was Adam — the human — that was tasked to name them.

you are proving that $n \cdot m$ is even if one of these two variables is even and the other is odd. It would be natural to do this in cases. Case 1: n is even and m is odd; Case 2: n is odd and m is even. One allowed trick is to say "without loss of generality, we will assume that n is even and m is odd." The reader can see that the proof of the second case is exactly the same, and thus you only have to write out the proof once. We will discuss this more in Chapter 6.

- We've begun to introduce some mathematical notation, and we will continue to introduce more throughout the text. Moreover, most courses you take will feature new notation, and even if you become a fully-fledged mathematical researcher, most research articles you pick up will invent some new notation to further their discussion.

  A word of warning: Mathematical notation can be tricky. For example, $(1, 2)$ means a point in the xy-plane... unless it means the interval of real numbers between 1 and 2. Raising something to the -1 power means "1 divided by that thing"... unless it means the inverse function. The square root of a negative number is undefined... until it's not. The symbol $\equiv$ means modular congruence... unless it means two functions are identical. The symbol $\cong$ means that two triangles are congruent... unless it means that two groups are isomorphic. And $\sim$ means so many different things it's just ridiculous. (You'll see one such meaning in Chapter 9!)

  Math is a big field, and we run out of symbols pretty quickly. You must use context clues to know what the symbols mean. Moreover, if you are confused by something you're reading, make sure you're not simply misinterpreting some of the notation. And if you read something which uses notation that you think you recognize but which doesn't seem to make sense, perhaps the notation has another meaning of which you are unaware.

  For each big, important symbol, each field of math tries to stick to one meaning (although there are exceptions). Meanwhile, other minor symbols are often redefined for each new paper, similar to how mathematicians keep redefining $f(x)$ to be something new each time, and everyone knows to look back a paragraph or two to see what its meaning is now.

- After proving something, get in the habit of asking, "is this true more generally?" If you just proved that if n is odd, then $3n^2 + 1$ is even, you might ask yourself whether the 3 is special there, or whether it could be replaced by any odd number. What about the "+1"? Could that 1 be replaced with any other odd number? These are ways to generalize the problem by weakening the assumptions in the problem

  You can also try to strengthen the result by seeing if it implies anything more. For example, while it true that $3n^2 + 1$ is always even, can we say what kind of "even" it is? For example, some even numbers are congruent to 0 (mod 4) and some are congruent to 2 (mod 4). Is $3n^2 + 1$ always one of these two? Or can it be both? What about 0, 2 or 4 (mod 6)? Is it always a positive even number, or can it be negative even number? It's good to engage with the material in these ways.

- When you take abstract algebra, you will learn that $\mathbb{Z}$ (with the addition and multiplication operations) is just one example of a more general algebraic object called a *ring*. You will learn that in a general ring, our definition of primality ($p \geq 2$ is prime if its only positive divisors are 1 and p) is actually the definition of an *irreducible element*. In a ring, an element p is called prime if $p \mid ab$ implies that $p \mid a$ or $p \mid b$. For the integers these are equivalent conditions, but for other rings they may not be.

- Learning how to prove things is difficult for beginners, because it really is a new way of thinking. Moreover, they have not yet accumulated the tips, tricks, strategies and techniques that come with experience. More than most classes, the time you put into this class will pay off in your future classes.

  For most classes, if a year after the class ended you went back and tried to retake the exams and redo the homework, you probably wouldn't do too well. There are a lot of important facts and definitions that are specific to that class and when you complete that course and move on to the later courses, you may forget many of those things (I know I did). This class is different. For most of what we do here, if you came back in a year and tried this material over again, you would find that you do *better*. Because many of the important ideas here you will be using again and again in later courses. Thus, it is all the more important to invest your time now.

  Continuing the "proof as a computer program" analogy from Chapter 1, experience proving things is like collecting program libraries. The knowledge you collect along the way is stored away, but can be quickly recalled when the need arises. The larger your "proof strategies" library, the easier your life will be, so work hard and your library will grow.

- A reminder: If possible, you should be discussing this material and solving problems with your friends. The best way to find the holes in your understanding is to try to explain it to someone else, and respond to their questions. You may also be surprised to discover that concepts which you feel confident about, others may think about in a very different way than you. By learning to see things from each others' perspective, you will enrich each others' understanding. This is sometimes called the *Feynman technique*.[33]

- In basically all of pure mathematics from here on out, including in the *Introduction to Number Theory* following this chapter, "log" refers to the natural log — the log with a base of e. In computer science, the base is always 2. It will never again be 10; those high school days are in the past.

[33]Relatedly, *Murphy's law* says that if you wish to find someone on the Internet to answer a question you have, the best approach is not to post your question, but to post a wrong answer to your question. Doing this guarantees that someone will come along to prove that you are wrong by telling you what the correct answer is — producing the exact answer that you sought. (In fact, this is not at all what Murphy's law says. Now, it is safe to print that, but be warned: If you were to post that on the Internet, then very quickly someone will show up to tell you what Murphy's law *really* says. This is guaranteed to occur, as I have said, according to Murphy's law.)

— Exercises[34] —

Exercise 2.1. List 5 skills that are important for someone to be successful in a college math class. Which skills seem most important for an upper-division math class? Which skills do you want to work to improve?

Exercise 2.2. The following are the squares of four numbers, each ending in a 5.

$$15^2 = 225$$
$$25^2 = 625$$
$$35^2 = 1225$$
$$45^2 = 2025.$$

Looking at these four squares, do you see anything interesting about their answers? Once you have noticed a pattern, answer the following.

(a) Write down a conjecture that explains what the answer is for the square of any integer ending in a 5.

(b) Give four more examples illustrating your conjecture.

(c) Prove your conjecture.

Exercise 2.3. For each of the following, give three examples of this property. Then, prove that it is true.

(a) The sum of an even integer and an odd integer is odd;

(b) The product of two even integers is even;

(c) The product of two odd integers is odd;

(d) The product of an even integer and an odd integer is even;

(e) An even integer squared is an even integer.

Exercise 2.4. For each of the following, give three examples of this property. Then, prove that it is true.

(a) If n is an even integer, then $-n$ is an even integer.

(b) If n is an odd integer, then $-n$ is an odd integer.

(c) If n is an even integer, then $(-1)^n = 1$. You may use standard properties of exponents.

---

[34] *Mo' chapters, mo' problems*

Exercise 2.5. Prove the following. For each, n is an integer.

(a) If n is odd, then $n^2 + 4n + 9$ is even.

(b) If n is odd, then n^3 is odd.

(c) If n is even, then $n + 1$ is odd.

Exercise 2.6. Prove the following. For each, m and n are integers.

(a) If m and n are odd, then $5m - 3n$ is even.

(b) If m and n are even, then $3mn$ is divisible by 4.

Exercise 2.7. Provide a second proof of Proposition 2.7 in which you first prove that $n(n + 1)$ is even, and then you apply Proposition 2.4.

Exercise 2.8. Give an example of each of the following properties. Then, prove that it is true.

(a) If n is an integer, then $n^2 + n$ is even.

(b) If n is an integer, then $3n^2 + 5n + 1$ is odd.

(c) If n is an integer, then $n^2 + 3n - 6$ is even.

(d) If m and n are integers of the same parity, then $7m - 3n$ is even.

Exercise 2.9. Determine conditions on integers m and n for which mn is even. Write down your conditions as a conjecture, and then prove that your conjecture is correct.

Exercise 2.10. Prove the following. For each, m, n and t are integers.

(a) If $m \mid n$, then $m^2 \mid n^2$.

(b) If $m \mid n$, then $m \mid (7n^3 + 13n^2 - n)$.

(c) If $m \mid n$ and $m \mid t$, then $m \mid (n + t)$.

(d) If $3 \mid 2n$, then $3 \mid n$.

(e) If $9 \mid 6n$, then $3 \mid n$.

(f) If $m^3 \mid n$ and $n^4 \mid t$, then $m^{12} \mid t$.

Exercise 2.11. Prove the following. For each, m, n and t are integers.

(a) $1 \mid n$.

(b) $n \mid n$.

(c) If $mn \mid t$, then $m \mid t$.

(d) If $mn \mid tn$, then $m \mid t$.

Exercise 2.12. Prove that if m and n are positive real numbers and $m < n$, then $m^2 < n^2$. You may use the fact that if $a < b$ and c is positive, then $ac < bc$.

Exercise 2.13. Define the absolute value of a real number x in this way:

$$|x| = \begin{cases} x & \text{if } x \geq 0 \\ -x & \text{if } x < 0. \end{cases}$$

Give three examples showing that if x and y are real numbers, then $|xy| = |x| \cdot |y|$. Then, prove that this is true.

Exercise 2.14. Prove that if m, n and t are integers, then at least one of $m - n$, $n - t$ and $m - t$ is even. Also, before your proof, write down three down examples of integers m, n and t, and show which of $m - n$, $n - t$ or $m - t$ are even.

Exercise 2.15.

(a) Prove that if n is a positive integer, then 4 divides $1 + (-1)^n(2n - 1)$.

(b) Prove that every multiple of 4 is equal to $1 + (-1)^n(2n - 1)$ for some positive integer n.

Exercise 2.16. For each pair of integers, find the unique quotient and remainder when a is divided by m.

(a) $a = 15$, $m = 4$

(b) $a = 4$, $m = 15$

(c) $a = -7$, $m = 3$

(d) $a = 65$, $m = 11$

(e) $a = -1$, $m = 15$

(f) $a = 0$, $m = 4$

Exercise 2.17. For each of the following pairs of numbers, list all of their common divisors (positive and negative!), and find $\gcd(a, b)$.

(a) $a = 12$, $b = 330$

(b) $a = -36$, $b = 64$

(c) $a = 7$, $b = -27$

Exercise 2.18. For each pair of integers, find $\gcd(a, b)$ and integers k and ℓ such that $\gcd(a, b) = ak + b\ell$. (Note: We know that such integers exist by Theorem 2.13.)

(a) $a = 3$, $b = 13$

(b) $a = 13$, $b = 3$

(c) $a = -25$, $b = 40$

(d) $a = -22$, $b = -14$

(e) $a = 62$, $b = 48$

(f) $a = 13$, $b = -50$

Exercise 2.19. Let a and b be positive integers, and suppose r is the nonzero remainder when b is divided by a. Prove that when $-b$ is divided by a, the remainder is $a - r$.

Exercise 2.20. Determine the remainder when 3^{302} is divided by 28, and show how you found your answer (without a calculator!).

Exercise 2.21. Assume that a, b, c, d and n are integers. Also assume that $a \equiv b \pmod{n}$ and $c \equiv d \pmod{n}$. Prove the following.

(i) $a - c \equiv b - d \pmod{n}$.

(ii) $a \cdot c \equiv b \cdot d \pmod{n}$.

Exercise 2.22. Assume that a is an integer and p and q are distinct primes. Prove that if $p \mid a$ and $q \mid a$, then $pq \mid a$. Also, before your proof, give three examples of this property.

Exercise 2.23. Prove that if abc is a multiple of 10, then at least one of ab, ac or bc is a multiple of 10. Also, before your proof, give three examples of this property.

Exercise 2.24. Assume that a, b and c are integers, $a^2 \mid b$ and $b^3 \mid c$. Prove that $a^6 \mid c$. Also, before your proof, give three examples of this property.

Exercise 2.25. Prove that for every integer n, either $n^2 \equiv 0 \pmod{4}$ or $n^2 \equiv 1 \pmod{4}$.

Exercise 2.26. Prove that if a, b and n are positive integers and $a \equiv b \pmod{n}$, then $a^2 \equiv b^2 \pmod{n}$. Also, before your proof, write down three examples of this property.

Exercise 2.27. The Pythagorean theorem involves integers a, b and c for which $a^2 + b^2 = c^2$. Prove that if three integers satisfy this relationship, then either a or b will be divisible by 3.

> **Note.** The next three exercises will ask you to prove that one thing is true *if and only if* something else is true. If one says "P if and only if Q," where P and Q are some mathematical statements, what this means is "If P, then Q" and also "If Q, then P."

Exercise 2.28. Prove that n is even if and only if n^2 is even. To do this, here are the two things that you should prove:

(a) If n is even, then n^2 is even.

(b) If n^2 is even, then n is even.

Exercise 2.29. Suppose that a, and b are positive integers, and $\gcd(a, b) = d$. Prove that $a \mid b$ if and only if $d = a$. To do this, here are the two things that you should prove:

(a) If $a \mid b$, then $d = a$.

(b) If $d = a$, then $a \mid b$.

Exercise 2.30. Prove that $m \equiv n \pmod{15}$ if and only if $m \equiv n \pmod 3$ and $m \equiv n \pmod 5$. To do this, here are the two things that you should prove:

(a) If $m \equiv n \pmod{15}$, then $m \equiv n \pmod 3$ and $m \equiv n \pmod 5$.

(b) If $m \equiv n \pmod 3$ and $m \equiv n \pmod 5$, then $m \equiv n \pmod{15}$.

Exercise 2.31. Suppose that a and b are positive integers and $d = \gcd(a, b)$.

(a) Prove that $\gcd\left(\frac{a}{d}, \frac{b}{d}\right) = 1$.

(b) Prove that $\gcd(an, bn) = dn$ for every positive integer n.

Exercise 2.32. Assume that a, b and c are integers for which $\gcd(a, b) = 1$ and $\gcd(a, c) = 1$. Prove that $\gcd(a, bc) = 1$.

Definition. For nonzero integers a and b, an integer n is a *common multiple* of a and b if $a \mid n$ and $b \mid n$. The *least common multiple* of a and b is the smallest positive integer m such that m is a common multiple of a and b. We denote this value by $\text{LCM}(a, b)$.

This definition will be used in the following problem.

Exercise 2.33. For the pairs of numbers in parts (a)–(d), determine $\gcd(a, b)$, $\text{LCM}(a, b)$, and ab.

(a) $a = 6$, $b = 8$

(b) $a = 3$, $b = 6$

(c) $a = 4$, $b = 6$

(d) $a = 5$, $b = 6$

(e) Based on your answers to parts (a)–(d), conjecture a relationship between $\gcd(a, b)$, $\text{LCM}(a, b)$, and ab. Then, prove that your conjecture is correct.

Exercise 2.34. If $\gcd(a, b) = 1$, then we say that $\frac{a}{b}$ is in *reduced form*. Prove that if n is an integer, then

$$\frac{21n + 4}{14n + 3}$$

is in reduced form.

Exercise 2.35. Prove that $3 \mid (4^n - 1)$ for every $n \in \mathbb{N}$ in two different ways.

(a) First, prove it using modular arithmetic.

(b) Second, prove it using the fact (which you do not have to prove) that

$$x^n - y^n = (x - y)(x^{n-1} + x^{n-2}y + x^{n-3}y^2 + \cdots + xy^{n-2} + y^{n-1})$$

for any real numbers x and y.

Exercise 2.36. Prove that every odd integer is the difference of two squares. (For example, $11 = 6^2 - 5^2$.)

Exercise 2.37. Prove that for every positive integer n, there exist a string of n consecutive integers none of which are prime.

> **Note.** Recall from Chapter 1 that a *counterexample* is a specific example showing that a conjecture is false.

Exercise 2.38. The following conjectures are all false. Prove that they are false by finding a counterexample to each.

(a) Conjecture 1: Let $f(n) = n^2 - n + 5$. If n is an integer, then $f(n)$ is a prime number.

(b) Conjecture 2: Suppose a, b and c are positive integers. If $a \mid bc$, then $a \mid b$ or $a \mid c$.

(c) Conjecture 4: Suppose a and b are integers. If $a \mid b$ and $b \mid a$, then $a = b$.

Exercise 2.39. Suppose n is an integer. Prove that if $n^2 \mid n$, then n is either -1, 0 or 1.

Exercise 2.40. As Evelyn Lamb pointed out,

Every prime larger than 3 is precisely 1 off from a multiple of 3!

The cool thing about this statement is that it is true whether the "!" symbol is an exclamation or a factorial! Prove this.

Exercise 2.41. After defining a prime number, Definition 2.16 stated that an integer $n \geq 2$ being "not prime" was equivalent to n being able to be written as $n = st$, where s and t are integers and $1 < s, t < n$. Prove that these are indeed equivalent. That is, prove that if $n \geq 2$ is not prime, then $n = st$ for some integers s and t where $1 < s, t < n$. And then prove that if $n = st$ for some integers s and t where $1 < s, t < n$, then n is not prime.

Exercise 2.42. Read the *Introduction to Number Theory* following this chapter. Then, encrypt your first name using the RSA algorithm, and then show how to decrypt it. Show every step of your procedure, including what you used for your encryption key, what your numerical message is, and every calculation along the way.

— Open Questions —

In this chapter we investigated divisibility, a topic that stretches back to the ancient Greeks. In fact, the Greeks looked closely at the divisors of the positive integers and noticed that some of them had some special properties. It will be easier to see if we instead discuss the *proper divisors* of a number: A positive integer d is a *proper divisor* of n if $d \mid n$ and $d \neq n$. For example, 10's divisors are -10, -5, -2, -1, 1, 2, 5 and 10. The proper divisors of 10 are all of its divisors which are positive and are not equal to 10. Thus, they are 1, 2 and 5. Here are data on more n-values:

n	2	3	4	5	6	7	8	9	10	11	12
Proper Divisors	1	1	1, 2	1	1, 2, 3	1	1, 2, 4	1, 3	1, 2, 5	1	1, 2, 3, 4, 6

When the Pythagoreans (i.e., people who studied under Pythagoras) looked at this table, they noticed something special when $n = 6$. Can you spot it? In fact, they also noticed that the same special property occurred in the $n = 28$ case, whose proper divisors are 1, 2, 4, 7 and 14. Again: Can you spot it?

What they noticed is that in these two cases, n is equal to the sum of its proper divisors! That is, $6 = 3 + 2 + 1$ and $28 = 14 + 7 + 4 + 2 + 1$. Divisors already correspond to the number in a multiplicative way, but having a summation property was even more intriguing to them. In fact, they were so smitten with them, that they defined a *perfect number* to be any number that is the sum of its proper divisors.

A couple hundred years later, in about 300BC, Euclid proved that if $2^n - 1$ is prime, then $2^{n-1}(2^n - 1)$ is a perfect number.[35] Two thousand years after that, Euler proved that if a perfect number is even, then it must be of this form.

Prime numbers of the form $2^n - 1$ are today called *Mersenne primes*, and the largest primes we know are almost all Mersenne primes — for both computational and theoretical reasons, numbers of that form are where we point our flashlights when we go searching for the next monster prime.

The Greeks knew the first four perfect numbers (6, 28, 496, and 8128), and Egyptian mathematician Ismail ibn Ibrahim ibn Fallus (1194-1239) found the next three (33550336, 8589869056 and 137438691328; yes, they are quite rare!). Only 49 perfect numbers are known today, but every time our super(-duper) computers find us a new Mersenne prime, we get another perfect number for free. Below are two questions regarding perfect numbers that nobody knows the answer to.

Open Question 1. Are there infinitely many perfect numbers?

Open Question 2. Are all perfect numbers even?

[35]Three examples: Since $2^2 - 1 = 3$ is prime, therefore $2^1(2^2 - 1) = 6$ must be perfect, which it is. And since $2^3 - 1 = 7$ is prime, therefore $2^2(2^3 - 1) = 28$ must be perfect, which it is. And since $2^5 - 1 = 31$ is prime, therefore $2^4(2^5 - 1) = 496$ must be perfect, which it is.

Introduction to Number Theory

Take a look at the following:

$$\begin{aligned} 4 &= 2+2 \\ 6 &= 3+3 \\ 8 &= 5+3 \\ 10 &= 5+5 \\ 12 &= 7+5 \\ 14 &= 7+7 \\ 16 &= 11+5 \\ 18 &= 13+5 \\ 20 &= 17+3 \\ 22 &= 11+11. \end{aligned}$$

What do you notice about this? The numbers on the left are the even numbers larger than 2, and we have written each as a sum of two other numbers. Do you notice anything special about the numbers in the sums?

Each is written as a sum of two primes! Could it possibly be true that *every* even number larger than 2 can be written as a sum of two prime numbers? The data above is very little. The first ten cases work, but there are still infinitely many to go. Also, for small numbers, primes are everywhere! Take a look at the sequence of primes:[36]

$$2\ ,\ 3\ ,\ 5\ ,\ 7\ ,\ 11\ ,\ 13\ ,\ 17\ ,\ 19\ ,\ 23\ ,\ 29\ ,\ 31,\ 37\ ,\ 41\ ,\ \ldots$$

Among the ten odd numbers between 3 and 21, seven of them are prime! So the fact that we can write all the evens from 4 to 22 as a sum of two primes could very well be a coincidence—with 70% of the odds being prime, including the first three, *the odds seem in our favor.*[37] But could this result hold true even when the primes start thinning out? The density of the even numbers remains constant, so when the density of the primes starts dropping, perhaps there simply are not enough primes out there to give this pattern a fighting chance...

[36] A fundamental fact: The sequence of primes is never-ending (we will prove this in Chapter 7).

[37] Some say that puns make you numb. But this section's puns make you number.

Before Christian Goldbach discussed this problem with others in 1742, he presumably checked at least the first 100 cases, and sure enough, each of these can be written as a sum of two primes. Goldbach then told Leonhard Euler,[38] a master of numerical computation, who undoubtedly checked hundreds more. And so far, so good.

In 1938, Nils Pipping earned a spot in the math history books the hard way: he checked, by hand, all the even numbers up to 100,000—and sure enough, he was able to write each and every one as a sum of two primes. Then, once computers were invented, Pipping's labor could be replicated in the blink of an eye, and could keep going at will.[39] As of this writing, the first 200,000,000,000,000,000 cases have been checked, and every single one is a sum of two primes.

But is it true forever? Or does there exist at least one even integer, way down the line, that is not the sum of two primes? And what do primes have to do with it anyways? It is common to think about multiplying primes together, since every natural number is a product of primes (a fact we will prove in Chapter 4); for example, $15 = 3 \cdot 5$, and $28 = 2 \cdot 2 \cdot 7$. But adding primes together completely loses their factorization; if p_1 and p_2 are different primes, then $p_1 + p_2$ is a product of primes other than p_1 and p_2. So the main way in which we think about primes is failing us.

Goldbach's Conjecture states that every even number larger than 2 can be written as the sum of two primes.[40] But as of today, it remains a conjecture, meaning we still do not know whether it is true. Yet due to its age, its simplicity to understand, its difficulty to prove, and its relationship with prime numbers, whose study remains as important today as ever before, it stands as one of the most famous unsolved problems in all of mathematics.

The Prime Number Theorem

The study of formal mathematics began with the study of geometry and number theory. And from early on, mathematicians realized that in order to understand numbers, one must study the primes. In Chapter 2, we recalled that a prime is an integer p which is at least 2 and whose only positive divisors are 1 and p. If a positive integer is at least 2 and is not prime, then it is called *composite*. Not only do the primes thin out as you reach higher and higher portions of the natural numbers, but the *rate* at which the primes thin out is pretty steady. Indeed, if we let $\pi(N)$ denote the number of primes up to an integer N (e.g., $\pi(10) = 4$, since there are four primes in $\{1, \mathbf{2}, \mathbf{3}, 4, \mathbf{5}, 6, \mathbf{7}, 8, 9, 10\}$), then we know exactly how fast $\pi(N)$ is growing as N gets larger and larger. The answer:

$$\pi(N) \sim \frac{N}{\log(N)}.$$

[38] Pro-Tip: "Euler" is pronounced "Oiler."

[39] And a computer can do so without his wife getting mad at him because he won't put down the damn pencil. I mean, seriously Nils, 50,000 ain't enough? You live in Finland in the 1930s for God's sake. Pick up an ax already; your house won't heat itself!

[40] ....And puns about Goldbach's conjecture make you *even number*.

First, as we discussed in the Chapter 2 Pro-Tips, "log" refers to the natural log. Next, the symbol "$\sim$" means that

$$\lim_{N\to\infty} \frac{\pi(N)}{N/\log(N)} = 1.$$

Intuitively, this means that the two functions are growing at the same rate. You can think about this as saying that $\frac{1}{\log(N)}$ is the proportion of the numbers up to N which are prime, roughly. Because we are taking a limit, this does not need to be the *exact* proportion — and indeed, it's not the exact proportion — but over time the error is a smaller and smaller percentage of the total. This fact is extremely important, and is known as the *prime number theorem*.

Cryptography

I now want to turn to a modern application of prime numbers: cryptography. The classic way to decode a message is a *substitution cipher* in which each letter is replaced by a symbol. If each letter is replaced by the letter which is, say, five positions further down the alphabet, then you have a *Caesar cipher*; it is named after Julius Caesar, who is said to have communicated with his generals using a Caesar cipher with a shift of three. For example, Caesar might send

L kdyh lqyhqwhg d qhz vdodg, whoo wkh Juhhnv

which means

I have invented a new salad, tell the Greeks.

Since then, much more sophisticated techniques have been developed to send secret messages. During World War II, Nazi Germany famously used the *Enigma machine* to encrypt messages. They are famous for using it because although it worked well for them for awhile, eventually superhero mathematicians like Alan Turing threw their minds at it.[41] Their work allowed the Allies to read substantial amounts of Axis communication, which western Supreme Allied Commander Dwight D. Eisenhower said was "decisive" to the Allied victory.

So yeah, it's pretty cool stuff. And it doesn't stop there. Modern life depends on the ability to send secure messages through public channels of communication, like the Internet. From banking to email to social media, keeping our electronic communications and transactions secure from eavesdroppers is an enormously important and difficult task. And as it happens, the most popular encryption technique relies entirely on the mathematics of primes and modular arithmetic.

The modular arithmetic that we learned in Chapter 2 was developed by the great number theorist Carl Friedrich Gauss in his book *Disquisitiones Arithmeticae*, which the precocious whiz kid wrote at just 21 years of age. Gauss realized that focusing on remainders is a powerful tool to understand numbers, and while much of his work seemed entirely theoretical and application-free, it is now one of the most applied advanced mathematical tools in the world.

[41]Turing's story is portrayed well in the movie *The Imitation Game*, which I highly recommend. But be warned, his story is a sad one.

The RSA Algorithm

The *RSA*[42] *algorithm* is a cryptosystem — a way to encode and decode messages. Let's examine this through an example. We begin with an encryption lock, which is a pair of numbers; in our example, it will be $(7, 33)$. This pair of numbers is published publicly, which my friend will use to decode my message (and, as you'll see, I will not care if anybody else knows which pair of numbers I chose). Next, we need a message that we want to send. Let's send "math", which we write as numbers using the standard $a = 1$, $b = 2$, $c = 3$, and so on, which turns "math" into "13 1 20 8". (Or, one could concatenate this to 131208, or a variant, which we won't do here.)

Encryption Key: $(7, 33)$
Message: math
Numerical Message: 13 1 20 8

Next, we encrypt the numerical message. But we do not simply shift these numbers like Julius Caesar would do — we turn each number into another number in a more sophisticated way. Using our encryption key of $(7, 33)$, we turn a number a into the remainder when a^7 is divided by 33. That is, we are looking for $a^7 \pmod{33}$:

$$\begin{array}{ccccccc} \text{m} & \to & 13 & \to & 13^7 \pmod{33} & \equiv & 7 \pmod{33} \\ \text{a} & \to & 1 & \to & 1^7 \pmod{33} & \equiv & 1 \pmod{33} \\ \text{t} & \to & 20 & \to & 20^7 \pmod{33} & \equiv & 26 \pmod{33} \\ \text{h} & \to & 8 & \to & 8^7 \pmod{33} & \equiv & 2 \pmod{33} \end{array}$$

You can see that $1^7 \equiv 1 \pmod{33}$, and your phone's calculator can confirm the rest; for example, that $8^7 \pmod{33} \equiv 2097152 \pmod{33} \equiv 2 \pmod{33}$. From this, we see that the encryption key $(7, 33)$ would encrypt "math" as "7 1 26 2".

Suppose I send "7 1 26 2" to my friend, and she can see that my encryption key is $(7, 33)$. How can she decode my super-secret message? The trick will be to use a *decryption key*, which in this problem will be $(3, 33)$. We will talk later about how to come up with this, but for now let me show you how it works. What my friend does is she turns each number b into the remainder when b^3 is divided by 33. That is, she is looking for $b^3 \pmod{33}$:

$$\begin{array}{ccccccc} 7 & \to & 7^3 \pmod{33} & \equiv & 13 \pmod{33} & \to & \text{m} \\ 1 & \to & 1^3 \pmod{33} & \equiv & 1 \pmod{33} & \to & \text{a} \\ 26 & \to & 26^3 \pmod{33} & \equiv & 20 \pmod{33} & \to & \text{t} \\ 2 & \to & 2^3 \pmod{33} & \equiv & 8 \pmod{33} & \to & \text{h} \end{array}$$

And with that, she has decoded my message!

[42]Named after its desigers, Ron **R**ivest, Adi **S**hamir, and Leonard **A**dleman.

Finding the Lock and Key

How did I come up with my encryption lock, and how did my friend come up with her decryption key? To begin, recall that two numbers are considered *relatively prime* if they share no prime factors. For example, 1, 3, 7 and 9 are all relatively prime to 10, while 2, 4, 5, 6, 8 and 10 are not relatively prime to 10. With that, here are the steps for the RSA algorithm:

1. Pick any two primes that only you and your correspondent know, and let N be their product.
   - In the above example, I chose $p = 3$ and $q = 11$, which gave $N = 33$.
2. Compute $\phi(N)$: If $N = pq$, then $\phi(N) = (p-1)(q-1)$. See footnote.[43]
   - In the above example, $\phi(N) = (3-1)(11-1) = 20$.
3. Choose any number t in $\{2, 3, \ldots, \phi(N)\}$ which is relatively prime to N *and* is also relatively prime to $\phi(N)$; this will always be possible.[44]

   ◦ In the above example, we want a number in $\{2, 3, 4, \ldots, 20\}$ which is relatively prime to both 33 and 20. The options are: 7, 13, 17 and 19. I chose to use 7.
4. The numbers in Steps 3 and 1 give the encryption lock: (t, N).
   - In the above example, the lock is $(7, 33)$.
5. Choose the first number d in $\{1, 2, 3, \ldots, \phi(N)\}$ such that $td \equiv 1 \pmod{\phi(N)}$; this will always be possible in practice.[45]

   ◦ In the above example, where in Step 3 we said $t = 7$, if we checked $d = 1, 2, 3, 4, 5, \ldots$ we would get

$$\begin{aligned} 7 &\equiv 7 \pmod{20} \\ 14 &\equiv 14 \pmod{20} \\ 21 &\equiv 1 \pmod{20} \\ 28 &\equiv 8 \pmod{20} \\ 35 &\equiv 15 \pmod{20} \\ &\vdots \end{aligned}$$

   So with $d = 3$ we have $td = 21$, and $td \equiv 1 \pmod{20}$.
6. The numbers in Steps 5 and 1 give the decryption key: (d, N).
   - In the above example, the key is $(3, 33)$.

---

[43]In general, $\phi(N)$ is defined to be the number of integers in $\{1, 2, 3, \ldots, N\}$ which are relatively prime to N. But when $N = pq$, this has a simple formula: $\phi(N) = (p-1)(q-1)$. (By the way, this ϕ function is called the *Euler totient function*.)

[44] See: Group theory.

[45]Note: In theory it is possible for this not to occur, but in practice we use huge primes which essentially guarantees this.

Why it Works

The mathematics for why this algorithm works is based on a generalization of Fermat's little theorem (Theorem 2.19). The generalization is due to Euler, and it says this:

Theorem.

Theorem 2.22 (*Euler's theorem*). If a and N are positive integers which are relatively prime, then

$$a^{\phi(N)} \equiv 1 \pmod{N}.$$

Now suppose we have such an a and N, just like in the RSA algorithm. We then found t and d such that $td \equiv 1 \pmod{\phi(N)}$. By applying the definition of modular congruence, this means $td = 1 + k \cdot \phi(N)$ for some integer k. And recall what we did in the RSA algorithm: First we encrypted the message by finding $a^t \pmod{N}$, then we decrypted it by taking this encrypted message and raising it to the power of d $\pmod{N}$. That is, $(a^t)^d \pmod{N}$. And the claim is that this will give us back a, our original message. And by applying a little algebra and the above theorem, it does:

$$(a^t)^d = a^{td} = a^{1+k\cdot\phi(N)} = a \cdot \left(a^{\phi(N)}\right)^k \equiv a \cdot 1^k \equiv a \pmod{N}.$$

That calculation shows why RSA works, but why is RSA secure from potential eavesdroppers? It is secure because in order to undo this procedure, you need to know more than N—you need to know p and q. If you know p and q, then you can easily find $\phi(N)$, and with $\phi(N)$ you can find d, and with d you can decrypt the message. But how do you find p and q? If $N = 33$, then it's easy: just factor N. But in practice, the primes used are not small, like 3 and 11 from our example. The primes are huge. And when I say huge, I really mean HUGE. The primes typically have *hundreds* of digits.

For instance, here is a single prime number containing 300 digits:

203956878356401977405765866929034577280193993314348263094772646453 2...

Let me fix:

2039568783564019774057658669290345772801939933143482630947726464532...

...8306272270127763293661606314408817331237288267712387953870940 01...

...5830656733832827915449969836607190676644003707421711780569087 27...

...9284814911202228633214487618337632651208357482164793399296124 99...

...17319836219304274280243803104015000563790123

Yes, that number is prime.[46] Now imagine you took another prime, this one with 400 digits, and multiplied them together. That's N. It's a gigantic number

[46]Fun fact: There is a much bigger prime that you can memorize. The number 11111...111, which is a number with only the digit 1, and which consists of 270,343 copies of 1 in a row, is a prime number. It's called a *repunit prime*.

with around 700 digits. If you were eavesdropping on a digital transaction between somebody's computer and Bank of America[47] and you wanted to decode the message, your first challenge would be to factor a 700-digit number. If you managed to factor N, then you have a fighting chance to decode the message, but factoring a 700-digit number with our current knowledge and tools is *hard*. Like, really freaking hard. And without that, you're sunk.[48]

The world's privacy and finances rely on the assumption that nobody will ever figure out how to factor really big numbers quickly. Of course, our ability to factor big numbers is steadily increasing simply because our computing capabilities are steadily increasing. But the response to that is easy: Instead of using 300- and 400-digit primes for p and q, just use 500- and 600-digit primes. Or 700- and 800-digit primes. As long as you can keep finding primes with another hundred digits, the fix is easy.

To really be vulnerable, someone would have to find a method that far surpasses our abilities to just pick larger primes. There would have to be an advancement far and above what mathematicians envision as plausible. In some ways, our world rests on the assumption that humans are not smart enough to solve that problem. And although such a solution would surely be fascinating... let's all hope we're not.

Before cryptography ran our world, and before all of our secrets were secured with algebraic locks, number theory was viewed as an exceptionally pure and abstract field, driven by theory and curiosity, not application. Numbers are everywhere in the real world, but the questions of number theory seemed to be on a higher plane. Gauss called number theory the "queen of mathematics."

In G. H. Hardy's book *A Mathematician's Apology* — a small book that every working mathematician should read — he writes about beauty in mathematics. To Hardy, math without applications is something to be praised, especially when the experimental sciences have too often had their work used for death and destruction.

Hardy celebrates number theory for remaining above the fray, saying "No one has yet discovered any warlike purpose to be served by the theory of numbers." Ironically, at that very moment German Nazis were secretly using number theory to encrypt their most classified communications to coordinate their military action in the second world war — the deadliest conflict in human history. Yet in a bit of mathematical justice, it was just a few years later that number theory was used to crack the German enigma codes, a pivotal moment that turned the tide against the Nazis.

Today, as we rely more and more on computer-run infrastructure, the pernicious threat of cyber-warfare requires a constant defense. I saw this in 2010 when interning at the NSA, before its threats began to regularly spill into the news and our national discourse. Thus, while academic research in number theory continues, work on its many applications do, too.

---

[47] Activities which we at Long-Form Mathematics do NOT condone.

[48] By the way, if you want to help find new big primes, check out the Great Internet Mersenne Prime Search.

Chapter 3: Sets

3.1 Definitions

We began with the most fundamental form of proof — the direct proof. Now we turn to one of the most fundamental objects in math — sets. Let's kick that off with some important definitions.

Definition.

Definition 3.1.

- A *set* is an unordered collection of distinct objects, which are called *elements*.[1]
- If x is an element of a set S, we write $x \in S$. This is read "x in S."

When possible, sets are often drawn with curly braces enclosing their elements, like $\{2, \pi, 6\}$. Let's record some important sets and their notation.

Definition.

Definition 3.2.

- The set of *natural numbers*, denoted $\mathbb{N}$, is the set $\{1, 2, 3, \dots\}$.
- The set of *integers*, denoted $\mathbb{Z}$, is the set $\{\dots, -3, -2, -1, 0, 1, 2, 3, \dots\}$.
- The set without any elements, denoted $\emptyset$ or $\{\}$, is called the *empty set*.

Another way to think about a set is as a box, possibly with some things inside. When you look into a box, the things inside do not have any particular order; the same can be said about the elements of a set. Indeed, consider the following.

[1]Alternative definition: Everything. Everything is a set. Almost no definition in the world is as general as that of a set. And while some day you may learn that "everything" is just slightly too broad, it is pretty dang close.

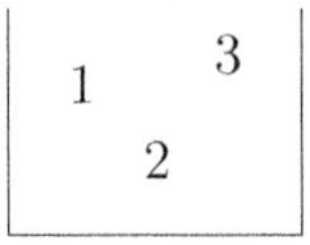

1 3 2

$$= \{2,3,1\}$$

The above box also corresponds to $\{1, 2, 3\}$, $\{1, 3, 2\}$, $\{2, 1, 3\}$, $\{3, 1, 2\}$ and $\{3, 2, 1\}$. Indeed, we will view all of these sets as being *equal* to each other, since they contain the exactly the same elements. For example, $\{1, 3, 2\} = \{3, 2, 1\}$.

Another important thing to note about sets is that the elements do not have to be numbers. The elements of a set can be *anything.*

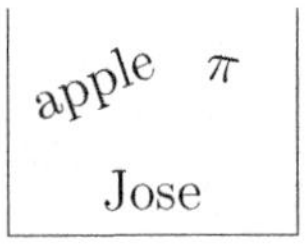

apple π Jose

$$= \{\text{apple, Jose, } \pi\}$$

Also, just as boxes can be empty, so can sets!

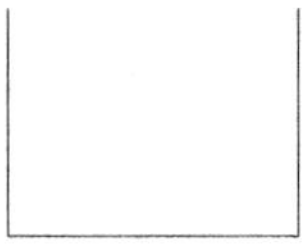

$$= \emptyset$$

Furthermore, it's certainly possible for one box to be inside another box. Likewise, it's certainly possible for one set to be a single element inside another set.

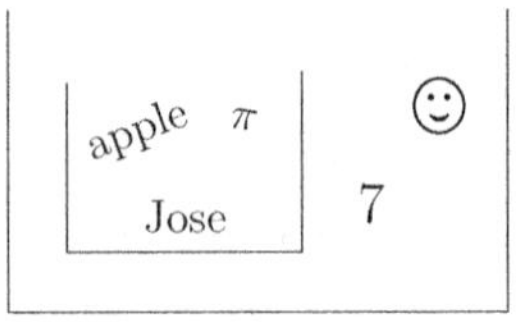

apple π Jose ☺ 7

$$= \{\{\text{apple, Jose, } \pi\}, 7, ☺\}$$

Notice that the above set has three elements in it: (1) a set (containing three specific elements), (2) the number 7, and (3) a smiley face. Your box could also have just one thing in it: a smaller box with nothing inside it. This looks like the following.

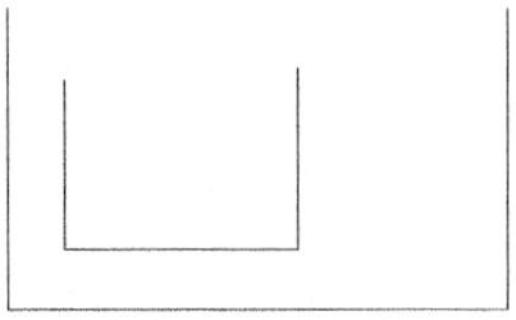

$$= \{\emptyset\}$$

In the last three pictures we have seen an example for $\emptyset$ and for $\{\emptyset\}$. Notice that these are different! The empty set is different than the set containing the empty set, just as an empty box is different than a box containing an empty box. It would be a mistake to think about $\emptyset$ as being nothing. It's something! It's a set! It doesn't have anything in it, but it's still a thing. In the same way, $\{\emptyset\}$ is a set containing one element — its element is a set which contains no elements, but it's still there and it's still a thing.

Definition 3.1 defined a set to be an unordered collection of distinct objects. So far, we have focused our discussion on what it means to be "unordered," what it means to be a "collection" and what counts as "objects." The final key word in the definition is "distinct." We require that the elements in a set be distinct — meaning, they are all different. For example, {apple, Jose, π} is a set where each of its three elements is distinct from the others.

This might bring up a point of confusion: if we write $\{1, 1, 2\}$, is this not a set? The decision mathematicians have made is to regard $\{1, 1, 2\}$ as being the same as the set $\{1, 2\}$. That is, we regard the duplicates as being automatically removed. Thus, $\{a, b, b, c\}$ and $\{a, a, a, b, c, c\}$ are both considered to be the set $\{a, b, c\}$. This convention will pay off in many instances, including one coming up soon.

All of our examples thus far have been sets written like $\{\dots\}$, where inside the braces is just a list of the elements, like $\{1, 2, 3\}$ or $\{1, 2, 3, 4, \dots\}$. Sometimes, though, they are defined by a rule; this is called *set-builder notation.* Set-builder notation either looks like this:

$$\{\text{elements} \ : \ \text{conditions used to generate the elements}\},$$

or perhaps like this:

$$\{\text{elements} \in S \ : \ \text{conditions used to generate the elements}\},$$

where S is some larger set in which the conditions are restricting. Let's discuss a couple examples of both of these forms. First, here are two examples of the first form:

- $\{n^2 : n \in \mathbb{N}\} = \{1, 4, 9, 16, 25, \dots\}$.
- $\{|n| : n \in \mathbb{Z}\} = \{0, 1, 2, 3, \dots\}$.

The first example uses the condition $n \in \mathbb{N}$, which means[2] that you should plug in $n = 1, 2, 3, 4, 5, \dots$ into n^2 to get the elements of the set. Next, here are two examples of the second form:

- $\{n \in \mathbb{Z} : \ n \text{ is even}\} = \{\dots, -6, -4, -2, 0, 2, 4, 6, \dots\}$
- $\{n \in \mathbb{N} : 6 \mid n\} = \{6, 12, 18, 24, 30, \dots\}$

Next[4] up, let's discuss one weird set and one important set. First, here is the weird set: $\{w : w \text{ is weird}\}$. And now, for the important set: the set of rational

[2]Recall that 0 is *not* considered a natural number.[3] I will defend this to my grave.

[3]The set $\{0, 1, 2, 3, \dots\}$ is denoted $\mathbb{N}_0$. (Fun fact: '0' was first discovered by an ancient Babylonian who asked how many of his friends wanted to talk about numbers with him.)

[4]Quick note: Be careful when you use dot-dot-dots. They are not rigorous — they are an informal way to say "and continue this pattern forever." It is fine to use them in your work *provided the pattern is clear.* For example, "$1, 2, \dots$" is not clear at all. Does this mean the arithmetic sequence $1, 2, 3, 4, 5, \dots$? Or the geometric sequence $1, 2, 4, 8, 16, \dots$? Or perhaps it is the sequence of factorials $1, 2, 6, 24, 120, \dots$? Or the sequence of Catalan numbers $1, 2, 5, 14, 42, \dots$? Make sure your pattern is very clear before throwing down the dot-dot-dots.

numbers, which is important enough to deserve a special symbol, and to have its definition be enclosed in a definition box.

Definition.

Definition 3.3. The set

$$\mathbb{Q} = \left\{ \frac{a}{b} : a, b \in \mathbb{Z}, b \neq 0 \right\}$$

is called the set of *rational numbers.*

The equation in Definition 3.3 is read like so:

$\mathbb{Q}$	$=$	$\{$	$\frac{a}{b}$	$:$	$a, b \in \mathbb{Z}$	,	$b \neq 0\}$
The rational numbers	are defined to be	the set of all	fractions of the form $\frac{a}{b}$	such that	a and b are integers	and	b is nonzero

You might notice that the definition of $\mathbb{Q}$ considers both $\frac{2}{3}$ and $\frac{4}{6}$ and $\frac{6}{9}$, and infinitely more representations of this same number. But remember that a set only keeps one of each element, so the duplicates of each rational number will be automatically removed, simply because they are part of a set.

The set of real numbers, denoted $\mathbb{R}$, is more difficult to define, so for now just rely on your intuition — real numbers are all the numbers you can write down with a decimal point. This includes integers like -4, finite-decimals like 12.439, and infinite-decimals like $3.14159\ldots$. To define them rigorously would literally take dozens of pages, which you would likely find much more confusing than enlightening.

Let's now use $\mathbb{R}$ and set-builder notation to generate other familiar sets. The set of 2×2 real matrices can be written

$$\left\{ \begin{bmatrix} a & b \\ c & d \end{bmatrix} : a, b, c, d \in \mathbb{R} \right\}.$$

The xy-plane represents the set of *ordered pairs* of real numbers. This set can be written

$$\mathbb{R}^2 = \{(x, y) : x \in \mathbb{R} \text{ and } y \in \mathbb{R}\}.$$

The unit circle (circle of radius 1 centered at the origin) is contained inside of $\mathbb{R}^2$, and can be defined as follows:

$$S^1 = \{(x, y) \in \mathbb{R}^2 : x^2 + y^2 = 1\}.$$

The closed interval $[a, b]$ can be defined as follows:

$$[a, b] = \{x \in \mathbb{R} : a \leq x \leq b\}.$$

The notation for the open interval is "(a, b)"—this looks the same as an ordered pair, and you must use context to determine which is which. Its definition is this: $(a, b) = \{x \in \mathbb{R} : a < x < b\}$, and it applies even if $a = -\infty$ and/or $b = \infty$. The definitions for the half open intervals, $(a, b]$ and $[a, b)$, are similar.

3.2 Proving $A \subseteq B$

Definition.

Definition 3.4. Suppose A and B are sets. If every element in A is also an element of B, then A is a *subset* of B, which is denoted $A \subseteq B$.

Just as definitions are human choices which at times provide intuition, notation is too. The notation "$A \subseteq B$" for sets A and B looks similar to "$x \leq y$" for numbers x and y, and shares many properties: If $A \subseteq B$, then B is bigger than A is some sense. And if $A \subseteq B$ and $B \subseteq C$, then $A \subseteq C$. Later on we will discuss other similarities.

Below are three standard examples and one subtle example.

Example 3.5.

- $\{1, 3, 5\} \subseteq \{1, 2, 3, 5, 7\}$, because 1, 3 and 5 are all in $\{1, 2, 3, 5, 7\}$.
- $\mathbb{N} \subseteq \mathbb{Z} \subseteq \mathbb{Q} \subseteq \mathbb{R}$. (If you know the complex numbers: $\mathbb{N} \subseteq \mathbb{Z} \subseteq \mathbb{Q} \subseteq \mathbb{R} \subseteq \mathbb{C}$.)
- $\{a, b, c\} \not\subseteq \{a, b, e, f, g\}$, because c is not in $\{a, b, e, f, g\}$.
- For *every* set B, it is true that $\emptyset \subseteq B$. Why does this satisfy Definition 3.4? To see it, first note that, because there are no elements in $\emptyset$, it would be true to say "for any $x \in \emptyset$, x is a purple elephant that speaks German." It's vacuously[5] true! You certainly can't disprove it, right? You can't present to me any element in $\emptyset$ that is *not* a purple elephant that speaks German.

  By this reasoning, I could switch out "is a purple elephant that speaks German" for *any other statement* and it would still be true! And this includes the subset criteria: If $x \in \emptyset$, then $x \in B$, which by definition means that $\emptyset \subseteq B$. Again, you certainly cannot present to me any $x \in \emptyset$ which is *not* also an element of B, can you?[6]

Notice that if $A = B$, then $A \subseteq B$. In the case that $A \subseteq B$ and $A \neq B$, we say that A is a *proper subset* of B. We will not use it in this text, but the correct notation for this is "$A \subset B$."[7]

[5]A *vacuum* in physics is a container in which the air inside has been sucked out, leaving nothing left. Likewise in math, saying something is *vacuously true* means that the set of elements that the statement is referring to is empty; therefore there is nothing to prove, and it's automatically true.

[6]Perhaps $\emptyset \subseteq B$ is true only due to a technicality—but this is a technical subject!

[7]Note: Some people use "$\subset$" to mean "is a subset of" and "$\subsetneq$" for "is a proper subset of." These people are wrong. We write $\leq$ and $<$ for our inequalities, and our subset notation should be likewise. I wrote this book mainly as a vehicle to push my opinions on mathematical notation, so don't let me down here. Go forth and spread the word.

Given a pair of sets A and B, Definition 3.4 tells us that in order to prove that $A \subseteq B$, what we would have to show is this:

"If $x \in A$, then $x \in B$."

Thus, here is the outline for a (direct) proof that a set A is a subset of a set B:

Proposition. $A \subseteq B$.

Proof. Assume $x \in A$.

$\langle\langle$ An explanation of what $x \in A$ means $\rangle\rangle$

⋮ apply algebra,
logic, techniques

$\langle\langle$ Oh hey, that's what $x \in B$ means $\rangle\rangle$

Therefore $x \in B$.

Since $x \in A$ implies that $x \in B$, it follows that $A \subseteq B$. □

Let's practice.

Proposition.

Proposition 3.6. It is the case that[8]

$$\{n \in \mathbb{Z} : 12 \mid n\} \subseteq \{n \in \mathbb{Z} : 3 \mid n\}.$$

Scratch Work. For a problem like this, where it is possible to write out more explicitly what sets we are dealing with, it's always a good idea to write out a few of the terms to make sure you believe that it is true. This may also help you prove the result. Here is the first set:

$$\{n \in \mathbb{Z} : 12 \mid n\} = \{\dots, -24, -12, 0, 12, 24, \dots\}.$$

And here is the second set:

$$\{n \in \mathbb{Z} : 3 \mid n\} = \{\dots, -15, -12, -9, -6, -3, 0, 3, 6, 9, 12, 15 \dots\}.$$

[8]Note: It is considered improper to start a sentence with mathematical notation. So we add a short statement like "It is the case that" or "We have" at the start of propositions like this. (Reminder: more advice on writing proofs is contained in Appendix C at the end of this book.)

So yes, it does seem to be checking out. It looks like the terms in the first set make up one fourth of the terms in the second set.

As for the proof, we will follow the outline above; here, $A = \{n \in \mathbb{Z} : 12 \mid n\}$ and $B = \{n \in \mathbb{Z} : 3 \mid n\}$. "An explanation of what $x \in A$ means" will basically just be an application of Definition 2.8 to explain what it means to say "$12 \mid x$." This brings our proof outline to this point:

Proof. Assume $x \in \{n \in \mathbb{Z} : 12 \mid n\}$.

Thus $x \in \mathbb{Z}$ and $12 \mid x$, which by Def 2.8 means $x = 12k$ for some $k \in \mathbb{Z}$.

⋮ apply algebra,
logic, techniques

Therefore, $x = 3m$ for some $m \in \mathbb{Z}$. Thus, by Definition 2.8, $3 \mid x$.

Therefore, $x \in \{n \in \mathbb{Z} : 3 \mid n\}$.

Since $x \in \{n \in \mathbb{Z} : 12 \mid n\}$ implies that $x \in \{n \in \mathbb{Z} : 3 \mid n\}$, it follows that $\{n \in \mathbb{Z} : 12 \mid n\} \subseteq \{n \in \mathbb{Z} : 3 \mid n\}$. □

Can you see how to bridge the gap? Think about it on your own, then check out the proof below.

Proof. Assume $x \in \{n \in \mathbb{Z} : 12 \mid n\}$. This means that $x \in \mathbb{Z}$ and $12 \mid x$, which by Definition 2.8 implies that $x = 12k$ for some $k \in \mathbb{Z}$. Equivalently,

$$x = 3 \cdot (4k).$$

And since $k \in \mathbb{Z}$, by Fact 2.1 it is also true that $4k \in \mathbb{Z}$. Thus, by the definition of divisibility (Definition 2.8), this means that $3 \mid x$. So, $x \in \{n \in \mathbb{Z} : 3 \mid n\}$.

Since $x \in \{n \in \mathbb{Z} : 12 \mid n\}$ implies that $x \in \{n \in \mathbb{Z} : 3 \mid n\}$, it follows that $\{n \in \mathbb{Z} : 12 \mid n\} \subseteq \{n \in \mathbb{Z} : 3 \mid n\}$. □

As we showed above,[9] to prove that $A \subseteq B$, we pick an $x \in A$ and prove that $x \in B$. It is really important to remember that this x has to be an *arbitrary* element of A. We do not pick a specific element, like 24, from A. Moreover, we are allowed to assume nothing about x beyond that it is in A.

If $A = \{n \in \mathbb{Z} : 12 \mid n\}$, like in the last example, the x we choose might be positive, negative or 0. It might be a big number or a smaller number. At no point in our proof did we make any assumptions about our x beyond that it is an integer that is divisible by 12. This is important, because by doing so, anything we prove about our *arbitrary* element of A will then apply to *every* element of A.

[9]FYI: Mathematicians use "above" a lot to simply mean "what we did earlier." At times, the "above" that is being referenced could even be on the previous page.

The next example looks a little different, but the same general principles apply.

Proposition.

Proposition 3.7. Let $A = \{-1, 3\}$ and $B = \{x \in \mathbb{R} : x^3 - 3x^2 - x + 3 = 0\}$. Then $A \subseteq B$.

Proof Idea. Remember that what we must show is that if $x \in A$, then $x \in B$. The trick here is to realize that $x \in A$ can only mean one of two things: either $x = -1$ or $x = 3$. Since there are just two distinct options, this suggests that perhaps using a proof by cases is the way to go.

Next, in each case, how do we show that $x \in B$? We must show that such an x satisfies $x^3 - 3x^2 - x + 3 = 0$; if it does then it's in B since that's literally how B is defined. This is how we proceed.

Proof. Assume $x \in A$. Then either $x = -1$ or $x = 3$. Consider these two cases separately.

Case 1: $x = -1$. Note that this x is a real number, and

$$(-1)^3 - 3(-1)^2 - (-1) + 3 = -1 - 3 + 1 + 3 = 0,$$

which by the definition of B implies $x \in B$.

Case 2: $x = 3$. Note that this x is a real number, and

$$(3)^3 - 3(3)^2 - (3) + 3 = 27 - 27 - 3 + 3 = 0,$$

which by the definition of B implies $x \in B$.

Since $x \in A$ implies that $x \in B$, it follows that $A \subseteq B$. □

Now, when writing a book, there are times where the book's formatting will be better if some part starts at the top of a page. This is one of those times. The curves below are solely to fill up space, so that my next part will start on a fresh page.

3.3 Proving $A = B$

Recall that, for sets A and B, to say that "$A = B$" is to say that these two sets contain *exactly* the same elements. Said differently, it means these two things:

1. Every element in A is also in B (which means $A \subseteq B$), and
2. Every element in B is also in A (which means $B \subseteq A$).

Indeed, a slick way to prove that $A = B$ is to prove both $A \subseteq B$ and $B \subseteq A$—both of which can be done using the approach discussed above.[10]

Thus, here is the outline for one way to prove that two sets are equal:

Proposition. $A = B$.

Proof. Assume $x \in A$.

⟪An explanation of what $x \in A$ means⟫

$\vdots$ apply algebra, logic, techniques

⟪Oh hey, that's what $x \in B$ means⟫

Therefore $x \in B$.

Since $x \in A$ implies that $x \in B$, it follows that $A \subseteq B$.

Next, assume $x \in B$.

⟪An explanation of what $x \in B$ means⟫

$\vdots$ apply algebra, logic, techniques

⟪Oh hey, that's what $x \in A$ means⟫

Therefore $x \in A$.

Since $x \in B$ implies that $x \in A$, it follows that $B \subseteq A$.

We have shown that $A \subseteq B$ and $B \subseteq A$. Therefore, $A = B$. □

We will do some examples of this in the next section.

[10] This is analogous to saying: If x and y are numbers for which $x \leq y$ and $y \leq x$, then $x = y$.

3.4 Set Operations

Next, we define some important operations for sets.

Definition.

Definition 3.8.

- The *union* of sets A and B is the set $A \cup B = \{x : x \in A \text{ or } x \in B\}$.
- The *intersection* of sets A and B is the set $A \cap B = \{x : x \in A \text{ and } x \in B\}$.
- Likewise, if $A_1, A_2, A_3, \ldots, A_n$ are all sets, then the union of all of them is the set $A_1 \cup A_2 \cup \cdots \cup A_n = \{x : x \in A_i \text{ for some } i\}$. This set is also denoted
$$\bigcup_{i=1}^{n} A_i.$$
- Likewise, if $A_1, A_2, A_3, \ldots, A_n$ are all sets, then the intersection of all of them is the set $A_1 \cap A_2 \cap \cdots \cap A_n = \{x : x \in A_i \text{ for all } i\}$. This set is also denoted
$$\bigcap_{i=1}^{n} A_i.$$

To test your understanding, think about what the union and intersection of two sets would look like from the box interpretation with which we began this chapter.[11] One answer: The union of two boxes A and B can be obtained by dumping everything in A and everything in B into a new box, and then removing any duplicate items. The intersection can be obtained by identifying everything in A that is also in B, and putting those items into a new box. The intersection can also be obtained by dumping everything in A and everything in B into a new box, and then removing one of each item (so if there are two of something, you remove just one of the two).

Next, if $A_1, A_2, A_3, \ldots, A_n$ are all boxes, think if you can now describe the following in terms of boxes.

- $\displaystyle\bigcup_{i=1}^{n} A_i = A_1 \cup A_2 \cup \cdots \cup A_n$
- $\displaystyle\bigcap_{i=1}^{n} A_i = A_1 \cap A_2 \cap \cdots \cap A_n$

[11]Or, for additional intuition, here's a popular meme involving the union and intersection of an interesting haircut with a balding man:

 $\cap$  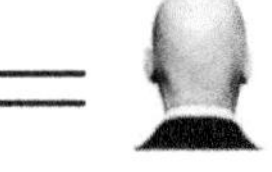 and  $\cup$   

Another helpful way to picture a collection of sets is via *Venn diagrams.* For example, below are $A \cup B$, $A \cap B$, $A \cup B \cup C$ and $A \cap B \cap C$, represented by the shaded region in the diagrams.

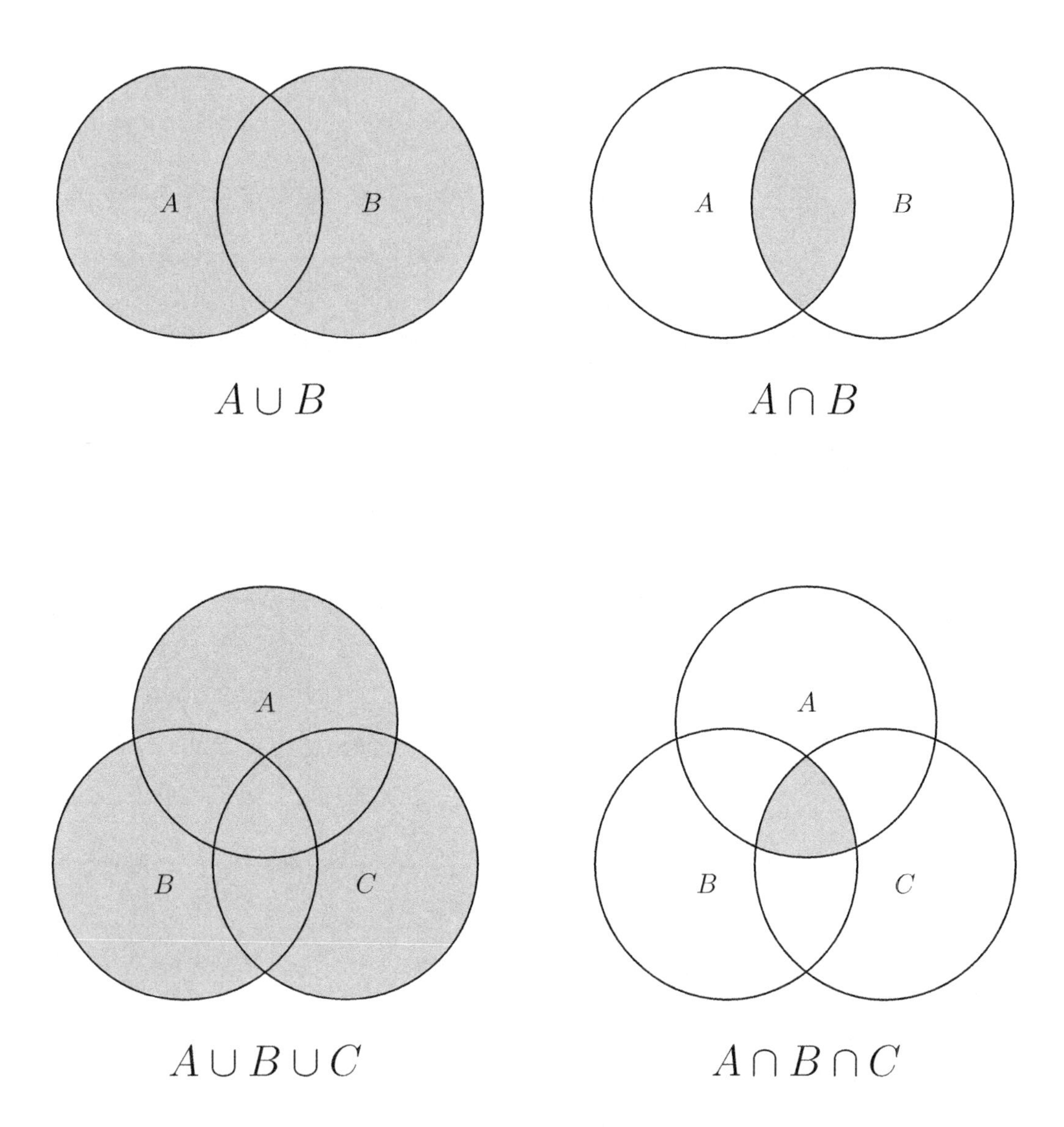

For numbers, there are the operations of addition, subtraction and multiplication, and each of these has an analogous operation in set theory. Indeed, for sets there are union, *set subtraction* and *Cartesian product* operations. Also, just as taking the absolute value of a number tells you how big it is, in set theory one can determine a set's *cardinality.* These are some of the major set operations left to discuss. We will (mostly) go through them two-at-a-time.

Subtraction and Complements

Definition.

Definition 3.9. Assume A and B are sets and "$x \notin B$" means that x is not an element of B.

- The *subtraction* of B from A is $A \setminus B = \{x : x \in A \text{ and } x \notin B\}$.
- If $A \subseteq U$, then U is called a *universal set* of A. The *complement*[12] of A in U is $A^c = U \setminus A$.

Intuitively, $A \setminus B$ means "all the elements in A that are not in B." You can find this set by starting with A, and then removing everything in it that is also in B. As for the complement, A^c intuitively means "everything that is not in A," with one caveat: When we say "everything" we are only referring to things in the universe U. Here are their Venn diagrams:

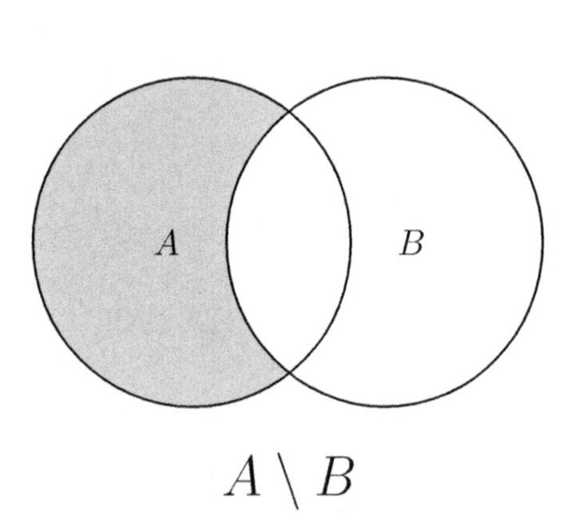

$A \setminus B$

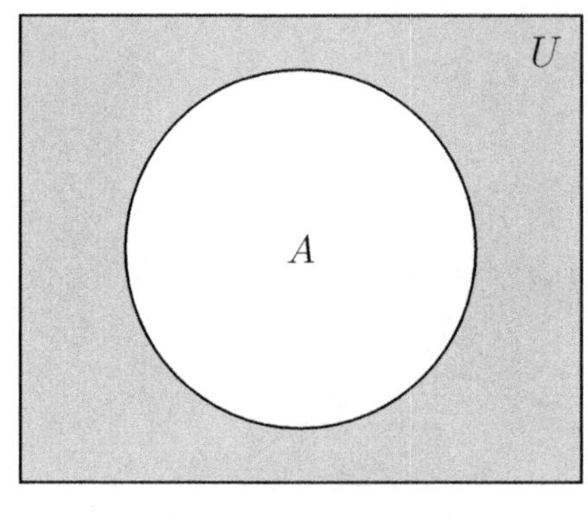

A^c in U

Note that $A \cup B = B \cup A$ and $A \cap B = B \cap A$ are always true, but $A \setminus B = B \setminus A$ is rarely true.[13] Here are some examples:

Example 3.10. Let A be the set of odd integers and B be the set of even integers.

- See if you can determine what $A \cup B$, $A \cap B$, $\mathbb{Z} \setminus A$ and $A \setminus B$ are. Only once you have a guess in mind, check out the answer in the footnote. (And to encourage you to try it first, it is upside down.)[14]
- If $\mathbb{Z}$ is the universal set, then $A^c = B$ and $B^c = A$ and $\emptyset^c = \mathbb{Z}$ and $\mathbb{Z}^c = \emptyset$.

---

[12] If is the universal set, then the complement of  is .

(Now you can say "looks good!" to more balding people. You're just giving them a complement!)

[13] Pop-quiz: When is it true?

[14] $A \cup B = \mathbb{Z}$, $A \cap B = \emptyset$, $\mathbb{Z} \setminus A = B$ and $A \setminus B = A$.

Power Sets and Cardinality

Next, here are two set operations that involve just a single set. The first is the power set, which takes in a set and outputs a much larger set. The second is the cardinality operator, which takes in a set and outputs a number.

Definition.

Definition 3.11. Assume A is a set.

- The *power set* of A is $\mathcal{P}(A) = \{X : X \subseteq A\}$.
- The *cardinality* of A is the number of elements in A, and is denoted $|A|$.

The power set of a set A is denoted $\mathcal{P}(A)$. Since $\mathcal{P}(A)$ is a set, what are the elements of $\mathcal{P}(A)$? First, every element of $\mathcal{P}(A)$ is itself a set.[15] And *which* sets have earned the honor of being an element of $\mathcal{P}(A)$? If X is a subset of A, then X is an element of $\mathcal{P}(A)$. (Read that last sentence as many times as needed for it to make sense.)

Example 3.12. Below are two examples of power sets.

- The power set[16] of $\{1,2,3\}$ is
$$\mathcal{P}(\{1,2,3\}) = \big\{\{1,2,3\},\{1,2\},\{1,3\},\{2,3\},\{1\},\{2\},\{3\},\emptyset\big\}.$$
- The power set $\mathcal{P}(\mathbb{N})$ is the set of all sets of natural numbers. Every set which contains only natural numbers—whether that set is infinite like the set of even natural numbers, or finite like $\{23, 74, 140\}$—is an element of $\mathcal{P}(\mathbb{N})$. Make sense?[17]

Most students find cardinality a little easier to grasp. It just tells you how many elements are in your set. For example, $|\{1,2,3\}| = 3$, and $|\{a,b,c\}| = 3$, and $|\{1,4,9,16,25,36,49,64,81,100\}| = 10$, and $|\mathbb{N}| = \infty$. The one tricky case to mention is when a set's elements include other sets. For example, $|\{\{1,2\},\{a,b,c\}\}| = 2$, since the two elements of this set are $\{1,2\}$ and $\{a,b,c\}$.

Cartesian Products

Our final set operation is the Cartesian product. This is once again an operation that combines *two* sets to create a new set.

[15] Remember, a box can contain anything, including other boxes!

[16] *"Don't forget your empty set!"* is the *"Don't forget your +C!"* of set theory.

[17] If so, now try to make sense of the set $\mathcal{P}(\mathcal{P}(\mathbb{N}))$. Got it?[18]

[18] If so, now try to make sense of the set $\mathcal{P}(\mathcal{P}(\mathcal{P}(\mathbb{N})))$. Got it?[19]

[19] If so, now try to... (Attn: I hereby define the footnote of a footnote to be a *toenote*.)

Definition.

Definition 3.13. Assume A and B are sets.

- The *Cartesian product* of A and B is $A \times B = \{(a, b) : a \in A \text{ and } b \in B\}$.

The Cartesian product is a way to "multiply" sets. The product of sets A and B is a set which is denoted $A \times B$. It is a set for which each of its elements is an ordered pair (like $(1, 2)$). Which ordered pairs have earned the right to be an element of $A \times B$? If a is an element of A, and b is an element of B, then (a, b) is an element of $A \times B$.

Example 3.14. Below are two examples of Cartesian products.

- The Cartesian product of $\{1, 2, 3\}$ and $\{☺, \pi\}$ is

$$\{1, 2, 3\} \times \{☺, \pi\} = \{(1, ☺), (2, ☺), (3, ☺), (1, \pi), (2, \pi), (3, \pi)\}.$$

These elements can be generated via a table:[20]

	1	2	3
π	$(1, \pi)$	$(2, \pi)$	$(3, \pi)$
☺	$(1, ☺)$	$(2, ☺)$	$(3, ☺)$

- When a set has two "dimensions" to it, it can often be viewed as a Cartesian product. For example, consider the integer points in the xy-plane.

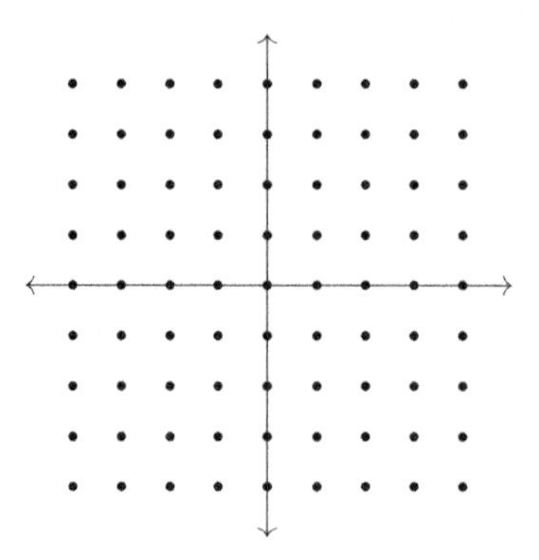

These points have a x-axis dimension and a y-axis dimension, so they should be a Cartesian product. And indeed, they are: This is a plot of the set $\mathbb{Z} \times \mathbb{Z}$.

We have now learned five new set operations. And if you're not tired of it yet, try to describe $A \setminus B$, A^c, $\mathcal{P}(A)$, $|A|$ and $A \times B$ in terms of boxes.

[20]In fact, this perspective helps one see that if A and B are finite sets, then $|A \times B| = |A| \cdot |B|$. And if you had fun with that, here's another important one: $|A \cup B| = |A| + |B| - |A \cap B|$.

– BEGIN INTERJECTION –

We interrupt your regularly-scheduled programming to bring you an important theorem, whose numbering could only fit in right now.

Theorem.

Theorem 3.14159... The number π is super cool.

Proof. Despite being defined in terms of circles, π has the property that

$$1 - \frac{1}{3} + \frac{1}{5} - \frac{1}{7} + \frac{1}{9} - \cdots = \frac{\pi}{4}.$$

And I think we can all agree that is super cool, thus completing the proof. □

– END INTERJECTION –

Back to sets, let's prove something! Next up is a proposition whose main goal is to test our understanding of power sets and subsets.

Proposition.

Proposition 3.15. Suppose A and B are sets. If $\mathcal{P}(A) \subseteq \mathcal{P}(B)$, then $A \subseteq B$.

Proof Idea. Recall the general structure for such a proof:

Proposition. $A \subseteq B$.

Proof. Assume $x \in A$.

≪ An explanation of what $x \in A$ means ≫

⋮ apply algebra,
logic, techniques

≪ Oh hey, that's what $x \in B$ means ≫

Therefore $x \in B$.

Since $x \in A$ implies that $x \in B$, it follows that $A \subseteq B$. □

This proof will basically come down to remembering the definitions of a subset and a power set. In fact, these are the important observations:

- If $x \in A$, then $\{x\} \subseteq A$. Also, conversely, if $\{x\} \subseteq A$, then $x \in A$.
- If $\{x\} \subseteq A$, then $\{x\} \in \mathcal{P}(A)$. Also, conversely, if $\{x\} \in \mathcal{P}(A)$, then $\{x\} \subseteq A$.

Before moving on to the proof, make sure both of these bullet points make sense to you; if they don't, then go back and stare at the definitions of a subset and a power set until they do. The proof will be a blur unless these are clear in your mind.

Proof. Assume that A and B are sets and $\mathcal{P}(A) \subseteq \mathcal{P}(B)$. Let $x \in A$. Note that this implies that $\{x\} \subseteq A$ by the definition of a subset (Definition 3.4), and so $\{x\} \in \mathcal{P}(A)$ by the definition of a power set (Definition 3.11). And since we assumed that $\mathcal{P}(A) \subseteq \mathcal{P}(B)$, this in turn means that $\{x\} \in \mathcal{P}(B)$, again by the definition of a subset. Finally, by each of these definitions one last time, $\{x\} \in \mathcal{P}(B)$ means that $\{x\} \subseteq B$, which in turn means that $x \in B$.

We showed that $x \in A$ implies $x \in B$, and so $A \subseteq B$. □

In math there is often more than one way to prove something. Proposition 3.15 is a good example of this. Below is a second proof.

Second Proof. Assume A and B are sets and $\mathcal{P}(A) \subseteq \mathcal{P}(B)$. To begin, observe that $A \subseteq A$; this is because $x \in A$ of course implies $x \in A$, which means that $A \subseteq A$ by the definition of a subset (Definition 3.4).

By the definition of the power set of A (Definition 3.11), the fact that $A \subseteq A$ means that $A \in \mathcal{P}(A)$. And since we assumed that $\mathcal{P}(A) \subseteq \mathcal{P}(B)$, this means that $A \in \mathcal{P}(B)$.

Finally, by the definition of the power set of B (Definition 3.11), having $A \in \mathcal{P}(B)$ means that $A \subseteq B$. This concludes the (second) proof. □

We just proved that if $\mathcal{P}(A) \subseteq \mathcal{P}(B)$, then $A \subseteq B$. In Exercise 3.17 you will be asked to prove the other direction: If $A \subseteq B$, then $\mathcal{P}(A) \subseteq \mathcal{P}(B)$.

It's now time for a result which you know is important because it's labeled a theorem, it has a name, and the result is called a law. Any one of these should cause you to sit up and pay attention. But all three?? This is a theorem to remember.

Theorem.

Theorem 3.16 (*De Morgan's laws*). Suppose A and B are subsets of a universal set U. Then,

$$(A \cup B)^c = A^c \cap B^c \qquad \text{and} \qquad (A \cap B)^c = A^c \cup B^c.$$

Proof Idea. We will prove the first identity and leave the second as an exercise. Let's see if the first identity makes sense based on its Venn diagram. To begin, here is the set $A \cup B$, inside the set U:

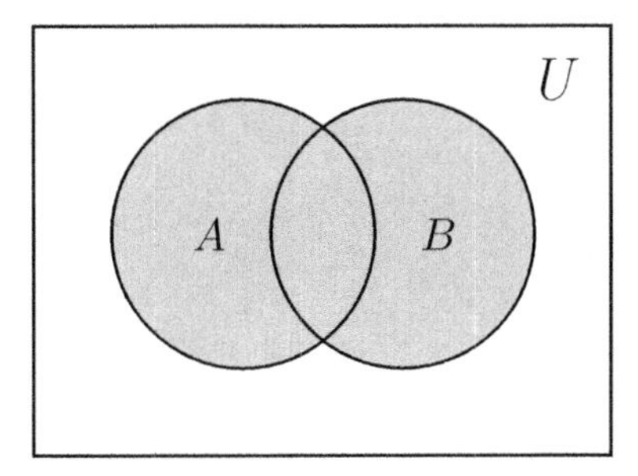

Taking the complement,[21] this is $(A \cup B)^c$:

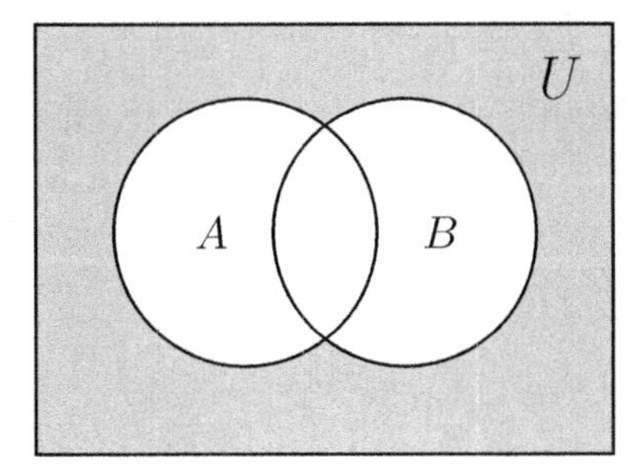

Meanwhile, here are A^c and B^c:

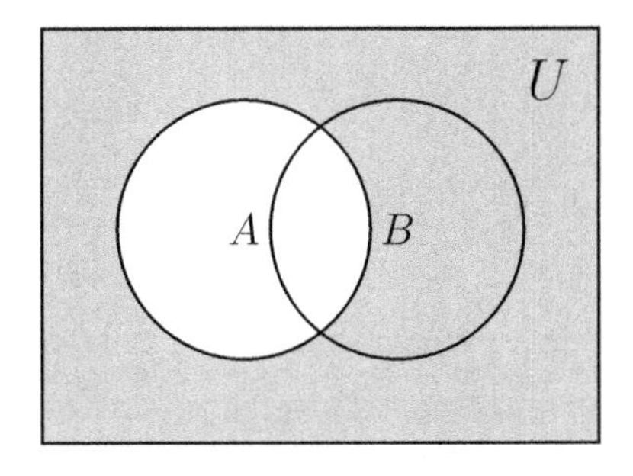

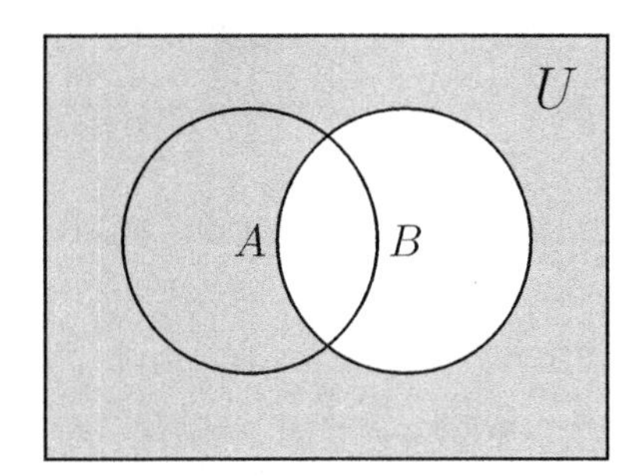

Since these are A^c and B^c, the Venn diagram of $A^c \cap B^c$ is the set of all points which are shaded in *both* of the above diagrams. Which is this:

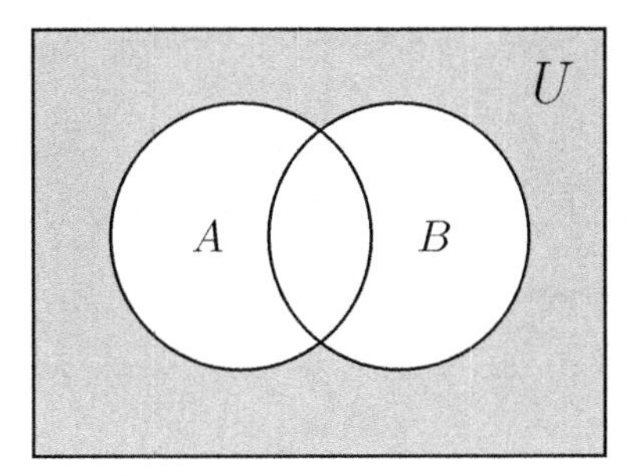

The Venn diagram for $(A \cup B)^c$ is the same as the Venn diagram for $A^c \cap B^c$!

[21]There are two types of people in this world. Those who understand complements and $\left(\text{those who understand complements}\right)^c$.

This is intuition for why these two are the same, but to prove it we will use the approach laid out in Section 3.3. That is, we will prove that

$$(A \cup B)^c \subseteq A^c \cap B^c \qquad \text{and} \qquad A^c \cap B^c \subseteq (A \cup B)^c.$$

Collectively, these will prove that

$$(A \cup B)^c = A^c \cap B^c.$$

Ok, let's do it.

Proof. Assume A and B are subsets of U and all complements are taken inside U. We will prove that $(A \cup B)^c \subseteq A^c \cap B^c$ and $A^c \cap B^c \subseteq (A \cup B)^c$. Together, this will prove that $(A \cup B)^c = A^c \cap B^c$.

First, we will prove that $(A \cup B)^c \subseteq A^c \cap B^c$. To this end, assume $x \in (A \cup B)^c$. Then, by the definition of the complement (in U), $x \in U$ and

$$x \notin (A \cup B).$$

By the definition of the union, x can be in neither A nor B. Said differently,[22]

$$x \notin A \text{ and } x \notin B,$$

which by the definition of the complement means

$$x \in A^c \text{ and } x \in B^c.$$

And hence, by the definition of an intersection,

$$x \in A^c \cap B^c.$$

We have shown that $x \in (A \cup B)^c$ implies $x \in A^c \cap B^c$, which means

$$(A \cup B)^c \subseteq A^c \cap B^c.$$

Next, we will prove that $A^c \cap B^c \subseteq (A \cup B)^c$. To this end, assume $x \in A^c \cap B^c$. Then, by the definition of the intersection,

$$x \in A^c \text{ and } x \in B^c.$$

By the definition of the complement (in U), $x \in U$ and

$$x \notin A \text{ and } x \notin B,$$

which by the definition of the union means

$$x \notin (A \cup B).$$

[22]Note: We will be studying the logic of this step in depth in Chapter 5.

And hence by the definition of the complement,

$$x \in (A \cup B)^c.$$

We have shown that $x \in A^c \cap B^c$ implies $x \in (A \cup B)^c$, which means

$$A^c \cap B^c \subseteq (A \cup B)^c.$$

We have shown that

$$(A \cup B)^c \subseteq A^c \cap B^c \quad \text{and} \quad A^c \cap B^c \subseteq (A \cup B)^c.$$

Together, this demonstrates that

$$(A \cup B)^c = A^c \cap B^c,$$

completing the proof. □

The above proof was longer and more challenging than others in this chapter.[23] But it was good practice for our subset proofs, and is a great reminder of how to work with unions, intersections and complements. But now that I've forced you to suffer through a page-long argument, I thought I'd mention that there is another way to prove some of these set equalities by manipulating set-builder notation. The above proof, for example, can be consolidated into just 4 lines:

$$\begin{aligned}
A^c \cap B^c &= \{x \in \mathbb{R} : x \in A^c \text{ and } x \in B^c\} && \text{(definition of intersection)}\\
&= \{x \in \mathbb{R} : x \notin A \text{ and } x \notin B\} && \text{(definition of complement)}\\
&= \{x \in \mathbb{R} : x \notin (A \cup B)\} && \text{(definition of union)}^{24}\\
&= (A \cup B)^c. && \text{(definition of complement)}
\end{aligned}$$

Some Final Things to Ponder

We will close the main content of this chapter with four miscellaneous topics.

Describing a Set by Listing its Elements

One way to define a finite set is by simply listing its elements: e.g., $\{1, 2, 3\}$. As we've discussed already, the order in which the elements are listed, or whether elements are listed multiple times, does not matter. For example,

$$\{1, 2, 3\} = \{3, 2, 1\} = \{1, 2, 2, 3, 3, 3\}. \qquad (\spadesuit)$$

[23] The Struggle $\in \mathbb{R}$.

[24] Again, this step will be studied in depth in Chapter 5.

In fact, every finite set can be be thought of in terms of set-builder notation. For example, "$\{1, 2, 3\}$," is the set

$$\{x : x = 1 \text{ or } x = 2 \text{ or } x = 3\}.$$

So why should this set equal the other two in Equation (♠)? Well, because the three statements

$$\begin{aligned} &\text{“}x = 1 \text{ or } x = 2 \text{ or } x = 3\text{”} \\ \text{and} \quad &\text{“}x = 3 \text{ or } x = 2 \text{ or } x = 1\text{”} \\ \text{and} \quad &\text{“}x = 1 \text{ or } x = 2 \text{ or } x = 2 \text{ or } x = 3 \text{ or } x = 3 \text{ or } x = 3\text{”} \end{aligned}$$

are all equivalent! (You can make this precise by using *truth tables*, which are described in Chapter 5.) Because the three sets have equivalent membership conditions, they must be equal.

Different Names For the Same Set

The first mathematical facts that you learned — for example, $3^2 = 9$ — asserted that two different mathematical objects were equal. But there aren't really "two different mathematical objects" here, since they are equal! That doesn't mean that $3^2 = 9$ is a vacuous statement; it asserts that 3^2 and 9 are different names for the same number.[25]

As some non-mathematical examples:

- The 50th state admitted to the United States is the only island state.
- The only player ever to have won the NBA Finals MVP with three different franchises is also the player with the record for most playoff points.

From a certain perspective, these are boring statements; the first is saying "Hawaii is Hawaii" (since both "the 50th state" and "the only island state" are both Hawaii), and the second is saying "LeBron James is LeBron James." But that perspective is lacking: Something really *is* being asserted when one says that the 50th US state is also the only island state.

The same principle applies to sets. We have already discussed how a set can be described by listing its elements. So the same set can be named in different ways by listing its elements in a different order, or listing the same element multiple times. There are more exciting examples of this phenomenon, like this one from trigonometry:

$$\{(x, y) : x^2 + y^2 = 1\} = \{(\cos(t), \sin(t)) : 0 \leq t \leq 2\pi\}.$$

And many open problems can be stated in terms of equality of sets. For example, if we write E for the set of even natural numbers, then a famous open problem asks if it is true that

$$\{n \in E : n > 2\} = \{n \in E : n \text{ is the sum of two primes}\}.$$

[25]In philosophical jargon, 3^2 and 9 are expressions with different *intentions* but the same *extension*.

And if those two sets are indeed shown to be equal, the above equality does not become a trivial statement, like $\{1, 2\} = \{1, 2\}$. On the contrary, showing that those sets are equal would be an extremely deep mathematical result.

Proving $a \in A$

The next topic is proving that an element a belongs to a set A. This is considered miscellaneous because, although one can create research-level problems of this type, for the examples we care about it is typically more-or-less clear whether or not a specific element is in a specific set; and when it is not clear, the methods are highly dependent on the specific element and set. For example, consider the set $\mathbb{Q}$. Given a rational number written in the standard way, like $\frac{-143}{14}$, it is clear that this number is in $\mathbb{Q}$ by the definition of $\mathbb{Q}$. And given a number not written in this way, such as π or e or $\sqrt{2}$, there is not a general method to determine whether each of these is in $\mathbb{Q}$. Moreover, any vague method we might articulate would likely not apply if $\mathbb{Q}$ and the elements are changed.

Nevertheless, what follows is a quick discussion about general strategies when A is written in set-builder notation. Consider the set

$$A = \{x \in S : P(x)\},$$

where $P(x)$ is some condition on x. For instance, if $A = \{x \in \mathbb{Z} : 6 \mid x\}$, then $P(x)$ is the condition "$6 \mid x$."

Given a set of this form, if you are presented with a specific a and you wish to prove that $a \in A$, then you must show that

(i) $a \in S$, and

(ii) $P(a)$ is true.

Below is an example of this.

Example 3.17. Let $A = \{(x, y) \in \mathbb{Z} \times \mathbb{N} : x \equiv y \pmod 5\}$. Then $(17, 2) \in A$.

Proof. First, note that $(17, 2) \in \mathbb{Z} \times \mathbb{N}$ since $17 \in \mathbb{Z}$ and $2 \in \mathbb{N}$. Next, observe that

$$17 - 2 = 5(3),$$

which by the definition of divisibility (Definition 2.8) means that

$$5 \mid (17 - 2),$$

which by the definition of modular congruence (Definition 2.14) means that

$$17 \equiv 2 \pmod 5.$$

Thus, $(17, 2) \in A$. □

Indexed Families of Sets

The final topic is a bit of notation that will pop up periodically throughout your mathematical career. It is notable in that it often causes students confusion. It will be used in the *Introduction to Topology* following this chapter, but will otherwise not be too important for the rest of this book. Still, when Future-You is eventually reintroduced to it in later classes, you can thank your Yester-Self for taking the time now to begin to understand it, while sets are fresh in your mind.[26]

We'll explain the idea through examples. Recall that a set can contain other sets, like the set $\mathcal{F} = \{\{1, -2, 3\}, \mathbb{N}, \{7, \pi, -22\}\}$. If every element of $\mathcal{F}$ is itself a set, then $\mathcal{F}$ is called a *family* of sets. Then, one can ask questions about such a family — like, what is the union of all of the sets in $\mathcal{F}$?[27] That is,

$$\bigcup_{S \in \mathcal{F}} S = \{x : x \in S \text{ for some } S \in \mathcal{F}\}.$$

For example, if $\mathcal{F} = \{\{2, 4, 6, 8, \dots\}, \{3, 6, 9, 12, 15, \dots\}, \{0\}\}$, then

$$\bigcup_{S \in \mathcal{F}} S = \{0, 2, 3, 4, 6, 8, 9, 10, 12, 14, 15, 16, \dots\}.$$

Likewise,

$$\bigcap_{S \in \mathcal{F}} S = \{x : x \in S \text{ for every } S \in \mathcal{F}\}.$$

For example, if $\mathcal{F} = \{\mathbb{N}, \{2, 4, 6, 8, 10, \dots\}, \{5, 10, 15, 20, 25, \dots\}\}$, then

$$\bigcap_{S \in \mathcal{F}} S = \{10, 20, 30, 40, \dots\}.$$

Two final examples are in Exercise 3.42, where you will be asked to ponder two families of sets, where each set in these families is a closed interval. The first family is here:

$$\mathcal{F}_1 = \left\{ \left[2 - \frac{1}{n}, 4 + \frac{1}{n}\right] : n \in \mathbb{N} \right\}.$$

And the second family is here:

$$\mathcal{F}_2 = \left\{ \left[2 + \frac{1}{n}, 4 - \frac{1}{n}\right] : n \in \mathbb{N} \right\}.$$

Indeed, you will be asked to determine what

$$\bigcup_{n \in \mathbb{N}} \left[2 - \frac{1}{n}, 4 + \frac{1}{n}\right] \qquad \text{and} \qquad \bigcap_{n \in \mathbb{N}} \left[2 - \frac{1}{n}, 4 + \frac{1}{n}\right]$$

are equal to, and then what

$$\bigcup_{n \in \mathbb{N}} \left[2 + \frac{1}{n}, 4 - \frac{1}{n}\right] \qquad \text{and} \qquad \bigcap_{n \in \mathbb{N}} \left[2 + \frac{1}{n}, 4 - \frac{1}{n}\right]$$

are equal to.

[26] Do you ever thank your past self for doing something so that your current self doesn't have to? Or fault them for leaving you annoying tasks? I call mine YesterJay.

[27] Which mathematicians *should* have called a family reunion.

3.5 Bonus Examples

The first bonus example expands on Proposition 3.6.

Proposition.

Proposition 3.18. It is the case that

$$\{n \in \mathbb{Z} : 12 \mid n\} = \{n \in \mathbb{Z} : 3 \mid n\} \cap \{n \in \mathbb{Z} : 4 \mid n\}.$$

Scratch Work. Let's make sure we believe the result. Here are the n such that $n \in \mathbb{Z}$ and $3 \mid n$:

$$\dots, -24, -21, -18, -15, -12, -9, -6, -3, 0, 3, 6, 9, 12, 15, 18, 21, 24, \dots$$

And here are the n such that $n \in \mathbb{Z}$ and $4 \mid n$:

$$\dots, -24, -20, -16, -12, -8, -4, 0, 4, 8, 12, 16, 20, 24, \dots$$

Which n are in both lists? These:

$$\dots, -24, -12, 0, 12, 24, \dots$$

Those are indeed the n such that $n \in \mathbb{Z}$ and $12 \mid n$, so it seems to be checking out.

Now, following the proof outline from Section 3.3, we will prove this proposition by showing that

$$\{n \in \mathbb{Z} : 12 \mid n\} \subseteq \{n \in \mathbb{Z} : 3 \mid n\} \cap \{n \in \mathbb{Z} : 4 \mid n\}$$

and

$$\{n \in \mathbb{Z} : 3 \mid n\} \cap \{n \in \mathbb{Z} : 4 \mid n\} \subseteq \{n \in \mathbb{Z} : 12 \mid n\}.$$

Let's jump right into it.

Proof. To make our proof more readable, let's define

$$\begin{aligned} A &= \{n \in \mathbb{Z} : 3 \mid n\}, \\ B &= \{n \in \mathbb{Z} : 4 \mid n\}, \text{ and} \\ C &= \{n \in \mathbb{Z} : 12 \mid n\}. \end{aligned}$$

Thus, our aim is to prove that $C \subseteq A \cap B$ and $A \cap B \subseteq C$; we begin with the former. To this end, assume $x \in C$. This means that $x \in \mathbb{Z}$ and $12 \mid x$, which by Definition 2.8 implies that $x = 12k$ for some $k \in \mathbb{Z}$. Equivalently,

$$x = 4 \cdot (3k).$$

And since $k \in \mathbb{Z}$, by Fact 2.1 it is also true that $3k \in \mathbb{Z}$. By Definition 2.8 this means that $4 \mid x$. Therefore, $x \in B$.

Since $x \in C$ implies that $x \in B$, it follows that $C \subseteq B$. The fact that $C \subseteq A$ is by Proposition 3.6, which by the definition of a subset means that if $x \in C$, then $x \in A$.

We have proven that $x \in A$ and $x \in B$, so by the definition of the intersection (Definition 3.8), this implies $x \in A \cap B$.

We have shown that if $x \in C$, then $x \in A \cap B$. This implies $C \subseteq A \cap B$, as desired.

Next, assume $x \in A \cap B$, which by the definition of the intersection means that $x \in A$ and $x \in B$. This means that $x \in \mathbb{Z}$, $3 \mid x$ and $4 \mid x$, which by Definition 2.8 implies that

$$x = 3k \qquad \text{and} \qquad x = 4\ell$$

for some $k, \ell \in \mathbb{Z}$. That is, $3k = 4\ell$. Since $k \in \mathbb{Z}$, by the definition of divisibility (Definition 2.8) this means $3 \mid 4\ell$. We now apply Lemma 2.17 part (iii);[28] since 3 is prime and $3 \mid 4\ell$, either $3 \mid 4$ or $3 \mid \ell$. Clearly $3 \nmid 4$, so it must be the case that $3 \mid \ell$. That is, $\ell = 3m$ for some $m \in \mathbb{Z}$.

We have shown that $x = 4\ell$ and $\ell = 3m$, where $\ell, m \in \mathbb{Z}$. Combined, this means that

$$x = 4(3m) = 12m$$

where $m \in \mathbb{Z}$, which by the definition of divisibility means $12 \mid x$. And so, $x \in C$.

We have proved that if $x \in A \cap B$, then $x \in C$. This implies $A \cap B \subseteq C$, as desired.

We have now shown that

$$C \subseteq A \cap B \qquad \text{and} \qquad A \cap B \subseteq C.$$

Combined, this implies that

$$C = A \cap B,$$

completing the proof. $\square$

The Cardinality of the Power Set

Suppose A is a set with n elements. How many subsets of A are there? Said differently, what is $|\mathcal{P}(A)|$? We could check the first few cases by hand.

A	$\lvert A\rvert$	Subsets of A	$\lvert\mathcal{P}(A)\rvert$
$\{1\}$	1	$\{1\}, \emptyset$	2
$\{1,2\}$	2	$\{1,2\}, \{1\}, \{2\}, \emptyset$	4
$\{1,2,3\}$	3	$\{1,2,3\}, \{1,2\}, \{1,3\}, \{2,3\}, \{1\}, \{2\}, \{3\}, \emptyset$	8
$\{1,2,3,4\}$	4	$\{1,2,3,4\}, \{1,2,3\}, \{1,2,4\}, \{1,3,4\}, \{2,3,4\}, \{1,2\}, \{1,3\}, \{1,4\}, \{2,3\}, \{2,4\}, \{3,4\}, \{1\}, \{2\}, \{3\}, \{4\}, \emptyset$	16

[28] *"Yo, lemma help you prove that proposition."*

There appears to be a pattern! It sure looks like if $|A| = n$, then $|\mathcal{P}(A)| = 2^n$. Why would this be true? There is actually a pretty slick way to see it. Every subset of $\{1, 2, 3\}$ can be generated by going through the elements of $\{1, 2, 3\}$ and asking whether or not each element is included in the subset. For example, $\{1, 3\}$ can be thought of as $\langle$yes, no, yes$\rangle$, since 1 was included, 2 was not, and 3 was. Likewise:

- $\{1\} \leftrightarrow \langle$yes, no, no$\rangle$
- $\{2, 3\} \leftrightarrow \langle$no, yes, yes$\rangle$
- $\emptyset \leftrightarrow \langle$no, no, no$\rangle$
- $\{1, 2, 3\} \leftrightarrow \langle$yes, yes, yes$\rangle$

Suppose you're trying to generate a subset of $\{1, 2, 3\}$. You could think about doing so by asking three yes/no questions, the answers to which uniquely determine your set. With 2 options for the first element, 2 for the second, and 2 for the third, in total there are $2 \times 2 \times 2 = 8$ ways to answer the three questions, and hence 8 subsets! In general, a subset of $A = \{1, 2, 3, \dots, n\}$ can be generated like this:

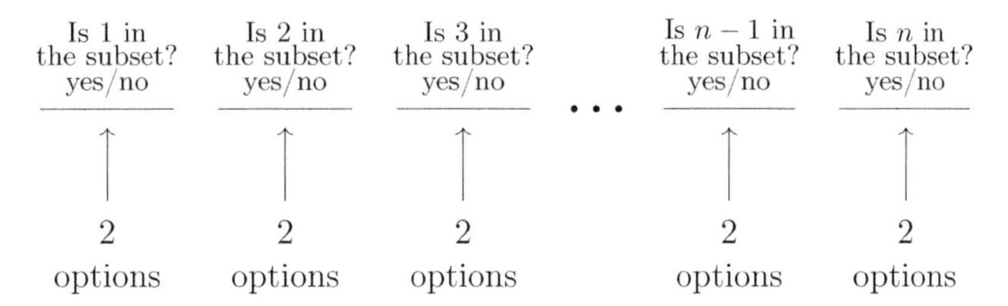

With n straight yes/no questions, there are $2 \times 2 \times \cdots \times 2 = 2^n$ ways to answer the questions, each corresponding uniquely to a subset of A. Thus, if $|A| = n$, then $|\mathcal{P}(A)| = 2^n$.

This property has the following neat consequence.

Proposition.

Proposition 3.19. Given any $A \subseteq \{1, 2, 3, \dots, 100\}$ for which $|A| = 10$, there exist two different subsets $X \subseteq A$ and $Y \subseteq A$ for which the sum of the elements in X is equal to the sum of the elements in Y.

For example, I asked a computer for 10 random numbers from $\{1, 2, 3, \dots, 100\}$, and here is what it spit out:

$$\{6, 23, 30, 39, 44, 46, 62, 73, 90, 91\}.$$

And sure enough, I was able to find two subsets X and Y which work. If we let

$$X = \{6, 23, 46, 73, 90\} \qquad \text{and} \qquad Y = \{30, 44, 73, 91\},$$

then the elements in both sets sum to 238:

$$6 + 23 + 46 + 73 + 90 = 238 = 30 + 44 + 73 + 91$$

(Now, since 73 was included in both sets, if we removed it from both we would have another pair of sets satisfying the theorem. Or if we added 39 to both, then again it would be satisfied.)

This seems like quite the amazing property! *Any* set of 10 elements from $\{1, 2, 3, \ldots, 100\}$ has this property. You might think there are just too many possible sums for such a thing to be guaranteed. But when you wonder whether there are "too many" of something to guarantee some property, your pigeoney senses should start tingling, since the pigeonhole principle is a great tool to determine whether or not there are enough of something.[29]

Proof. Suppose $A \subseteq \{1, 2, 3, \ldots, 100\}$ and $|A| = 10$. The smallest possible subset sum would be with the subset $\emptyset$, whose elements sum to 0, since there are no elements.[30] Meanwhile, the largest possible subset sum would correspond to the subset $\{91, 92, 93, 94, 95, 96, 97, 98, 99, 100\}$, whose sum is

$$91 + 92 + 93 + 94 + 95 + 96 + 97 + 98 + 99 + 100 = 955.$$

Thus, there are certainly no more than 956 possible subset sums of A. Imagine a box for each possible sum.

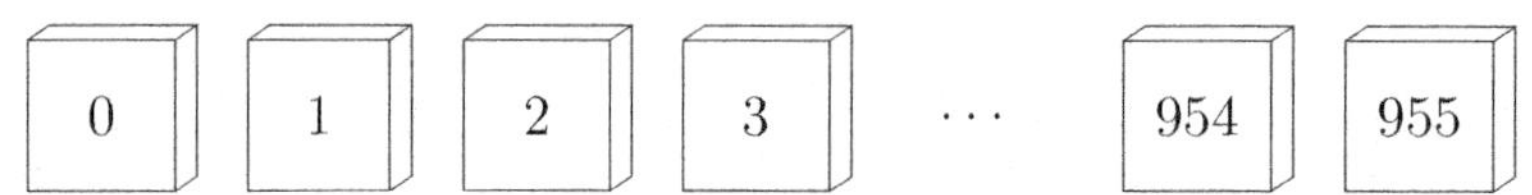

How many subsets of A are there, if $|A| = 10$? Before this proof we showed that the answer is $2^{10} = 1024$. For each subset of A, place it into the box corresponding to its sum. We are placing 1024 objects into 956 boxes, so by the pigeonhole principle (Principle 1.5) there must be a box containing two subsets of A—which means these two subsets have the same sum. □

[29] A pigeoney sense is similar to a spidey sense: *With great power sets comes great responsibility.*

[30] Alternatively, you don't lose anything by focusing on nonempty subsets, in which case the smallest possible subset sum is 1, corresponding to the subset $\{1\}$.

— Chapter 3 Pro-Tips —

- Typically when taking the complement of A in some universal set U, the set U is clear from the context. If you're taking real analysis and your professor discusses the complement of the open interval $(1, 5)$, what she means is $(-\infty, 1] \cup [5, \infty)$, because the complement is assumed to be in the reals. When you're reading a research article on combinatorics on the integers and the author writes $\{\dots, -9, -6, -3, 0, 3, 6, 9, \dots\}^c$, what they mean is the set of integers which are not divisible by 3, because it is assumed that the universal set is the integers. Context clues help relax the writing in advanced mathematics. If one wishes to refer to the unit circle in the xy-plane, perhaps they would define S to be this set of points: $S = \{(x, y) \in \mathbb{R}^2 : x^2 + y^2 = 1\}$. Then, if later they wish to refer to the points in the xy-plane which are *not* on the unit circle, they might write S^c, and the understanding would be that the universal set is $\mathbb{R}^2$.

- If you're trying to write a xi (ξ), a three (3) or a right set brace ($\}$), and they all look like one of these:

  then I know how you feel. I had a professor in undergrad who loved to use xi as his variable, and I spent so much mental energy just trying to draw them that it's literally the only thing I remember from that class. But this is a chapter on sets, let's focus on how to write set braces. Try this: write a 2 (2)

  and then add an S (S) right below it:

  That will give you a right set brace. As for the left, write an S and then a 2: S + 2 = ξ . It'll take a little practice to get the curves right, but that should help. (But if your prof starts using xi in every proof, my best advice is to just drop the class.)

  Also, since we are on the topic, putting a short horizontal line through the middle of your 7s is smart (7), to distinguish them from 1s. Likewise, crossing your Zs is smart (Ƶ), to distinguish them from 2s. And adding a "tail" to your Us is smart (u), to distinguish them from Vs. And it's good keep your Ss curvy on top, but your 5s straight. And now that we are dealing with the empty set, your zeros should not be crossed (0) because your empty sets are (∅).

- Up until now, we have been very careful to always justify every small step in every proof. We started each proof by stating our assumptions, we said when nearly every definition was used, and we worked out every little bit of algebra. From this point on, we will begin, ever so slightly, to pull back from this meticulousness. And in your later courses your professors will probably pull back a little more. And if you go to graduate school in math, or read math research papers, even more will be held back. While I firmly believe that research papers and advanced math books should say a *lot* more than they do... it is practical to not cite every last definition and work out the details of every small algebraic step.

  Here's how I think about it. When my dad taught me how to drive, he insisted that I do everything *perfectly*. Hand placement, mirrors, speed limit, spacing, signs, blinkers, lights, focus, radio, ...every last thing should be done perfectly. It's not that being soooo meticulous is crucial; less so would still be plenty safe. It's because everyone relaxes this alertness eventually — and if you start by driving perfectly, then once you relax you will still end up in a great place. This was my dad's reasoning.

  I believe the same holds with proof writing. If your proofs begin with surgical precision, then once you inevitably relax a bit, you won't do so to a point that mistakes are introduced or your readers are confused.[31] Moreover, even if certain aspects of your proof-writing relax, it is important that your thinking always stays sharp. This approach will not only improve your proof-writing, but will help train your brain to think as rigorously as possible.

- When possible, try to find multiple proofs of a theorem. A proof shows you a single perspective — like a picture does. If you want to learn what a castle looks like, one picture of it is certainly much better than zero. But one is not enough to get a full understanding of what the castle is like, and each additional picture will improve your perception of it.

- If you're trying to prove a theorem but are stuck, try to prove a special case. For example, if you are trying to prove that something holds for all even integers n, try it in the case that n is positive, or n is a multiple of 4. Or, even more specifically, when $n = 2$ or 4 or 6. Likewise, if you are trying to prove something holds for any function, try proving it for a specific function, like $f(x) = x^2$. After you understand why it is true in these special cases, start peeling back your added assumptions to see if you can get closer to proving the general case.

[31] When my grandma taught my dad to drive, her advice was simpler: "Assume every other driver is an idiot." While I did consider making this the lesson for your proof writing... I ultimately chose to go with my dad's more wholesome take.

— Exercises —

Exercise 3.1. Below are pictures of three boxes: a small box inside a medium box inside a large box. Inside some of those boxes are up to three black cats: Tofu, Porco and Dragon. These are my brother's cats, and they received a considerable amount of edible compensation for their work on this project.

For each picture, write down the set representing each of these, as discussed in the chapter. For example, one answer might be $\{\text{Tofu}, \{\{\text{Porco}\}\}\}$.

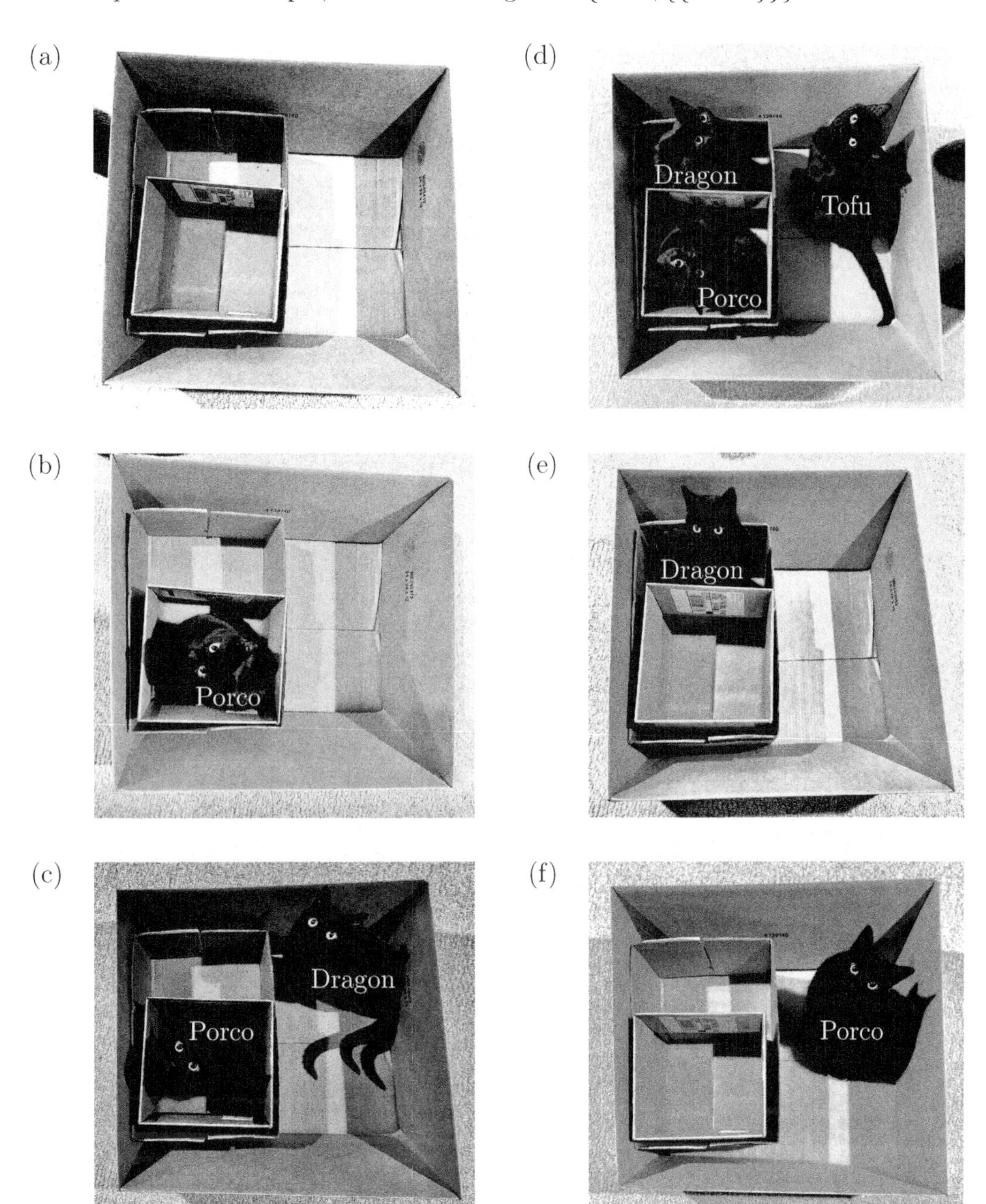

Exercise 3.2. Suppose A and B are two boxes (possibly with things inside). Describe the following in terms of boxes: $A \setminus B$, $\mathcal{P}(A)$, and $|A|$.

Exercise 3.3. Rewrite each of the following sets by listing their elements between braces. If it is an infinite set, write out enough elements for the reader to see the pattern, then use ellipses.

(a) $\{n \in \mathbb{Z} : -5 \leq n < 4\}$

(b) $\{n \in \mathbb{N} : -5 \leq n < 4\}$

(c) $\{3n : n \in \mathbb{Z} \text{ and } |2n| < 8\}$

(d) $\{1, 3, 4, 5\} \times \{☺, \text{math}\}$

(e) $\emptyset \times \{1, 2, 3\}$

(f) $\{1, 2\} \times \left(\{a, b, d\} \times \{☺\}\right)$

(g) $\mathcal{P}(\{a, 2, \square\})$

(h) $\mathcal{P}(\{\{1, 2\}, \{a, b\}\})$

(i) $\{A \in \mathcal{P}(\{a, b, c\}) : |A| < 2\}$

(j) $\{\frac{m}{n} \in \mathbb{Q} : |\frac{m}{n}| < 1 \text{ and } 1 \leq n \leq 4\}$

(k) $\{x \in \mathbb{R} : x^2 + 5x + 6 = 0\}$

(l) $\{5n + 3 : n \in \mathbb{Z}\}$

Exercise 3.4. Suppose $A = \{1, 2, 3, 4, 5\}$, $B = \{3, 4, 5, 6, 7\}$ and $C = \{1, 3, 5, 7\}$, with universal set $\{1, 2, 3, 4, 5, 6, 7, 8\}$. Determine the following.

(a) $A \cup B$

(b) $B \cap C$

(c) $A \setminus C$

(d) $C \setminus A$

(e) $(C \setminus A)^c$

(f) $(A \cup B \cup C)^c$

(g) $\mathcal{P}(A) \setminus \mathcal{P}(B)$

(h) $(A \cap B) \times (B \cap C)$

(i) $(A \setminus B) \times (B \setminus C)$

(j) $\mathcal{P}(A \cap C)$

(k) $\mathcal{P}(A) \cap \mathcal{P}(C)$

(l) $(A \setminus A)^c$

Exercise 3.5. Rewrite each of the following sets in set-builder notation.

(a) $\{3, 5, 7, 9, 11, \ldots\}$

(b) $\left\{\ldots, -\frac{3\pi}{2}, -\pi, -\frac{\pi}{2}, 0, \frac{\pi}{2}, \pi, \frac{3\pi}{2}, \ldots\right\}$

(c) $\{-2, -1, 0, 1, 2, 3, 4, 5\}$

(d) $\left\{\ldots, -\frac{8}{27}, -\frac{4}{9}, -\frac{2}{3}, 1, \frac{2}{3}, \frac{4}{9}, \frac{8}{27}, \ldots\right\}$

Exercise 3.6. Find the cardinality of each of the following sets.

(a) $\{a, b, d\}$

(b) $\{\{1\}, 3, \{\{1\}, 3\}\}$

(c) $\{\{1, 2, 3\}\}$

(d) $\{s, e, t\} \times \{t, h, e, o, r, y\}$

(e) $\mathcal{P}(\{1, 2, 3\})$

(f) $\mathcal{P}(\mathcal{P}(\{a, b\}))$

Exercise 3.7. Sketch the following sets as points/arcs/regions in the xy-plane.

(a) $\{(x, y) : x \in [1, 3] \text{ and } y \in [2, 4]\}$

(b) $\{(x, y) : x \in \mathbb{Z} \text{ and } y \in \mathbb{R}\}$

(c) $\{(x, y) : x^2 + y^2 = 4\}$

(d) $\{(x, y) : x^2 + y^2 \leq 4\}$

(e) $\{1, 2, 3\} \times \{-1, 1\}$

(f) $[-1, 2] \times [1, \pi]$

Exercise 3.8. Determine whether each of the following is true or false.[32]

(a) $1 \in \{1, \{1\}\}$

(b) $1 \subseteq \{1, \{1\}\}$

(c) $1 \in \mathcal{P}(\{1, \{1\}\})$

(d) $\{1\} \in \{1, \{1\}\}$

(e) $\{1\} \subseteq \{1, \{1\}\}$

(f) $\{1\} \in \mathcal{P}(\{1, \{1\}\})$

(g) $\{\{1\}\} \in \{1, \{1\}\}$

(h) $\{\{1\}\} \subseteq \{1, \{1\}\}$

(i) $\{\{1\}\} \in \mathcal{P}(\{1, \{1\}\})$

(j) $\emptyset \in \mathbb{N}$

(k) $\emptyset \subseteq \mathbb{N}$

(l) $\emptyset \in \mathcal{P}(\mathbb{N})$

(m) $\mathbb{Q} \times \mathbb{Q} \subseteq \mathbb{R} \times \mathbb{R}$

(n) $\mathbb{R}^2 \subseteq \mathbb{R}^3$

(o) $\emptyset \subseteq \{1, 2, 3\} \times \{a, b\}$.

Exercise 3.9. Write down all subsets of each of the following.

(a) $\{1, 2, 3\}$

(b) $\{\mathbb{N}, \mathbb{Q}, \mathbb{R}\}$

(c) $\{\mathbb{N}, \{\mathbb{Q}, \mathbb{R}\}\}$

(d) $\emptyset$

Exercise 3.10. The set $\{5a + 3b : a, b \in \mathbb{Z}\}$ is equal to a familiar set. By examining which elements are possible, determine the familiar set.

Exercise 3.11. Suppose A, B and C are sets. Is there a difference between $(A \times B) \times C$ and $A \times (B \times C)$? Explain your answer.

Exercise 3.12. Prove the second identity in De Morgan's Law (Theorem 3.16). That is, suppose A and B are subsets of $\mathbb{R}$. Using U as our universal set,

$$(A \cap B)^c = A^c \cup B^c.$$

Also, before your proof, pick three specific sets A, B and U and show that this property holds for your chosen sets.

Exercise 3.13. For sets A, B and C, and a universal set U, draw the Venn diagram representing each of the following.

(a) $(A \setminus B)^c$

(b) $A \cup (B \setminus C)$.

(c) $(A \cap B) \setminus C$

(d) $A^c \cap (B \setminus C)$

[32]For part (n), note that $\mathbb{R}^2$ is defined to be $\mathbb{R} \times \mathbb{R}$, which is the set of ordered pairs of real numbers. Meanwhile, $\mathbb{R}^3$ is the set of ordered triples of real numbers.

Exercise 3.14. For each of the following Venn diagrams, write down an expression which would describe that Venn diagram. There are multiple correct answers.

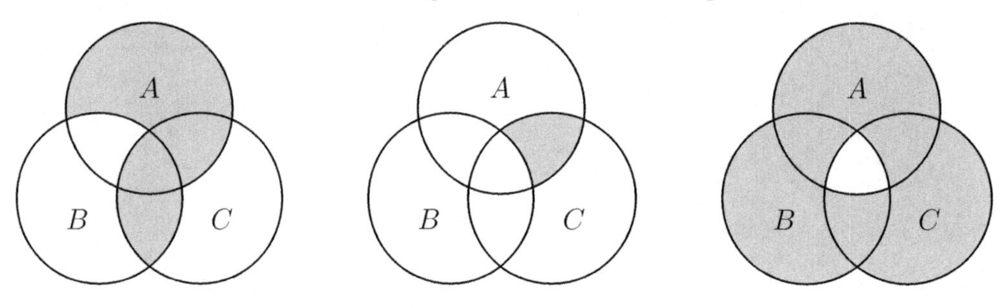

Exercise 3.15. Suppose A and B are sets. Prove that

$$\mathcal{P}(A) \cup \mathcal{P}(B) \subseteq \mathcal{P}(A \cup B).$$

Also, before your proof, write down an example of sets A and B, find their union and the powersets of all three, and check that this property holds in your example.

Exercise 3.16. Let $A = \{n \in \mathbb{Z} : 2 \mid n\}$, $B = \{n \in \mathbb{Z} : 3 \mid n\}$, and $C = \{n \in \mathbb{Z} : 6 \mid n\}$. Prove that $A \cap B = C$ by proving that $A \cap B \subseteq C$ and that $C \subseteq A \cap B$.

Exercise 3.17. Let A and B be sets. Prove that if $A \subseteq B$, then $\mathcal{P}(A) \subseteq \mathcal{P}(B)$. Also, before your proof, write down an example of sets A and B, find their powersets, and check that this property holds in your example.

Exercise 3.18. Suppose A, B and C are sets with $C \neq \emptyset$.

(a) Prove that if $A \times C = B \times C$, then $A = B$.

(b) Explain why the condition "$C \neq \emptyset$" is necessary.

Exercise 3.19. Suppose A, B and C are sets. Prove that

$$A \cap (B \cup C) = (A \cap B) \cup (A \cap C).$$

Exercise 3.20. Give examples of sets A, B, C and D where the following hold.

(a) $A \cup B \setminus B = A$

(b) $C \cup D \setminus D \neq C$

Exercise 3.21. Prove that

$$\{n \in \mathbb{Z} : 2 \mid n\} \cap \{n \in \mathbb{Z} : 9 \mid n\} \subseteq \{n \in \mathbb{Z} : 6 \mid n\}.$$

Exercise 3.22. Prove that

$$\{(m, n) \in \mathbb{Z} \times \mathbb{Z} : m \equiv n \pmod 6\} \subseteq \{(m, n) \in \mathbb{Z} \times \mathbb{Z} : m \equiv n \pmod 2\}.$$

Exercise 3.23. Prove that

$$\{n \in \mathbb{Z} : n \equiv 1 \pmod 4\} \not\subseteq \{n \in \mathbb{Z} : n \equiv 1 \pmod 8\}.$$

Exercise 3.24. Prove the following.

(a) $\{5k + 1 : k \in \mathbb{Z}\} = \{5k + 6 : k \in \mathbb{Z}\}$

(b) $\{12a + 3b : a, b \in \mathbb{Z}\} = \{3k : k \in \mathbb{Z}\}$

(c) $\{8a + 17b : a, b \in \mathbb{Z}\} = \mathbb{Z}$

> **Note.** Recall from Chapter 1 that a *counterexample* is a specific example showing that a conjecture is false.

Exercise 3.25. For each of the following conjectures, either prove it is true or find a counterexample demonstrating that it is false. For each, suppose A, B, C and D are sets.

(a) Conjecture 1: If $A \subseteq B \cup C$, then $A \cup B = B$ or $A \cup C = C$.

(b) Conjecture 2: If $A \subseteq B \cup C$, then $A \cap B \subseteq B \cap C$.

(c) Conjecture 3: If $A \subseteq B \cup C$, then $A \cap B \subseteq C$.

(d) Conjecture 4: If $A = B \setminus C$, then $B = A \cup C$.

(e) Conjecture 5: $A \setminus (B \cap C) = (A \setminus B) \cap (A \setminus C)$.

(f) Conjecture 6: $(A \times B) \cup (C \times D) = (A \cup C) \times (B \cup D)$.

(g) Conjecture 7: $(A \times B) \cap (C \times D) = (A \cap C) \times (B \cap D)$.

(h) Conjecture 8: $\mathcal{P}(A) \cap \mathcal{P}(B) = \mathcal{P}(A \cap B)$.

(i) Conjecture 9: $\mathcal{P}(A) \setminus \mathcal{P}(B) = \mathcal{P}(A \setminus B)$.

(j) Conjecture 10: $(A \times B) \times C = A \times (B \times C)$.

Exercise 3.26. Suppose someone conjectured that $A \cup (B \cap C) = (A \cup B) \cap C$ for any sets A, B and C. Find three sets which are a counterexample to this conjecture. Also, draw Venn diagrams for each to see why they do not align. (Note: This shows that parentheses matter for these set operations! Writing "$A \cup B \cap C$" is ambiguous.)

Exercise 3.27. Suppose someone conjectured that, for any sets A and B which contain finitely many elements, we have

$$|A \cup B| = |A| + |B|.$$

Determine whether this conjecture is true or false. If it is true, prove it. If it is false, give a counterexample demonstrating that it is false, and then conjecture a different formula for $|A \cup B|$.

Exercise 3.28.

(a) Give an example of three sets A, B and C for which $A \cup B = A \cup C$, but $B \neq C$.

(b) Give an example of three sets A, B and C for which $A \cap B = A \cap C$, but $B \neq C$.

(c) Let A, B and C be sets. Prove that if $A \cup B = A \cup C$ *and* $A \cap B = A \cap C$, then $B = C$.

Exercise 3.29. Suppose A, B and C are sets. Prove that if $A \subseteq B$, then $A \setminus C \subseteq B \setminus C$. Also, before your proof, give an example of sets A, B and C and verify that this property holds in your example.

Exercise 3.30. Suppose A and B are sets, with universal set U. Prove that $A \setminus B = A \cap B^c$. Also, before your proof, give an example of sets A and B and verify that this property holds in your example.

Exercise 3.31. Suppose A, B and C are sets. Prove the following.

(a) $A \cup (B \cap C) = (A \cup B) \cap (A \cup C)$.

(b) $A \cap (B \cup C) = (A \cap B) \cup (A \cap C)$.

(c) $A \setminus (B \cap C) = (A \setminus B) \cup (A \setminus C)$.

(d) $A \setminus (B \cup C) = (A \setminus B) \cap (A \setminus C)$.

(e) $A \times (B \cap C) = (A \times B) \cap (A \times C)$.

(f) $A \times (B \cup C) = (A \times B) \cup (A \times C)$.

Exercise 3.32. Describe in words what each of the equalities in Exercise 3.31 means.

> **Note.** The next two exercises will ask you to prove that one thing is true *if and only if* something else is true. If one says "P if and only if Q," where P and Q are some mathematical statements, what this means is "If P, then Q" and also "If Q, then P."

Exercise 3.33. Suppose A and B are sets. Prove that $A \subseteq B$ if and only if $A \setminus B = \emptyset$. To do this, here are the two things that you should prove:

(a) If $A \subseteq B$, then $A \setminus B = \emptyset$.

(b) If $A \setminus B = \emptyset$, then $A \subseteq B$.

Exercise 3.34. Suppose A and B are sets. Prove that $A \subseteq B$ if and only if $A \cap B = A$. To do this, here are the two things that you should prove:

(a) If $A \subseteq B$, then $A \cap B = A$.

(b) If $A \cap B = A$, then $A \subseteq B$.

Exercise 3.35. Prove that $(\mathbb{N} \times \mathbb{Z}) \cap (\mathbb{Z} \times \mathbb{N}) = \mathbb{N} \times \mathbb{N}$.

Exercise 3.36. If $\mathbb{R} \times \mathbb{R}$ is our universal set, describe the elements in the set

$$(\mathbb{Q} \times \mathbb{Q})^c.$$

Exercise 3.37. Let $A = \{a, b\}$. Write out the set $A \times \mathcal{P}(A)$.

Exercise 3.38. Take another look at footnote 12 on page 108. Below, we will switch out the "universal set" guy for a guy with shoulder-length hair.

(a) Sketch a picture of the concluding picture with this setup:

If  is the universal set, then the complement of  is ______ .

(b) Describe what is difficult about sketching a picture of the concluding picture with this setup:

If  is the universal set, then the complement of  is ______ .

Exercise 3.39. Let A and B be sets. Prove that $\mathcal{P}(A \cap B) = \mathcal{P}(A) \cap \mathcal{P}(B)$. Also, before your proof, give an example of sets A and B and verify that this property holds in your example.

Exercise 3.40. Let C be any set.

(a) Prove that there is a unique set $A \in \mathcal{P}(C)$ such that for every $B \in \mathcal{P}(C)$ we have $A \cup B = B$.

(b) Prove that there is a unique set $A \in \mathcal{P}(C)$ such that for every $B \in \mathcal{P}(C)$ we have $A \cup B = A$.

(c) Prove that there is a unique set $A \in \mathcal{P}(C)$ such that for every $B \in \mathcal{P}(C)$ we have $A \cap B = B$.

(d) Prove that there is a unique set $A \in \mathcal{P}(C)$ such that for every $B \in \mathcal{P}(C)$ we have $A \cap B = A$.

Definition. The *symmetric difference* of sets A and B is the set

$$A \triangle B = (A \cup B) \setminus (A \cap B).$$

Exercise 3.41. Use the above definition to complete the following.

(a) Draw a Venn diagram representing the symmetric difference.

(b) For sets A, B and C, prove that $(A \triangle B) \cup C = (A \cup C) \triangle (B \setminus C)$.

(c) For sets A, B and C, prove that $(A \triangle B) \cap C = (A \cap C) \triangle (B \cap C)$.

(d) For sets A, B and C, prove that $(A \cup B) \triangle C = (A \triangle C) \triangle (B \setminus A)$.

(e) For sets A, B and C, prove that $(A \triangle B) \triangle C = (A \triangle C) \triangle (A \setminus B)$.

Exercise 3.42.

(a) Make a conjecture as to what you think

$$\bigcup_{n \in \mathbb{N}} \left[2 - \frac{1}{n}, 4 + \frac{1}{n}\right] \qquad \text{and} \qquad \bigcap_{n \in \mathbb{N}} \left[2 - \frac{1}{n}, 4 + \frac{1}{n}\right]$$

are equal to. You do not need to formally prove your answer, but you should explain your reasoning.

(b) Make a conjecture as to what you think

$$\bigcup_{n \in \mathbb{N}} \left[2 + \frac{1}{n}, 4 - \frac{1}{n}\right] \qquad \text{and} \qquad \bigcap_{n \in \mathbb{N}} \left[2 + \frac{1}{n}, 4 - \frac{1}{n}\right]$$

are equal to. You do not need to formally prove your answer, but you should explain your reasoning.

Exercise 3.43. Read the *Introduction to Topology* following this chapter. Then, complete the following.

(a) Prove Theorem 3.22.

(b) On the final two pages of that section, it is described how to make a Möbius strip out of paper. Make a Möbius strip strip and cut it down the middle the long way, so that you are (presumably) dividing it up into two skinny Möbius strips; take a picture of what you got and describe what happened.

Next, make something very similar to a Möbius strip: Begin with a strip of paper, but instead of giving it a "half twist" and taping the ends like before, give it a full twist and tape the ends. Now cut it down the middle the long way, take a picture of what you got, and describe what happened.

— Open Question —

One of the most important properties of $\mathbb{Q}$ is that it is *dense* in $\mathbb{R}$. What this means is that, given any two distinct real numbers, x and y, there is guaranteed to be a rational number between x and y. Why is this true? Let's break it down into cases. The first case is that one of these numbers is negative (maybe $x = -2.1485$) and one is positive (maybe $y = \pi$). Then, 0 is guaranteed to be rational number that is between x and y. So that case is done.

Case 2 is that x and y are both positive. Note that either x or y is the bigger one, so let's just suppose $x < y$. We aim to find a rational number $\frac{m}{n}$ between x and y. That is, with $x < \frac{m}{n} < y$.

First, think about the numbers $1, \frac{1}{2}, \frac{1}{3}, \frac{1}{4}, \frac{1}{5}$, and so on. Does it make sense that this sequence gets closer and closer to 0? In fact, if you pick any small positive number, like 0.001, there is an n for which $\frac{1}{n} < 0.001$. And if you had instead chosen 0.0000000001, then it would take a larger n to achieve it, but there is still some n for which $\frac{1}{n} < 0.0000000001$. This is sometimes called *the Archimedean principle.*

This is a useful observation, because the fact that $x < y$ means that $y - x$ is some positive number! It may be 0.001 or 0.0000000001 or 12543.455453425 or something else, but it is certainly a positive number. And so, by the Archimedean principle there must be some $n \in \mathbb{N}$ for which $\frac{1}{n} < y - x$.

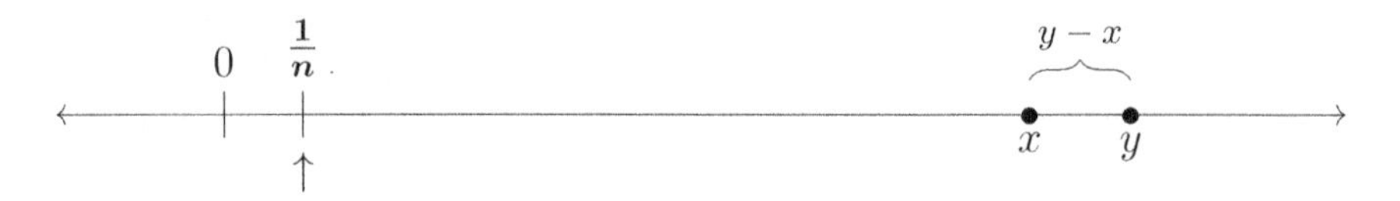

Now think about all integer multiples of this $\frac{1}{n}$.

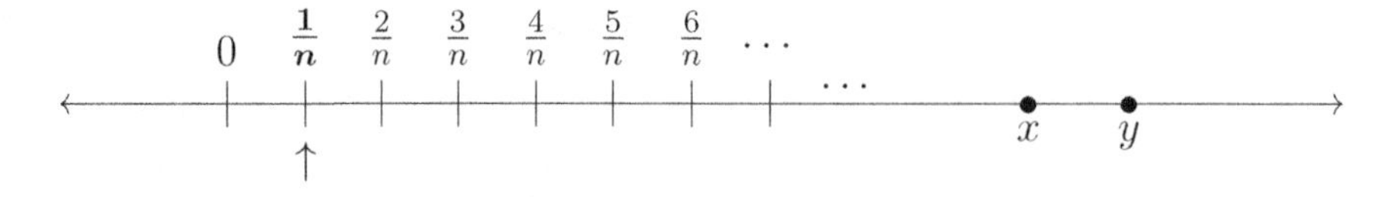

Note that since each of these multiples is $\frac{1}{n}$ away from the next one, but $y - x > \frac{1}{n}$, it is impossible for these dashes to completely hop over the interval between x and y. That is, at least one of these must fall between x and y.

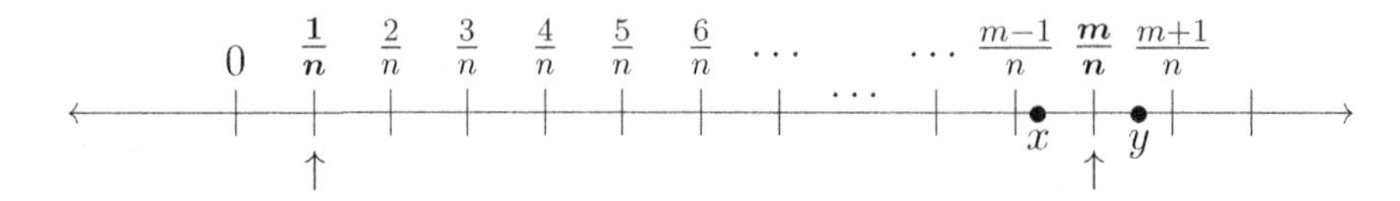

This is the intuition that shows that if x and y are positive, there must be a rational number between them.

Case 3 is that they are both negative, and a very similar argument shows that there is a rational number in-between. Thus, $\mathbb{Q}$ is dense in $\mathbb{R}$.

What other sets are dense in $\mathbb{R}$? If you took $\mathbb{Q}$ but removed the integers, the remaining set would still be dense in $\mathbb{R}$. That is, for any $x, y \in \mathbb{R}$, there exists some $\frac{m}{n}$ from $\mathbb{Q} \setminus \mathbb{Z}$ which is between x and y. Basically, even if $x = 7.99999$ and $y = 8.00001$, you don't *need* to choose $\frac{8}{1}$ as your $\frac{m}{n}$; you could instead choose $\frac{800000001}{100000000} = 8.00000001$, for instance.

We can also ask about density within other sets. Recall that $[0, 1]$ is the set of real numbers between 0 and 1. The set S is dense inside of $[0, 1]$ if, given any $x, y \in [0, 1]$, there exists some element $s \in S$ which is between x and y.

An interesting question is to consider the set

$$\left\{\frac{3}{2}, \left(\frac{3}{2}\right)^2, \left(\frac{3}{2}\right)^3, \left(\frac{3}{2}\right)^4, \left(\frac{3}{2}\right)^5, \left(\frac{3}{2}\right)^6 \dots\right\},$$

but for each number remove the integer portion;[33] for example, $(3/2)^3 = 3.375$ would be reduced to 0.375.

Thus, if we start with the above set

$$\left\{\frac{3}{2}, \left(\frac{3}{2}\right)^2, \left(\frac{3}{2}\right)^3, \left(\frac{3}{2}\right)^4, \left(\frac{3}{2}\right)^5, \left(\frac{3}{2}\right)^6 \dots\right\},$$

and then write each of these as a decimal, we get

$$\{1.5, 2.25, 3.375, 5.0625, 7.59375, 11.3906, \dots\}.$$

Finally, we remove the integer portions to get

$$S = \{0.5, 0.25, 0.375, 0.0625, 0.59375, 0.3906, \dots\}.$$

Every element of S is inside of $[0, 1]$. Is S dense in $[0, 1]$? That is, if you pick any two numbers $x, y \in [0, 1]$, are we guaranteed that some $s \in S$ is squished between x and y? Nobody knows. There is some evidence to think so, but yet a proof remains elusive.[34]

Open Question. Is the set S dense in $[0, 1]$?

---

[33] If you know about floor functions and can imagine generalizing modular congruence to non-integer values, then here is another way to say it: "$(3/2)^n \pmod 1$" or "$(3/2)^n - \lfloor (3/2)^n \rfloor$."

[34] In fact, some guess that S is *uniformly distributed* in $[0, 1]$ (meaning that S doesn't "clump up" in any one region more than any other region), which is a stronger condition. If you have an infinite set like S, and it is uniformly distributed throughout $[0, 1]$, then that would imply it is dense. But it is also possible that it is dense without it being uniformly distributed.

Introduction to Topology

There is a terrible joke about topology that goes: "A topologist is someone who is unable to tell the difference between their coffee cup and their doughnut." And, like many terrible jokes, it must first be explained. The good news is, there is a lot of interesting math that we can discuss while we work to understand this joke!

The set of real numbers has a lot of *structure*. The numbers in $\mathbb{R}$ can be added together, multiplied, subtracted, and usually divided. Any two real numbers have a distance between them, and some numbers are closer together than others are. This may seem like a pretty basic level of structure, but it is actually a lot; it is enough to do algebra and calculus and much else on the reals! The field of *topology* asks what happens when you peel back some of this structure.

Discussing the topology of the real numbers can be confusing, though, because $\mathbb{R}$ is viewed as a number line and it is hard to separate the set from the arithmetic. Instead, let's discuss a sphere.

Sure, you can take two points on this sphere and ask for the distance between them, and it is even possible to create an "arithmetic" of sorts on this sphere. But topology strips all that away. Instead, think of this sphere as being made out of play-dough. With your play-dough sphere, it is possible to knead it into different shapes. For example, you could knead it into a cube.

(Although, let's be honest, you'd probably knead it into a long snake.)

In topology, the sphere, the cube and the long snake are all considered topologically equivalent. In fact, any other shape that you are able to knead the play-dough into would also be considered topologically equivalent, provided you follow these rules:

1. You are not allowed to rip the play-dough into two pieces or poke any holes into it;

2. You are not allowed to make any new connections that cannot be obtained with simple kneading.

For example, it would be impossible to create a torus (doughnut shape) with your sphere of play-dough.

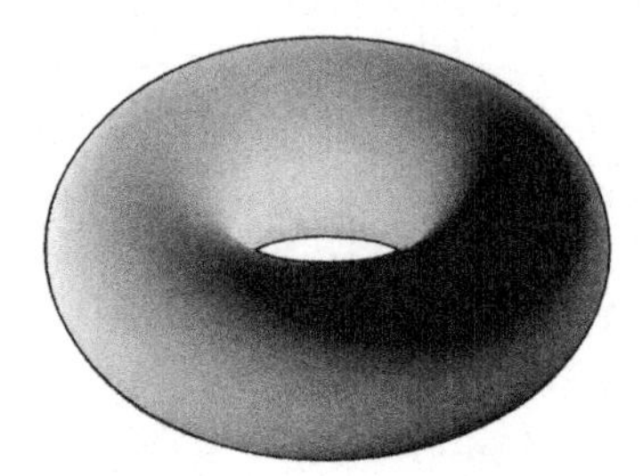

The only two ways to create this shape are to take a sphere, flatten it some, and then poke a hole in the middle — but that's prevented by Rule 1 above; or you could knead the sphere into a snake and connect the ends together — but that's prevented by Rule 2. In the same way, the above doughnut can not be legally transformed into either of the SUPER DOUGHNUTS below. Moreover, the two SUPER DOUGHNUTS below can not be transformed into each other:

Topologists talk about the *genus* of a surface, where if a shape has n holes in it, then it has genus n. If two (orientable[35]) surfaces have the same genuses, then one can be legally transformed into the other; if they have different genuses, then they can not be legally transformed into the other.[36] The two rules ensure this.

This finally explains the terrible joke. A doughnut and a coffee mug each have genus 1, so if they were made out of play-dough you could knead one into the other. Thus, they are topologically equivalent. Get it? *Ba dum tss!*

[35]This is a technical term to make this statement correct. Ignore it for now; all surfaces you're used to apply. (The last two pages of this Introduction will show you non-orientable surfaces.)

[36]Good question for a fun dinner conversation: Does a straw have one hole or two?

The genus of a surface is one example of a *topological invariant*, which are numbers that describe the shape or structure of your play-dough object, and which do not change when you knead or deform your object. Finding topological invariants is particularly important, because when your objects are mushable, one needs to be able to know whether the two objects they are looking at are the same or different.

The formal study of topology began in the late 1800s as a way to take the principle ideas from analysis, but free them of the notion of distance. There is no distance preservation when you knead play-dough, and two play-dough molecules that started near each other on the sphere might become really far apart on the long snake. Despite formally starting in the late 1800s, some early insights into topology were being realized in the mid-1700s.

Leonhard Euler, whose name you will hear about 1000 more times before your mathematics education is complete, discovered an important invariant for convex polyhedra. This is different than our play-dough discussion, but it proved to be an important step to a really important topological invariant which was named the *Euler characteristic* in his honor.

A polyhedron is comprised of (flat) faces which meet at edges, and edges which meet at vertices. Consider the following table which counts how many vertices, edges and faces some of the simplest polyhedra have.

Vertices	4	8	6	12
Edges	6	12	12	30
Faces	4	6	8	20
$V - E + F$	2	2	2	2

What Euler noticed, and indeed was able to prove, is that given *any* convex polyhedron, if V is the number of its vertices, F is the number of its faces, and E is the number of its edges, then $V - E + F = 2$.

First and foremost: This is really cool. It works for any convex polyhedron,[37] and when you realize how different such polyhedra can look, its simplicity is wonderful.

Notice that the formula $V - E + F$ starts with the 0-dimensional object (vertices), then subtracts the 1-dimensional object (edges), then adds the 2-dimensional object (faces). In a more general topological space called a *CW-complex*, if k_n is the number of n-dimensional "cells" in the CW-complex, then the alternating sum

$$k_0 - k_1 + k_2 - k_3 + \dots$$

is always a fixed constant. This constant is the topological invariant called the Euler characteristic.

[37] A polyhedron is convex if all of its vetices point outward; none are "caving in."

Now, the reason that topology is being introduced after our chapter on sets is that if you take a course on topology, you will likely have many weeks — and perhaps a whole semester — in which you will primarily be doing set theory. Topology was grounded in the notion of sets. And kneading play-dough is the idea of applying a continuous function to those sets; indeed, the idea of *continuous deformations* are central to topology. But when you stare at the fundamental definitions in topology, you basically just see a lot of sets.

The most fundamental object of study is that of an *open set*, and here is what it means to be an open subset of $\mathbb{R}$:

Definition.

Definition 3.20. A set $V \subseteq \mathbb{R}$ is *open* if every point in V has an open interval around it that is also contained entirely in V. That is, for every $x \in V$, there is a number $\delta > 0$ such that the open interval $(x - \delta, x + \delta) \subseteq V$.

Intuitively, there are no "edge points."[38] Below are some examples and non-examples.

Example 3.21.

(i) The open interval (a, b) is open. To see this, pick any $x \in (a, b)$. Let δ be whichever is smaller between $x - a$ and $b - x$. Then, $(x - \delta, x + \delta) \subseteq (a, b)$.

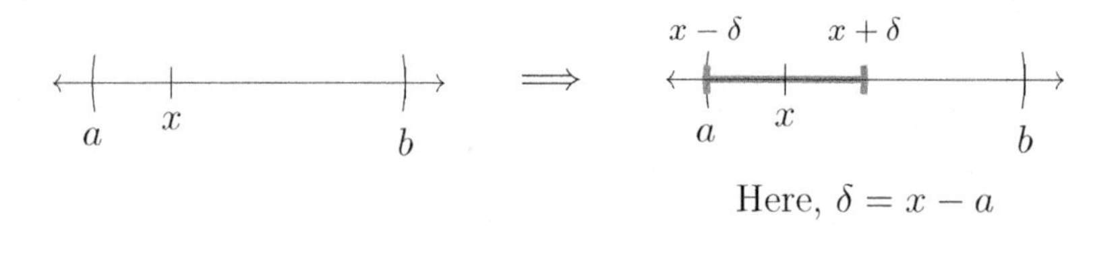

Here, $\delta = x - a$

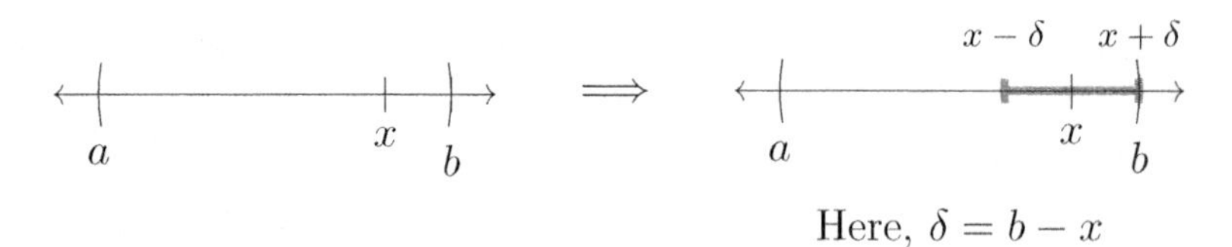

Here, $\delta = b - x$

(ii) The intervals $(-\infty, a)$ and (b, ∞) are open. To see this, pick any $x \in (-\infty, a)$ and let $\delta = a - x$; or pick any $x \in (b, \infty)$ and let $\delta = x - b$.

(iii) A non-example: The closed interval $[3, 7]$ is *not* an open set. To see this, note that $3 \in [3, 7]$, but for any $\delta > 0$ we have

$$(3 - \delta, 3 + \delta) \not\subseteq [3, 7].$$

[38] In fact, it is possible to develop topology using this intuition, called the *Kuratowski closure axioms*.

By including numbers a little less than 3, the set $(3-\delta, 3+\delta)$ can *not* be a subset of $[3,7]$. For example, $3-\frac{\delta}{2}$ is such a number, since $3-\frac{\delta}{2} \in (3-\delta, 3+\delta)$, but $3-\frac{\delta}{2} \notin [3,7]$. So we dissatisfied the definition "for all $x \in V$, there is..." by exhibiting a single x for which it fails. Therefore, $[3,7]$ is *not* open.

(iv) The set $\mathbb{R}$ is open. To see this, note that for any $x \in \mathbb{R}$, if you let $\delta = 17$ then $(x-17, x+17) \subseteq \mathbb{R}$. (Of course, there is nothing special about 17, pick any other positive δ you want and this same property holds.)

(v) The empty set $\emptyset$ is open. To see this, note that, because there are no elements in $\emptyset$, it's true to say "if $x \in \emptyset$, then x is a yellow kangaroo that juggles fire." Or, more to the point, "if $x \in \emptyset$, then $(x-3, x+3) \subseteq \emptyset$." Like we discussed in Chapter 3, this statement is vacuously true, because it begins by choosing an x from the empty set.

To address a confusing point right at the top: The field of topology studies what are also called *topologies*. It's like how people play the game of basketball with a ball that is also called a basketball. The word "topology" has two meanings.

In fact, there is one other confusing point, also involving a double-meaning word. Given a set X, a *topology on* X is a family $\mathcal{T}$ of subsets of X (with a few other conditions, which we will discuss later). The confusing part is that the sets inside $\mathcal{T}$ are always called "open sets."[39] If $X = \mathbb{R}$, then there is something called the *Euclidean topology* on X for which the open sets in $\mathcal{T}$ are precisely the open sets as defined in Definition 3.20. But for other topologies, the term "open set" is reused to refer to whatever is inside $\mathcal{T}$.

One reason for this is that a topology aims to *generalize* the notion of openness that we discussed in Definition 3.20. For example, just like how the entire set $\mathbb{R}$ and the empty set $\emptyset$ were open sets according to Definition 3.20, we will be demanding that every topology on a set X include X as an "open set," and $\emptyset$ as an "open set." This is the first of the conditions mentioned in the previous paragraph.

Before we can write down the formal definition of a topology, there are two more very important properties we must identify. Think again about the open subsets of $\mathbb{R}$ that we discussed above. Notice that if you take any two open sets and union them together, you will get back another open set. For example, the intervals $(1,2)$ and $(4,5)$ are open, and you can check on your own that

$$(1,2) \cup (4,5)$$

is also open, according to Definition 3.20. In fact, if you union together *any* (possibly infinite!) family of open sets, what you get back will be another open set. For intersections, though, the rules change: You may intersect any *finite* family of open sets and be assured that the result is another open set, but with infinitely many sets this is not guaranteed. This is what the next theorem is all about.

[39]Example: If $X = \{a,b,c\}$, then one topology on X will be $\mathcal{T} = \{\{a,b,c\}, \emptyset, \{a,b\}, \{b,c\}, \{b\}\}$. And the "open sets" in this topology are $\{a,b,c\}$, $\emptyset$, $\{a,b\}$, $\{b,c\}$ and $\{b\}$.

Theorem.

Theorem 3.22. Assume all of the following sets are subsets of $\mathbb{R}$, and "open" is as defined in Definition 3.20.

(i) If $\{V_\alpha\}$ is a family of open sets, then $\bigcup_\alpha V_\alpha$ is also an open set.

(ii) If $\{V_1, V_2, \ldots, V_n\}$ is a family of finitely many open sets, then $\bigcap_{k=1}^{n} V_k$ is also an open set.

With this, we can now state the formal definition of a *topology*, which is any family of sets which satisfies these main properties of open sets that we have laid out.

Definition.

Definition 3.23. Let X be a set. A *topology* on X is a family $\mathcal{T}$ of subsets of X having the following properties:

(i) $\emptyset$ and X are in $\mathcal{T}$;

(ii) The union of the sets in any subfamily of $\mathcal{T}$ is also in $\mathcal{T}$; and

(iii) The intersection of the sets in any finite subfamily of $\mathcal{T}$ is also in $\mathcal{T}$.

Let's turn from the Euclidean topology to some examples where X is simpler.

Example 3.24. Let $X = \{a, b, c\}$. Below are nine topologies on X. Check on your own that each topology satisfies Definition 3.23: It contains $\emptyset$ and X, and the union and intersection of any collection of sets from the topology is also in the topology.

1. $\mathcal{T}_1 = \big\{\{a, b, c\}, \emptyset\big\}$.
2. $\mathcal{T}_2 = \big\{\{a, b, c\}, \emptyset, \{a, b\}, \{a\}\big\}$.
3. $\mathcal{T}_3 = \big\{\{a, b, c\}, \emptyset, \{a, b\}, \{b, c\}, \{b\}\big\}$.
4. $\mathcal{T}_4 = \big\{\{a, b, c\}, \emptyset, \{b\}\big\}$.
5. $\mathcal{T}_5 = \big\{\{a, b, c\}, \emptyset, \{a\}, \{b, c\}\big\}$.
6. $\mathcal{T}_6 = \big\{\{a, b, c\}, \emptyset, \{a, b\}, \{b, c\}, \{b\}, \{c\}\big\}$.
7. $\mathcal{T}_7 = \big\{\{a, b, c\}, \emptyset, \{a, b\}\big\}$.
8. $\mathcal{T}_8 = \big\{\{a, b, c\}, \emptyset, \{a, b\}, \{a\}, \{b\}\big\}$.
9. $\mathcal{T}_9 = \big\{\{a, b, c\}, \emptyset, \{a, b\}, \{b, c\}, \{a, c\}, \{a\}, \{b\}, \{c\}\big\}$.

And here is a schematic for each:

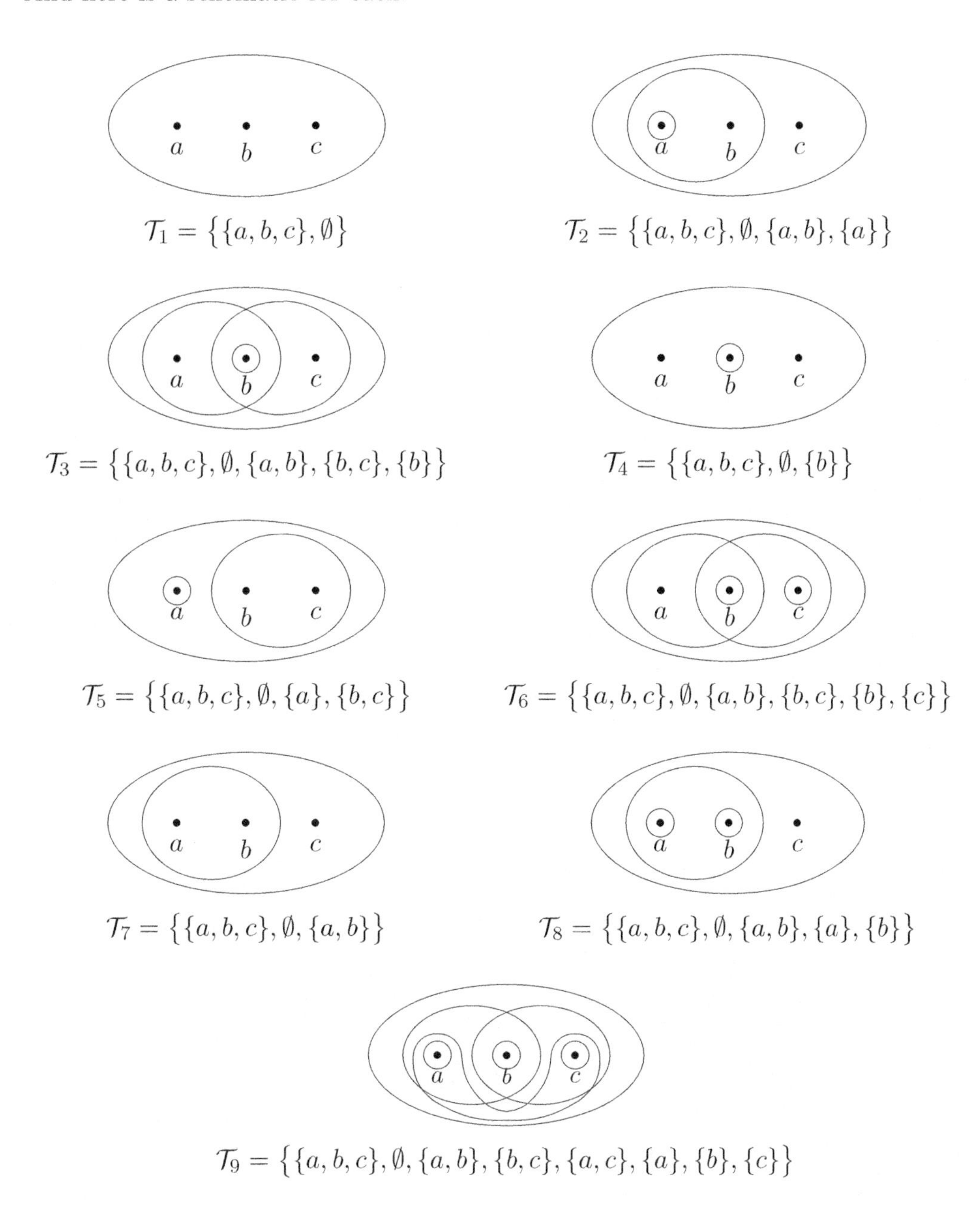

$\mathcal{T}_1 = \{\{a, b, c\}, \emptyset\}$

$\mathcal{T}_2 = \{\{a, b, c\}, \emptyset, \{a, b\}, \{a\}\}$

$\mathcal{T}_3 = \{\{a, b, c\}, \emptyset, \{a, b\}, \{b, c\}, \{b\}\}$

$\mathcal{T}_4 = \{\{a, b, c\}, \emptyset, \{b\}\}$

$\mathcal{T}_5 = \{\{a, b, c\}, \emptyset, \{a\}, \{b, c\}\}$

$\mathcal{T}_6 = \{\{a, b, c\}, \emptyset, \{a, b\}, \{b, c\}, \{b\}, \{c\}\}$

$\mathcal{T}_7 = \{\{a, b, c\}, \emptyset, \{a, b\}\}$

$\mathcal{T}_8 = \{\{a, b, c\}, \emptyset, \{a, b\}, \{a\}, \{b\}\}$

$\mathcal{T}_9 = \{\{a, b, c\}, \emptyset, \{a, b\}, \{b, c\}, \{a, c\}, \{a\}, \{b\}, \{c\}\}$

You see? It's a whole lot of set theory when you start out. It's worth noting, though, that topology after that point is really different. After taking a course in this set-theoretic topology (which is called *point-set topology*), you might take a course in *algebraic topology*, which will feel *very* different than the point-set stuff. And geometric topology feels different than both.

Let's talk a little more about geometric topology. In particular, about an important structure called a *manifold*.

Manifolds

Topologies are simply a family of sets which satisfy a few properties — a very general definition which took mathematicians a long time to settle on. Given this generality, topologies can be quite weird, abstract and seemingly-detached from anything concrete. They can also look exactly like spaces we think about all the time, like the real numbers, the xy-plane, xyz-space, and so on; these are called *Euclidean spaces of dimension 1, 2, 3*, and so on. Or... they can be halfway between. One important area of topology is the study of so-called *manifolds*, which are topological spaces with a few more bells and whistles.

- If the topological space locally resembles real n-dimensional Euclidean space, then it is called a *topological manifold.*[40]
- If that topological manifold is also equipped with a differentiable structure, allowing one to do calculus on the manifold just like how we do calculus on $\mathbb{R}$, then it is called a *differentiable manifold.*
- If the differentiable manifold is also equipped with a function that allows you to determine distance and angles (and hence notions of area, volume and "hyper-volume"), then it is called a *Riemannian manifold.*

Consider the *Möbius strip*, which is pictured below. (Or, better yet, make one on your own! Cut a long strip of paper. While holding the two ends, give it a half twist and tape the two ends together. *Voilà!*)

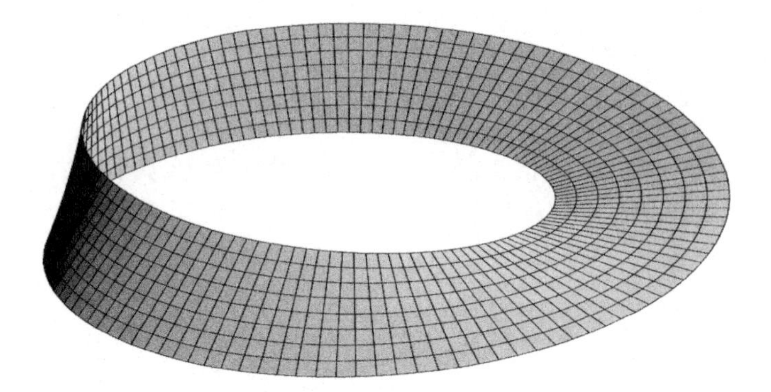

Suppose you steal your nearsighted little brother's glasses, shrink him down, and place him (away from an edge) on a gigantic Möbius strip. When he stands still and looks around, it will look to him like he was placed on a flat plane. Sure, if you zoomed out or gave him really good glasses he might see a difference, but his nearsighted eyes can only see what is happening *locally*. And locally it looks like a small region of the xy-plane.[41] Thus, the Möbius strip is a topological manifold.

[40] These are spaces which, when you focus on the region right around a point (i.e., *locally*), they look very much like Euclidean space, even if they don't overall (i.e., *globally*).

[41] Likewise, a sphere (like the Earth) looks flat locally. Which must be what the Flat Earthers have been trying to tell us all along! LEARN TOPOLOGY, SHEEPLE!

The Möbius strip is a very strange surface. If you made a paper version of it, go ahead and move your finger along one edge of the strip. Notice anything interesting? There *is* only one edge! Trace your finger along one side. You get to "both" sides because there is only one side![42] Now, cut your Möbius strip right down the middle the long way, so that you are (presumably) dividing it up into two skinny Möbius strips. See what happened? Cool, huh?![43]

Another geometric shape which (in theory) your shrunken, robbed, scared, helpless, nearsighted little brother could tell us looks like the xy-plane, is the Klein bottle. Although, while the Möbius strip can be constructed neatly in our three-dimensional world, the Klein bottle would require a fourth dimension to be properly created;[44] drawn in three dimensions, the bottle is forced to incorrectly intersect itself. Nevertheless, here is a 2D picture of a 3D attempt to visualize a 4D object:

In fact, there is a connection between Möbius strips and Klein bottles: A Klein bottle is simply two Möbius strips glued together! This is told in limerick form, essentially matching a poem by Leo Moser:

A mathematician named Klein
Thought Möbius strips were divine.
Said he: "If you glue
The edges of two,
You'll get a weird bottle, like mine."

[42] Why did the chicken cross the Möbius strip? To get to the same side.

[43] Next, make something very similar to a Möbius strip: Begin with a strip of paper, but instead of giving it a "half twist" and taping the ends like before, give it a full twist and tape the ends. Now cut it down the middle the long way. Super cool, huh?!

[44] Another benefit if we lived in four dimensions: our cables would be far less likely to get tangled. A downside: It would be impossible to tie our shoes. Want to know more? There's an area of topology called *knot theory*, which contains some pretty fun stuff.

Chapter 4: Induction

4.1 Dominoes, Ladders and Chips

Consider a line of dominoes, perfectly arranged, just waiting to be knocked over.

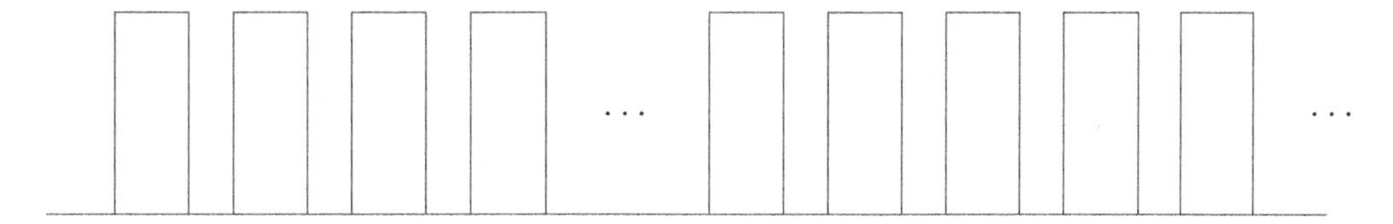

Dominoes stacked up like this have the following properties:

- If you give the first domino a push, it will fall (in particular, it will fall into the second domino, knocking it over).
- Moreover, *every* domino, when it's knocked over, falls into the next one and knocks it over.

Given these two properties, it must be the case that if you knock over the first domino, then every domino will eventually fall. The first property gets the process going: it implies that the first domino will fall. The second property keeps it going: it implies that the falling first domino will cause the second domino to fall. Applying the second property again implies that the second falling domino will cause the third domino to fall. Applying the second property again implies that the third falling domino will cause the fourth domino to fall. And so on.

A similar thing works in mathematics. For example, take a look at the following.

$$\begin{aligned} 1 &= 1 = 1^2 \\ 1+3 &= 4 = 2^2 \\ 1+3+5 &= 9 = 3^2 \\ 1+3+5+7 &= 16 = 4^2 \\ 1+3+5+7+9 &= 25 = 5^2 \\ 1+3+5+7+9+11 &= 36 = 6^2 \\ 1+3+5+7+9+11+13 &= 49 = 7^2 \end{aligned}$$

It sure looks like the sum of the first n odd numbers is n^2. What a neat property![1] But how can we prove that it's true for every one of the infinitely many n? The trick is to use the domino idea. Imagine one domino for each of the above statements.

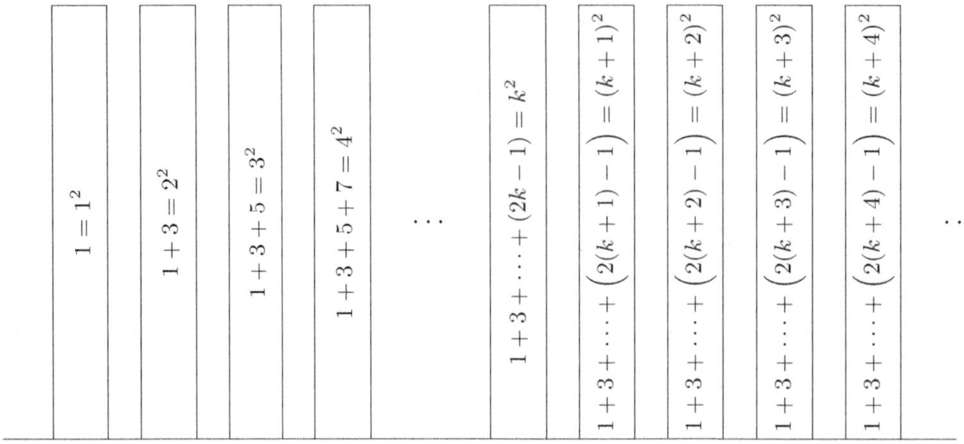

$1 = 1^2$ | $1+3 = 2^2$ | $1+3+5 = 3^2$ | $1+3+5+7 = 4^2$ | $\cdots$ | $1+3+\cdots+(2k-1) = k^2$ | $1+3+\cdots+\big(2(k+1)-1\big) = (k+1)^2$ | $1+3+\cdots+\big(2(k+2)-1\big) = (k+2)^2$ | $1+3+\cdots+\big(2(k+3)-1\big) = (k+3)^2$ | $1+3+\cdots+\big(2(k+4)-1\big) = (k+4)^2$ | $\cdots$

Suppose we do the following:

- Show that the first domino is true (this is trivial, since obviously $1 = 1^2$).
- Show that *any* domino, if true, implies that the following domino is true too.

Given these two, we may conclude that *all* the dominoes are true. It's exactly the same as noting that all the dominoes from earlier will fall. This is a slick way to prove infinitely many statements all at once, and it is called *the principle of mathematical induction*, or, when among friends, it is simply called *induction*.[2]

Principle.

Principle 4.1 (*Induction*)**.** Consider a sequence of mathematical statements, $S_1, S_2, S_3, \ldots$.

- Suppose S_1 is true, and
- Suppose, for each $k \in \mathbb{N}$, if S_k is true then S_{k+1} is true.

Then, S_n is true for every $n \in \mathbb{N}$.

[1]There is also a pleasant way to visualize this fact. Here's the case $5^2 = 1+3+5+7+9$:

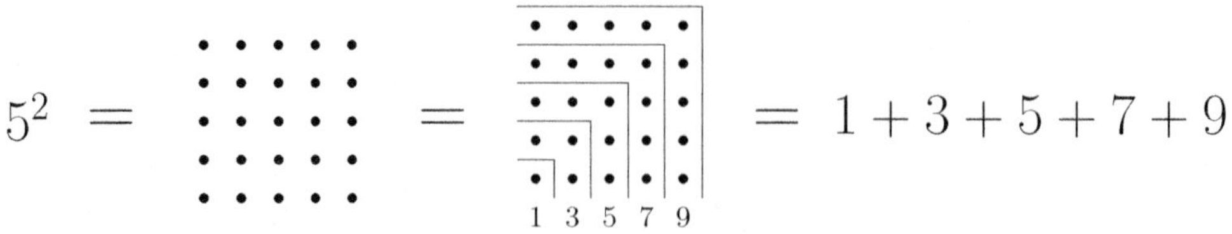

$$5^2 = \begin{matrix}\bullet&\bullet&\bullet&\bullet&\bullet\\\bullet&\bullet&\bullet&\bullet&\bullet\\\bullet&\bullet&\bullet&\bullet&\bullet\\\bullet&\bullet&\bullet&\bullet&\bullet\\\bullet&\bullet&\bullet&\bullet&\bullet\end{matrix} = \begin{matrix}\bullet&\bullet&\bullet&\bullet&\bullet\\\bullet&\bullet&\bullet&\bullet&\bullet\\\bullet&\bullet&\bullet&\bullet&\bullet\\\bullet&\bullet&\bullet&\bullet&\bullet\\\bullet&\bullet&\bullet&\bullet&\bullet\\1&3&5&7&9\end{matrix} = 1+3+5+7+9$$

[2]The principle of induction, like the pigeonhole principle, will be considered true without proof.

This is modeled by the following picture.

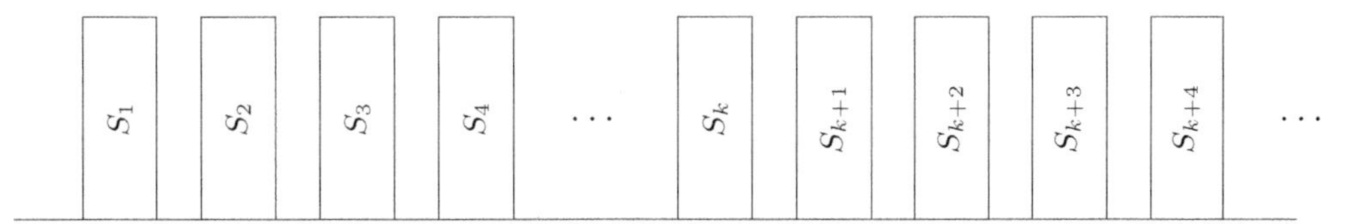

The above also suggests a general framework for how to use induction.

Proposition. S_1, S_2, S_3, ... are all true.

Proof. ≪General setup or assumptions, if needed≫

Base Case. ≪Demonstration that S_1 is true≫

Inductive Hypothesis. Assume that S_k is true.

Induction Step. ≪Proof that S_k implies S_{k+1}≫

Conclusion. Therefore, by induction, all the S_n are true. □

Before we get into examples, why is this section called Dominoes, *Ladders and Chips*? First, there is another popular metaphor for induction that uses ladders. And in case you're not falling for the domino metaphor, perhaps this next one will elevate your understanding.

Assume there is a ladder that rests on the ground but climbs upwards forever. Assuming you can step on the first rung, and assuming that you can always step from one rung to the next, then sky's (not even) the limit! You can climb upward forever![3]

And in case dominoes *and* ladders aren't doing it for you, I came up with one final metaphor for you—one that really resonates in my soul. Assume you have an endless bag of potato chips. Assuming you eat a first chip, and assuming that eating a chip always makes you want to eat another chip, then you will want to eat chips forever.

4.2 Examples

The example that we have discussed thus far will be saved for Exercise 4.1, but fear not, there are many more beautiful results for us to tackle. I want to go simpler than adding up the first n odd natural numbers—let's simply sum the first n natural numbers: $1+2+3+4+\cdots+n$. These sums are called the *triangular numbers* since they can be pictured as the number of balls in the following triangles.

[3]Between these two metaphors, I prefer dominoes, although some prefer the latter.

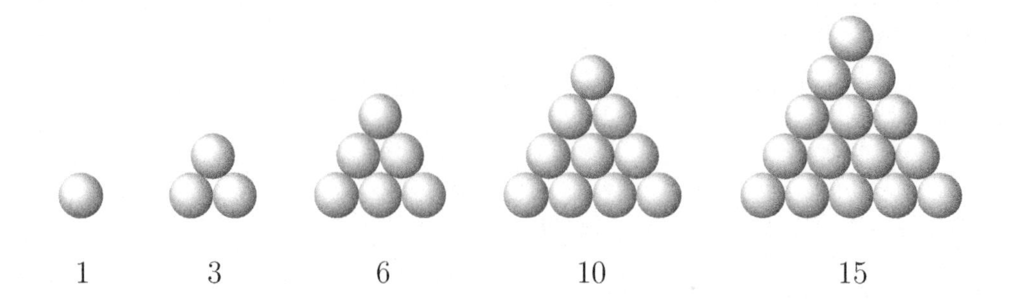

These sums also have a wonderfully simple formula.

Proposition.

Proposition 4.2. For any $n \in \mathbb{N}$,

$$1 + 2 + 3 + \cdots + n = \frac{n(n+1)}{2}.$$

Principle 4.1 was phrased in terms of a sequence of statements. In this proposition, for example, S_3 is the statement $1 + 2 + 3 = \frac{3(3+1)}{2}$, and S_8 is the statement $1 + 2 + 3 + 4 + 5 + 6 + 7 + 8 = \frac{8(8+1)}{2}$. We aim to prove all S_n are true.

Proof Idea. Since we are aiming to prove something for all $n \in \mathbb{N}$, it makes sense to consider induction. The base case will be fine: If $n = 1$ in the formula in Proposition 4.2, the left side is just 1, and the right side is $\frac{1(1+1)}{2}$. Since these are indeed equal, the statement S_1 has been shown to be true.

Next up is our inductive hypothesis, in which we assume the kth step (S_k) is true. That is, we *assume* that

$$1 + 2 + 3 + \cdots + k = \frac{k(k+1)}{2}.$$

Here, k is some fixed natural number; we don't know what it is — perhaps $k = 1$ or $k = 2$ or $k = 174$. Our assumption is independent of the choice, but we do assume it is fixed. It's like assuming the kth domino will at some point fall, and all you're wondering is whether it is guaranteed to knock over the $(k+1)^{\text{st}}$ domino.[4]

[4]Think back to Chapter 2 where we referred to an arbitrary odd integer as $n = 2a + 1$ where a is some integer. It wasn't that n was *all* the odd integers at once, but at the same time it wasn't guaranteed to be 7 or 23 or 101 either. It was a fixed odd integer, but it was also an arbitrary odd integer. Thus, every thing we did to it (like finding $n^2 = (2a+1)^2 = 4a^2 + 4a + 1$) would apply equally to every odd integer. Indeed, our proof of Proposition 2.6 proceeded by showing that if n is an arbitrary odd number, then n^2 is also odd. By proving it for a fixed-but-arbitrary odd integer, we could conclude that it holds for every odd integer! In the same way, the kth domino is fixed but arbitrary. Our induction step will prove that this arbitrary domino must knock over the next one, and because k was arbitrary this in turn means that *every* domino will knock over the next one.

Ok, so we have stated our assumption, and we wish to use it to prove that the $(k+1)^{\text{st}}$ step must also be true:[5]

$$1 + 2 + 3 + \cdots + (k+1) = \frac{(k+1)((k+1)+1)}{2}.$$

How do we do it?[6] And how do we make use of the assumption that we know what $1 + 2 + \cdots + k$ is equal to? If I told you that $1 + 2 + \cdots + 60 = 1830$, and then I asked you to tell me what $1 + 2 + \cdots + 61$ was equal to, what would you do? You wouldn't start at the beginning, you would simply take $1830 + 61 = 1891$, and that's the answer! The same trick works here: The sum of the first $k + 1$ natural numbers begins with the sum of the first k natural numbers:

$$1 + 2 + 3 + \cdots + (k+1) = 1 + 2 + 3 + \cdots + k + (k+1).$$

Makes sense? Instead of writing, say, "$1 + 2 + \cdots + 7$" we wrote "$1 + 2 + \cdots + 6 + 7$." These are certainly both ways to represent "$1 + 2 + 3 + 4 + 5 + 6 + 7$."

This new representation is helpful, though, because it helps us realize how we can apply our assumption. Now that we have a $1 + 2 + \cdots + k$ appearing, and since we know by our inductive hypothesis that $1 + 2 + 3 + \cdots + k = \dfrac{k(k+1)}{2}$, we can now use this!

$$1 + 2 + 3 + \cdots + (k+1) = \underbrace{1 + 2 + 3 + \cdots + k}_{= \frac{k(k+1)}{2}, \text{ by induc. hyp.}} + (k+1)$$

After some algebra, this approach will work out.

Proof. We proceed by induction.

Base Case. The base case is when $n = 1$, and

$$1 = \frac{1(1+1)}{2},$$

as desired.

Inductive Hypothesis. Let $k \in \mathbb{N}$, and assume that

$$1 + 2 + 3 + \cdots + k = \frac{k(k+1)}{2}.$$

Induction Step. We aim to prove that the result holds for $k + 1$. That is, we wish to show that

$$1 + 2 + 3 + \cdots + (k+1) = \frac{(k+1)((k+1)+1)}{2}.$$

[5]This was obtained by looking at the proposition and plugging in $k + 1$ for n.

[6]Whenever you begin the induction step in one of your own induction proofs, I suggest you ask yourself: "How am I going to use the inductive hypothesis to prove this?" If you didn't need the inductive hypothesis, then there is no point to using induction. Moreover, the inductive hypothesis is a massive assumption! You are assuming the kth domino has fallen! Use that!

Written slightly differently, we wish to show

$$1+2+3+\cdots+k+(k+1)=\frac{(k+1)(k+2)}{2}.$$

To do this, we begin with the expression on the left, we apply the inductive hypothesis to the sum of the first k numbers, and after three further steps of algebra we will obtain the expression on the right. Indeed, by the inductive hypothesis we see that

$$1+2+3+\cdots+k+(k+1)=\frac{k(k+1)}{2}+(k+1).$$

Finding a common denominator and simplifying, the above

$$\begin{aligned}&=\frac{k^2+k}{2}+\frac{2(k+1)}{2}\\&=\frac{k^2+3k+2}{2}\\&=\frac{(k+1)(k+2)}{2},\end{aligned}$$

as desired.

Conclusion. Therefore, by induction, $1+2+3+\cdots+n=\frac{n(n+1)}{2}$ for all $n\in\mathbb{N}$. $\square$

Another way to visualize this proposition is the following. If we let T_n be the sum of the first n natural numbers,

$$T_n=1+2+3+\ldots+(n-2)+(n-1)+n,$$

then it is of course also true that

$$T_n=n+(n-1)+(n-2)+\ldots+3+2+1,$$

since adding up the same n numbers in a different order does not change its sum.[7] Next, look what happens when we add these two sums together:

$$\begin{array}{rcccccccccccc} T_n &=& 1 &+& 2 &+& 3 &+\ldots+& (n-2) &+& (n-1) &+& n \\ T_n &=& n &+& (n-1) &+& (n-2) &+\ldots+& 3 &+& 2 &+& 1 \\ \hline 2T_n &=& (n+1) &+& (n+1) &+& (n+1) &+\ldots+& (n+1) &+& (n+1) &+& (n+1) \end{array}$$

[7]Remarkable fact: It actually is important that we are only adding up finitely many numbers here. If you add up infinitely many numbers, changing the order in which you add them *can* change the result! Reference: real analysis.

Since there are n copies of $n+1$, this shows that

$$2T_n = n(n+1),$$

and hence

$$T_n = \frac{n(n+1)}{2}.$$

Neat![8] However, we should be a little careful here. When you see the ellipses (the dot-dot-dots), there is implicitly an induction going on. I showed you 6 pairs that added to $n+1$ and just asserted that the other $n-6$ pairs will also add to $n+1$. Now, you might think that it's clear that this pattern will continue and all the terms will add up to $n+1$, but formally that leap should be proven by induction.

Example 2: Sums of Triangular Numbers

Using this formula for T_n and pushing these ideas farther, notice the following:

n	T_n	$T_n + T_{n+1}$
1	1	4
2	3	9
3	6	16
4	10	25
5	15	36
6	21	49

It sure looks like $T_n + T_{n+1} = (n+1)^2$. Neat! Let's use induction to prove this fact!

Proposition.

Proposition 4.3. Let T_n be the sum of the first n natural numbers. Then, for any $n \in \mathbb{N}$,

$$T_n + T_{n+1} = (n+1)^2.$$

We will prove this proposition twice. The first proof is by induction, and the second will be a direct proof.

[8] One of history's most accomplished number theorists was Carl Friedrich Gauss. He passed away in 1855, and the following year his biographer recorded a story which he says Gauss used to tell late in his life. As the story goes, when Gauss was seven he was in arithmetic class and his teacher told the class that he would give them a problem to solve; as soon as a student found the answer, they were to place their slate on one of the tables. The problem was to find the sum $1+2+3+\cdots+100$. As his biographer wrote, "The problem was barely stated before Gauss threw his slate on the table with the words (in the low Braunschweig dialect): 'There it lies.'" According to the elder Gauss, he solved it with a similar trick to what we discussed. His approach: $1+2+3+\cdots+100 = (1+100)+(2+99)+(3+98)+\cdots+(50+51) = 50\times(101) = 5050$.

First Proof (Induction Proof). We proceed by induction.

Base Case. The base case is when $n = 1$, and

$$T_1 + T_2 = 1 + 3 = 4 = (1+1)^2,$$

as desired.

Inductive Hypothesis. Let $k \in \mathbb{N}$, and assume that[9]

$$T_k + T_{k+1} = (k+1)^2.$$

Induction Step. We aim to prove that the result holds for $k+1$. That is, we wish to show that

$$T_{k+1} + T_{k+2} = (k+2)^2.$$

To do this,[10] we will use the fact that since T_{k+1} is the sum of the first $k+1$ natural numbers, you can write T_{k+1} as $T_k + (k+1)$, as we did before.[11] Likewise, $T_{k+2} = T_{k+1} + (k+2)$. Using this and the inductive hypothesis,

$$\begin{aligned}
T_{k+1} + T_{k+2} &= \big(T_k + (k+1)\big) + \big(T_{k+1} + (k+2)\big) \\
&= T_k + T_{k+1} + 2k + 3 \\
&= (k+1)^2 + 2k + 3 && \text{(by the inductive hypothesis)} \\
&= (k^2 + 2k + 1) + 2k + 3 \\
&= k^2 + 4k + 4 \\
&= (k+2)^2.
\end{aligned}$$

Conclusion. Therefore, by induction, the proposition must hold for all $n \in \mathbb{N}$. □

Before giving you the second proof, a quick note: For some proof techniques, adding a sentence at the end of your proof is nice but not required. For induction,

[9]Reminder: You should think about this k as being a fixed natural number. It is an arbitrary choice, so it could be any natural number, but it is a single, fixed choice. Thus, this inductive hypothesis is not asserting something about *all* the natural numbers (as in the statement of the proposition), but rather about this one particular natural number, k. Back to the metaphor: we are assuming here that the kth domino falls down, and in the induction step we will show that the $(k+1)^{\text{st}}$ domino will also fall. When we say the kth domino knocks over the $(k+1)^{\text{st}}$, we imagine a single fixed (but arbitrary) domino knocking over the next one. That's what's happening here.

[10]Reminder: At this point you should be asking yourself: "How are we going to use the inductive hypothesis to prove this?" The inductive hypothesis is our most powerful tool right now, so we will definitely use it. In this, we will find a way to turn $T_{k+1} + T_{k+2}$ into something involving $T_k + T_{k+1}$, because our inductive hypothesis tells us what $T_k + T_{k+1}$ is equal to.

[11]If you're confused by this, let's clarify—this is exactly what we did in the last problem:

$$\begin{aligned}
T_{k+1} &= 1 + 2 + 3 + \cdots + (k+1) \\
&= 1 + 2 + 3 + \cdots + k + (k+1) \\
&= T_k + (k+1).
\end{aligned}$$

though, it really is required. You can prove that the first domino will fall, and you can prove that each domino — if fallen — will knock over the next domino, but why does this mean they all fall? Because induction says so! Until you say "by induction..." your work will not officially prove the result.

Let's now discuss the direct proof of Proposition 4.3. Although most propositions like Proposition 4.3 are best proved by induction, in this case there is a slick way to do it as a direct proof. This sometimes happens with the "first examples" for a new topic. Usually there won't be a trick like this to avoid induction.

Second Proof (Direct Proof). By Proposition 4.2,

$$\begin{aligned} T_n + T_{n+1} &= \frac{n(n+1)}{2} + \frac{(n+1)(n+2)}{2} \\ &= \frac{1}{2}(n^2 + n + n^2 + 3n + 2) \\ &= \frac{1}{2}(2n^2 + 4n + 2) \\ &= n^2 + 2n + 1 \\ &= (n+1)^2. \end{aligned}$$

□

And there you have it! Apply the previous proposition and do a little algebra and the conclusion pops right out.

We started this chapter by talking about how the sum of the first n odd natural numbers is equal to n^2. We just now proved Proposition 4.3, which shows that $T_n + T_{n+1} = (n+1)^2$. That is, $T_n + T_{n+1}$ is equal to the sum of the first $n+1$ odd natural numbers. So there should be some connection between $T_n + T_{n+1}$ and the sum of odd numbers. See if you can find the connection on your own!

Example 3: An Example with Products

We have seen examples involving sums; what about an example involving products? As you may have learned in a previous course, the *factorial* of a positive integer n is denoted $n!$, and is defined to be $n \cdot (n-1) \cdot (n-2) \cdots 3 \cdot 2 \cdot 1$. For example, $4! = 4 \cdot 3 \cdot 2 \cdot 1 = 24$.

Proposition.

Proposition 4.4. For every $n \in \mathbb{N}$, the product of the first n odd natural numbers equals $\frac{(2n)!}{2^n n!}$. That is,

$$1 \cdot 3 \cdot 5 \cdot \ldots \cdot (2n-1) = \frac{(2n)!}{2^n n!}.$$

Scratch Work. When presented with a problem like this, it is a good idea to immediately do an example. This helps convince yourself that it is true, and also might suggest a reason *why* it is true — hence suggesting a path to prove it. Below we check it for $n = 1$, 2 and 3.

- For $n = 1$, note that $1 = 1$, and also $\dfrac{(2 \cdot 1)!}{2^1 1!} = \dfrac{2!}{2 \cdot 1} = \dfrac{2}{2} = 1.$ ✓
- For $n = 2$, note that $1 \cdot 3 = 3$, and also $\dfrac{(2 \cdot 2)!}{2^2 2!} = \dfrac{4!}{4 \cdot 2} = \dfrac{24}{8} = 3.$ ✓
- For $n = 3$, note that $1 \cdot 3 \cdot 5 = 15$, and also $\dfrac{(2 \cdot 3)!}{2^3 3!} = \dfrac{6!}{8 \cdot 6} = \dfrac{720}{48} = 15.$ ✓

As these cases show, factorials quickly become large, and so the numbers quickly become hard to work with. Nevertheless, we were able to check the first few cases.

Now, to prove this by induction, we will need to show that the base case works, and hey, what do you know, we just did — the first bullet point above is the base case. (Scratch work FTW!)

For the inductive hypothesis, we will be assuming that, for some $k \in \mathbb{N}$,

$$1 \cdot 3 \cdot 5 \cdot \ldots \cdot (2k-1) = \frac{(2k)!}{2^k k!}.$$

Our goal in the induction step will be to show that

$$1 \cdot 3 \cdot 5 \cdot \ldots \cdot (2(k+1)-1) = \frac{(2(k+1))!}{2^{k+1}(k+1)!}. \qquad (☺)$$

Now, do you remember how in the last couple examples it was really beneficial to note

$$1 + 2 + \cdots + (k+1)$$

is really

$$(1 + 2 + \cdots + k) + (k+1)?$$

That allowed us to apply the inductive hypothesis to turn knowledge about the kth step into knowledge about the $(k+1)^{\text{st}}$ step.

Is there a similar trick we can use here? Starting on the left side of (☺), how can we write

$$1 \cdot 3 \cdot 5 \cdot \ldots \cdot (2(k+1)-1)$$

to include the penultimate term? Each term in the above is 2 bigger than the previous. And $2(k+1) - 1$ simplifies to $2k + 1$. Thus, the above is the same as

$$1 \cdot 3 \cdot 5 \cdot \ldots \cdot (2k-1) \cdot (2k+1)$$

What about the right side of (☺)? How do we work with the fatorials? Notice that $(k+1)!$ is this:

$$(k+1)! = 1 \cdot 2 \cdot \ldots \cdot (k+1).$$

Right before that $(k+1)$ in the product must have been a k, and so

$$(k+1)! = (1 \cdot 2 \cdot \ldots \cdot k) \cdot (k+1),$$

which you may notice means that $(k+1)! = k! \cdot (k+1)$. Likewise, $(2k+2)! = (2k)! \cdot (2k+1) \cdot (2k+2)$. Finally, note that the 2^{k+1} term is just $k+1$ copies of 2 multiplied together, and hence $2^{k+1} = 2^k \cdot 2$. You see, with a little algebra we are able to turn information about the $(k+1)^{\text{st}}$ step into knowledge about the kth step. Using this new knowledge, we can put together a proof.

Proof. We proceed by induction.

Base Case. The base case is when $n = 1$, and

$$1 = \frac{2!}{2^1 \cdot (1!)},$$

as desired.

Inductive Hypothesis. Assume that for some $k \in \mathbb{N}$ we have

$$1 \cdot 3 \cdot 5 \cdot \ldots \cdot (2k-1) = \frac{(2k)!}{2^k k!}.$$

Induction Step. We aim to prove that the result holds for $k+1$. That is, we wish to show that

$$1 \cdot 3 \cdot 5 \cdot \ldots \cdot (2(k+1)-1) = \frac{(2(k+1))!}{2^{k+1}(k+1)!}.$$

Written slightly differently, we wish to show

$$1 \cdot 3 \cdot 5 \cdot \ldots \cdot (2k-1) \cdot (2k+1) = \frac{(2k+2)!}{2^{k+1}(k+1)!}.$$

To do this, we begin on the left side, and notice that the "$1 \cdot 3 \cdot 5 \cdot \ldots \cdot (2k-1)$" portion can be replaced by $\frac{(2k)!}{2^k k!}$, according to the inductive hypothesis. Indeed, by doing this, and then some algebra, we can arrive at the right side.

$$\begin{aligned}
1 \cdot 3 \cdot 5 \cdot \ldots \cdot (2k-1) \cdot (2k+1) &= \frac{(2k)!}{2^k k!} \cdot (2k+1) \\
&= \frac{(2k+1)!}{2^k k!} \\
&= \frac{(2k+1)! \cdot (2k+2)}{2^k k! \cdot (2k+2)} \\
&= \frac{(2k+2)!}{2^k k! \cdot 2(k+1)} \\
&= \frac{(2k+2)!}{2^{k+1}(k+1)!},
\end{aligned}$$

as desired.

Conclusion. Therefore, by induction, the equality must hold for all $n \in \mathbb{N}$. □

Example 4: A Tiling Problem

Next up is another tiling problem, harking back to the very first pages of this book. This time, though, we are not tiling with dominoes, we are tiling with ⊞–shaped tiles:

Moreover, we are going to try to perfectly cover chessboards of size $2^n \times 2^n$; that is, a 2×2 board, a 4×4 board, an 8×8 board, a 16×16 board, and so on. Now, as stated, this is impossible. The ⊞–shaped tiles cover three squares at a time, but since a $2^n \times 2^n$ chessboard has $2^n \cdot 2^n = 2^{2n} = 4^n$ squares, and $3 \nmid 4^n$ for any $n \in \mathbb{N}$, it is impossible to cover all the squares. However, according to Exercise 2.35, it *is* true that $3 \mid (4^n - 1)$ for every $n \in \mathbb{N}$. Therefore, if we remove one square from a $2^n \times 2^n$ chessboard, it is *possible* that such a board can be perfectly covered by ⊞–shaped tiles — divisibility alone does not prevent a perfect covering.

Now, "it is possible" is no guarantee. Perhaps it is impossible for some other reason, and perhaps whether it can be perfectly covered depends on which square you remove, as it did in Chapter 1. And yet, surprisingly, such a board can always be perfectly covered *no matter which square you remove!*

For example, consider a 4×4 chessboard. If the square we remove is the one marked on the left, then a perfect covering is shown on the right:

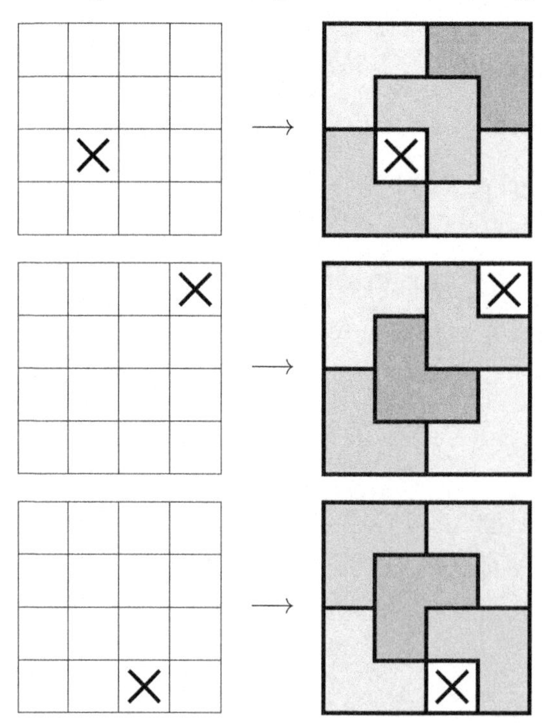

With a small board like a 4×4, trial and error (and thinking about the corners) can get you pretty far, but for a 64×64 board — or a $2^{294} \times 2^{294}$ board — it's not so easy. Nevertheless, induction will get us there.

Proposition.

Proposition 4.5. For every $n \in \mathbb{N}$, if any one square is removed from a $2^n \times 2^n$ chessboard, the result can be perfectly covered with ⊞-shaped tiles.

Proof Idea. Again, recall that the tiles cover three squares and look like this:

Since the proposition refers to something being true "for every $n \in \mathbb{N}$," that's a pretty good indication that induction is the way to proceed. The base case (when $n = 1$) will be fine. For the inductive hypothesis, we will be assuming that any $2^k \times 2^k$ board, with one square removed, can be perfectly covered by ⊞-shaped tiles.

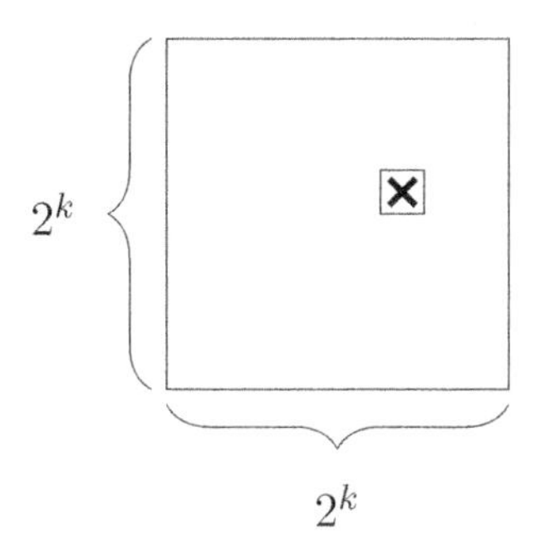

In the induction step we are going to consider a $2^{k+1} \times 2^{k+1}$ board — a board that is twice as big in each dimension — with one square missing.

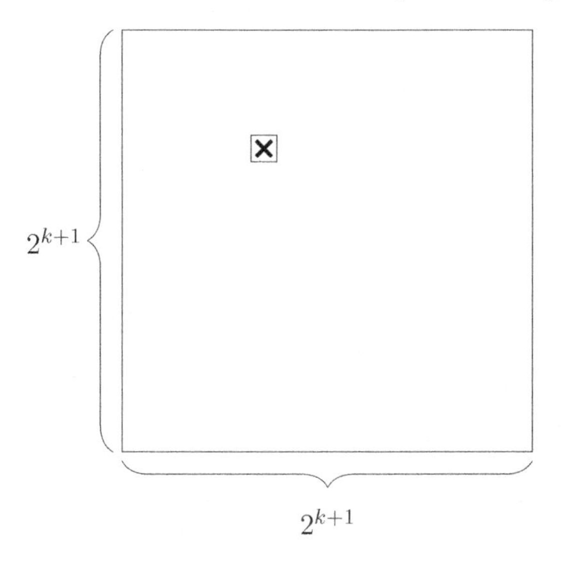

As always, the biggest question we should ask ourselves is: "How are we going to use the inductive hypothesis to prove this?" The inductive hypothesis deals with

$2^k \times 2^k$ boards, so to have any chance of applying it we need to find some $2^k \times 2^k$ chessboards somewhere! Do you see them? If not, think about a concrete example, like if $k = 2$. Can you spot the 4×4 chessboards inside the 8×8 chessboard?[12]

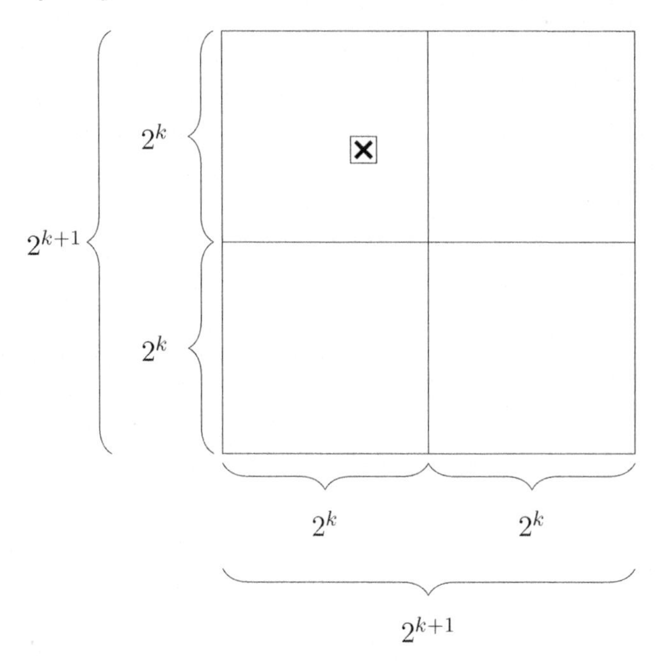

You see? There are four $2^k \times 2^k$ boards in the $2^{k+1} \times 2^{k+1}$ board! Now, one of these four $2^k \times 2^k$ chessboards contains the removed square, and hence by the inductive hypothesis it can be perfectly covered by ⊞−shaped tiles. Perhaps like this:

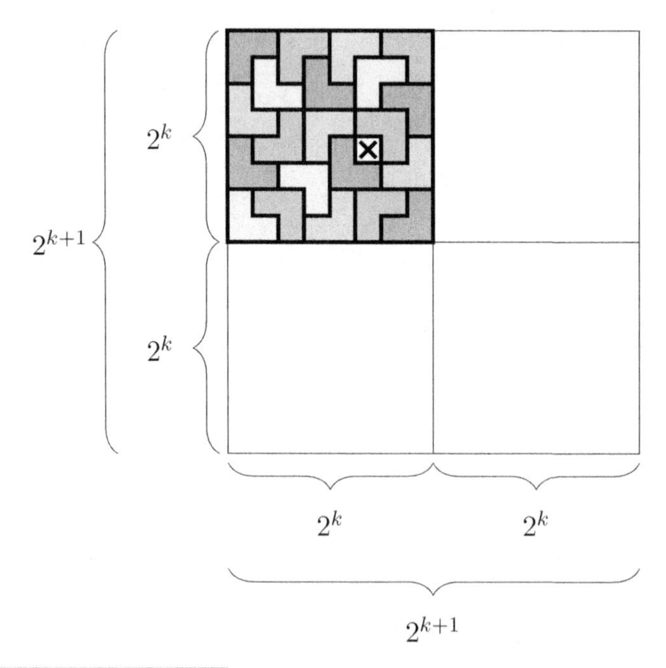

[12]Since $n + n = 2n$, and $2^a \cdot 2^b = 2^{a+b}$, we have $2^k + 2^k = 2 \cdot 2^k = 2^1 \cdot 2^k = 2^{1+k} = 2^{k+1}$.

But what about the other three $2^k \times 2^k$ boards? They don't have any squares removed, so we can't apply the inductive hypothesis to them. And if we picked a random square from each to remove, then sure we could cover the rest, but those three squares would be left uncovered by a tile.

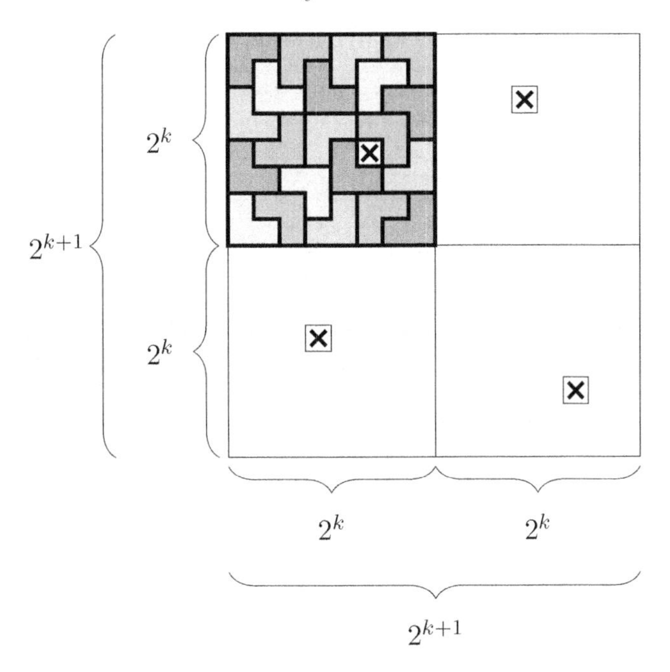

The trick is to remember that the inductive hypothesis says that if *any* square is removed, then a perfect covering exists. So we don't have to imagine that the squares are randomly chosen — we can choose them! For example, we could choose these three squares:

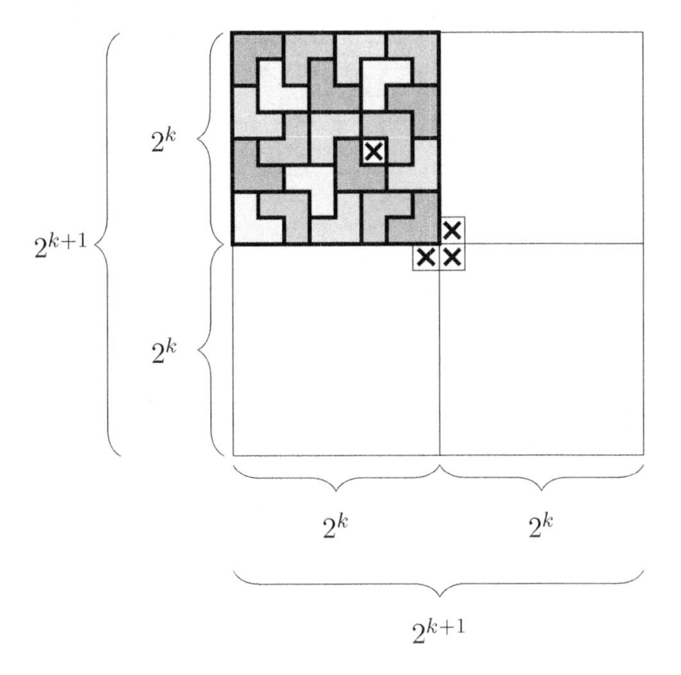

Then, two things happen at the same time. First, by the inductive hypothesis, the three remaining $2^k \times 2^k$ boards can be perfectly covered:

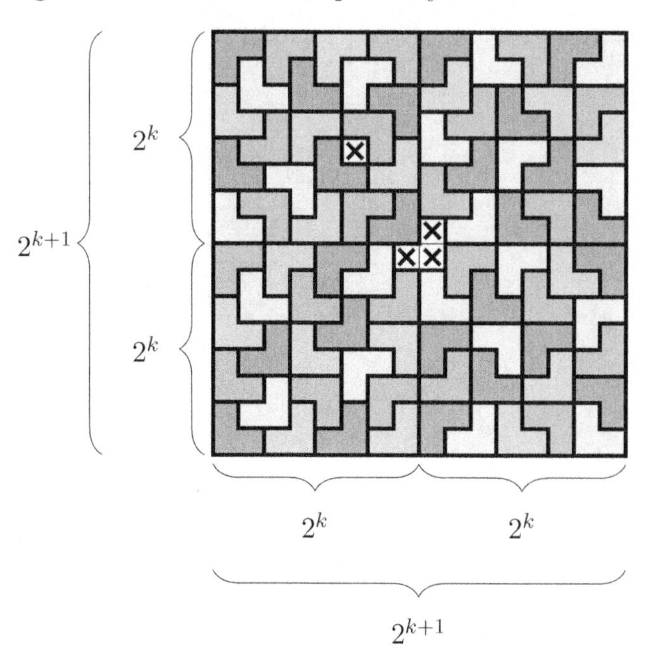

And second, those middle three squares that we crossed out can be covered by a single tile:

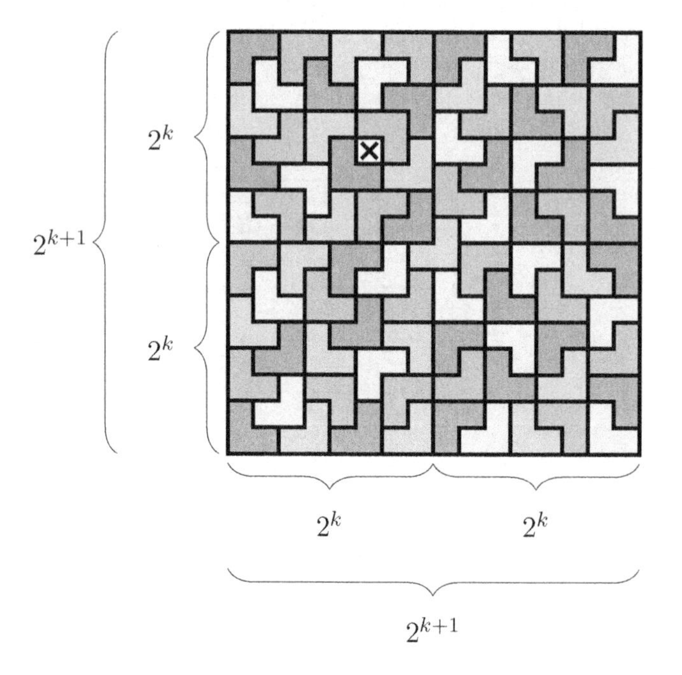

And there it is![13] A tiling of the entire $2^{k+1} \times 2^{k+1}$ board. Whew. Ok, that's the idea, now here's the formal proof.

[13] So cool that it made the cover.

***Proof*.** We proceed by induction.

Base Case. The base case is when $n = 1$, and among the four possible squares that one can remove from a 2×2 chessboard, each leaves a chessboard which can be perfectly covered by a single ⊞–shaped tile:

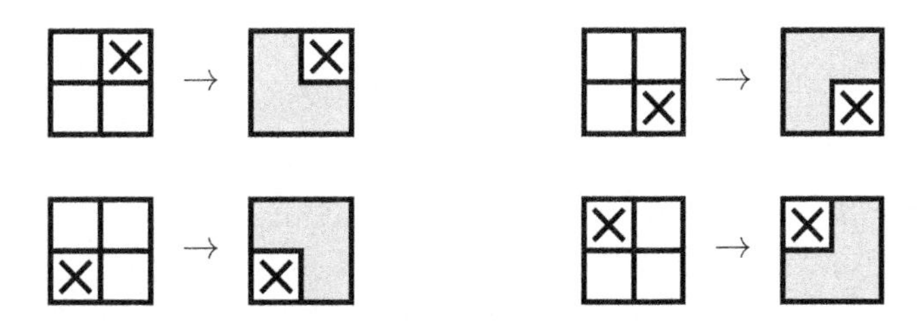

Inductive Hypothesis. Let $k \in \mathbb{N}$, and assume that if any one square is removed from a $2^k \times 2^k$ chessboard, the result can be perfectly covered with ⊞–shaped tiles.

Induction Step. Consider a $2^{k+1} \times 2^{k+1}$ chessboard with any one square removed. Cut this chessboard in half vertically and horizontally to form four $2^k \times 2^k$ chessboards. One of these four will have a square removed and hence by the induction hypothesis can be perfectly covered.

Next, place a single ⊞–shaped tile so that it covers one square from each of the other three $2^k \times 2^k$ chessboards, as shown in the picture below.

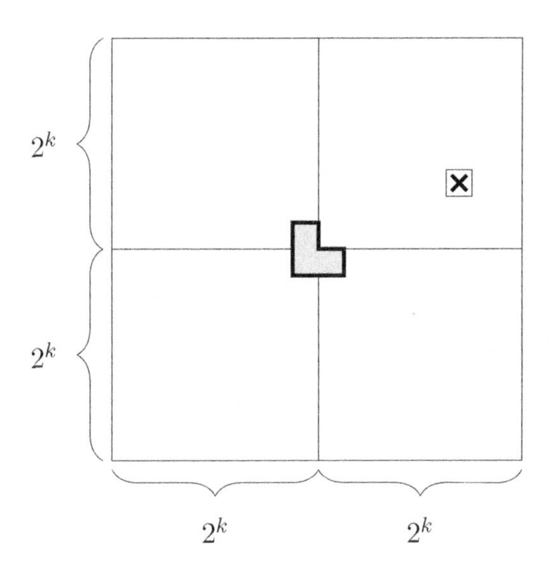

Each of these other three $2^k \times 2^k$ chessboards can be perfectly covered by the inductive hypothesis, and hence the entire $2^{k+1} \times 2^{k+1}$ chessboard can be perfectly covered.

Conclusion. By induction, for every $n \in \mathbb{N}$, if any one square is removed from a $2^n \times 2^n$ chessboard, the result can be perfectly covered with ⊞–shaped tiles. □

Note 4.6. So far, in all of our examples we proved that a statement holds for all $n \in \mathbb{N}$. The base case was $n = 1$ and in the inductive hypothesis we assumed that the result holds for some $k \in \mathbb{N}$.

There are times where one instead wants to prove that a statement holds for only the natural numbers past some point. For example, it is possible to prove the p-test by induction, a result that you might remember from your calculus class:

$$\sum_{i=1}^{\infty} \frac{1}{i^n} \quad \text{converges for all integers } n \geq 2.$$

To prove this result, the base case would be $n = 2$ and in the inductive hypothesis we would assume that the result holds for some k in $\{2, 3, 4, 5, \dots\}$.

At other times, you may want to prove that a result holds for more than just the natural numbers. For example, a result from combinatorics is that

$$\sum_{i=0}^{n} \binom{n}{i} = 2^n \quad \text{holds for all integers } n \geq 0.$$

Here, the base case is $n = 0$, and the inductive hypothesis is the assumption that this holds for some k in $\{0, 1, 2, 3, \dots\}$.

4.3 Strong Induction

The idea behind *strong induction* is that at the point when the 100th domino is the next to get knocked down, you know for sure that all of the first 99 dominoes have fallen, not just the 99th. Likewise, when you are proving some sequence of statements $S_1, S_2, S_3, S_4, \dots$, instead of just assuming that S_k is true in order to prove S_{k+1}, why not assume that $S_1, S_2, \dots, S_k$ are *all* true in order to prove S_{k+1}? After all, by the time that it is S_{k+1}'s turn to be proven to be true, $S_1, S_2, \dots, S_k$ must already have been shown to be true! Why not use that information?

Principle.

Principle 4.7 (*Strong induction*)**.** Consider a sequence of mathematical statements, $S_1, S_2, S_3, \dots$.

- Suppose S_1 is true, and
- Suppose, for any $k \in \mathbb{N}$, if $S_1, S_2, \dots, S_k$ are all true, then S_{k+1} is true.

Then, S_n is true for every $n \in \mathbb{N}$.

In regular induction, you essentially use S_1 to prove S_2, and then S_2 to prove S_3, and then S_3 to prove S_4, and so on. With strong induction, you use S_1 to prove S_2, and then S_1 and S_2 to prove S_3, and then S_1, S_2 and S_3 to prove S_4, and so on.

Below is the general structure for a strong induction proof, which is just slightly different than the structure of a regular induction proof.

Proposition. S_1, S_2, S_3, ... are all true.

Proof. ≪ General setup or assumptions, if needed ≫

Base Case. ≪ Demonstration that S_1 is true ≫

Inductive Hypothesis. Assume that $S_1, S_2, \ldots, S_k$ are all true.

Induction Step. ≪ Proof that $(S_1, S_2, \ldots, S_k)$ implies S_{k+1} is true ≫

Conclusion. Therefore, by induction, all the S_n are true. □

For our first example, recall from Definition 2.16 that if n is an integer and $n \geq 2$, then n is either prime or composite. An integer p is *prime* if $p \geq 2$ and its only positive divisors are 1 and p. A positive integer $n \geq 2$ that is not prime is called *composite*, and is therefore one that can be written as $n = st$, where s and t are integers smaller than n but larger than 1. And with that, it is time for a really big and important result.

Theorem.

Theorem 4.8 (*Fundamental theorem of arithmetic*). Every integer $n \geq 2$ is either prime or a product of primes.[14]

Examples: $21 = 3 \cdot 7$, and $24 = 2 \cdot 2 \cdot 2 \cdot 3$, and $25 = 5 \cdot 5$, and 31 is prime.

Proof Idea. The base case will be $n = 2$, which is prime and hence satisfies the theorem. The inductive hypothesis will be that each of $2, 3, 4, \ldots, k$ is either prime or a product of primes. How do we prove that $k + 1$ is also prime or a product of primes? Regular induction does not seem helpful at all here — if you know that k is prime or a product of a couple primes, then that may tell you something useful about, say, $2k$ or $3k$. But what does it say about $k + 1$? Seemingly very little! This is why regular induction is faltering. But as you'll see, strong induction is just what the doctor[15] ordered.

Note that $k + 1$ is an integer larger than 1, and hence must be either prime or composite (i.e., a product of primes). We will consider these two cases separately. If

[14] Furthermore, each such n can be written as a prime or product of primes in just one way. You will be asked to prove this in Exercise 4.21.

[15] (of philosophy)

$k+1$ is prime, then that's fantastic—it satisfies the theorem! What about if it is composite? Being composite, that would mean $k+1 = st$ for some smaller numbers s and t. Do you see why this is exactly what we need? By *strong* induction, both s and t will satisfy the theorem. And if s and t are both either prime or a product of primes, their product will be too.

Here's a quick summary:

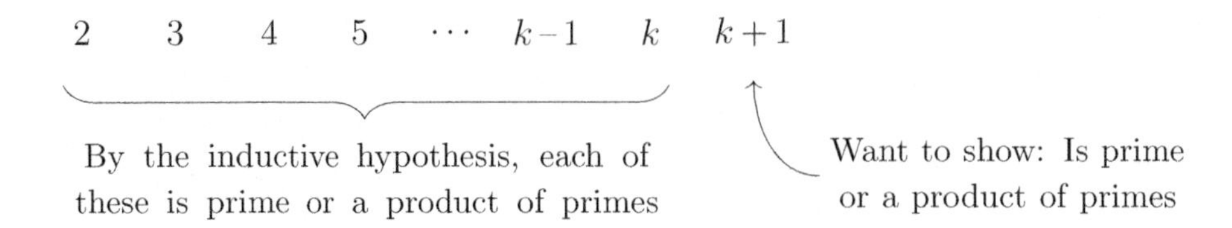

If $k+1$ is prime, then we're done. Otherwise, $k+1 = st$, and both s and t are in the range of numbers covered by the inductive hypothesis.

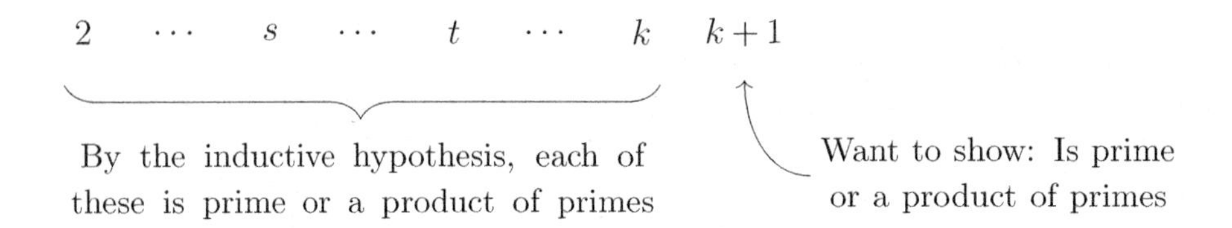

And if s and t are both primes or products of primes, then so must be st, which is $k+1$. Ok, now here's the proof.

Proof. We proceed by strong induction.

Base Case. Our base case is when $n = 2$, and since 2 itself is prime, we are done.

Inductive Hypothesis. Let k be a natural number such that $k \geq 2$, and assume that each of the integers $2, 3, 4, \dots, k$ is either prime or a product of primes.

Induction Step. Next, we consider $k+1$ and we aim to show that $k+1$ is either prime or a product of primes. Since $k+1$ is an integer larger than 1, it is either prime or composite. Consider these two cases separately. Case 1 is that $k+1$ is prime. Since our goal is to show that $k+1$ is either prime or a product of primes, we are immediately done.

Case 2 is that $k+1$ is composite; that is, $k+1$ has positive factors other than 1 and itself. Say, $k+1 = st$ where s and t are positive integers, and

$$1 < s < k+1 \qquad \text{and} \qquad 1 < t < k+1.$$

By the inductive hypothesis, s and t can both be written as a product of primes. Say,

$$s = p_1 \cdot p_2 \cdots p_m \qquad \text{and} \qquad t = q_1 \cdot q_2 \cdots q_\ell$$

where each p_i and q_j is prime.[16] Then,

$$k + 1 = st = (p_1 \cdot p_2 \cdots p_m)(q_1 \cdot q_2 \cdots q_\ell)$$

is written as a product of primes.

Conclusion. By strong induction, every positive integer larger than 2 can be written as a product of primes. □

Chocolate Bar Example

Let's change gears a bit; next up is an example involving chocolate bars. A chocolate bar is typically a grid of squares, which enables it to be broken into smaller pieces. For example, here is a bar broken in two:

Suppose you had a chocolate bar and you wanted to break it up completely, so that each piece is only one square of chocolate. How many breaks will be required to break it all up? To answer this question, there is another question we should ask first: Does the answer depend on how you break it up? Is there an efficient way to break it all up and a slow way to break it all up? Or will the answer be the same no matter how you do it?

The rules are simple: No tricks. You can't stack pieces to break them together and only count that as one break. No tricky ways to hold many pieces at once. The simplest way to think about it is that after you break a piece into two, you then have to work on those two new pieces separately.

Start thinking about this on your own, and at least have a guess in mind of whether the answer depends on how you do it, or whether all breaking sequences are the same number of steps. And then, when you're ready, here's a very small example:

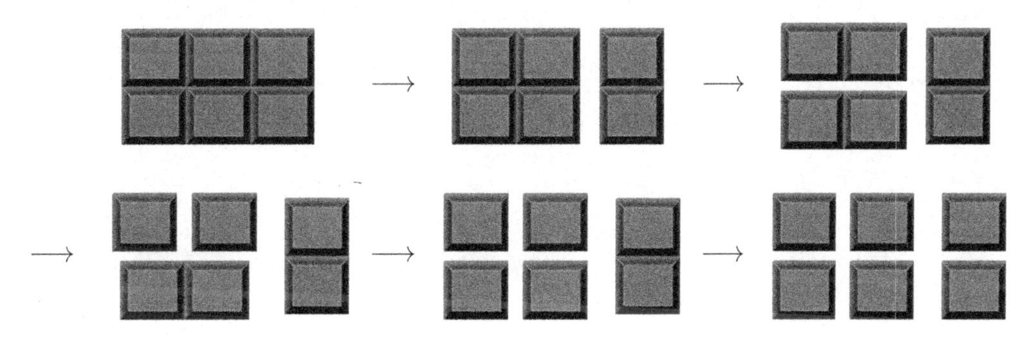

[16]Note that if, say, s is prime, then $m = 1$ and the expression for s is simply $s = p_1$. So this includes the cases in which s and/or t are prime.

So with this sequence of breaks, it took 5 breaks to break a 2×3 chocolate bar into individual squares.

And here is the answer to the first question: It does not matter how you break it up, the answer will always be the same. This holds even for ~~extra delicious~~ very large chocolate bars, where there are loads of different ways to break it all up. Moreover, the number of breaks required follows a very simple formula: It is always equal to one less than the number of squares.[17] For the example above, it had 6 squares and required 5 breaks.

Proposition.

Proposition 4.9. Suppose you have a chocolate bar that is an $m \times n$ grid of squares. The entire bar, or any smaller rectangular piece of that bar, can be broken along the vertical or horizontal lines separating the squares.

The number of breaks to break up that chocolate bar into individual squares is precisely $mn - 1$.

Here is what a 4×7 chocolate bar looks like:

Let's now sketch the proof.

Proof Sketch. The base case will deal with the 1×1 chocolate bar, and will work out fine. So let's turn our attention to the inductive hypothesis and the induction step. The inductive hypothesis will say that all bars with at most k squares satisfy the result, and we wish to prove that any chocolate bar with $k + 1$ squares satisfies the result, too.

Now, due to the fact that we are a grid of squares, thinking in terms of a single variable k makes it more confusing. So instead we will phrase the problem in terms of a grid. Instead of showing it is true for $k + 1$, where $k + 1$ is some $m \times n$, for now let's talk about it in terms of that $m \times n$.

[17]Just to be safe, I recommend you go to the grocery store right now and buy lots of different-sized chocolate bars and try this on your own. For science.

With that perspective, our inductive hypothesis will be that all chocolate bars with fewer than mn squares satisfies the proposition, and we will aim to prove that the $m \times n$ chocolate bar satisfies the result.

Consider the first break of an $m \times n$ chocolate bar, which will break the bar into two pieces. There are many ways to make this first break, but here's one vertical break:

Suppose the first of the two pieces has a squares and the second has b squares. Since the original had mn squares, this means $a + b = mn$. Moreover, notice that what we have essentially done is produce two new smaller rectangular chocolate bars! In fact, since both of these have fewer than mn squares, we can apply the inductive hypothesis to each of them! Here's what that gives us: The bar with a squares can be completely broken up with $a - 1$ breaks, and the bar with b squares can be completely broken up with $b - 1$ breaks. Combined, this tells us how many breaks it takes for the original bar.

Before we write out the formal proof, notice that we really do need strong induction. With regular induction, when proving the $(k+1)^{\text{st}}$ case you are only permitted to use the previous case—the kth case. Now, if the first break of a bar with $(k+1)$ squares was guaranteed to produce a bar with k squares, then you could use regular induction—but this is not the case. Typically, the first break produces two bars which have fewer than k pieces, and thus we need strong induction.[18]

Proof. We proceed by strong induction.

Base Case. Our base case is for a chocolate bar with just 1 square; the only bar like this is the 1×1 bar. And the number of breaks required to break the 1×1 bar into individual squares is clearly 0, as it is already an individual square. This satisfies the result, as $0 = 1 \cdot 1 - 1$ is one less than the number of squares in the bar.

Inductive Hypothesis. Let $k \in \mathbb{N}$, and assume that all bars with at most k squares satisfy the proposition.

[18]Go hit the gym regular-induction, we need some  up in here.

Induction Step. Consider now any chocolate bar with $k+1$ squares;[19] suppose this bar has dimensions $m \times n$. Any sequence of breaks begins with a first break which breaks the bar into two smaller bars. Consider an arbitrary first break, and suppose the two smaller bars have a squares and b squares, respectively. Note that we must have $a + b = mn$, because the number of squares in the smaller bars must add up to the number of squares in the original $m \times n$ bar.

By the inductive hypothesis, the bar with a squares will require $a - 1$ breaks to completely break it up, and the bar with b breaks will require $b - 1$ breaks. Therefore, to break up the $m \times n$ bar, we must make a first break, followed by $(a-1) + (b-1)$ additional breaks. The total number of breaks is then

$$1 + (a-1) + (b-1) = (a+b) - 1 = mn - 1.$$

And $mn - 1$ is indeed one less than the number of squares in the $m \times n$ bar.

Conclusion. By strong induction, a chocolate bar of any size requires one break less than its number of squares to break it up into individual squares. □

When you prove a result, it is good practice to ask yourself, "were all the assumptions in the problem necessary?" Proposition 4.9 assumed that the bar was an $m \times n$ grid of squares, and then concluded that $mn - 1$ breaks were needed. Said differently, the number of breaks was 1 less than the number of squares.

What if the pieces were in the shape of a triangle? If a bar was comprised of T triangular pieces, would it still require $T - 1$ breaks?

What about chocolate bars of other shapes? What if a bar started off with some "missing" pieces in the middle (like the above triangular bar, with the middle piece gone)? Interestingly, none of that matters. If the bar has T pieces, then it requires $T - 1$ breaks, no matter the bar's shape, and even if has holes in the middle. As long as each of your "breaks" divides one chunk into two, then $T - 1$ is the answer.

Here is some intuition for that: No matter the shape, the bar starts out as a single "chunk" of chocolate, and after your sequence of breaks the bar is broken into T chunks of chocolate — the T individual pieces. How many breaks does it take to move from 1 chunk to T chunks? Notice that every break increases the number of chunks by 1. So after 1 break, there will be 2 chunks. After 2 breaks, there will be 3 chunks. And so on. Thus, after $T - 1$ breaks there will be T chunks, which is why $T - 1$ breaks is guaranteed to be the answer, no matter which shape you started with.

[19]By the way, note that there could be many different bars with $k+1$ squares, but there is guaranteed to be at least one: the long and skinny bar with dimensions $(k+1) \times 1$.

Multiple Base Cases

When proving the $(k+1)^{\text{st}}$ case within the induction step, strong induction allows us to apply not just the kth step, but any of the steps 1, 2, 3, ..., k. In the previous two examples, we had no idea which earlier steps we would need, so it was vital that we assumed them all. At times, though, we really only need, say, the previous *two* steps. The kth step is perhaps not enough, but the $(k-1)^{\text{st}}$ step and the kth step is guaranteed to be enough.

If we rely on the two previous steps, then that is analogous to saying that it takes the previous *two* dominoes to knock over the next one. Thus, if we knock over dominoes 1 and 2, then they will collectively knock over the third. Then, since the second and third have fallen, those two will collectively knock over the fourth. Then the third and fourth will knock over the fifth. And so on. Thus, the induction relies on two base cases, because without knocking over the first two dominoes, the third won't fall and the process won't begin.

Here is a chart showing how this situation differs from strong induction.

	Strong Induction (Typical Form)	Strong Induction (Two-Step Form)	For Both Forms, After This Step We Know
Step $n=1$:	1 is a base case	1 is a base case	1 is true
Step $n=2$:	1 implies 2	2 is a base case	1 and 2 are true
Step $n=3$:	(1 and 2) imply 3	(1 and 2) imply 3	1, 2 and 3 are true
Step $n=4$:	(1, 2 and 3) imply 4	(2 and 3) imply 4	1, 2, 3 and 4 are true
Step $n=5$:	(1, 2, 3 and 4) imply 5	(3 and 4) imply 5	1, 2, 3, 4 and 5 are true

In the same way, if each step relies on the previous three steps, then you must prove three base cases. If each step relies on the previous four, then you must prove four base cases. And so on. But let's not get too crazy, below is an example relying on just two base cases.

Proposition.

Proposition 4.10. Every $n \in \mathbb{N}$ with $n \geq 11$ can be written as $2a + 5b$ for some natural numbers a and b.

As an example of this, note that $n = 41$ can be written as $2 \cdot 3 + 5 \cdot 7$.

Scratch Work. First, note that this proposition asserts that this property holds for $n = 11, 12, 13, 14, \dots$. Since the process starts at $n = 11$, this will be a base case. So we will have to find an $a, b \in \mathbb{N}$ for which $11 = 2a + 5b$. I think $a = 3$ and $b = 1$ works. But again, just to get our feet wet, let's write out the first few cases. There are at times multiple ways of doing so, but remember that $a, b \in \mathbb{N}$, so these numbers cannot be negative or zero.

- $11 = 2 \cdot 3 + 5 \cdot 1$
- $12 = 2 \cdot 1 + 5 \cdot 2$
- $13 = 2 \cdot 4 + 5 \cdot 1$
- $14 = 2 \cdot 2 + 5 \cdot 2$
- $15 = 2 \cdot 5 + 5 \cdot 1$
- $16 = 2 \cdot 3 + 5 \cdot 2$

Writing out some examples is often the best way to discover a proof. Do you see anything interesting about the numbers? In particular, do you see a pattern between the $n = 11$, 13 and 15 cases? And perhaps you can spot a pattern between the $n = 12$, 14 and 16 cases?

To move from the $n = 13$ case to the $n = 15$ case, for example... all you need is an extra 2! So $2 \cdot \mathbf{4} + 5 \cdot 1$ simply turns into $2 \cdot \mathbf{5} + 5 \cdot 1$, and that's it! This is how we will prove it. Each case relies on two cases back. How do you show that there is a way to write $(k+1)$ in this way? Well, by the inductive hypothesis for strong induction, it is possible to write $(k-1)$ in such a way, and now you just tack on another 2. Let's do it.

Proof. We proceed by strong induction.

Base Cases. In the induction step we will need two cases prior, so we show two base cases here: $n = 11$ and $n = 12$. Both of these can be written as asserted:

$$11 = 2 \cdot 3 + 5 \cdot 1$$
$$12 = 2 \cdot 1 + 5 \cdot 2.$$

Inductive Hypothesis. Assume that for some integer $k \geq 12$, the result holds for

$$n = 11, 12, 13, \dots, k.$$

Induction Step. We aim to prove the result for $k + 1$. By the inductive hypothesis,

$$k - 1 = 2a + 5b$$

for some $a, b \in \mathbb{N}$. Adding 2 to both sides of the above,

$$k + 1 = 2(a + 1) + 5b.$$

Observe that $(a + 1) \in \mathbb{N}$ and $b \in \mathbb{N}$, proving that this is indeed a representation of $(k + 1)$ in the desired form.

Conclusion. Therefore, by strong induction, every integer $n \geq 11$ can be written as the proposition asserts. $\square$

To close out this section, I will note that while there are many instances where regular induction is not enough and strong induction is needed, you will discover that regular induction comes up far more often than strong induction — usually the kth case is enough to prove the $(k+1)^{\text{st}}$ case. And the most common instances in which you need strong induction are ones like the above, where you need a fixed number of prior cases to prove the next.

4.4 Non-Examples

What if instead of doing induction properly, you make only a teeny-tiny mistake that's super hard to notice? Then what could we prove? Lots of things! Behold, a fun non-example!

Fake Proposition.

Fake Proposition 4.11. Everyone on Earth has the same name.

Fake Proof. We will consider groups of n people at a time, and by induction we will "prove" that for every $n \in \mathbb{N}$, every group of n people must have everyone with the same name.

Base Case. If $n = 1$, then of course everyone in the group has the same name, since there's only one person in the group!

Inductive Hypothesis. Let $k \in \mathbb{N}$, and assume that any group of k people all have the same name.

Induction Step. Consider a group of $k + 1$ people.

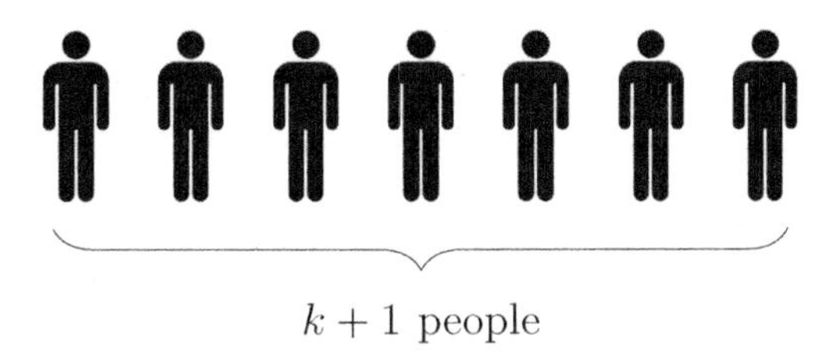

But notice that we can look at the first k of these people and then the last k of these people, and to each of these groups we can apply the inductive hypothesis:

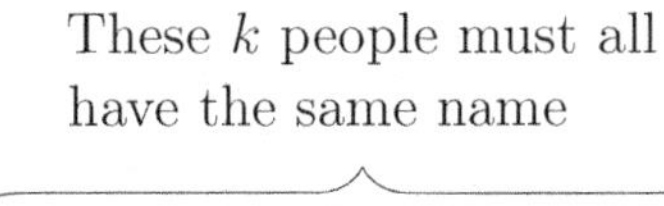

And these k people must all have the same name

And the only way that this can all happen, is if all $k + 1$ people have the same name.

Conclusion. This "proves" by induction that for every $n \in \mathbb{N}$, every group of n people must have the same name. So if you let n be equal to the number of people on Earth, this "proves" that everyone has the same name. ☒

This must, of course, be flawed somewhere. To find the mistake, think about how the above argument moves from the $n = 1$ case to the $n = 2$ case... Exercise 4.10 asks for an explanation of the error.

Let's do one more. In calculus you probably learned that the *harmonic series* diverges. That is,

$$1 + \frac{1}{2} + \frac{1}{3} + \frac{1}{4} + \cdots = \infty.$$

You may have learned this on its own, or perhaps within the discussion of the series p-test. And your calc professor did not lie to you — what you learned was completely true, so only a Fake Proof could assert otherwise. See if the Fake Proof below does the job.

Fake Proposition.

Fake Proposition 4.12. The harmonic series converges. That is,

$$1 + \frac{1}{2} + \frac{1}{3} + \frac{1}{4} + \cdots < \infty.$$

Fake Proof. We proceed by induction. In the notation of the principle of mathematical induction, we will let S_n be the statement

$$1 + \frac{1}{2} + \frac{1}{3} + \cdots + \frac{1}{n} < \infty.$$

Base Case. If $n = 1$, then of course $1 < \infty$.

Inductive Hypothesis. Let $k \in \mathbb{N}$, and assume that

$$1 + \frac{1}{2} + \frac{1}{3} + \cdots + \frac{1}{k} < \infty.$$

Induction Step. By the inductive hypothesis, $1 + \frac{1}{2} + \frac{1}{3} + \cdots + \frac{1}{k}$ is some finite number, which we will call F. Then,

$$\begin{aligned} 1 + \frac{1}{2} + \frac{1}{3} + \cdots + \frac{1}{k+1} &= 1 + \frac{1}{2} + \frac{1}{3} + \cdots + \frac{1}{k} + \frac{1}{k+1} \\ &= F + \frac{1}{k+1}. \end{aligned}$$

And since a finite number plus another finite number is finite, $F + \frac{1}{k+1}$ is finite. This means that

$$1 + \frac{1}{2} + \frac{1}{3} + \cdots + \frac{1}{k+1} < \infty.$$

Conclusion. By induction, this means that

$$1 + \frac{1}{2} + \frac{1}{3} + \frac{1}{4} + \cdots < \infty,$$

completing the "proof."

Once again, this must be flawed somewhere. To find the mistake, think about what conclusion is actually being reached by the first three stages of this proof... Exercise 4.11 asks for an explanation of the error.

4.5 Bonus Examples

We are going to call our first bonus example a lemma, since it will be used in the second bonus example. A quick reminder before we begin: $\mathbb{N}_0$ is the set $\{0, 1, 2, 3, \dots\}$.

Lemma.

Lemma 4.13. For every $n \in \mathbb{N}_0$,

$$1 + 2 + 4 + 8 + \cdots + 2^n = 2^{n+1} - 1.$$

Scratch Work. Let's do some examples to convince ourselves that this seems true.

$$\begin{aligned} 1 &= 2^1 - 1 \quad \checkmark \\ 1 + 2 &= 2^2 - 1 \quad \checkmark \\ 1 + 2 + 4 &= 2^3 - 1 \quad \checkmark \\ 1 + 2 + 4 + 8 &= 2^4 - 1 \quad \checkmark \end{aligned}$$

Seems to check out! The inductive hypothesis will be $1+2+4+8+\cdots+2^k = 2^{k+1}-1$. See if you can see a way to use this to prove that $1+2+4+8+\cdots+2^{k+1} = 2^{k+2}-1$, which is the induction step. Then check out the proof below.

Proof. We proceed by induction.

Base Case. The base case is when $n = 0$, and

$$1 = 2^{0+1} - 1,$$

as desired.

Inductive Hypothesis. Assume that for some $k \in \mathbb{N}_0$, we have

$$1 + 2 + 4 + 8 + \cdots + 2^k = 2^{k+1} - 1.$$

Induction Step. We aim to prove that the result holds for $k+1$. That is, we wish to show that

$$1 + 2 + 4 + 8 + \cdots + 2^{k+1} = 2^{(k+1)+1} - 1.$$

Written slightly differently, we wish to show

$$1 + 2 + 4 + 8 + \cdots + 2^k + 2^{k+1} = 2^{k+2} - 1.$$

Starting with the inductive hypothesis, we can add 2^{k+1} to both sides, and then do a little algebra, to get

$$\begin{aligned} 1 + 2 + 4 + 8 + \cdots + 2^k + 2^{k+1} &= 2^{k+1} - 1 + 2^{k+1} \\ &= 2 \cdot 2^{k+1} - 1 \\ &= 2^{k+2} - 1 \end{aligned}$$

as desired.

Conclusion. Therefore, by induction, $1 + 2 + 4 + 8 + \cdots + 2^n = 2^{n+1} - 1$ holds for all $n \in \mathbb{N}_0$. □

Our next bonus example deals with these same powers of 2. Just to be clear: by *powers of 2* we mean $2^0 = 1$, $2^1 = 2$, $2^2 = 4$, $2^3 = 8$, $2^4 = 16$, and so on. These are important when discussing a number's *binary* representation — which, along with *beep boop*, is the language of computers. A number like 13 can be represented by sums of powers of 2 like this:

$$13 = 1 \cdot 8 + 1 \cdot 4 + 0 \cdot 2 + 1 \cdot 1.$$

This written in binary as 1101, representing how many 8s, 4s, 2s, and 1s you need. In fact, every $n \in \mathbb{N}_0$ can be represented in binary using only 0s and 1s, and moreover this representation is unique. That is, every $n \in \mathbb{N}_0$ can be represented in precisely one way as a sum of distinct powers of 2. Here are the first representations; check them each on your own.

- $0 \to 0$
- $1 \to 1$
- $2 \to 10$
- $3 \to 11$
- $4 \to 100$
- $5 \to 101$
- $6 \to 110$
- $7 \to 111$
- $8 \to 1000$

Let's[20] now use strong induction to prove that every $n \in \mathbb{N}$ has a unique binary representation.

[20]Given the third example, it would be a dereliction of duty not to include the classic binary joke: "There are 10 types of people in the world: Those who understand binary and those who don't."

Theorem.

Theorem 4.14. Every $n \in \mathbb{N}$ can be expressed as a sum of distinct powers of 2 in precisely one way.

Once you include 0 as the binary representation of 0, this theorem also tells us that every $n \in \mathbb{N}_0$ has a unique binary representation.

Proof. We proceed by strong induction.

Base Case. Our base case is when $n = 1$. Note that 1 can be written as 2^0, and this is the only way to write 1 as a sum of distinct powers of 2, because all other powers of 2 are larger than 1.

Inductive Hypothesis. Let $k \in \mathbb{N}$. Assume that each of the integers $1, 2, 3, \dots, k$ can be expressed as a sum of distinct powers of 2 in precisely one way.

Induction Step. We now aim to show that $k+1$ can be expressed as a sum of distinct powers of 2 in precisely one way.

Let 2^m be the largest power of 2 such that $2^m \leq k+1$. We now consider two cases: the first case is that $2^m = k+1$, and the second case is that $2^m < k+1$.

Case 1: $2^m = k+1$. If this occurs, then 2^m is itself a way to express $k+1$ as a (one-term) sum of distinct powers of 2. Moreover, there is no other way to express $k+1$ as a sum of distinct powers of 2, because by Lemma 4.13 all smaller powers of 2 sum to $2^m - 1 = k$. Thus, even by including all smaller powers of 2, we are unable to reach $k+1$. So, in Case 1, there is precisely one such expression for $k+1$.

Case 2: $2^m < k+1$. In order to apply the inductive hypothesis, we will consider $(k+1) - 2^m$. First, note that $(k+1) - 2^m$ is less than 2^m, because otherwise $k+1$ would have two copies of 2^m within it, implying that $2^m + 2^m \leq k+1$. However, since $2^m + 2^m = 2 \cdot 2^m = 2^{m+1}$, this would mean $2^{m+1} \leq k+1$. This can't be, since 2^m was chosen to be the *largest* power of 2 that is at most $k+1$. Thus, it must be the case that $(k+1) - 2^m < 2^m$.

Next, by the inductive hypothesis, $(k+1) - 2^m$ can be expressed as a sum of distinct powers of 2 in precisely one way, and since $(k+1) - 2^m < 2^m$, this unique expression for $(k+1) - 2^m$ will not contain a 2^m. Thus, by adding a 2^m to it, we obtain an expression for $k+1$ as a sum of powers of 2. And this expression is unique because the $(k+1) - 2^m$ is unique according to the inductive hypothesis, and the 2^m portion is unique because, again by Lemma 4.13, even if you summed all of the smaller powers of 2, you will not reach 2^m.

Conclusion. By strong induction, every $n \in \mathbb{N}$ can be expressed as a sum of distinct powers of 2 in precisely one way. □

Our next goal is to use induction to provide a second proof of Fermat's little theorem, in the case that $a \in \mathbb{N}$. The proof is going to rely on a theorem we have not yet discussed called the *binomial theorem*, but which you may have seen in some form in an earlier course.

Theorem.

Theorem 4.15 (*The Binomial Theorem*). For $x, y \in \mathbb{R}$ and $n \in \mathbb{N}_0$,

$$(x+y)^n = \sum_{m=0}^{n} \binom{n}{m} x^m y^{n-m}.$$

Here, when $n \geq m$, the binomial coefficient $\binom{n}{m}$ is defined to be $\frac{n!}{m!(n-m)!}$, which one can show is always an integer.[21] The binomial coefficients can also be defined combinatorially: $\binom{n}{m}$ is equal to the number of ways to choose m elements from an n-element set; in fact, $\binom{n}{m}$ is read "n choose m." For example, $\binom{4}{2} = 6$ because there are six subsets of the set $\{1,2,3,4\}$ containing two elements:

$$\{1,2\}\ ,\ \{1,3\}\ ,\ \{1,4\}\ ,\ \{2,3\}\ ,\ \{2,4\}\ ,\ \{3,4\}.$$

Binomial coefficients can be computed iteratively using *Pascal's rule*, which says that

$$\binom{n}{r} = \binom{n-1}{r-1} + \binom{n-1}{r},$$

as well as the fact that $\binom{n}{0} = 1$ and $\binom{n}{n} = 1$ for all $n \in \mathbb{N}_0$. A beautiful way to combine these facts is called *Pascal's triangle*, which begins like this:

$$\begin{array}{ccccccccccc} &&&&& \binom{0}{0} &&&&& \\ &&&& \binom{1}{0} && \binom{1}{1} &&&& \\ &&& \binom{2}{0} && \binom{2}{1} && \binom{2}{2} &&& \\ && \binom{3}{0} && \binom{3}{1} && \binom{3}{2} && \binom{3}{3} && \\ & \binom{4}{0} && \binom{4}{1} && \binom{4}{2} && \binom{4}{3} && \binom{4}{4} & \\ \binom{5}{0} && \binom{5}{1} && \binom{5}{2} && \binom{5}{3} && \binom{5}{4} && \binom{5}{5} \end{array} \quad = \quad \begin{array}{ccccccccccc} &&&&& 1 &&&&& \\ &&&& 1 && 1 &&&& \\ &&& 1 && 2 && 1 &&& \\ && 1 && 3 && 3 && 1 && \\ & 1 && 4 && 6 && 4 && 1 & \\ 1 && 5 && 10 && 10 && 5 && 1 \end{array}$$

[21] Note: We define 0! to be equal to 1.

Indeed, we can even prove the binomial theorem by induction, by making use of Pascal's rule. Here is a sketch of that proof:

Proof Sketch. The base case is when $n = 0$, and indeed $(x+y)^0 = 1$. The next couple cases are more interesting, and you can check that $(x+y)^1 = x + y$ and $(x+y)^2 = x^2 + 2xy + y^2$ do indeed match the theorem. The inductive hypothesis will be

$$(x+y)^k = x^k + \binom{k}{1}x^{k-1}y + \binom{k}{2}x^{k-2}y^2 + \cdots + \binom{k}{k-1}xy^{k-1} + y^k.$$

For the induction step, we perform easy algebra, then apply the inductive hypothesis, then perform hard algebra, and then apply Pascal's rule:

$$\begin{aligned}
&(x+y)^{k+1} \\
&= (x+y)(x+y)^k \\
&= (x+y)\cdot\left[x^k + \binom{k}{1}x^{k-1}y + \binom{k}{2}x^{k-2}y^2 + \cdots + \binom{k}{k-1}xy^{k-1} + y^k\right] \\
&= x^{k+1} + \left[\binom{k}{0} + \binom{k}{1}\right]x^k y + \left[\binom{k}{1} + \binom{k}{2}\right]x^{k-1}y^2 \\
&\qquad + \cdots + \left[\binom{k}{k-1} + \binom{k}{k}\right]xy^k + y^{k+1} \\
&= x^{k+1} + \binom{k+1}{1}x^k y + \binom{k+1}{2}x^{k-1}y^2 + \cdots + \binom{k+1}{k}xy^k + y^{k+1}.
\end{aligned}$$

And that — a few boring algebraic details omitted — is the proof. □

The binomial theorem tells us that in order to expand $(x+y)^5$ you can just look at the 5th row of Pascal's triangle (where the top element counts as the 0th row, so the 5th row is 1 5 10 10 5 1):

$$(x+y)^5 = 1x^5 + 5x^4y + 10x^3y^2 + 10x^2y^3 + 5xy^4 + 1y^5$$

Moreover, by plugging in special values for x and y, all sorts of neat identities pop out. There are loads of examples of this,[22] but here are just three:

- By plugging in $x = 1$, $y = 1$, we prove $2^n = \sum_{k=0}^{n}\binom{n}{k}$.
- By plugging in $x = 2$, $y = 1$, we prove $3^n = \sum_{k=0}^{n}\binom{n}{k}2^k$.
- By plugging in $x = -1$, $y = 1$, we prove $0 = \sum_{k=0}^{n}(-1)^k\binom{n}{k}$.

[22] *"A theorem that launched a thousand corollaries!"*

But let's move on to the main event. The binomial theorem is a means to provide a second proof of (the positive case of) *Fermat's little theorem*, which we first discussed while studying modular arithmetic in Chapter 2. Here is that theorem, written just slightly differently by multiplying each side of the congruence by a, which can also be undone by using the cancellation law (Proposition 2.18).

Theorem.

Theorem 2.19 (*Fermat's little theorem*)**.** If a is a natural number and p is a prime which does not divide a, then

$$a^p \equiv a \pmod{p}.$$

Quick note: The proof below will begin by saying "Fix a prime p." What this means is that, throughout the proof, p is a single prime number. However, we are not saying that it is 7 or 11 or any specific prime. Indeed, the fact that it is an arbitrary prime is important: If our proof goes through without saying which prime it is, then that means that it could have been *any* prime, and the proof would have worked! Thus, by beginning with a fixed (but arbitrary) prime p, it in effect proves it for *every* prime. Ok, let's prove it.

Proof. Fix a prime p. We will prove the theorem by inducting on a.

Base Case. Our base case is when $a = 1$, and indeed

$$\begin{aligned} a^p &= 1^p \\ &= 1 \\ &\equiv a \pmod{p}, \end{aligned}$$

as needed.

Inductive Hypothesis. Let $k \in \mathbb{N}$, and assume that

$$k^p \equiv k \pmod{p}.$$

Induction Step. We aim to prove that $(k+1)^p \equiv k+1 \pmod{p}$. To do this, we make use of the binomial theorem, which says that

$$(k+1)^p = k^p + \binom{p}{1}k^{p-1} + \binom{p}{2}k^{p-2} + \cdots + \binom{p}{p-1}k + 1.$$

Next, note that $\binom{p}{k} = \frac{p!}{k!(p-k)!}$ is an integer where the numerator is divisible by p but the denominator is not (since the denominator is a product of numbers smaller than p, and hence contains only primes smaller than p). Therefore, every term except for the first and the last is congruent to 0 (mod p). That is,

$$\begin{aligned}(k+1)^p &\equiv k^p + \binom{p}{1}k^{p-1} + \binom{p}{2}k^{p-2} + \cdots + \binom{p}{p-1}k + 1 \pmod{p}\\ &\equiv k^p + 0 + 0 + \cdots + 0 + 1 \pmod{p}\\ &\equiv k^p + 1 \pmod{p}\\ &\equiv k + 1 \pmod{p},\end{aligned}$$

where in the final step we used that $k^p \equiv k \pmod{p}$, which was given to us by our inductive hypothesis. We have successfully shown that $(k+1)^p \equiv k+1 \pmod{p}$, completing the induction step.

Conclusion. Therefore, for any fixed p we have shown that, by induction, Fermat's little theorem holds for all $a \in \mathbb{N}$. And since p was arbitrary, this theorem holds for any prime p. □

We just proved Fermat's little theorem in the case that $a \in \mathbb{N}$, but in Chapter 2 we proved that the theorem applies to any $a \in \mathbb{Z}$. The $a = 0$ case is clear enough, but does the theorem for $a \in \mathbb{N}$ imply the theorem for negative integers?

If p is an odd prime (meaning, $p \neq 2$), then by multiplying both sides by -1 we can turn $a^p \equiv a \pmod{p}$ into $-a^p \equiv -a \pmod{p}$, and because p is odd this means $(-a)^p \equiv -a \pmod{p}$, showing that the negative case is satisfied when p is an odd prime.

What about if $p = 2$? Here, things are even simpler. If $p = 2$, then observe that $1 \equiv -1 \pmod{p}$, and so having a negative sign or not makes no difference. So $a^p \equiv a \pmod{p}$ is the same as $a^p \equiv -a \pmod{p}$. And because $a^2 = (-a)^2$ by basic algebra, $a^p \equiv -a \pmod{p}$ is the same as $(-a)^p \equiv -a \pmod{p}$, showing that the negative case is satisfied when $p = 2$.

Thus, with a little more work, our induction proof could quickly be amended to account for the general $a \in \mathbb{Z}$ case.

Mantel's Theorem

As a final example, here is an important result from graph theory[23] called *Mantel's theorem*. It is best phrased in terms of graphs with an even number of vertices, so we will refer to the number of vertices as $2n$. The question is, how many edges must we have in order to *guarantee* that the graph contains a triangle (three vertices for which the three possible edges between them are all present). For example, here is a graph with 4 vertices, 4 edges, and which contains a triangle.

[23]Please review the short introduction to graphs on pages 24 and 25 if the idea of a graph, vertex or edge is unclear.

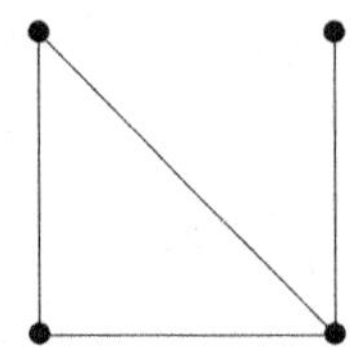

However, having 4 vertices and 4 edges does not *guarantee* a triangle, because the following is a graph with these statistics which does not have a triangle.

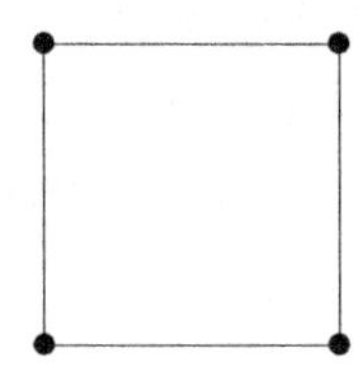

However, it turns out that with 4 vertices and 5 edges, a triangle cannot be avoided. A proof sketch of this: With four vertices, the maximum number of edges is 6, so with 5 edges there can only be one missing edge. So you could just draw all six options and note that each one contains a triangle. Or you could note this: By symmetry, removing any one edge essentially results in the same graph as if you had removed any other edge. So the picture is essentially always this:

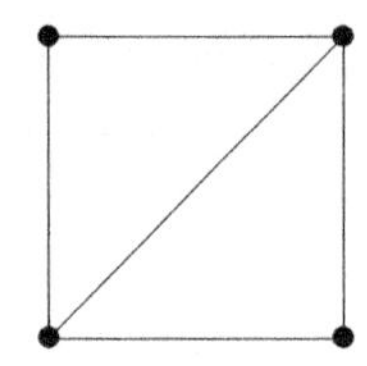

And this graph does indeed have a triangle.

For a general graph with $2n$ vertices, how many edges are needed to guarantee a triangle? The *complete bipartite graph* is the graph with $2n$ vertices which is best drawn by placing n vertices on the left, n vertices on the right, and drawing in all possible edges from the left to the right — but adding no edge between any two vertices on the left, or any two on the right. For example, here is the complete bipartite graph on 8 vertices:

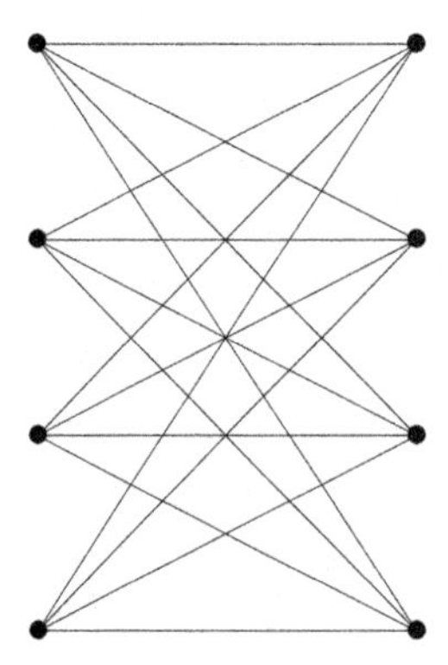

It is an example of a graph with $4^2 = 16$ edges, and observe that it contains no triangles! In general, the complete bipartite graph on $2n$ vertices contains n^2 edges and no triangles. Can we do better than this? What if our graph on $2n$ vertices had $n^2 + 1$ edges? Could such a graph also contain no triangles? The answer is no; the complete bipartite graph is the best we can possibly do. This is what Mantel's theorem says. Let's state and prove this theorem now.

Theorem.

Theorem 4.16 (*Mantel's theorem*)**.** If a graph G has $2n$ vertices and $n^2 + 1$ edges, then G contains a triangle.

Proof. We proceed by induction.

Base Case. Our base case is when $n = 1$, giving $2 \cdot 1 = 2$ vertices and $1^2 + 1 = 2$ edges. And because there are no graphs on 2 vertices and 2 edges, the conclusion is vacuously true.[24]

Inductive Hypothesis. Let $k \in \mathbb{N}$, and assume that every graph on $2k$ vertices and $k^2 + 1$ edges contains a triangle.

Induction Step. We aim to prove that every graph on $2(k+1)$ vertices and $(k+1)^2+1$ edges contains a triangle. Among our $2k + 2$ vertices, choose any two which are connected by an edge, and call these u and v. The other $2k$ vertices form a graph of their own (let's call this graph H), with a certain number of edges going between these vertices.

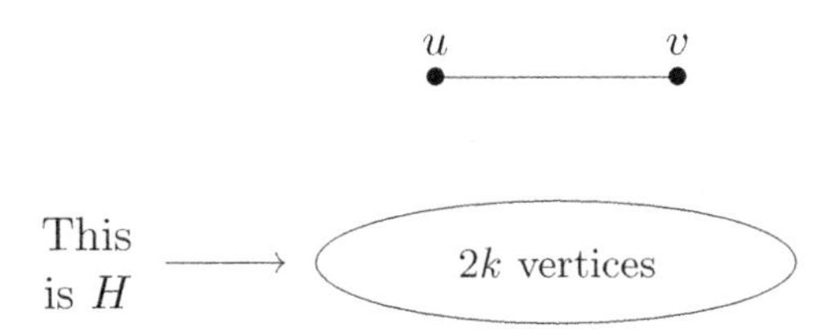

If the $2k$ vertices in H have at least $k^2 + 1$ edges between them, then by the inductive hypothesis there must be a triangle among these vertices!

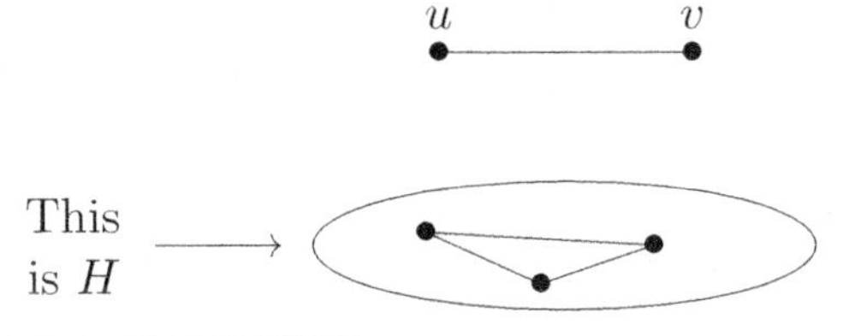

[24] If this is unsatisfying, then we could instead add the condition "for $n \geq 2$" to the theorem so that our base case starts at $n = 2$. And in this case, we are asking about 4 vertices and 5 edges. The fact that this guarantees a triangle was the example that we discussed one page ago.

If this happens, we are done! We have found our triangle! What if there are not $k^2 + 1$ edges among these $2k$ vertices? Then there are at most k^2 edges down there. We had assumed our graph contained $(k+1)^2 + 1$ edges, so there must be at least

$$(k+1)^2 + 1 - k^2 = 2k + 2$$

edges to go. One of these is between u and v, so the remaining $2k+1$ must look like this: one of their vertices is u or v, and the other is one of the $2k$ vertices in H. And by the pigeonhole principle, with $2k+1$ edges like this, but only $2k$ vertices in H, at least one vertex in H (call it w) must be a part of *two* of these edges — which can only mean that both u and v are connected to w. Thus, u, v and w form a triangle.

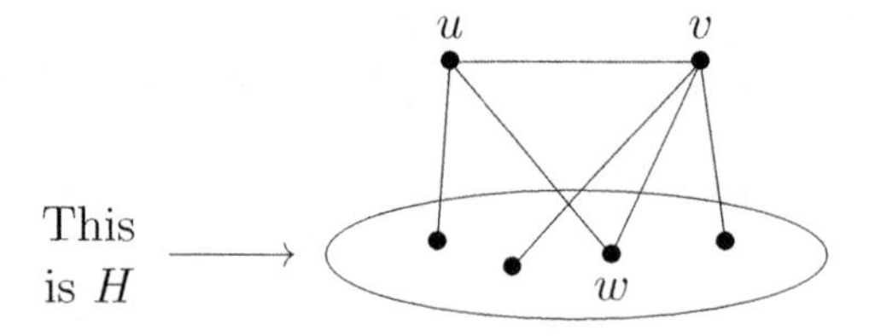

Conclusion. Therefore, by induction, Mantel's theorem holds for every $n \in \mathbb{N}$. $\square$

— Chapter 4 Pro-Tips —

- While it is common at this point of your math journey to carefully label your base case, inductive hypothesis, induction step and conclusion, you'll notice that in later courses that some of these habits will be relaxed. And in a math research paper you would never see "Inductive Hypothesis" and such. In fact, it's even the case that in math papers for which the base case is trivial, that the base case isn't even mentioned. I am aware of one very good combinatorics researcher, who has used induction in a lot of his papers, and who told me once that it is always his goal to set up his induction just right so that the base case is vacuously true — not even trivially true, he aims for *vacuously* true each time.[25] I imagine that getting that just right each time is more effort than simply working out his base cases... but I very much respect his conviction.

- There are some generalizations and extensions of induction. We discussed strong induction and multiple base cases, but here are more:

  - If a theorem holds for every $n \in \mathbb{Z}$, then you may have to perform two inductions: One for the positive values and one for the negative values. For instance, if you prove that S_0 holds, and you prove that $S_k \Rightarrow S_{k+1}$ for every $k \in \{0, 1, 2, 3 \dots\}$, and you prove that $S_k \Rightarrow S_{k-1}$ for every $k \in \{0, -1, -2, -3, \dots\}$, then combined this would prove that S_n holds for every $n \in \mathbb{Z}$. Cool![26]
  - It turns out that this "backwards" technique does not only work with negative numbers, it also provides a second way to prove the positive case. Here's how: Suppose you wish to prove a result for every $n \in \mathbb{N}$. If you can prove that it is true for an infinite sequence of base cases, like for the $n \in \{1, 2, 4, 8, 16, 32, 64, 128, \dots\}$, and you can prove that for every $k \in \mathbb{N}$ with $k \geq 2$, that $S_k \Rightarrow S_{k-1}$, then every case must hold. For example, why is the $n = 60$ case true? Well, the $n = 64$ case was one of the base cases that was proven to hold, and therefore by backwards induction, $S_{64} \Rightarrow S_{63} \Rightarrow S_{62} \Rightarrow S_{61} \Rightarrow S_{60}$, which shows that the $n = 60$ case holds. As long as you can show some infinite sequence works, then you can backwards induct to any case. This technique is called *backwards induction.*
  - Suppose you want to prove something only for all $n \in \{1, 2, 3, \dots, 100\}$. Can you still use induction? You can! Your base case would be $n = 1$, in your inductive hypothesis you would assume the result for any $k \in \{1, 2, 3, \dots, 99\}$, and then in your induction step you would show that $S_k \Rightarrow S_{k+1}$. In this way, induction can also be used to prove a result in finitely many cases.
  - There are also times when you have to perform two or more inductions within the same proof where each induction is on a different variable. And

[25] As a reminder of what it means to be *vacuously true*, see Footnote 5 on page 101

[26] Think about that: A doubly infinite sequence of dominoes!

to answer your first question, yes, it can get really confusing. There are also many ways to do these, but here is perhaps the simplest: If $S_{1,1}$ is true, and $S_{m,n} \Rightarrow S_{m+1,n}$ for any $m, n \in \mathbb{N}$, and $S_{m,n} \Rightarrow S_{m,n+1}$ for any $m, n \in \mathbb{N}$, then $S_{m,n}$ is true for all $m, n \in \mathbb{N}$. Cool![27] I once saw an instance in which the proof of the base case was itself a proof by induction!

- There is a fascinating extension of induction called *transfinite induction.* One of its early successes was in 1904 when Ernst Zermelo proved that every set can be well-ordered, which is one of the really cool results that every mathematician should know. Go look it up! (Its statement may read a little confusing at this point in your math career... but that will serve as motivation to take more math courses!)

- Pictures can go a long way in aiding your proofs, but you should still appreciate their limitations. Maybe your picture is capturing only the "nicest" scenario of your problem? Your proofs should highlight the main ideas of your proof — they should not replace them.

- Sometimes in math, it is easier to prove a stronger result. For example, suppose you wish to prove by induction that

$$\sum_{\ell=1}^{n} \frac{1}{\ell^2} \leq 2$$

for every $n \in \mathbb{N}$. The base case boils down to $1 \leq 2$, which you'll be interested to learn is true. The inductive hypothesis will be the assumption that

$$\sum_{\ell=1}^{k} \frac{1}{\ell^2} \leq 2,$$

for some $k \in \mathbb{N}$, and is used in the induction step to produce this:

$$\begin{aligned}\sum_{\ell=1}^{k+1} \frac{1}{\ell^2} &= \left(\sum_{\ell=1}^{k} \frac{1}{\ell^2}\right) + \frac{1}{(k+1)^2} \\ &\leq 2 + \frac{1}{(k+1)^2}.\end{aligned}$$

But that's not good enough! We need $\sum_{\ell=1}^{k+1} \frac{1}{\ell^2} \leq 2$ for a successful induction...

Ok, so let's make our job seemingly harder. Let's prove by induction the stronger result that

$$\sum_{\ell=1}^{n} \frac{1}{\ell^2} \leq 2 - \frac{1}{n}$$

[27]Think about that: An infinite grid of dominoes!

for every $n \in \mathbb{N}$. The base case still works: it boils down to $1 \leq 2 - 1$, which is again true. The inductive hypothesis now will be the assumption that

$$\sum_{\ell=1}^{k} \frac{1}{\ell^2} \leq 2 - \frac{1}{k}$$

for some $k \in \mathbb{N}$, and is used in the induction step to produce this:

$$\begin{aligned}
\sum_{\ell=1}^{k+1} \frac{1}{\ell^2} &= \left(\sum_{\ell=1}^{k} \frac{1}{\ell^2}\right) + \frac{1}{(k+1)^2} \\
&\leq \left(2 - \frac{1}{k}\right) + \frac{1}{(k+1)^2} \\
&= 2 - \frac{(k+1)^2}{k(k+1)^2} + \frac{k}{k(k+1)^2} \\
&= 2 - \frac{(k^2+2k+1) - k}{k(k+1)^2} \\
&= 2 - \frac{k^2+k+1}{k(k+1)^2} \\
&< 2 - \frac{k^2+k}{k(k+1)^2} \\
&= 2 - \frac{k(k+1)}{k(k+1)^2} \\
&= 2 - \frac{1}{k+1}.
\end{aligned}$$

Thus, by induction, we have succeeded in showing that $\sum_{\ell=1}^{n} \frac{1}{\ell^2} \leq 2 - \frac{1}{n}$ for every $n \in \mathbb{N}$. □

Why did this work? How could we have failed to prove something easier, and then did the exact same thing with a harder problem and succeeded? It comes down to the inductive hypothesis; proving a harder result allows us to assume more in the inductive hypothesis, which was needed in the induction step. There are other examples of this in mathematics, and it's certainly not limited to proofs by induction. Sometimes adding additional criteria can help you see what's really going on.

By the way, this result also implies that the infinite version of this sum, $\sum_{\ell=1}^{\infty} \frac{1}{\ell^2}$, must also be at most 2. But what does it equal? At age 24, the great Leonhard Euler proved the remarkable answer:

$$\sum_{\ell=1}^{\infty} \frac{1}{\ell^2} = \frac{\pi^2}{6}.$$

— Exercises —

Exercise 4.1. Prove that the sum of the first n odd natural numbers equals n^2 by induction or strong induction.

Exercise 4.2. Provide three proofs that if $n \in \mathbb{N}$, then $n^2 - n$ is even.

(a) Prove it by cases, by considering the "n is even" and "n is odd" cases.

(b) Prove it by applying Proposition 4.2 to the sum $1 + 2 + 3 + \cdots + (n-1)$.

(c) Prove it by induction.

Exercise 4.3. Use induction or strong induction to prove that the following hold for every $n \in \mathbb{N}$.

(a) $3 \mid (4^n - 1)$

(b) $6 \mid (n^3 - n)$

(c) $9 \mid (3^{4n} + 9)$

(d) $5 \mid (n^5 - n)$

(e) $6 \mid (5^{2n} - 1)$

(f) $5 \mid (6^n - 1)$

Exercise 4.4. Prove that each of the following hold for every $n \in \mathbb{N}$.

(a) $1^2 + 2^2 + 3^2 + \cdots + n^2 = \dfrac{n(n+1)(2n+1)}{6}$

(b) $1^3 + 2^3 + 3^3 + \cdots + n^3 = \dfrac{n^2(n+1)^2}{4}$

(c) $1 \cdot 2 + 2 \cdot 3 + 3 \cdot 4 + \cdots + n \cdot (n+1) = \dfrac{n(n+1)(n+2)}{3}$

(d) $1 \cdot 3 + 2 \cdot 4 + 3 \cdot 5 + \cdots + n \cdot (n+2) = \dfrac{n(n+1)(2n+7)}{6}$

(e) $1^3 + 2^3 + 3^3 + \cdots + n^3 = (1 + 2 + 3 + \cdots + n)^2$

(f) $1 \cdot 1! + 2 \cdot 2! + 3 \cdot 3! + \cdots + n \cdot n! = (n+1)! - 1$

(g) $2^0 + 2^1 + 2^2 + 2^3 + \cdots + 2^n = 2^{n+1} - 1$

(h) $3^1 + 3^2 + 3^3 + \cdots + 3^n = \dfrac{3^{n+1} - 3}{2}$

(i) $4^0 + 4^1 + 4^2 + 4^3 + \cdots + 4^n = \dfrac{4^{n+1} - 1}{3}$

(j) $\dfrac{1}{2!} + \dfrac{2}{3!} + \dfrac{3}{4!} + \cdots + \dfrac{n}{(n+1)!} = 1 - \dfrac{1}{(n+1)!}$

(k) $\dfrac{1}{1 \cdot 2} + \dfrac{1}{2 \cdot 3} + \dfrac{1}{3 \cdot 4} + \cdots + \dfrac{1}{n \cdot (n+1)} = \dfrac{n}{n+1}$

Exercise 4.5. Prove that each of the following holds for every $n \in \mathbb{N}$. Also, before each proof, pick three specific n-values and verify that the result holds for those values.

(a) $n + 2 < 4n^2$

(b) $\frac{1}{\sqrt{1}} + \frac{1}{\sqrt{2}} + \frac{1}{\sqrt{3}} + \cdots + \frac{1}{\sqrt{n}} \leq 2\sqrt{n} - 1$

(c) $1 + \frac{n}{2} \leq \frac{1}{1} + \frac{1}{2} + \frac{1}{3} + \frac{1}{4} + \frac{1}{5} + \cdots + \frac{1}{2^n - 1} + \frac{1}{2^n}$

(d) $2^n \leq 2^{n+1} - 2^{n-1} - 1$

(e) $1 + 2^n \leq 3^n$

(f) $4^{n+4} \geq (n+4)^4$

Exercise 4.6. Make a conjecture as to which $n \in \mathbb{N}$ have the property that $2^n < n!$. Then prove your conjecture by induction or strong induction.

Exercise 4.7. Prove that $n^2 < 3^n$ for every $n \in \{0, 1, 2, 3, \dots\}$.

Exercise 4.8. The *Fermat number* $\tilde{F}_n$ is defined to be $\tilde{F}_n = 2^{2^n} + 1$ for $n \geq 0$. For example, $\tilde{F}_3 = 2^{2^3} + 1 = 2^8 + 1 = 256 + 1 = 257$. The first five are $\tilde{F}_0 = 3$, $\tilde{F}_1 = 5$, $\tilde{F}_2 = 17$, $\tilde{F}_3 = 257$ and $\tilde{F}_4 = 65537$. These are all prime numbers, which lead Pierre de Fermat to conjecture that all the Fermat numbers are prime. As it turns out, $\tilde{F}_5$ is not prime (as Euler showed), and so far there is no known prime Fermat number after $\tilde{F}_4$. (Oops.)

Prove that, for every $n \in \mathbb{N}$,

$$\tilde{F}_1 \cdot \tilde{F}_2 \cdot \tilde{F}_3 \cdots \tilde{F}_n = \tilde{F}_{n+1} - 2.$$

Exercise 4.9. If your friend Lexi asked you to explain the difference between deductive reasoning and inductive reasoning, what would you tell her? Feel free to look up definitions online before writing your explanation.

Exercise 4.10. Explain the error in the "proof" of Fake Proposition 4.11.

Exercise 4.11. Explain the error in the "proof" of Fake Proposition 4.12.

Exercise 4.12. Determine the smallest M such that every $n \in \mathbb{N}$ with $n \geq M$ can be written as $3a + 7b$ for some natural numbers a and b. Then, prove that you are correct.

Exercise 4.13. Find a formula for the sum

$$2 + 4 + 6 + \cdots + 2n,$$

where $n \in \mathbb{N}$. Then, prove that your formula works in two different ways. First, by using Proposition 4.2. Second, by induction.

Exercise 4.14. Find a formula for the sum

$$m + (m+1) + (m+2) + \cdots + n,$$

where $n \in \mathbb{N}$. Then, prove that your formula works in two different ways. First, by using Proposition 4.2. Second, by induction.

Exercise 4.15. Use induction to prove that if A is a set and $|A| = n$, then $|\mathcal{P}(A)| = 2^n$.

Exercise 4.16. Suppose $x \in \mathbb{R}$ with $x > -1$. Prove that, for every $n \in \mathbb{N}$,

$$1 + nx \leq (1+x)^n.$$

Exercise 4.17. Prove that, for every $n \in \mathbb{N}$, there are n distinct natural numbers $a_1, a_2, \ldots, a_n$ such that $a_1^2 + a_2^2 + \cdots + a_n^2$ is a perfect square.

Exercise 4.18. Suppose $A_1, A_2, A_3, \ldots, A_n$ are subsets of some universal set U. Prove that the following hold for every $n \in \mathbb{N}$. You may use the fact that unions and intersections are associative.

(a) $\left(A_1 \cap A_2 \cap A_3 \cap \cdots \cap A_n\right)^c = A_1^c \cup A_2^c \cup A_3^c \cup \cdots \cup A_n^c$

(b) $\left(A_1 \cup A_2 \cup A_3 \cup \cdots \cup A_n\right)^c = A_1^c \cap A_2^c \cap A_3^c \cap \cdots \cap A_n^c$

Exercise 4.19. Where is the mistake in the following "fake proof" that $2n = 0$ for all $n \in \{0, 1, 2, 3, 4, \ldots\}$?

Fake Proof. The base case is when $n = 0$, and indeed $2n = 2(0) = 0$, as desired, when $n = 0$.

Since we are using strong induction, our inductive hypothesis is the assumption that $2m = 0$ for all $m \in \{0, 1, 2, \ldots, k\}$, and we wish to show that $2(k+1) = 0$.

In the induction step, we choose to write $k + 1 = a + b$ for some smaller a and b from $\{0, 1, 2, \ldots, k\}$. For example, you could use $a = k$ and $b = 1$, or you could use any other a and b that work; this is just like in the proof of Proposition 4.9, where we broke up a chocolate bar with $k + 1$ pieces into two parts, containing a and b pieces, respectively. But no matter how you break it up, since a and b are smaller, the inductive hypothesis tells us that $2a = 0$ and $2b = 0$, and hence

$$2(k+1) = 2(a+b) = 2a + 2b = 0 + 0 = 0.$$

Thus, by strong induction, $2n = 0$ for all $n \in \{0, 1, 2, 3, \ldots\}$.

Exercise 4.20.

(a) Find a way to place the numbers $1, 2, 3, 4, 5, 6$ into the boxes below, so that the displayed inequalities are all correct.

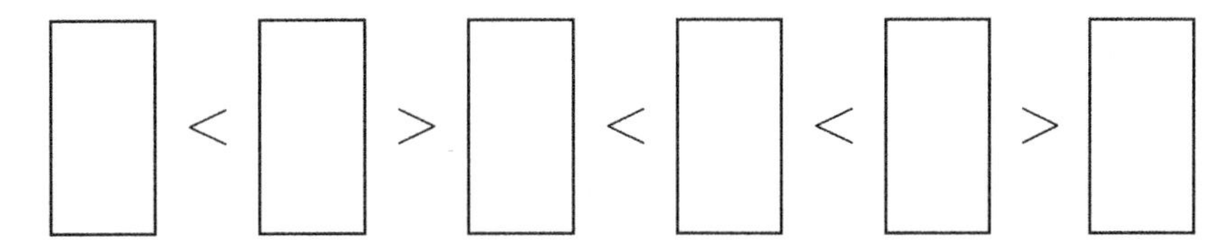

(b) Suppose $n \in \mathbb{N}$. Prove that if one has n boxes like this, with any $n-1$ inequalities between them, then it is always possible to place the numbers $1, 2, 3, \ldots, n$ into these boxes so that the inequalities are all correct.

Exercise 4.21.

(a) Suppose $n \in \mathbb{N}$, $a_1, a_2, \ldots, a_n$ are positive integers at least 2, and p is a prime. Use Lemma 2.17 and induction to prove that if

$$p \mid (a_1 \cdot a_2 \cdot a_3 \cdot \ldots \cdot a_n),$$

then $p \mid a_i$ for some i.

(b) In Theorem 4.8 we proved the fundamental theorem of arithmetic. Prove that if $n \geq 2$ is an integer, then it has a *unique* prime factorization in the sense that if

$$n = p_1 p_2 \cdots p_k \qquad \text{and} \qquad n = q_1 q_2 \cdots q_\ell$$

where each p_i and q_j is a prime, then there are the same number of primes in each list ($k = \ell$) and in fact the primes $p_1, p_2, \ldots, p_k$ are the same as the primes $q_1, q_2, \ldots, q_\ell$, perhaps just in a different order.

Exercise 4.22. Some forms of the fundamental theorem of arithmetic include the assertion that every integer $n \geq 2$ can be written *uniquely* as a product of primes. Explain why, if we included 1 as a prime number, this would no longer be true.

Exercise 4.23. In this exercise you will use strong induction to study sequences which are defined *recursively*.

(a) Define a sequence $a_1, a_2, a_3, \ldots$ recursively where $a_1 = 1$, $a_2 = 3$, and for $n \geq 3$, $a_n = 2a_{n-1} - a_{n-2}$. Prove that $a_n = 2n - 1$ for all $n \in \mathbb{N}$.

(b) Define a sequence $a_1, a_2, a_3, \ldots$ recursively where $a_1 = 1$, $a_2 = 4$, and for $n \geq 3$, $a_n = 2a_{n-1} - a_{n-2} + 2$. Through scratch work, conjecture a formula for a_n, and then prove that your conjecture is correct.

(c) Define a sequence $a_1, a_2, a_3, \ldots$ recursively where $a_1 = 1$, $a_2 = 2$, and for $n \geq 3$, $a_n = a_{n-1} + 2a_{n-2}$. Through scratch work, conjecture a formula for a_n, and then prove that your conjecture is correct.

Note. Recall from Chapter 1 that a *counterexample* is a specific example showing that a conjecture is false.

Exercise 4.24. The following conjecture is false. Prove that it is false by finding a counterexample.

(a) Conjecture: For every positive integer n,

$$1 + \frac{1}{2} + \frac{1}{3} + \frac{1}{4} + \cdots + \frac{1}{n} < 3.$$

Exercise 4.25. Let $n \in \mathbb{N}$. Suppose that an equilateral triangle is cut into 4^n congruent equilateral triangles, but with the top corner removed. For example, the $n = 1$ and $n = 2$ cases are here:

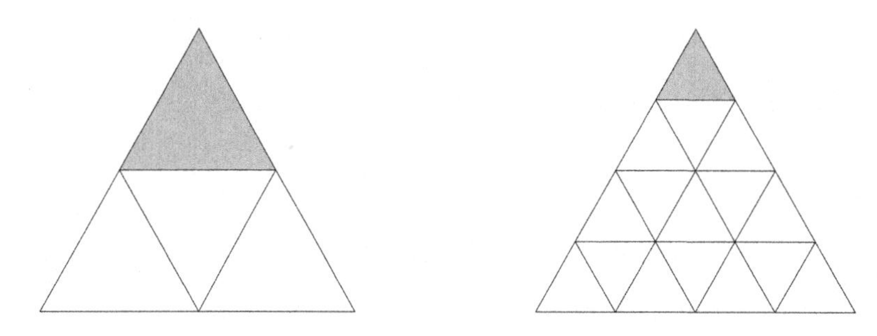

Prove that the remaining $4^n - 1$ triangles can be perfectly covered using tiles of this shape: ⏢. As usual, you are allowed to rotate (and flip) these tiles as you please. Also, just to be clear, these tiles will be properly sized for each n, so that they cover three triangles. For $n = 1$, a single tile will cover all of the non-removed squares, and for $n = 2$ you will need five tiles.

Exercise 4.26. Let P be any polygon in the plane. Prove that it is possible to divide P into triangles, all of whose vertices are vertices of P. For example:

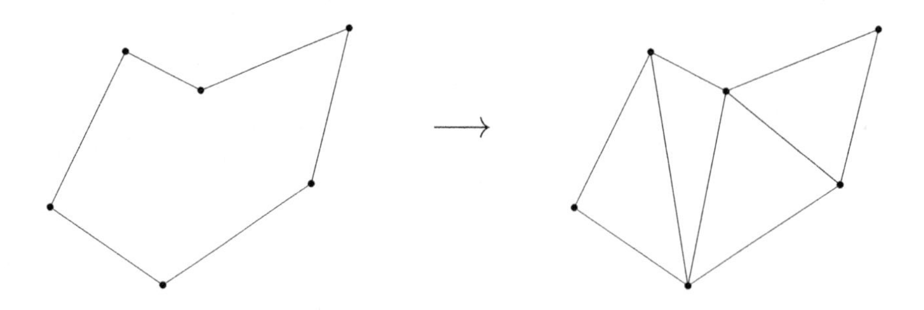

Exercise 4.27. Prove that for any natural numbers a and b, there exists a natural number m such that $mb > a$. This is a version of the so-called *Archimedean principle.*

Exercise 4.28. In chess, a rook attacks all the squares in its row and column. Consider the problem of placing n non-attacking rooks on an $n \times n$ chessboard; that is, n rooks such that none attack any other. One way to do this is to place the rooks on a single diagonal.

But that's boring. Prove that for every $n \geq 4$, it is possible to place n non-attacking rooks on the $n \times n$ chessboard so that none of the rooks are on either (or both) of the two diagonals.

Exercise 4.29. A *magic square* is an $n \times n$ matrix where the sum of the entries in each row, column and diagonal equal the same value. For example,

8	1	6
3	5	7
4	9	2

is a 3×3 matrix whose three rows, three columns, and two diagonals each sum to 15. Thus, this is a magic square.

An *antimagic square* is an $n \times n$ matrix where each row, column and diagonal sums to a distinct value. For example,

9	4	5
10	3	-2
6	9	7

is a 3×3 matrix whose rows sum to 18, 11 and 22, columns sum to 25, 16 and 10, and diagonals sum to 19 and 14. Notice that all eight of these numbers are different than the rest, showing that this is an antimagic square.

Prove that, for every integer $n \geq 2$, there exists an $n \times n$ antimagic square all of whose entries are positive integers.

Exercise 4.30. Read the introduction to graphs on Page 24. A graph is called a *tree* if it can be drawn so that it branches upwards and none of its branches intersect. Here are two examples:

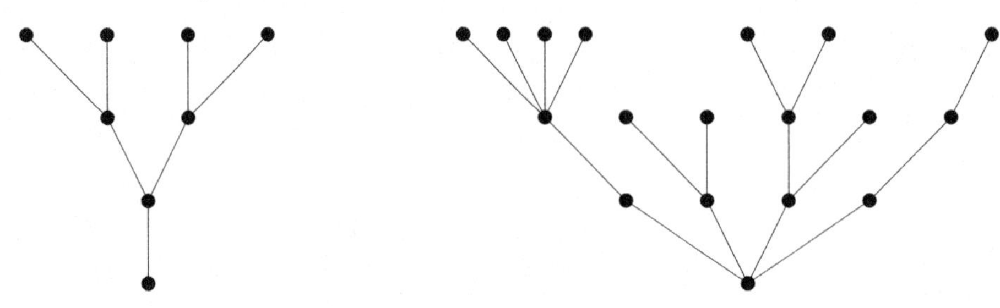

Prove that if a tree has n vertices, then it has $n-1$ edges.

Exercise 4.31. Read the *Introduction to Sequences* following this chapter, and prove the following hold for every $n \in \mathbb{N}$.

(a) $F_1 + F_2 + F_3 + \cdots + F_n = F_{n+2} - 1$

(b) $F_1 + F_3 + F_5 + \cdots + F_{2n-1} = F_{2n}$

(c) $(F_{n+1})^2 - F_{n+1}F_n - (F_n)^2 = (-1)^n$

(d) If $a = F_n F_{n+3}$, $b = 2F_{n+1}F_{n+2}$, and $c = (F_{n+1})^2 + (F_{n+2})^2$, then $a^2 + b^2 = c^2$

(e) $F_n = \dfrac{1}{\sqrt{5}}\left(\dfrac{1+\sqrt{5}}{2}\right)^n - \dfrac{1}{\sqrt{5}}\left(\dfrac{1-\sqrt{5}}{2}\right)^n$

(f) F_{3n} is even, F_{3n+1} is odd, and F_{3n+2} is odd

(g) $\gcd(F_n, F_{n+1}) = 1$

(h) $F_{n+6} = 4F_{n+3} + F_n$

— Open Question —

This chapter reintroduced you to graphs, and Exercise 4.30 introduced you to *trees*, which are graphs which can be drawn so that they branch upwards and none of their branches intersect. Here are two examples:

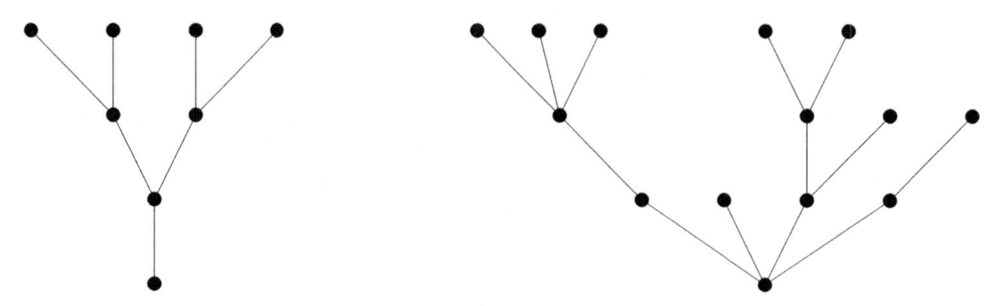

Let's now decorate the two trees above. Suppose we label the vertices and edges like this:

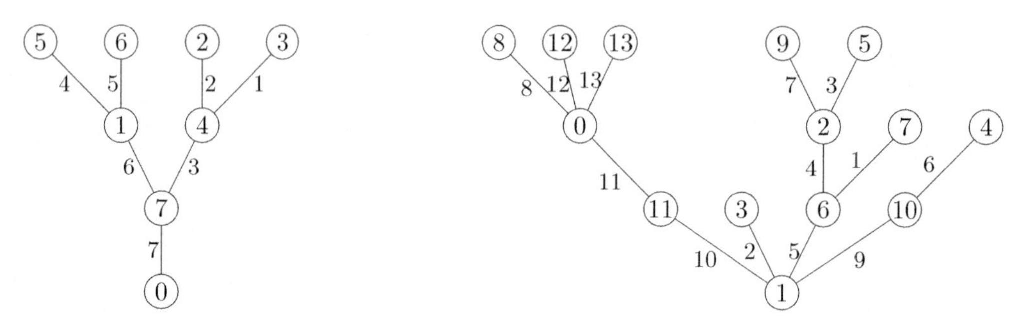

We labeled these graphs by following three rules.

1. If the graph has m edges, then we label the vertices with the numbers from the set $\{0, 1, 2, 3, \ldots, m\}$, in such a way that each vertex gets a different label.[28]

2. If an edge connects vertices with labels a and b, then we label that edge with the number $|a - b|$. (Look at each graph above and verify that this was done.)

3. We must choose these labels so that no two edges receive the same label.

Rule 3 is the crucial one. Anyone could assign numbers to vertices and find their differences, but can you do so in such a way that no two edges receive the same label? That's tough. A labeling following these three rules is called a *graceful labeling*.

Not every graph has a graceful labeling. For example, there is no graceful labeling of the 5-cycle graph, which looks like this: . Let's prove this. This graph has 5 edges, so the vertex labels must come from $\{0, 1, 2, 3, 4, 5\}$; thus, let's suppose we have labeled the vertices using numbers a, b, c, d and e, which all came from this set. Recall also that if two vertices are labeled x and y, then the edge between them is

[28]Note that the first graph has 7 edges, so its vertex labels came from the set $\{0, 1, 2, 3, 4, 5, 6, 7\}$. And the second graph has 13 edges, so its vertex labels came from the set $\{0, 1, 2, 3, \ldots, 13\}$.

labeled $|x - y|$. Adding in these edge labels too, here is the picture:

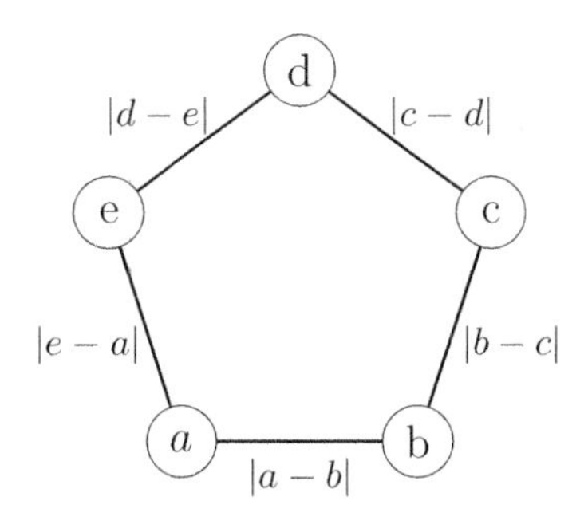

To show that this graph can not possibly have a graceful labeling, we must show that it is impossible for all of those five edge labels to be different. To do so, let's consider the sum of these edge labels:

$$|a - b| + |b - c| + |c - d| + |d - e| + |e - a|.$$

At the end, we will only care whether this sum is even or odd, so let's think about this sum modulo 2. Note that if n is an integer, then $n \equiv -n \pmod 2$ (because $2 \mid (n - (-n))$). Since negative numbers are equivalent to their positive values when considered modulo 2, this means that $|n| \equiv n \pmod 2$. Applying this to each edge label,

$$\begin{aligned} |a - b| + |b - c| &+ |c - d| + |d - e| + |e - a| \\ &\equiv a - b + b - c + c - d + d - e + e - a \pmod 2 \\ &\equiv 0 \pmod 2. \end{aligned}$$

This shows that the sum of all of the edge labels must be congruent to 0 modulo 2, which means that the sum of the edge labels is an even number.

Next, note that since the vertex labels are from $\{0, 1, 2, 3, 4, 5\}$, the biggest possible edge label is 5 (if a vertex labeled 5 connects to one labeled 0), and the smallest possible edge label is 1 (if a vertex labeled 1 connects to a vertex labeled 2, or a 2-vertex to a 3-vertex, etc.). So if a graceful labeling did exist, the five edges would be labeled with the numbers 1, 2, 3, 4 and 5. Thus, if the labeling were a graceful labeling, then adding up the edge labels is simply

$$1 + 2 + 3 + 4 + 5 = 15.$$

We concluded that the sum of the edge labels had to be an even number, and then we showed that if the labeling is graceful, the sum had to equal 15. Since 15 is not even, this shows that a graceful labeling of ⬠ is impossible.

We just showed that not every graph has a graceful labeling. Are there certain types of graphs which are guaranteed to have a graceful labeling? Do all trees have a graceful labeling? This is unknown, and is our open question.

Open Question. Does every tree have a graceful labeling?

Introduction to Sequences

A *sequence* of numbers is simply an infinite list of numbers, like

$$1\ ,\ 4\ ,\ 9\ ,\ 16\ ,\ 25\ ,\ 36\ ,\ 49\ ,\ 64\ ,\ 81\ ,\ \ldots.$$

You may be surprised to learn that the study of sequences of numbers is an important area of mathematics. A combinatorist[29] cares deeply about sequences of finite discrete structures, which in turn means they care deeply about numerical sequences. Your study of real analysis, in which you will prove the soundness of everything you did in calculus, will likely begin with the study of numerical sequences; they are particularly important when studying series and continuity. In topology, sequences are used to describe *compactness*, a very important idea. In abstract algebra, you may study *exact sequences*, although these are sequences of structures, not of numbers. And in set theory you may study generalizations of sequences which are *ordinal*-indexed. So yes, the simple idea of a sequence is surprisingly useful.

Sequences are also really interesting! Let's discuss a sequence that has been discovered an rediscovered many times, including from the study of a simple biological model. Suppose you have a pair of newly-born rabbits, one female and one male.

This simple model allows for something to happen each month: The rabbits mate once a month, beginning at one month old,[31] which results in the female rabbit giving birth to a pair of new rabbits each month beginning on her second month. Each new pair will contain one male and one female, which repeat this process, and will start producing a pair of new rabbits every month beginning on their second month.[32] So,

[29] I am on a lifetime mission to make the term for someone-who-does-combinatorics be "combinatorist", rather than the incorrect, yet sadly more common, "combinatorialist". There are not algebraicists or topologicalists, and there should also not be combinatorialists. If a compromise is needed, another acceptable title is "combinatron." Other <u>un</u>acceptable titles include "combinator," "combiner," "counter," and "cucumber."[30]

[30] But if we let the combinatorists to go by something as cool as "combinatrons," then it's only fair to let the number theorists be called "primatologists" and the algebraists be called "groupies."

[31] Rabbits actually begin mating at between 3 and 6 months of age, which is still... really young.

[32] While writing this I have also learned that rabbit inbreeding is not uncommon, although you probably shouldn't breed siblings. Also, a male rabbit is a buck, while a female is doe—just like with deer. A father rabbit is a sire, while a mother rabbit is a dam—just like with horses.

using each level of the tree to represent one month, we get this reproduction tree:

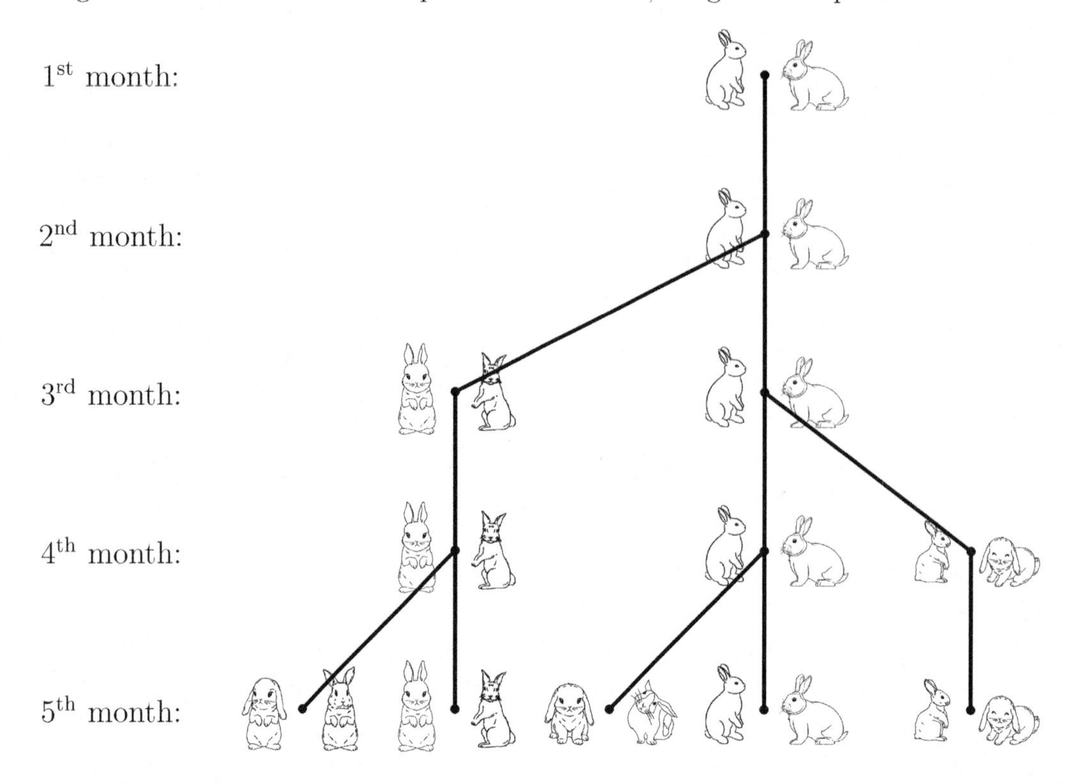

At the start there was 1 pair of rabbits. After the first month, there is still 1 pair. After the second month, 2 pairs. After the third, 3 pairs. Then 5 pairs. And if we extended this tree another few months, we would find 8 pairs, then 13 pairs, then 21 pairs. If we assume the rabbits never die (and never get tired...), then this pattern generates a sequence known as the *Fibonacci sequence*. Here are the first terms:

$$1\ ,\ 1\ ,\ 2\ ,\ 3\ ,\ 5\ ,\ 8\ ,\ 13\ ,\ 21\ ,\ 34\ ,\ 55\ ,\ \dots.$$

Interestingly, the Fibonacci sequence first appeared in the study of poetry. Indian mathematicians asked how many forms a certain class of poems can take, and the answer was a sequence of integers — the Fibonacci sequence. The sequence is named after Fibonacci[33] who independently discovered it when he sought to answer the exact rabbit problem we discussed above.

The Fibonacci sequence has many interesting properties[34] which do a good job of highlighting some of the fascinating paths on which a single sequence can guide you.

[33]Fun fact: Fibonacci's name was actually Leonardo Pisano (the surname just indicating that he was from Pisa). So why is he called Fibonacci? It's actually an accident of history. In 1202, "Leonardo" wrote an enormously influential book called *Liber Abaci*, but because this work predated the printing press, it was entirely handwritten. Now, Leonardo's father's name was Guglielmo Bonaccio, and in the title of *Liber* he included the words "filius Bonacci," meaning "son of Bonaccio." But when scholars were later studying this text, they misinterpreted that as being his surname. Eventually, "filius Bonacci" was reduced to "Fibonacci," and so, long after his death, his name was changed for history.

[34]And it also has many interesting jokes! For example, this Fibonacci joke is as funny as the last two Fibonacci jokes you've heard combined.

Patterns Inside Patterns

The Fibonacci sequence is denoted by $F_1, F_2, F_3, F_4, \ldots$, where F_1 and F_2 both equal 1, and every term thereafter is the sum of the previous two:

$$F_n = F_{n-1} + F_{n-2}$$

for $n = 3, 4, 5, 6, \ldots$.

By its very definition, this sequence has a summation property. Here's one way to map that, where we highlight the fact that 5 is the sum of the previous two Fibonacci numbers, 2 and 3.

$$1\ ,\ 1\ ,\ 2\ ,\ 3\ ,\ \boxed{5}\ ,\ 8\ ,\ 13\ ,\ 21\ ,\ 34\ ,\ 55\ ,\ \ldots$$

$$1\ ,\ 1\ ,\ \boxed{2+3}\ ,\ 5\ ,\ 8\ ,\ 13\ ,\ 21\ ,\ 34\ ,\ 55\ ,\ \ldots$$

Now, for a sequence defined by one property, you might not expect other properties would pop up too. But this is math after all, so usually when one interesting thing happens, many more follow in its wake. For example, what happens when we compare the Fibonacci sequence to the square of the Fibonacci sequence?

$$1\ ,\ 1\ ,\ 2\ ,\ 3\ ,\ 5\ ,\ 8\ ,\ 13\ ,\ 21\ ,\ 34\ ,\ 55\ ,\ \ldots$$

$$1\ ,\ 1\ ,\ 4\ ,\ 9\ ,\ 25\ ,\ 64\ ,\ 169\ ,\ 441\ ,\ 1156\ ,\ 3025\ ,\ \ldots$$

For a sequence defined by *sums*, introducing *products* into the mix seems suspect. And yet, some really surprising stuff happens when we do. For example, what do we get if we add together the first k terms of the square sequence (for various k)? For example,

$$\begin{aligned}
&1 + 1 + 4 = 6 \\
&1 + 1 + 4 + 9 = 15 \\
&1 + 1 + 4 + 9 + 25 = 40 \\
&1 + 1 + 4 + 9 + 25 + 64 = 104.
\end{aligned}$$

If you squint long enough at 6, 15, 40 and 104, you might notice some Fibonacci numbers hidden within. Do you see them?

$$\begin{aligned}
&6 = 2 \times 3 \\
&15 = 3 \times 5 \\
&40 = 5 \times 8 \\
&104 = 8 \times 13.
\end{aligned}$$

For example:

$$1\ ,\ 1\ ,\ 2\ ,\ 3\ ,\ 5\ ,\ (8\times 13)\ ,\ 21\ ,\ 34\ ,\ 55\ ,\ \ldots$$

$$(1+1+4+9+25+64)\ ,\ 169\ ,\ 441\ ,\ 1156\ ,\ 3025\ ,\ \ldots$$

But why is this the case? Let's focus on the example above, that

$$1+1+4+9+25+64 = 8\times 13.$$

One way to picture the squares of the Fibonacci sequence (1, 1, 4, 9, 25, 64, ...) is by imagining geometric squares of dimensions

$$1\times 1\ ,\ 1\times 1\ ,\ 2\times 2\ ,\ 3\times 3\ ,\ 5\times 5\ ,\ 8\times 8\ ,\ \ldots.$$

Now, both the Fibonacci sequence as well as the square of the Fibonacci sequence are appearing: The side lengths of these shapes are from the actual Fibonacci sequence, while the *areas* of these shapes are from the square of the Fibonacci sequence.

Start by placing two 1×1 squares next to each other.

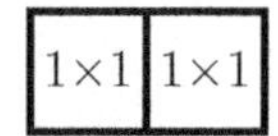

Then, since the next term in the Fibonacci sequence is $1+1=2$, the 2×2 square can sit perfectly flush against them. This then allows for the 3×3 square to fit perfectly to the right of these squares.

1×1 1×1
3×3
2×2

This in turn allows the 5×5 square to fit below them, and then the 8×8 square to the right.

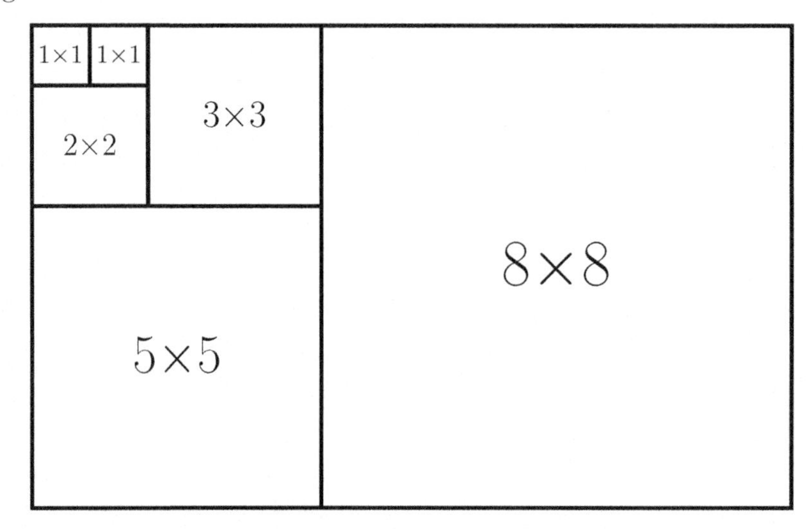

What is the area of this entire shape? On one hand, it is the sum of the areas of each of these squares, which is

$$1 + 1 + 4 + 9 + 25 + 64.$$

But on the other hand, like any rectangle, its area is width times height, which is $8 \cdot 13$. Thus, since these both answer the same question, they must be equal to each other:

$$1 + 1 + 4 + 9 + 25 + 64 = 8 \cdot 13.$$

This is only one example, but I expect you can envision how what we did is completely generalizable. Indeed, using the above as a template, one can prove that

$$(F_1)^2 + (F_2)^2 + (F_3)^2 + \cdots + (F_n)^2 = F_n \cdot F_{n+1}$$

for every $n \in \mathbb{N}$. And the patterns don't stop there. Given that sequences are indexed by the natural numbers, many of these properties can be proven by induction. Although the above has this nice geometric proof that helps us understand why it's true, the induction proof is quicker. The base case is simply that $(F_1)^2 = F_1 \cdot F_2$, which is true. In the inductive hypothesis we would assume that, for some $k \in \mathbb{N}$,

$$(F_1)^2 + (F_2)^2 + (F_3)^2 + \cdots + (F_k)^2 = F_k \cdot F_{k+1}.$$

In the induction step we would aim to show that it also holds for $k+1$. This can be done by simply applying the inductive hypothesis, factoring out an F_{k+1}, and then using the Fibonacci property that $F_k + F_{k+1} = F_{k+2}$:

$$\begin{aligned}(F_1)^2 + (F_2)^2 + (F_3)^2 + \cdots + (F_k)^2 + (F_{k+1})^2 &= F_k \cdot F_{k+1} + (F_{k+1})^2 \\ &= F_{k+1} \cdot (F_k + F_{k+1}) \\ &= F_{k+1} \cdot F_{k+2}.\end{aligned}$$

There are literally dozens of other patterns hidden in this simple sequence, but let's now turn to another use of sequences.

Generating Functions

You might recall from Calculus II that the geometric series $1 + x + x^2 + x^3 + x^4 + \ldots$ is equal to $\frac{1}{1-x}$, provided $|x| < 1$, which is a much more compact expression. This infinite sum is sometimes called a *power series*. One way to think about this power series is

$$1 + 1x + 1x^2 + 1x^3 + 1x^4 + \ldots.$$

It may seem strange to tack on a 1 to each term, but doing so emphasizes that for the rather boring sequence $1, 1, 1, 1, 1, \ldots$, if you wanted to express this sequence as *the coefficients of a power series*, then the above is how you would do so.

I remember how surprised I was to learn that doing this is amazingly powerful. It seemed to me that taking an infinite list of numbers, like $1, 1, 1, 1, 1, 1, \ldots$, and making them the coefficients of the power series, like $1 + 1x + 1x^2 + 1x^3 + \ldots$, would only be making things more complicated — but because this power series can be written succinctly as $\frac{1}{1-x}$, without any dot-dot-dots, this proves to be remarkably useful, in an almost magical way.

As a more interesting example, let's turn the Fibonacci sequence into a power series. This is the sequence $1, 1, 2, 3, 5, 8, 13, \ldots$, so making these the coefficients of a power series produces

$$1 + x + 2x^2 + 3x^3 + 5x^4 + 8x^5 + 13x^6 + \ldots.$$

Does this also have a compact form? It does! It relies on a little trickery. Let's call $f(x)$ the above power series. Then think about $f(x)$ and $x \cdot f(x)$ and $x^2 \cdot f(x)$:

$$\begin{aligned} f(x) &= 1 + x + 2x^2 + 3x^3 + 5x^4 + 8x^5 + 13x^6 + \ldots \\ xf(x) &= x + x^2 + 2x^3 + 3x^4 + 5x^5 + 8x^6 + 13x^7 + \ldots \\ x^2f(x) &= x^2 + x^3 + 2x^4 + 3x^5 + 5x^6 + 8x^7 + 13x^8 + \ldots \end{aligned}$$

Aligning these terms by the power of x, we can see the cancellation:

$$\begin{array}{rcccccccc} f(x) = & 1 + & x & + 2x^2 & + 3x^3 & + 5x^4 & + 8x^5 & + 13x^6 + & \ldots \\ xf(x) = & & x & + x^2 & + 2x^3 & + 3x^4 & + 5x^5 & + 8x^6 + & \ldots \\ x^2f(x) = & & & x^2 & + x^3 & + 2x^4 & + 3x^5 & + 5x^6 + & \ldots \end{array}$$

Indeed, subtracting the second and the third equations from the first, the Fibonacci property kicks in and we get a whole lot of cancellation. It reduces to just

$$f(x) - xf(x) - x^2f(x) = 1.$$

Remember, $f(x)$ is what we are looking for, and with just a little bit of algebra, we can now find it. By factoring out an $f(x)$, we get $f(x) \cdot (1 - x - x^2) = 1$, and so by dividing we obtain

$$f(x) = \frac{1}{1 - x - x^2}.$$

Isn't that crazy?! The function $\dfrac{1}{1 - x - x^2}$, if written out as a power series, has coefficients which are exactly the Fibonacci sequence:

$$1 + x + 2x^2 + 3x^3 + 5x^4 + 8x^5 + 13x^6 + \ldots.$$

And yet, the function $\dfrac{1}{1 - x - x^2}$ is so simple! We can manipulate it easily with algebra or calculus, which can then tell us new things about the Fibonacci sequence or other sequences. It would take us too far afield to get into the details now, but you will see this magic if you study generating functions in a future course.

Let's now turn to a final topic related to sequences — studying how quickly they grow.

The Golden Ratio

Sequences like 2^n grow *exponentially*. And exponential growth is much, much faster than, say, polynomial growth. For small values of n, a polynomial like n^2 and an exponential like 2^n may not look so different, but zoom out and you'll see the blistering speed of an exponential. Even an exponential like $(1.1)^n$ will, eventually, surpass a polynomial like n^{100}, and over time it will really pull away.[35]

Does the Fibonacci sequence grow exponentially? Notice that for exponential functions like 2^n, the base number is the ratio of consecutive terms: $\frac{2^{n+1}}{2^n} = 2$. For functions which grow like an exponential but aren't precisely of this form, the ratio of consecutive terms *approaches* some fixed number. Let's see what happens when we look at the ratio of consecutive terms of the Fibonacci sequence. As a reference, here's the sequence again:

$$1\ ,\ 1\ ,\ 2\ ,\ 3\ ,\ 5\ ,\ 8\ ,\ 13\ ,\ 21\ ,\ 34\ ,\ 55\ ,\ 89\ ,\ 144\ ,\ 233\ ,\ 377\ ,\ \ldots$$

And looking at the ratios:

$$\begin{array}{lll} \frac{1}{1} = 1 & \frac{8}{5} = 1.6 & \frac{55}{34} = 1.6176\ldots \\ \frac{2}{1} = 2 & \frac{13}{8} = 1.625 & \frac{89}{55} = 1.6181\ldots \\ \frac{3}{2} = 1.5 & \frac{21}{13} = 1.6153\ldots & \frac{144}{89} = 1.6179\ldots \\ \frac{5}{3} = 1.6666\ldots & \frac{34}{21} = 1.6190\ldots & \frac{233}{144} = 1.6180\ldots \end{array}$$

It sure looks like it is stabilizing around 1.618. What this means is that each term is roughly 1.618 times larger than the previous, and suggests that the Fibonacci sequence is growing basically as fast as the exponential function $(1.618)^n$ is growing. This constant, $\approx 1.6180\ldots$, that is so intimately related to the Fibonacci sequence is important enough that people have given it a symbol and a name: It is denoted ϕ, and is called *the golden ratio*.[36] It is indeed true that the Fibonacci ratios converge to this constant ratio; that is,

$$\phi = \lim_{n\to\infty} \frac{F_{n+1}}{F_n},$$

Let's try to find an exact value for ϕ. How do we do this? It's really clever. First, note that if we look at $\frac{F_{n+1}}{F_n}$ for $n = 1, 2, 3, 4, \ldots$, compared to $\frac{F_n}{F_{n-1}}$ for $n = 2, 3, 4, 5, \ldots$, we are looking at the exact same thing! In particular, as $n \to \infty$, these two ratios will certainly equal the same thing. That is:

$$\lim_{n\to\infty} \frac{F_{n+1}}{F_n} = \lim_{n\to\infty} \frac{F_n}{F_{n-1}}.$$

[35]Fun question: What's a function that eventually grows faster than any polynomial, but slower than any exponential? One answer is contained in the footnote on the next page.

[36]An amazing equation combining e, i, π, ϕ, 0 and 1: $(e^{\phi} + i^{\pi})(1-1) = \sqrt{0}$. Incredible.

Next,[37] way back in Calculus I you may have learned what are called the *limit laws* — rules for things you do to limits. One of these laws says that since

$$\lim_{n\to\infty} \frac{F_n}{F_{n-1}} = \phi, \qquad (👽)$$

if we instead inverted the terms in our limit $\left(\frac{F_{n-1}}{F_n}, \text{ instead of } \frac{F_n}{F_{n-1}}\right)$, it would converge to the inverse of the previous answer $\left(\frac{1}{\phi}, \text{ instead of } \phi\right)$. That is,

$$\lim_{n\to\infty} \frac{F_{n-1}}{F_n} = \frac{1}{\phi}. \qquad (🐦)$$

Finally, recall that the Fibonacci sequence satisfies $F_{n+1} = F_n + F_{n-1}$. Combining this fact with the equations in lines (👽) and (🐦), gives this:

$$\phi = \lim_{n\to\infty} \frac{F_{n+1}}{F_n} = \lim_{n\to\infty} \frac{F_n + F_{n-1}}{F_n} = \lim_{n\to\infty} \left(1 + \frac{F_{n-1}}{F_n}\right) = 1 + \frac{1}{\phi}.$$

That is, by using a little bit of trickery with limits we were able to show that whatever ϕ is, it must satisfy the equation

$$\phi = 1 + \frac{1}{\phi}.$$

And here is one of those wonderful moments when the key to finishing off a problem relies only on the math you learned in high school. If ϕ must satisfy the above equation, then we can solve that equation to determine what ϕ must be! Rewriting the above,

$$\phi^2 = \phi + 1.$$

Moving everything to the left,

$$\phi^2 - \phi - 1 = 0.$$

And how do you solve a equation like this that doesn't easily factor? The quadratic formula! The solutions are

$$\frac{-(-1) \pm \sqrt{(-1)^2 - 4(1)(-1)}}{2(1)} = \frac{1 \pm \sqrt{5}}{2}.$$

Clearly the Fibonacci sequence is increasing, which means ϕ has to be at least 1; and so we must have

$$\phi = \frac{1 + \sqrt{5}}{2}.$$

This ϕ is the precise value of the golden ratio. There are many mathematical and real-world applications of the golden ratio, but one of my favorites is to distance. Recall that 1 mile is about 1.61 kilometers which, just by a silly coincidence, means that 1 mile is roughly ϕ kilometers. So if you want to convert from miles to kilometers, you can use the Fibonacci sequence! For example, 3 miles is about 5 kilometers (as runners know), and 5 miles is about 8 kilometers, and 8 miles is about 13 kilometers, and 13 miles is about 21 kilometers, and so on.[38]

[37] In Footnote 35 I asked a question. Here's one answer to that question: $n^{\log(n)}$.

[38] *"That's gold, Jerry! Gold!"*

Chapter 5: Logic

It is common for Chapter 1 of an intro to proofs book to be on logic, and for good reason: Proofs rely entirely on logic. I decided to move it later for a few reasons. First, the logic needed to begin discussion on the pigeonhole principle, direct proofs and induction is not sophisticated, and you all mastered it naturally years ago. The logic thus far has been about 60% common sense and 30% hard work.[1] The problem is that when you first learn formal logic, it is really easy to get confused. When you come through the other side things will feel a lot more natural than when you're in the middle of it, but while you're in the midst, it is easy to lose track of your intuition.

Plus, advanced logic is legitimately weird. It is hard to grasp and requires really careful thinking. The great logician Bertrand Russell defined research-level logic as "The subject in which nobody knows what one is talking about, nor whether what one is saying is true." But fear not, we will not venture too far off the beaten path.

Mathematical proofs are tough to learn, but one of your best tools is your natural intuition and ingenuity. I feared that starting with logic before you'd ever seen a proof would send the wrong message—that you need to start warping your mind in order to reason through a proof. You don't! You have now proven dozens of results without any fancy logic, and make sure not to lose that. Formal logic will teach us some necessary things, and will open the door to some fundamental proof techniques, but your natural logic will still be far more important than anything we cover here, and your intuition is indispensable.

5.1 Statements

Logic is the process of deducing information correctly—it is *not* the process of deducing correct information. For example,

1. Socrates is a Martian
2. Martians live on Pluto
3. Therefore, Socrates lives on Pluto

...is *logically* correct, even though all three statements are false. And if I said "Socrates is a Martian and Martians live on Pluto, therefore $2 + 2 = 4$," then what I

[1] The other half is intelligence.

said was logically incorrect, even though the conclusion is correct.[2] In mathematics, we state axioms and then use logic to prove the necessary consequences of those axioms. Mathematicians search for some form of truth, but don't confuse correct logic for correct information, and this chapter focuses on the logic.

Statements

The building blocks of logic are *statements*. We have used this term many times in this book (including in the previous paragraph), but let's formally define them now.

Definition.

Definition 5.1. A *statement* is a sentence or mathematical expression that is either true or false.

By *sentence*, I mean a traditional sentence in English or another language. Not all sentences are statements, though, as they need to have *a truth value* (they are either true of false).

Example 5.2. Here are six examples of a statement:

True statements:

1. $\sqrt{2} \in \mathbb{R}$
2. All polynomials are continuous
3. $2 + 3 = 5$

False statements:

4. $\sqrt{2} \in \mathbb{Z}$
5. All integers are even
6. $3 + 4 = 15$

These, meanwhile, are not statements:

7. $8 + 9$
8. $\mathbb{Q}$
9. $x + 7$
10. Are polynomials differentiable?

The last one is a question—its answer is yes or no, not true or false. Also, whenever a sentence is ambiguous, it is not a statement.[3] Note also:

- Every theorem/proposition/lemma/corollary is a (true) statement;

[2] If the logic is valid *and* the statements are true, then it is called *sound*.

[3] Example: "I ran" would be too ambiguous to be considered a sentence (it is neither true nor false). "I ran to my car this morning" is better, although we might need to define "ran" and "car" and "morning" to be completely confident that it is either true or false.

Math expressions can also be ambiguous. Indeed, half the time that "math" goes viral on social media is due to some post asking for the value of something like $8 \div 2(2 + 2)$. I advise against engaging with such posts.

- Every conjecture is a statement (of unknown truth value); and
- Every incorrect calculation is a (false) statement.

A related notion is that of an *open sentence*, which are sentences or mathematical expressions which (1) do not have a truth value, (2) depend on some unknown, like a variable x or an arbitrary function f, and (3) when the unknown is specified, then the open sentence becomes a statement (and so has a truth value). Their truth value depends on which value of x or f one chooses.

Example 5.3. Here are four examples of open sentences:

1. $x + 7 = 12$
2. $3 \mid x$
3. f is continuous
4. x is even

For number 2, this open sentence is true if $x = 6$, but false if $x = 8$. For number 3, this open sentence is true if $f(t) = t^2$, but false if $f(t) = 1/t$ (with domains of $\mathbb{R}$ and $\mathbb{R} \setminus \{0\}$).

Note, though, that simply using unknowns does not mean something is an open sentence; an open sentence must not only use unknowns, but also have no truth value. So, "for each $x \in \mathbb{R}$, we have $x - x = 0$" is a (true) statement, while "$x + x = 2$" is an open sentence (true when $x = 1$, false otherwise). Indeed, as you were told years ago, the Pythagorean theorem is true and hence is a statement, even though it contains variables and the equation $a^2 + b^2 = c^2$ at the end.

Typically, we use capital letters for statements, like P, Q and R. Open sentences are often written the same, or perhaps like $P(x), Q(x)$ and $R(x)$ when one wishes to emphasize the variable. Below is some notation that is used often in logic, which turn one or more statements into a single new statement.

Notation.

Notation 5.4. Let P and Q be statements or open sentences.

1. $P \wedge Q$ means "P and Q"[4]
2. $P \vee Q$ means "P or Q" (or both)[5]
3. $\sim P$ means "not P"

[4]Way to remember this one: $\wedge$ is the start of $\wedge$nd, which kind of looks like And.

[5]Reminder: In math, 'or' is always an *inclusive or*, as compared to an *exclusive or*. An 'exclusive or' means that one or the other is true, but not both, like *"The light is on or off."* Meanwhile, an 'inclusive or' allows the possibility that both are true; in everyday language, people sometimes say "and/or" to emphasize that they mean an inclusive or. Notice that Lemma 2.17 part (iii) would be false if math used an 'exclusive or'. For example, if $p = 5$, $a = 10$ and $b = 15$, then $p \mid ab$, but it's not true that p only divides one of the two—it divides both!

Again, if P and Q are statements, then $P \wedge Q$ and $P \vee Q$ and $\sim P$ are all statements, too. (This is like saying, if x and y are integers, then $x + y$ is an integer, too. Sure, $2 + 3$ is a sum of integers, but it also equals 5, which is an integer in its own right.) Let's do some examples.

Example 5.5. Consider the following statements:

- P: The number 3 is odd
  → This is true
- Q: The number 4 is even
  → This is true
- R: The number 5 is even
  → This is false
- S: The number 6 is odd
  → This is false

Then,

1. $P \wedge Q$: 3 is odd and 4 is even
   → This is true[6]

   $P \wedge R$: 3 is odd and 5 is even
   → This is false
2. $P \vee R$: 3 is odd or 5 is even
   → This is true[7]

   $R \vee S$: 5 is even or 6 is odd
   → This is false
3. $\sim P$: 3 is not odd
   → This is false

   $\sim S$: 6 is not odd
   → This is true
4. $P \wedge \sim Q$: 3 is odd and 4 is not even
   → This is false

   $S \vee \sim S$: 6 is odd or 6 is not odd
   → This is true[8]

If a mom tells her son, "in order to go out, you must do the dishes and take out the trash," then the boy better do both. If instead she said, "in order to go out, you must do the dishes or take out the trash," then he can do either (or can do both!) and he would be allowed to go out. Now, let's talk implications.

Notation.

Notation 5.6. Let P and Q be statements or open sentences.

1. $P \Rightarrow Q$ means "P implies Q"
2. $P \Leftrightarrow Q$ means "P if and only if Q"

If P and Q are statements or open sentences, then $P \Rightarrow Q$, and $P \Leftrightarrow Q$, are statements or open sentences. That is, $P \Rightarrow Q$ and $P \Leftrightarrow Q$ must have truth values

[6] If I ever drop a rap album on math logic, my rapper name will be $m \wedge m$.

[7] Stranger at the store: "What a cute baby! Is your baby a boy or a girl?" Logician: "Yes."

[8] Shakespeare: To be $\vee$ $\sim$(To be). Literary critics: *Applause* Logicians: "True."
By the way, notice that $S \vee \sim S$ would be true for <u>any</u> statement S. This is called a *tautology*. (Meanwhile, $S \wedge \sim S$ is *false* for any statement S. This is called a *contradiction*.)

(they are either true or false). In fact, we have seen this many times, since most of our propositions and theorems are of the form $P \Rightarrow Q$. For example, "If n is odd, then n^2 is odd" is a (true) statement. In this way, not only are $P \wedge Q$ and $P \vee Q$ ways to turn a pair of statements into a new statement, but $P \Rightarrow Q$ and $P \Leftrightarrow Q$ are, too.

Let's now discuss a subtle aspect of implications: Translating them to and from English. Language can be complicated,[9] and we in fact have many different ways in English to say "P implies Q." Here are some more:

- If P, then Q
- Q if P
- P only if Q
- Q whenever P
- Q, provided that P
- Whenever P, then also Q
- P is a sufficient condition for Q
- For Q, it is sufficient that P
- For P, it is necessary that Q

For example, "If it is raining, then the grass is wet" has the same meaning as "The grass is wet if it is raining." These also mean the same as "The grass is wet whenever it is raining" or "For the grass to be wet, it is sufficient that it is raining."

Next, here are some ways to say "P if and only if Q":[10]

- P is a necessary and sufficient condition for Q
- For P, it is necessary and sufficient that Q
- P is equivalent to Q
- If P, then Q, and conversely
- P implies Q and Q implies P
- Shorthand: P iff Q[11]
- Symbolically: $(P \Rightarrow Q) \wedge (Q \Rightarrow P)$

Note that if we used our wet grass and rainy weather example from before, then these would all be false statements. This is because having wet grass does not imply it is raining—perhaps the sprinkler is on, or there is an awesome water balloon fight going down. To find an example where all the statements are true, we need to find statements P and Q which are equivalent—each implies the other.

For example, suppose that Jessica wears sunglasses whenever it is sunny, and never wears them when it is not sunny. That is, if it is sunny out, then Jessica wears sunglasses; and if Jessica wears sunglasses, then it is sunny out. This means that "Jessica wears sunglasses if and only if it is sunny out." This also means that

[9] Or "rich," if you're a linguist.

[10] And for each of these, you can also switch 'P' and 'Q' around. For example, "Q is a necessary and sufficient condition for P" is another way to say the same thing.

[11] More shorthand, in case you need it:
if = if
iff = if and only if
ifff = iff and only iff = if and only if and only if and only if
iffff = ifff and only ifff = iff and only iff and only iff and only iff = if and only if and only if and only if and only if and only if and only if and only if
etcetera etcetera etcetera

"Jessica wearing sunglasses is a necessary and sufficient condition for it to be sunny," or "Jessica wearing sunglasses is equivalent to it being sunny," or "If Jessica is wearing sunglasses, then it is sunny, and conversely."

As a math example, suppose $n \in \mathbb{Z}$. Then, "n is even if and only if $n \equiv 0 \pmod 2$" is the same as "n being even is equivalent to $n \equiv 0 \pmod 2$" or "n being even implies $n \equiv 0 \pmod 2$ and $n \equiv 0 \pmod 2$ implies n is even."

The fact that "P implies Q" is the same as "If P, then Q" or "Q if P" is sometimes intuitive to students. But the fact that these are all the same as "P only if Q" is often confusing. Most people's guts tell them that "P implies Q" should be the same as "Q only if P." What does your gut say?

Should	• $P \Rightarrow Q$ • If P, then Q • Q if P	be the same as	• P only if Q *or* • Q only if P	?

The answer is "P only if Q", and the way to think about it is that "P implies Q" means that whenever P is true, Q must also be true. And "P only if Q" means that P can *only be true* if Q is true. . . that is, whenever P is true, *it must be the case* that Q is also true. . . that is, $P \Rightarrow Q$.

Now, if P and Q are statements, then "$P \Rightarrow Q$" and "$P \Leftrightarrow Q$" are also statements, meaning they must also be either true or false. The statement $P \Rightarrow Q$ is called a *conditional statement*, whereas $P \Leftrightarrow Q$ is called a *biconditional statement*. These are minor definitions, but the following is an important definition.

Definition.

Definition 5.7. The *converse* of $P \Rightarrow Q$ is $Q \Rightarrow P$.

If $P \Rightarrow Q$, it is not necessarily the case that $Q \Rightarrow P$.[12] For example, "If $x = 2$, then x is even" is true, but its converse is "If x is even, then $x = 2$," which is false. There's also the classic example from 5th grade: "A square is a rectangle, but a rectangle is not necessarily a square." This could be rephrased as "If S is a square, then S is a rectangle," which is true; meanwhile, its converse is "If S is a rectangle, then S is a square," which is false.

Or, if you'd like an example from the real world: If person A likes person B, it's not always the case that person B likes person A. Just ask a mathematician.

[12] When a mathematician writes a sentence like this, what they mean is: If "$P \Rightarrow Q$" is a true statement, then it is not necessarily the case that "$Q \Rightarrow P$" is a true statement. (The converse certainly exists and is a statement; what is being communicated is that it could either be true or false.)

Intuition From Set Theory

If A and B are sets, then $A \cap B$ is the set of elements which are in A and in B. This is similar to how $P \wedge Q$ is true if P and Q are true. Likewise, $A \cup B$ are the elements that are in A or in B (or both), and $P \vee Q$ is true if P or Q is true (or both). Indeed, you could even write the definitions of $A \cup B$ and $A \cap B$ using our new notation.

$$A \cap B = \{x : x \in A \ \wedge \ x \in B\} \qquad \text{and} \qquad A \cup B = \{x : x \in A \ \vee \ x \in B\}.$$

This is especially nifty because $\cap$ and $\wedge$ look a lot alike, and $\cup$ and $\vee$ look a lot alike.

The similarities do not stop there. Notice that A^c for sets is analogous to $\sim P$ for statements. The former is asking what elements are outside of A, while $\sim P$ is asking what logical possibilities are outside of P. (In fact, some use $\overline{A}$ to denote A^c, and some use $\overline{P}$ to refer to $\sim P$.)[13]

Lastly, you can think about $P \Rightarrow Q$ as analogous to $A \subseteq B$. An implication like "If you live in Los Angeles, then you live in California" is true, because the "P" (the set of residents of LA) is smaller than the "Q" (the set of residents of CA). Likewise, $A \subseteq B$ if the "A" is smaller than the "B." In the same way, $Q \Rightarrow P$ is analogous to $B \subseteq A$, and thus $P \Leftrightarrow Q$ is analogous to $A = B$. Here is an example of all this:

- Sets: Suppose $A = \{x : x \text{ is an even integer}\}$ and $B = \mathbb{Z}$. Then, $A \subseteq B$.
  Logic: Suppose P is the open sentence "x is even" and Q is the open sentence "x is an integer." Then, $P \Rightarrow Q$.

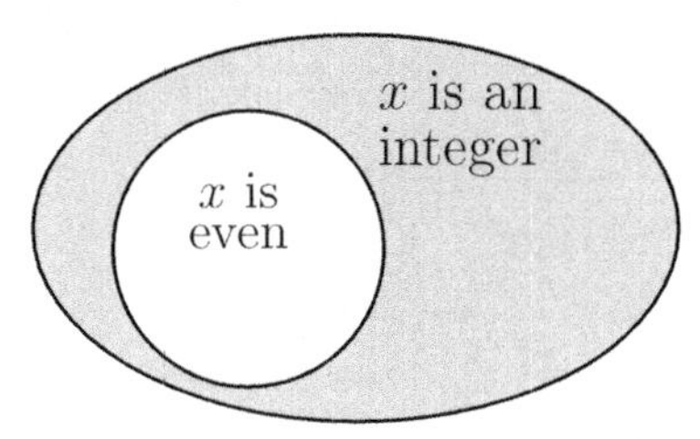

This example also shows that if your universal set is B, then A^c is the set of odd integers — the shaded portion above. And, again if your universe is the integers, then $\sim P$ is the statement "x is an odd integer." Another example:

- Sets: Suppose that $A = \{x \in \mathbb{Z} : 2 \mid x\}$ and $B = \{x \in \mathbb{Z} : 3 \mid x\}$. Then, $A \cap B = \{x \in \mathbb{Z} : 6 \mid x\}$.
  Logic: Suppose P is the open sentence "$x \in \mathbb{Z}$ and $2 \mid x$," and Q is the open sentence "$x \in \mathbb{Z}$ and $3 \mid x$." Then, $P \wedge Q$ is the open sentence "$x \in \mathbb{Z}$ and $6 \mid x$."

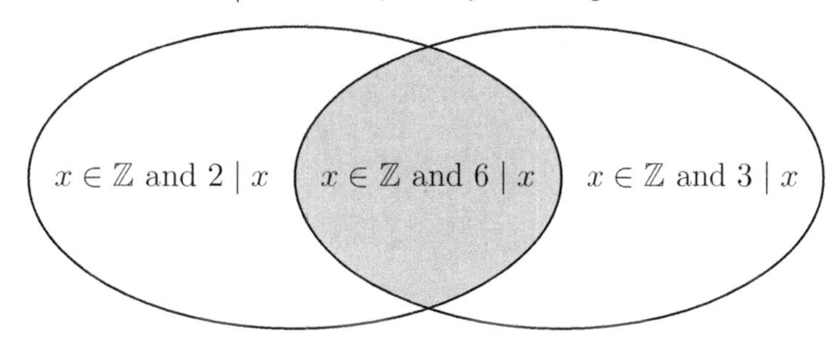

[13]In fact, while $\wedge$ and $\vee$ are very standard, the "not" symbol is, well, not. In addition to $\sim P$ and $\overline{P}$, you may also see $\neg P$ and $!P$.

5.2 Truth Tables

A truth table models the relationship between the truth values of one or more statements, and that of another. Let's first look at how the truth values of P and of Q affect the truth value of $P \wedge Q$.

P	Q	$P \wedge Q$
True	True	True
True	False	False
False	True	False
False	False	False

To the left of the double line are the possible truth value combinations of P and Q: They could be True/True, True/False, False/True or False/False.[14] To the right of the double line is what we are deducing. Refer back to Example 5.5 to see some concrete examples of these deductions. What those examples illustrate is that in order for "P and Q" to be a true statement, *both* P and Q must be independently true.[15]

For instance, the second row of the above truth table is telling us that if P is true but Q is false, then the statement $P \wedge Q$ is false. An example of this from Example 5.5: "3 is odd and 5 is even" is a false statement.

Next, here's how the truth values of P and of Q affect the truth value of $P \vee Q$.

P	Q	$P \vee Q$
True	True	True
True	False	True
False	True	True
False	False	False

Again, refer to Example 5.5 for concrete examples. Here, in order for "P or Q" to be a true statement, it is sufficient that either P is true or that Q is true (or both).

Finally, here is how the truth values of P affects that of $\sim P$.

P	$\sim P$
True	False
False	True

In order for "not P" to be true, it is required that P be false. By applying this reasoning twice, this also implies that $\sim\sim P$ and P always have the same truth value.[16] Using our intuition about sets, this is like how $(A^c)^c = A$.

[14]If you had three propositions, P, Q and R, you would need 8 rows to cover all the possible combinations.

[15]Teacher: Please gather around, boys and girls. Baby logician: *Doesn't gather around*

[16]Linguistics prof: "In English, a double negative forms a positive. However, in some languages, such as Russian, a double negative remains a negative. But there is no language where a double positive can form a negative." Heckler from the back of the room: "Yeah, right..."

Example 5.8. This example is more complicated. Here we will find the truth values of $(P \vee Q) \wedge \sim(P \wedge Q)$, given the four possible truth value combinations for P and Q. How do we do this? Well, to find the truth values of $(P \vee Q) \wedge \sim(P \wedge Q)$ we need the truth values of $(P \vee Q)$ and of $\sim(P \wedge Q)$, and for the latter we will need the truth values of $(P \wedge Q)$. This is how we proceed.

Truth values of P and of Q $\rightarrow$ Truth values of $(P \vee Q)$ and of $(P \wedge Q)$ $\rightarrow$ Truth values of $(P \vee Q)$ and of $\sim(P \wedge Q)$ $\rightarrow$ Truth values of $(P \vee Q) \wedge \sim(P \wedge Q)$

Indeed, in the truth table below, our first two columns are the four possible truth value combinations for P and Q. These are then used to deduce columns three and four. Column four is used to deduce column five. And columns three and five are used to deduce column six.

P	Q	$P \vee Q$	$P \wedge Q$	$\sim(P \wedge Q)$	$(P \vee Q) \wedge \sim(P \wedge Q)$
True	True	True	True	False	False
True	False	True	False	True	True
False	True	True	False	True	True
False	False	False	False	True	False

De Morgan's Logic Laws

Our next example is the logic form of De Morgan's law (Theorem 3.16). Take a look at the truth table for $\sim(P \wedge Q)$ and the truth table for $\sim P \ \vee \sim Q$, side by side:

P	Q	$P \wedge Q$	$\sim(P \wedge Q)$
True	True	True	False
True	False	False	True
False	True	False	True
False	False	False	True

P	Q	$\sim P$	$\sim Q$	$\sim P \ \vee \sim Q$
True	True	False	False	False
True	False	False	True	True
False	True	True	False	True
False	False	True	True	True

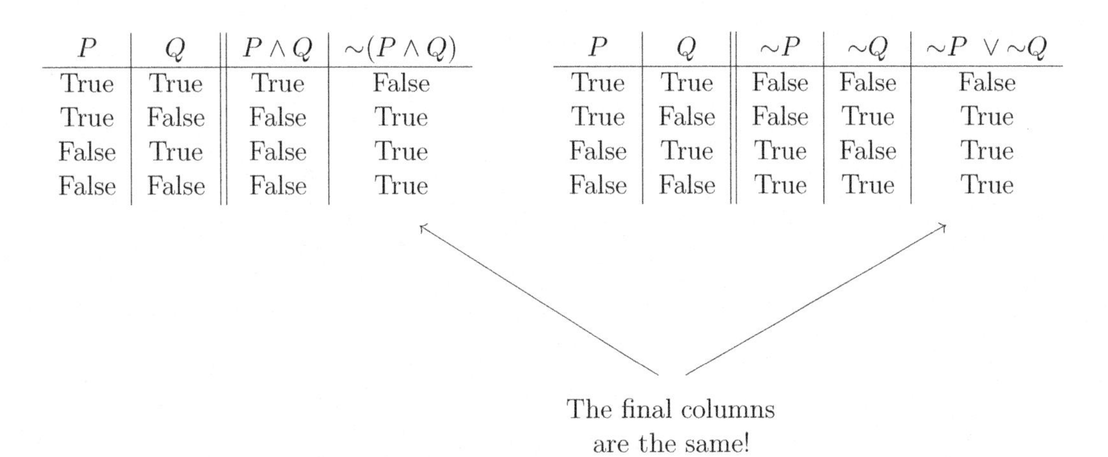

Since the final columns are the same, if one is true, the other is true; if one is false, the other is false; that is, there is no way to select P and Q without these two agreeing. This was basically a proof by cases, where we checked all the true/false combinations individually — and then consolidated all of the work into a single table.

When two statements have the same final column in their truth tables, like in the example above, they are said to be *logically equivalent* (one is true *if and only if* the other is true), which we denote with an "$\Leftrightarrow$" symbol. De Morgan's logic law, for example, can be written like this:

$$\sim(P \wedge Q) \;\Leftrightarrow\; \sim P \vee \sim Q.$$

In words, $\sim(P \wedge Q) \;\Leftrightarrow\; \sim P \vee \sim Q$ says this: "P and Q are not both true" is the same as "P is false or Q is false."[17]

As with De Morgan's laws for sets, there is also a second De Morgan law for logic: $\sim(P \vee Q) \;\Leftrightarrow\; \sim P \wedge \sim Q$. You can prove this in a similar way to the above, by simply checking the truth tables for both. This is asked of you in Exercise 5.22. Below we record these results.

Theorem.

Theorem 5.9 (*De Morgan's logic laws*). If P and Q are statements, then

$$\sim(P \wedge Q) \;\Leftrightarrow\; \sim P \vee \sim Q \qquad \text{and} \qquad \sim(P \vee Q) \;\Leftrightarrow\; \sim P \wedge \sim Q.$$

Truth Tables with Implications

In math, we deal with theorems, which are statements which typically contain an implication. This chapter began with the example of Socrates living on Pluto, which we said was logically correct, even if the components were false. This is what we wish to investigate now: Since $P \Rightarrow Q$ is a statement, it will have a truth table just like $P \wedge Q$ and $P \vee Q$ had truth tables, but what is it? Given the truth values of P and Q, what does that tell us about the truth value of $P \Rightarrow Q$? This is more subtle than you might think. For starters, consider this implication:

"If $n = 2$, then n is even."

This is a true statement. Indeed, here is a very small table which lays out the truth value of this statement:

n	$n = 2$	n is even	If $n = 2$, then n is even
2	True	True	True

Perhaps you can see where I am going with this. The above is an example of a $P \Rightarrow Q$ statement, in which P and Q are both true, and where $P \Rightarrow Q$ was in turn true. This makes sense, as $P \Rightarrow Q$ means "If P is true, then Q is also true," and both are indeed true. If the above were a (rather simple) theorem, then you would say that it is a true theorem, no sweat. You can perhaps even imagine that *whenever* P and Q are true, the implication $P \Rightarrow Q$ will be a true implication.

[17]By the way, De Morgan's logic laws also show that *technically* we were redundant when we defined all three of $\wedge$, $\vee$ and $\sim$. For example, if we had only defined $\wedge$ and $\sim$, then that would be enough to do all our logic, because $P \vee Q$ is the same as $\sim(\sim P \wedge \sim Q)$.

But suppose we added a few more rows to the table, imagining other values of n:

n	$n = 2$	n is even	If $n = 2$, then n is even
2	True	True	True
3	False	False	True
4	False	True	True

We had already established that the final column is true — the implication "if $n = 2$, then n is even" is absolutely true — and imagining any other possible values of n does not change the fact that this implication is true all by itself. However, the second row of this truth table does suggest that if P is false and Q is false, then $P \Rightarrow Q$ should be considered true. And the third row suggests that if P is false and Q is true, then $P \Rightarrow Q$ should be considered true. We still have the "True $\Rightarrow$ False" row to add, and as you might expect, this is a false implication.

Intuition is important, but we should say right at the top that this truth table we are generating is not up for dispute because this is how we are *defining* the implication:

P	Q	$P \Rightarrow Q$
True	True	True
True	False	False
False	True	True
False	False	True

The row we just added is now the second row in this table, and I think this one makes sense. The implication $P \Rightarrow Q$ means "If P is true, then Q is true," and the second row clearly does not satisfy this requirement. So for the second row, "If P is true, then Q is true" is false.

The "$n = 2 \Rightarrow n$ is even" example provided motivation for the last two rows, but I still expect these to be the hardest to think about. Why is the implication true if the assumption, P, is false? It's kind of like how we said that this is true: "If $x \in \emptyset$, then x is a purple elephant that speaks German." Since there is nothing in the empty set, if you suppose $x \in \emptyset$, you can then claim anything you want about x and it is inherently true — you certainly cannot present to me any element in the empty set that is *not* a purple elephant that speaks German. In the set theory chapter (on page 101), we called such a claim *vacuously true*.

Likewise, in a universe where P is true, the statement $P \Rightarrow Q$ has some real meaning that needs to be proven or disproven: Does P being true imply Q is true, or not? But in a universe where P is not true, it claims nothing, and hence $P \Rightarrow Q$ is vacuously true.

"If unicorns exist, then they can fly" can certainly *not* be considered false, because unicorns do not exist,[18] so any claim about them is considered vacuously true. Indeed, the way to falsify that proposition would be to locate a unicorn that cannot fly, which

[18] ☹

is impossible to do. Every unicorn in existence can indeed fly! Also, every unicorn in existence cannot fly! Neither can be disproven!

– $P \Rightarrow Q$ is false if $P \Rightarrow Q$ is a lie –

One final way to think about it is this: If I said to you, "If unicorns exist, then they can fly," would you say that I lied to you? We only label an implication as false if you would regard it as a lie, but I don't think most people would consider that implication a lie.

As another example, suppose I said "If you get an A on your final, then you will get an A in the class." And then suppose you get a B on the final and a B in the class. Would you say I lied to you? Of course not! And if you got a B on the final and an A in the class, I still did not lie. I said (A on final) $\Rightarrow$ (A in class); the two examples I gave are then "False $\Rightarrow$ False" and "False $\Rightarrow$ True." But in neither of these would you say I lied, so both "False $\Rightarrow$ False" and "False $\Rightarrow$ True" should be considered as true.[19] Compare this intuition with the truth table:

Grade on Final	Grade in Class	(A on Final) $\Rightarrow$ (A in Class)
A	A	Did not lie
A	B	LIE
B	A	Did not lie
B	B	Did not lie

P	Q	$P \Rightarrow Q$
True	True	True
True	False	False
False	True	True
False	False	True

(And if you still think it is weird, that's ok. Remember that, either way, the above is how it is because we are *defining* it to be so.)[20]

– Truth table for $P \Leftrightarrow Q$ –

Finally, let's combine the above knowledge about $P \Rightarrow Q$ and the corresponding truth table for $Q \Rightarrow P$ to generate the truth table for $P \Leftrightarrow Q$. Here are the two truth

[19] Now, if it were possible to give `Error 404` as an answer, then perhaps we could get out of calling it true. But since the only options are true and false, and it definitely ain't false, it's gotta be true.

[20] According to an old story, the great logician Bertrand Russell, in a lecture on logic, was asked about this strangeness by one of his students. In fact, the precocious student challenged Russell to prove that "if $1 = 0$, then you are the Pope." This is of the form $P \Rightarrow Q$ where P is false, so this should be a true implication, right?

Russell immediately replied, "Add 1 to both sides of the equation: then we have $2 = 1$. The set containing just me and the Pope has 2 members. But $2 = 1$, so it has only 1 member; therefore, I am the Pope."

(I like to imagine that at this point he held out has arm, dropped his mic, threw on some shades, said "Russell out," and exited the room to gasps and cheers.)

tables for $P \Rightarrow Q$ and $Q \Rightarrow P$:

P	Q	$P \Rightarrow Q$
True	True	True
True	False	False
False	True	True
False	False	True

P	Q	$Q \Rightarrow P$
True	True	True
True	False	True
False	True	False
False	False	True

Remember, $P \Leftrightarrow Q$ is true when both $P \Rightarrow Q$ is true *and* $Q \Rightarrow P$ is true. Thus, the truth table for $P \Leftrightarrow Q$ is this:

P	Q	$P \Leftrightarrow Q$
True	True	True
True	False	False
False	True	False
False	False	True

In closing, it is also useful at this point to reflect on the fact that the truth values of P and of Q are one thing that we have looked at, and the truth values of $P \Rightarrow Q$ and $P \Leftrightarrow Q$ are another, and as truth tables illustrate, these do not match. Think back to the first example of the chapter, with Socrates and Martians; correct logic (the implication) does not need to match correct information (the component statements). Make sure you distinguish these in your mind.

5.3 Quantifiers and Negations

Before discussing quantifiers, here is a quick riddle that we will come back to later. Suppose you saw this sign at a restaurant:

Good food is not cheap
Cheap food is not good

Here is the question: Are these two sentences saying the same thing, or different things? I'll let you mull that one over while we discuss quantifiers and negations.

— Quantifiers —

The (open) sentence

"n is even"

is not a statement as defined in Definition 5.1, because it is neither true nor false. One way to turn a sentence like this into a statement is to give n a value. For example,

"If $n = 5$, then n is even" *and* "If $n = 6$, then n is even"

are each bona fide statements. What I'd like to discuss now are two other basic ways to turn "n is even" into a statement: add *quantifiers*. A quantifier is an expression which indicates the number (or quantity) of our objects. For example:

"For all $n \in \mathbb{N}$, n is even."

Saying "for all $n \in \mathbb{N}$" means that we are asserting that all of the infinitely many n-values in $\mathbb{N}$ have the property. Another example:

There exists some $n \in \mathbb{N}$ such that n is even.

Saying "there exists some $n \in \mathbb{N}$" means that we are asserting that at least one n-value in $\mathbb{N}$ has the property.

Both of these are now statements—each is either true or false. "For all $n \in \mathbb{N}$, n is even" is a false statement, because it is not true that *all* $n \in \mathbb{N}$ are even (for example, $n = 5$). "There exists some $n \in \mathbb{N}$ such that n is even" is a true statement, because of course there *exists* such an n (for example, $n = 6$).

The phrases "for all" and "there exists" are the two most important quantifiers in math. Now, because language is complicated, there are many equivalent ways to say these quantifiers. And, just for good measure, mathematicians gave them names and symbols as well:

- The symbol $\forall$ means "for all" or "for every" or "for each", and is called the *universal quantifier.*
- The symbol $\exists$ means "there exists" or "for some", and is called the *existential quantifier.*[21]

Example 5.10. Here are some true statements:

- $\exists n \in \mathbb{N}$ such that $\sqrt{n} \in \mathbb{N}$

  Translation: There exists an n in the natural numbers such that $\sqrt{n}$ is also in the natural numbers.[22]

- $\forall n \in \mathbb{N}$, $\sqrt{n} \in \mathbb{R}$

  Translation: For all n in the natural numbers, $\sqrt{n}$ is in the reals.[23]

- $\nexists n \in \mathbb{N}$ where $\sqrt{n} = n + 1$

  Translation: There does not exist an n in the natural numbers such that $\sqrt{n} = n + 1$.

[21] Likewise, $\nexists$ means "there does not exist," and $\exists!$ means "there exists a unique." The symbol is a backwards E, which seems clear enough, but the fact that the $\forall$ symbol is an upside down 'A' took me like 5 years to realize, and exploded my mind when I did. Notwithstanding, the symbol now provides a nice way to write "math for all" in support of sharing math with more people: M$\forall$TH.

[22] Second translation: There exists a perfect square.

[23] Second translation: Every natural number has a square root.

Example 5.11. Here are some false statements:

- $\exists\, n \in \mathbb{N}$ such that $n = -n$

  Translation: There exists a natural number n such that $n = -n$.

- $\forall\, n \in \mathbb{N}$, $\sqrt{n} \in \mathbb{N}$

  Translation: For all n in the natural numbers, $\sqrt{n}$ is in the naturals.[24]

- $\nexists\, n \in \mathbb{N}$ where $\sqrt{n} = n + 2$

  Translation: There does not exist an n in the natural numbers such that $\sqrt{n} = n + 2$.

Note that the final example is false because 4 is in the natural numbers, and $n = 4$ satisfies the equation. So such an n does exist.

Example 5.12. Here is a statement that uses both quantifiers, followed by the three ways you can permute those quantifiers:

- $\forall\, x \in \mathbb{R}$, $\exists\, y \in \mathbb{R}$ such that $x^2 = y$ (this is true)[25]

  Translation: For all x in the reals, there exists some y in the reals such that $x^2 = y$.

- $\exists\, x \in \mathbb{R}$ such that $\forall\, y \in \mathbb{R}$, $x^2 = y$ (this is false)[26]

  Translation: There exists some x in the reals such that, for all y in the reals, we have $x^2 = y$.

- $\forall\, y \in \mathbb{R}$, $\exists\, x \in \mathbb{R}$ such that $x^2 = y$ (this is false)[27]

  Translation: For all y in the reals, there exists some x in the reals such that $x^2 = y$.

- $\exists\, y \in \mathbb{R}$ such that $\forall\, x \in \mathbb{R}$, $x^2 = y$ (this is false)[28]

  Translation: There exists some y in the reals such that, for all x in the reals, we have $x^2 = y$.

As these bullet points show, the order of your quantifiers is really important. Not only does the ordering affect their meaning, but it affects their truth value, too.

As a final note, although I used the $\forall$ and $\exists$ symbols in the above, these should not be used in formal proofs. For class notes, sure. For scratch work on your homework, sure. On an exam when you are crunched for time, sure. But when you are writing up any mathematics formally, be it your homework or a research paper, it is very rare to use such symbols. Use words instead.

---

[24] Second translation: Every natural number is a perfect square.
[25] Second translation: Every real number can be squared.
[26] Second translation: There is a real number which is every real number's square root.
[27] Second translation: Every real number has a square root in the reals.
[28] Second translation: There is a real number which is every real number's square.

— Negations —

We saw earlier that $\sim P$ is notation for "not P." Below are some examples, the first two of which use De Morgan's laws for logic (Theorem 5.9).

1. P: Socrates was a dog and Aristotle was a cat[29]

   $\sim P$: Socrates was not a dog or Aristotle was not a cat

2. Q: Plato was a walrus or a chimp[30]

   $\sim Q$: Plato was not a walrus and not a chimp

3. R: All Cretans are liars[31]

   $\sim R$: Not all Cretans are liars[32]

   $\sim R$: There exists a Cretan who is not a liar[33]

4. S: Someone in this room is sleepy[34]

   $\sim S$: No one in this room is sleepy

   $\sim S$: All people in this room are not sleepy

In the first example, the "and" in P turned into an "or" in $\sim P$. In the second example, it was the opposite. In the third example, a "for all" in R turned into a "there exists" in $\sim R$. In the fourth example, it was the opposite. This suggests some rules of thumb for negating statements:

- $\sim\wedge = \vee$
- $\sim\vee = \wedge$
- $\sim\forall = \exists$
- $\sim\exists = \forall$

[29]Note that you could write this as, say, $P \wedge Q$ where P is "Socrates was a dog" and Q is "Aristotle was a cat." But remember, $P \wedge Q$ is its own statement too! It's like how you could write 5 is the sum of two numbers, $2 + 3$, but its also a number by itself. So if we want to express "Socrates was a dog and Aristotle was a cat" as a single statement with a single letter, that is certainly fine.

[30]It is common to use the OG logicians in examples like this. I used to think it was a nice tribute to include them, but is it really if I call them walruses or chimps??

[31]This could be stated as: "For all people p, if p is a Cretan then p is a liar."

[32]Make sure you convince yourself that the negation should be "Not all Cretans are liars," and not "All Cretans are not liars." The negation of R should capture all the cases where R is false. And having, say, half the Cretans being liars is certainly one way that R could be false, and hence it is one of the cases that $\sim R$ should capture.

[33]This $\sim R$ means the exact same thing as the $\sim R$ right above it. Convince yourself of this.
P.S. This is not to be insulting to the people of Crete. This is a reference to the *Epimenides paradox*.
P.P.S. If this book sells a zillion copies I might just go live in Crete forever. It looks *beautiful*.

[34]This could be stated as: "There exists a person p in this room such that p is sleepy."

Example 5.13. Below are two examples applying these rules of thumb.

1. Applying a negation to every term, and the given rules of thumb,

   $\sim(P \wedge Q)$ is equivalent to $\sim P \sim\wedge \sim Q$ is equivalent to $\sim P \vee \sim Q$.

   This gives us a new way to understand De Morgan's law for logic (Page 215):

   $$\sim(P \wedge Q) \Leftrightarrow \sim P \vee \sim Q.$$

2. R: For every real number x, there is some real number y such that $y^3 = x$.

   In symbols,

   R: $\forall x \in \mathbb{R}$, $\exists y \in \mathbb{R}$ such that $y^3 = x$.

   Then,[35]

   $\sim R$: $\sim(\forall x \in \mathbb{R},\ \exists y \in \mathbb{R}$ such that $y^3 = x)$ is equivalent to[36]

   $\sim R$: $\exists x \in \mathbb{R}$ such that $\forall y \in \mathbb{R},\ y^3 \neq x$.

Notice that we negated every part of the statement R. Or did we? Notice that "$\forall x \in \mathbb{R}$" turned into "$\exists x \in \mathbb{R}$"—it did *not* turn into "$\exists x \notin \mathbb{R}$." Why is this? Recall that the negation swaps the truth values: If R was true, then $\sim R$ is false, and vice versa. And the statement R is saying that every real number x satisfies some property. Thus, the negation would be that not every real number x satisfies that property. But that is still referring to real numbers! Makes sense?

For example, if I made the statement "Every NBA player can dunk," then the negation would be "Someone in the NBA is unable to dunk." Indeed, in order to show "Every NBA player can dunk" is false, you would have to show that "someone in the NBA is unable to dunk" is true. Using quantifiers more explicitly, the negation of "For all players p in the NBA, player p can dunk" would be "There exists an NBA player p such that p cannot dunk."

Notice that in this example, the negation still referred to NBA players! If I said that every NBA player can dunk and you said "nuh uh, what about Serena Williams??", then your argument is flawed, because Serena Williams is not an NBA player (she plays tennis). If the universe of people I'm talking about are NBA players, then the negation remains in that universe. Getting back to our original example with R, if the universe of x-values that I am referring to is $\mathbb{R}$, then the negation remains in that universe. That's why the negation is "$\exists x \in \mathbb{R}$," rather than "$\exists x \notin \mathbb{R}$."

[35]Note that R is true, as it is saying that every real number has a cubed root. And since R is true, $\sim R$ will be false; indeed, $\sim R$ is saying that there exists a real number which is not the cubed root of *any* real number.

[36]Note: The fact that the words "such that" moved is because of the English, not the math. Include those words where they make sense linguistically. Typically, "there exists" is followed by a "such that." Meanwhile, "for all" is often followed by just a comma, or a phrase like "we have" or "it is the case that."

— Negations with Implications[37] —

If P and Q are statements, then $P \wedge Q$, $P \vee Q$ and $P \Rightarrow Q$ are statements, too. We have discussed negating the first two of these; let's now discuss how to negate $P \Rightarrow Q$. To think about this, first recall the truth table for $P \Rightarrow Q$:

P	Q	$P \Rightarrow Q$
True	True	True
True	False	False
False	True	True
False	False	True

The only way for $P \Rightarrow Q$ to be false is for both P to be true *and* for Q to be false. This shows that

$$\sim(P \Rightarrow Q) \;\Leftrightarrow\; P \wedge \sim Q.$$

This will be used in the following example.

Example 5.14. Let S be this statement: For every natural number n, if $3 \mid n$, then $6 \mid n$. (Note: This is false, so $\sim S$ will be true)

In symbols,

$$S\text{: } \forall n \in \mathbb{N},\ (3 \mid n) \Rightarrow (6 \mid n).$$

Then, by distributing the $\sim$ and using $\sim(P \Rightarrow Q) \;\Leftrightarrow\; P \wedge \sim Q$,

$$\begin{aligned}
\sim S\text{: } &\sim\Big(\forall n \in \mathbb{N}, (3 \mid n) \Rightarrow (6 \mid n)\Big)\\
&= \exists n \in \mathbb{N} \text{ such that } \sim\Big((3 \mid n) \Rightarrow (6 \mid n)\Big)\\
&= \exists n \in \mathbb{N} \text{ such that } (3 \mid n) \wedge \sim(6 \mid n)\\
&= \exists n \in \mathbb{N} \text{ such that } (3 \mid n) \wedge (6 \nmid n)
\end{aligned}$$

The negation of S, in words, is this: "There is some natural number n which is divisible by 3 but not by 6."[38]

---

[37] Also the name of my next punk rock album. 🤘

[38] Going to steal an opportunity to point out that while "P if and only if Q" is interesting and important, "P if or only if Q" is distinctly boring and confusing and that nobody should construct the truth table for "$(P \Rightarrow Q) \vee (Q \Rightarrow P)$" unless they are willing to risk chalking up all of logic to symbolic gobbledygook.

— The Contrapositive —

Remember when I told you that the *converse* was a very important definition? Well look alive, because here is another biggie.

Definition.

Definition 5.15. The *contrapositive* of $P \Rightarrow Q$ is $\sim Q \Rightarrow \sim P$.

Given the truth values of P and Q, let's build up the corresponding truth values of its contrapositive: $\sim Q \Rightarrow \sim P$. As always, the first two columns represent the four possible combinations of truth values for P and Q, and the last three columns are what we have deduced.

P	Q	$\sim Q$	$\sim P$	$\sim Q \Rightarrow \sim P$
True	True	False	False	True
True	False	True	False	False
False	True	False	True	True
False	False	True	True	True

Does that final column look familiar... It is the same as the final column as in the $P \Rightarrow Q$ truth table!

P	Q	$\sim Q$	$\sim P$	$\sim Q \Rightarrow \sim P$
True	True	False	False	True
True	False	True	False	False
False	True	False	True	True
False	False	True	True	True

P	Q	$P \Rightarrow Q$
True	True	True
True	False	False
False	True	True
False	False	True

The final columns are the same!

This shows that

$$(P \Rightarrow Q) \quad \Leftrightarrow \quad (\sim Q \Rightarrow \sim P).$$

This is an important result that we devote the entire next chapter to, so let's record it as a theorem.

Theorem.

Theorem 5.16. An implication is logically equivalent to its contrapositive. That is,

$$(P \Rightarrow Q) \quad \Leftrightarrow \quad (\sim Q \ \Rightarrow \ \sim P).$$

With this, let's discuss the riddle from the start of this chapter, which asked whether the two statements in the following sign are saying the same thing or different things.

Good food is not cheap
Cheap food is not good

To our ear, they do seem to be saying different things. The first is asserting something about good food, while the second is asserting something about cheap food. With our perspective of the contrapositive, though, there may be more going on. Here is a mathy way to write these, where F represents some food:

$$F \textit{ is good} \Rightarrow F \textit{ is not cheap}$$
$$F \textit{ is cheap} \Rightarrow F \textit{ is not good}$$

By Theorem 5.16, an implication is logically equivalent to its contrapositive, and so:

$$\begin{aligned}\Big(F \textit{ is good} \Rightarrow F \textit{ is not cheap}\Big) &\Leftrightarrow \Big(\sim(F \textit{ is not cheap}) \Rightarrow \sim(F \textit{ is good})\Big)\\ &\Leftrightarrow \Big(F \textit{ is cheap} \Rightarrow F \textit{ is not good}\Big).\end{aligned}$$

Logically, the two statements are equivalent! Take another look at the two statements, and see if you can convince yourself that they are both saying this: There is no food that is both good and cheap. Moreover, try to convince yourself that this is *all* that they are saying.

As mentioned, using contrapositives will be the focus of the following chapter, so we will soon pick back up this discussion.

5.4 Proving Quantified Statements

Here we provide a brief discussion of existence proofs (of the "there exists" variety) and universal proofs (of the "for all" variety).

Existential Proofs

To prove an existence statement, it suffices to exhibit an example satisfying the criteria. For example, in the opening pages of this book we proved that "there exists a perfect covering of a chessboard," and we did so by drawing one out. Likewise, if you were asked to prove that "there exists an integer with exactly three positive divisors," you could just find an integer which satisfies this; 9 is such an integer.

The above strategy is called a constructive proof—you literally construct an example. There are also non-constructive ways to prove something exists. Often (but not always!) non-constructive proofs make use of some other theorem. For example, in real analysis you will study the *intermediate value theorem*, which tells us that if f is a continuous function (such as a polynomial), and $f(a)$ is negative while $f(b)$ is positive, then there must be some point between a and b, call it c, for which $f(c) = 0$.

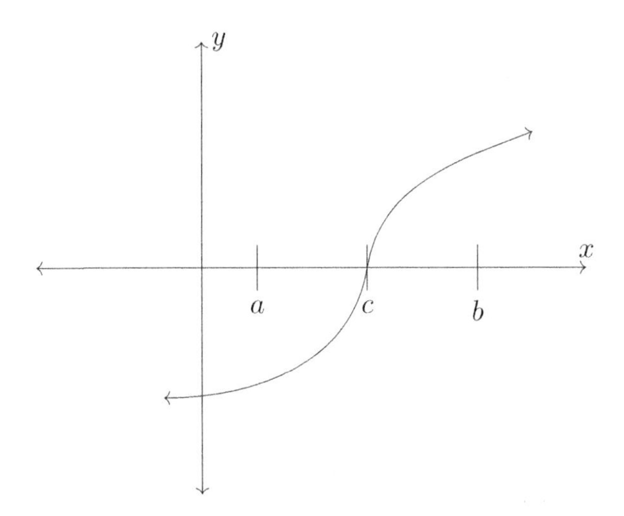

Now suppose you wish to prove that there exists an $c \in \mathbb{R}$ for which

$$c^7 = c^2 + 1.$$

Without the intermediate value theorem, this would be an existential crisis! But with it, it's a breeze. Indeed, we simply let $f(x) = x^7 - x^2 - 1$, which means that we are trying to find some c for which $f(c) = 0$. Note that $f(1)$ is negative while $f(2)$ is positive. Therefore, since f is continuous, the intermediate value theorem guarantees for us some c between 1 and 2 for which $f(c) = 0$.

Rather than finding the explicit c that works, we indirectly showed that such a c must exist. The pigeonhole principle worked in a similar way. It never told you which box had two objects in it, just that such a box must exist.

Universal Proofs

To prove a universal statement, it suffices to choose an arbitrary[39] case and prove it works there. We have seen several examples of this. For instance, if you were asked to prove that "For every odd number n, it must be that $n+1$ is even," your proof wouldn't explicitly check 1 and 3 and 5 and so on. Rather, you would say "Since n is odd, $n = 2a+1$ for some $a \in \mathbb{Z}$." Then you would note that $n+1 = (2a+1)+1 = 2(a+1)$ is even. The point here is that by letting $n = 2a + 1$, you were essentially selecting an arbitrary odd number, and operating on that. Every odd number can be written in that form, and every odd number can have 1 added to it and then factored like we did. Since our n was completely arbitrary, everything we did could be applied to any particular odd number. Proving something holds for an *arbitrary* element of a set, proves that it in turn holds for *every* element in that set.

The field of *real analysis* has a lot of definitions which make good use of quantifiers. For example, here is the definition of what it means for a sequence to *converge*:

> A sequence $a_1, a_2, a_3, \ldots$ *converges* to $a \in \mathbb{R}$ if for all $\varepsilon > 0$ there exists some N such that $|a_n - a| < \varepsilon$ for all integers $n > N$.

This is a universal statement, since it begins with a "for all," but it also includes an existential condition. Working with definitions like these is a great way to sharpen your knowledge of quantifiers. To give you a head start, in this chapter's Bonus Examples section, I include a 5-page introduction to proving sequence convergence. Then, following this chapter is an *Introduction to Real Analysis*, whose primary goal is to highlight a fun aspect of the field—the wealth of bizarre and exciting examples it contains.

5.5 Paradoxes

To close out the main content of this chapter on logic, let's steal a few pages to talk about paradoxes. It would be natural to start off by giving you a formal definition of a *paradox*, but the term is used in several distinct ways. It has been used to mean "something counterintuitive," such as when Derek Jeter had a worse batting average than David Justice in the 1995 season and again in the 1996 season, but yet, over the two years combined, had a better batting average.[40] Or how, from the year 2000

[39]Reminder: This means that it is fixed, but we know nothing else about it.

[40]This phenomenon is called *Simpson's paradox*, and it comes down to the sample sizes:

	1995	1996	Combined
Jeter	.250 (12/48)	.314 (183/582)	**.310** (195/630)
Justice	**.253** (104/411)	**.321** (45/140)	.270 (149/551)

This is similar to how a US presidential candidate can win the popular vote but lose the electoral college. How you divide up your at-bats—or your votes—matters.

to 2015, the median income fell for every educational group, but yet incomes rose when combining all of those incomes and considering the country as a whole.[41]

In these examples, something counterintuitive occurs, but logic itself is not bending or breaking. Not only are these consistent with logic, they occurred in the real world! And while these are fun to think about, they are not self-defeating paradoxes. They lead to no logical failing, and they have perfectly reasonable explanations.

Another class of "paradoxes" are of the magic-trick variety. For example, one of the first exercises in Chapter 1 of this book was to find the error in the following paradoxical "proof" that $2 = 1$.

Let $x = y$. Then,

$$\begin{aligned} x^2 &= xy \\ x^2 - y^2 &= xy - y^2 \\ (x+y)(x-y) &= y(x-y) \\ x + y &= y \\ 2y &= y \\ 2 &= 1. \end{aligned}$$

You will notice that $x = y$ implies that $x - y = 0$, which means that we were not allowed to divide by $(x - y)$ to move from the third line to the fourth. This "proof" was a careful bit of sleight-of-hand, but it is far from a math breaker. Most visual or animated "paradoxes" work this way. Like a magic trick, somewhere there is a small lie which produces the effect. Here's a final (and more subtle!) example which "proves" that $1 = -1$. Recalling that the imaginary number $i = \sqrt{-1}$,

$$1 = \sqrt{1} = \sqrt{(-1)(-1)} = \sqrt{-1}\sqrt{-1} = i \cdot i = i^2 = (\sqrt{-1})^2 = -1.$$

There are also paradoxes which, even under careful inspection, really seem to contradict math and logic. I kick off my real analysis book by talking about *Zeno's paradox*, which asks you to consider a hypothetical race between Achilles and a tortoise. Zeno argued that if the tortoise was given any amount of a head start, that Achilles cannot possibly win. It was a laughable conclusion, but his logic seemed airtight. It took the development of calculus to really pin down why Zeno's argument was flawed.

None of these so-called paradoxes have dealt a fatal blow to mathematics, but they have highlighted misconceptions among mathematicians. The age of rigorous,

[41] And because we are considering the median, this is *not* due to the super wealthy gaining so much more money. It is due to social mobility!

Highest Education Attained	Median Income Change (2000 - 2015)
High school dropout	Incomes fell by 7.9%
High school diploma	Incomes fell by 4.7%
Some college	Incomes fell by 7.6%
At least one college degree	Incomes fell by 1.2%
Everyone combined	Incomes rose by 0.9%

axiomatic-based math is in-part a campaign to guard ourselves against these pitfalls.[42]

A famous example of this comes from set theory. It is hard to get more simple than the notion of a set, but the mathematical world did a double take in 1901 when Bertrand Russell discovered a paradoxical set. Recalling that the elements of a set can themselves be sets, he considered the set

$$R = \{x : x \notin x\}.$$

Symbolically, there was no reason to disallow such a set definition, but yet you can work out the strange contradiction that $R \in R$ if and only if $R \notin R$; indeed, if R is not a member of itself, then according to its definition it must be a member of itself, and if it does contain itself, then it contradicts its own definition.[43] How un*set*tling! Russell's paradox demonstrated a hole in our theory, but that hole is now patched.[44]

In the above example, it was natural to assume that since the set R was able to be succinctly defined, that it must exist. But this was false, and no such R exists. The issue comes down to the fact that the set's definition, in some way, referred back to itself. No word in the dictionary uses itself in its definition, but in a less direct way language can also fall into this trap. As a fun example, a word is called *autological* if it is an example of itself. For instance, *word* is a word, and *unhyphenated* has no hyphens. Here are some more:

- A *pentasyllabic* word is one having five syllables, and pen-ta-syl-la-bic does indeed have five syllables.

- The word *hellenic* means to be to be of Greek origin. And this word is itself hellenic, as it comes from *Ελληνικός* (ellhnikos), which is the ancient Greek word for "Greek."

- The word *oxymoron* refers to a term that is self-contradicting. The word has two parts to it: The 'oxy' part comes from a Greek word meaning 'sharp', and the 'moron' part meaning 'dull'. The word itself is self-contradicting, so it itself is an oxymoron![45]

- The word *repetitive* does sounds kind of repetitive.[46]

On the other hand, there are also words that, in some way, are not examples of themselves. For instance, the word *verb* is a noun, and *monosyllabic* means to have one syllable, but the word itself has five syllables. There are other examples which take this to an an amusing degree. *Hippopotomonstrosesquippedaliophobia*, for instance, is the fear of long words.

---

[42]Admittedly, doing so does spoil some of our fun. We're just trying to have our minds blown, but the drabby-clothed logicians keep stepping in and saying *"Ack-tually, it was perfectly fine the whole time!"*

[43]Related: If Pinocchio said "my nose will grow right now," what would happen?

[44]In fact, they went so deep into the rabbit hole to find the patch, that the set theory we learned in Chapter 3 is now called *naive set theory*! How rude!

[45]Likewise with 'preposterous': combining 'pre' and 'post' in a single word is indeed preposterous!

[46]Related: Because of how it is spelled, the word "parallel" contains an example of what it means.

A word that does not describe itself—meaning it is *not* autological—is called *heterological.* So here's a question: is the word "heterological" itself heterological? Or is it autological? Well, if it is autological, then it describes itself, but what heterological means is something that doesn't describe itself—so if we assume it describes itself then we can conclude that it does not describe itself. So that can't be right!

So "heterological" must be heterological? Well, if it does not describe itself, and it itself means "does not describe itself," then that means it *does* describe itself. So if we assume it does not describe itself, then we conclude that it does describe itself. Again, that can't be right!

So it can't be anything?? This phenomenon is known as the Grelling-Nelson paradox and is another example of a self-referential paradox where the bind comes not from the logic, but a faulty definition.[47]

Indeed, there are surprising results that often get called paradoxes — especially in math fields for which we all have real-world intuition, like statistics, probability and geometry. And some logical tangles seemed paradoxical for centuries until we found sophisticated ways to unwind them — like work involving infinity. But now that we *have* proved/resolved them, it seems strange to me to keep calling them paradoxes. Simpson's paradox and Zeno's paradox are called paradoxes because they are counterintuitive, not because there is anything contradictory about them.

The remaining "paradoxes" in math are of the false-at-their-core type.[48] There is some fundamental flaw which is causing the inconsistency, like the basic misuse of an idea or object.[49] Sometimes, these are benign paradoxes which we learn to live with, like the Grelling-Nelson paradox. Other times, these are things which highlight an error in our theory, and we are forced to correct our work in some way, whether it be a definition, a piece of logic, or an axiom. Russell's paradox is an example of this.

The goal of rigorous, axiomatic mathematics is to drive out each of these "real" paradoxes (which are sometime called *antinomies*). In some ways this makes math beautiful and pure. In other ways, it loses something exciting. Paradoxes of relativity and quantum mechanics — genuine collisions of ideas — drove much of 20th century physics. How exciting to wonder about a cat in a closed box, your twin soaring through space near the speed or light, a box filled with light, or particle-like waves! Just a single strange idea, caught between competing theories, can spawn a hundred papers and a thousand YouTube videos!

To math's credit, there is something to be said for purity and knowable truth. And while physics is constrained by the laws of our universe, math's free-rein has allowed it to grow wider and freer and wilder than physics ever could. Our field contains so much depth for those willing to submerge. And plus, everyone can watch their YouTube videos.

[47]A classic riddle works along the same lines: "A barber cuts the hair of everyone in his town who does not cut their own hair. Does the barber cut his own hair?"

[48]Unless you count unprovable statements as paradoxes, which I do not.

[49]Unless you're thinking beyond math in which case, let's be honest, they usually involve time travel.

5.6 Bonus Examples

Truth tables can involve more than just two statements, P and Q. Let's do an example with three statements, P, Q and R. With three statements, we will have three columns before the double line, and these first three columns must contain all possible true and false combinations for P, Q and R. There are eight such combinations:

- True/True/True
- True/True/False
- True/False/True
- True/False/False
- False/True/True
- False/True/False
- False/False/True
- False/False/False

Thus, our table will begin like this:

P	Q	R		
True	True	True		
True	True	False		
True	False	True		
True	False	False		
False	True	True		
False	True	False		
False	False	True		
False	False	False		

Example 5.17. The truth table for the statement $(\sim P) \Leftrightarrow (Q \vee R)$ is below.

P	Q	R	$\sim P$	$Q \vee R$	$(\sim P) \Leftrightarrow (Q \vee R)$
True	True	True	False	True	False
True	True	False	False	True	False
True	False	True	False	True	False
True	False	False	False	False	True
False	True	True	True	True	True
False	True	False	True	True	True
False	False	True	True	True	True
False	False	False	True	False	False

If we had a truth table with four statements, P, Q, R and S, then it would require 16 rows.[50]

[50] One way to think about these is by taking the numbers $0, 1, 2, 3, \ldots, 15$, writing them in binary, and then turning 0s to "True" and 1s to "False." Thus:

- $0 \to 0000 \to$ TTTT
- $1 \to 0001 \to$ TTTF
- $2 \to 0010 \to$ TTFT
- $3 \to 0011 \to$ TTFF
- $4 \to 0100 \to$ TFTT
- $5 \to 0101 \to$ TFTF
- $6 \to 0110 \to$ TFFT
- $7 \to 0111 \to$ TFFF
- $8 \to 1000 \to$ FTTT

$\cdots$

P	Q	R	S			
True	True	True	True			
True	True	True	False			
True	True	False	True			
True	True	False	False			
True	False	True	True			
True	False	True	False			
True	False	False	True			
True	False	False	False			
False	True	True	True			
False	True	True	False			
False	True	False	True			
False	True	False	False			
False	False	True	True			
False	False	True	False			
False	False	False	True			
False	False	False	False			

Proving a Sequence Converges

As promised, let's practice using quantifiers by turning to an example which combines the "for all" and "there exists" quantifiers: sequences of numbers, and what it means for a sequence to converge. We discussed sequences in the last *Introduction to*, but as a quick reminder, a sequence is an infinite list of numbers, like

$$1\ ,\ \frac{1}{2}\ ,\ \frac{1}{3}\ ,\ \frac{1}{4}\ ,\ \frac{1}{5}\ ,\ \dots .$$

You may notice that the above sequence is getting closer and closer to 0. In fact, we say that this sequence *converges* to 0. How should we define convergence? Saying that it gets "closer and closer" is not sufficiently precise since, for example, the sequence

$$1.1\ ,\ 1.01\ ,\ 1.001\ ,\ 1.0001\ ,\ 1.00001\ ,\ \dots$$

also gets closer and closer to 0, in the sense that every term is closer to 0 than the last. But this is clearly not what we mean, since all the terms are farther than 1 away. We want the terms to get *arbitrarily close* to 0. The sequence, $1, \frac{1}{2}, \frac{1}{3}, \frac{1}{4}, \dots,$ not only is getting closer and closer to 0, but there is no bound on how close it gets.

Consider an arbitrary sequence, which we denote

$$a_1\ ,\ a_2\ ,\ a_3\ ,\ a_4\ ,\ a_5\ ,\ \dots .$$

If this sequence is going to converge to a number a, it must be that after some point, all the terms of the sequence are within 0.5 of a. And it must be that after some later point, all the terms of the sequence are within 0.1 of a. And, later on, it must be that after some later point, all the terms of the sequence are within 0.0001 of a. And so on.

In general, if ε is any number larger than 0 (maybe $\varepsilon = 0.27$ or $\varepsilon = 0.01$ or something else a little bigger than 0), then the terms of the sequence must eventually get within ε of a, and from that point on they must remain within ε of a. Here's what that looks like:

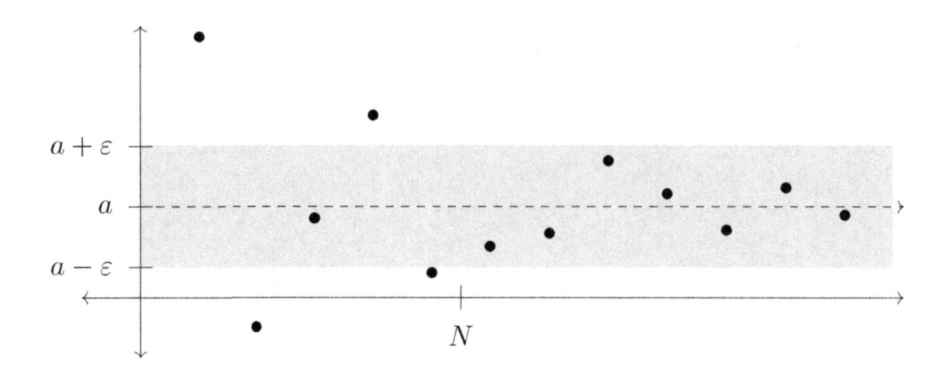

In this picture, the "N" represents the point after which all the terms of the sequence remain within ε of a. Now, if you chose a smaller value of ε, which we will call ε_2, then you may need to choose a larger value of N, which we will call N_2, to mark the point after which all the terms are within ε_2 of a.

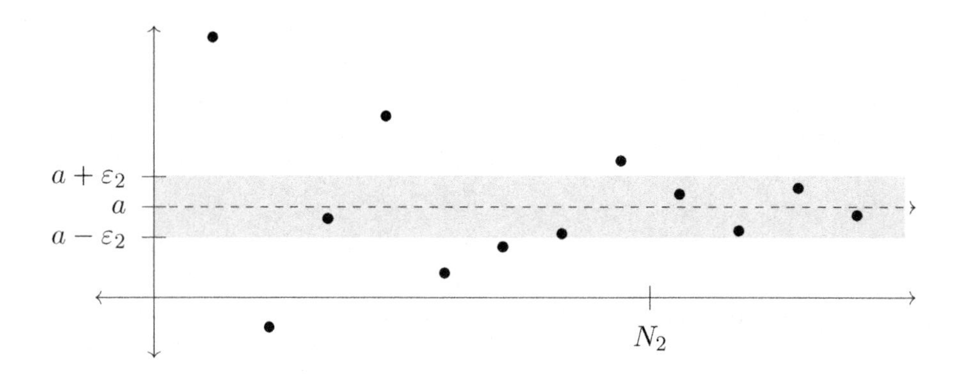

With this intuition, see if you can parse the following precise definition for sequence convergence.

Definition.

Definition 5.18. This definition has two parts:

- A sequence $a_1, a_2, a_3, \ldots$ *converges* to $a \in \mathbb{R}$ if for all $\varepsilon > 0$ there exists some $N \in \mathbb{R}$ such that $|a_n - a| < \varepsilon$ for all integers $n > N$.
- When this happens, we write $a_n \to a$ and call a the *limit* of this sequence.

Because the sequence must be getting arbitrarily close to a, it must be the case that, for all $\varepsilon > 0$, eventually its terms get within ε of a. But for any such ε, all we require is that there exists just one N to mark the point after which all the terms of the sequence are within ε of a. And, finally, a_n being within ε of a simply means $|a_n - a| < \varepsilon$.

Given a specific sequence $a_1, a_2, a_3, \ldots$ and a real number a which this sequence converges to, here is the outline for how we will prove that this sequence converges to a.

Outline.

Outline 5.19. To show that $a_n \to a$, begin with preliminary work:

0. Scratch work: Start with $|a_n - a| < \varepsilon$ and unravel to solve for n. This tells you which N to pick for Step 2 below.

Now for your actual proof:

1. Let $\varepsilon > 0$.
2. Let N be the final value of n you got in your scratch work, and let $n > N$.
3. Redo scratch work (without ε's, and[51] in the opposite order), but at the end use N to show that $|a_n - a| < \varepsilon$.

In short: The strategy is to start at the end ($|a_n - a| < \varepsilon$), and then unwind that until you reach something you know to be true. The actual proof will then be in the reverse of your scratch work. By beginning at the end, you learn how to begin![52]

Let's do some examples!

[51]To grammar nazis: I am aware that the contraction does not pluralize. However, in math where we use letters as variables, "εs" can be misread as a product between 'ε' and 's'. Occasionally I add an apostrophe before the 's' when pluralizing a variable in order to avoid mathematical confusion, even if it is grammatically problematic... Also to grammar nazis: Apologies for all the mistakes that I have no idea I'm making. Please email me all my typos and errors. Also to grammar nazis: Good job learning proofs!

[52] From *Alice in Wonderland* (written by mathematician Lewis Carroll): "'Begin at the beginning,' the King said gravely, 'and go on till you come to the end: then stop.'" This is good advice, provided you can find your way from the beginning to the end. But if you find yourself going in loops without reaching the end, instead try to find your way from the end to the beginning. Sometimes that's much easier, just ask an 8 year old trying to solve one of those pencil maze puzzles.[53]

[53]Fun Fact: Essentially all[54] of those mazes can be solved by simply following the wall on your right wherever it goes.

[54]Technical condition is that it must be "simply connected." If it fails, though, then you may have to start over.[52] Oh, and since I defined a *toenote* earlier, let's also take this opportunity to declare that the plural of 'footnote' is 'feetnote'.

Proposition.

Proposition 5.20. Let $a_1, a_2, a_3, \ldots$ be the sequence where $a_n = \dfrac{1}{n}$. Then, $a_n \to 0$.

Scratch Work. Here is the corresponding picture:

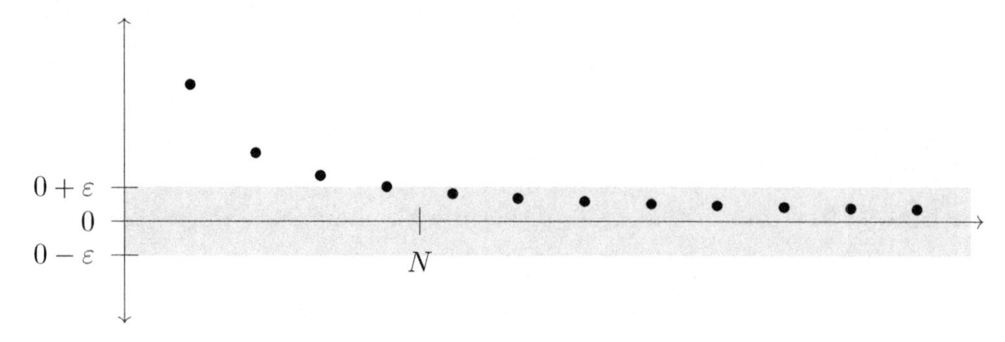

We will follow Outline 5.19. Given an arbitrary $\varepsilon > 0$, we will find a specific N which guarantees that, for every $n > N$, we have $|a_n - 0| < \varepsilon$. For example, if $\varepsilon = \frac{1}{2}$, then $N = 2$ works. If $\varepsilon = \frac{1}{3}$, then $N = 3$ works. If $\varepsilon = \frac{1}{4}$, then $N = 4$ works. You see the pattern, but as practice for how to approach harder problems, here is how we might come about it in general: you start at where you want to reach ($|a_n - a| < \varepsilon$), and then unwind this until we discover an N which would guarantee this.

We want the following:

$$|a_n - a| < \varepsilon$$
$$\left|\frac{1}{n} - 0\right| < \varepsilon$$
$$\frac{1}{n} < \varepsilon$$
$$\frac{1}{\varepsilon} < n.$$

So as long as we choose $N = \dfrac{1}{\varepsilon}$, then for any $n > N$ we will have $n > \dfrac{1}{\varepsilon}$, which by the above will imply that $\frac{1}{n} \to 0$, as desired.

(As you saw, in the above we essentially did the important steps of Outline 5.19 in reverse order. We started with Step 3, undoing a bunch of algebra to find an N that will work for Step 2.) Ok, now here is the formal proof of the proposition.

Proof. Fix any $\varepsilon > 0$. Set $N = \frac{1}{\varepsilon}$. Then for any $n > N$ (implying $\frac{1}{n} < \frac{1}{N}$),

$$|a_n - a| = \left|\frac{1}{n} - 0\right| = \frac{1}{n} < \frac{1}{N} = \frac{1}{1/\varepsilon} = \varepsilon.$$

That is, $|a_n - a| < \varepsilon$. So by Definition 5.18 we have shown that $\dfrac{1}{n} \to 0$. $\square$

Proposition.

Proposition 5.21. Let $a_1, a_2, a_3, \ldots$ be the sequence where $a_n = 2 - \frac{1}{n^2}$. Then, $a_n \to 2$.

We follow Outline 5.19.

Scratch Work. Here is the corresponding picture:

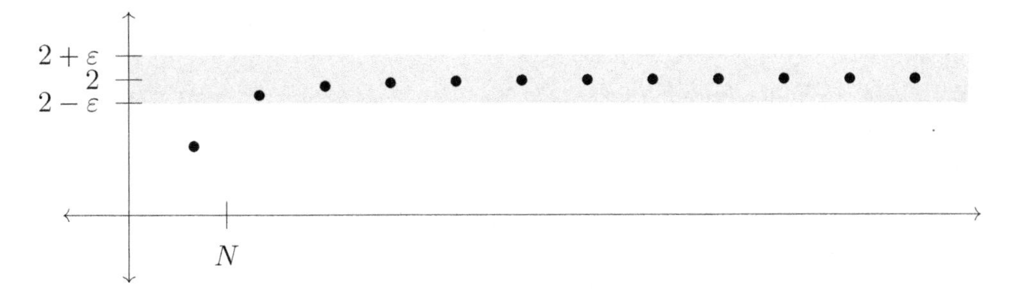

Again, we first begin with our conclusion (that $|a_n - a| < \varepsilon$), and do some algebra to figure out which values of n would give this.

We want the following:

$$\begin{aligned}
|a_n - a| &< \varepsilon \\
\left|\left(2 - \frac{1}{n^2}\right) - 2\right| &< \varepsilon \\
\left|-\frac{1}{n^2}\right| &< \varepsilon \\
\frac{1}{n^2} &< \varepsilon \\
\frac{1}{\varepsilon} &< n^2 \\
\frac{1}{\sqrt{\varepsilon}} &< n.
\end{aligned}$$

So as long as we choose $N = \frac{1}{\sqrt{\varepsilon}}$, then for any $n > N$ we will have $n > \frac{1}{\sqrt{\varepsilon}}$, which by the above will imply that $2 - \frac{1}{n^2} \to 2$, as desired. Below is the formal proof.

Proof. Fix any $\varepsilon > 0$. Set $N = \frac{1}{\sqrt{\varepsilon}}$. Then for any $n > N$,

$$|a_n - a| = \left|\left(2 - \frac{1}{n^2}\right) - 2\right| = \frac{1}{n^2} < \frac{1}{N^2} = \frac{1}{1/(\sqrt{\varepsilon})^2} = \frac{1}{1/\varepsilon} = \varepsilon.$$

That is, $|a_n - a| < \varepsilon$. So by Definition 5.18 we have shown that $2 - \frac{1}{n^2} \to 2$. □

This next one looks a bit trickier, but the same procedure works.

Proposition.

Proposition 5.22. Let $a_1, a_2, a_3, \dots$ be the sequence where $a_n = \dfrac{3n+1}{n+2}$. Then, $a_n \to 3$.

Scratch Work. Again, we first play around. We start with where we want to get to (that $|a_n - a| < \varepsilon$), and then do some algebra to figure out which values of n would give this.

We want the following:

$$\begin{aligned}
|a_n - a| &< \varepsilon \\
\left|\frac{3n+1}{n+2} - 3\right| &< \varepsilon \\
\left|\frac{3n+1}{n+2} - \frac{3(n+2)}{n+2}\right| &< \varepsilon \\
\left|\frac{3n+1-3n-6}{n+2}\right| &< \varepsilon \\
\left|\frac{-5}{n+2}\right| &< \varepsilon \\
\frac{5}{n+2} &< \varepsilon \\
\frac{5}{\varepsilon} &< n+2 \\
\frac{5}{\varepsilon} - 2 &< n
\end{aligned}$$

So as long as we choose $N = \dfrac{5}{\varepsilon} - 2$, then for any $n > N$ we will have $n > \dfrac{5}{\varepsilon} - 2$, which by the above will imply that $\frac{3n-1}{n+2} \to 3$, as desired.

Proof. Fix any $\varepsilon > 0$. Set $N = \frac{5}{\varepsilon} - 2$. Then for any $n > N$,

$$\begin{aligned}
|a_n - a| &= \left|\frac{3n+1}{n+2} - 3\right| = \left|\frac{3n+1}{n+2} - \frac{3n+6}{n+2}\right| \\
&= \frac{5}{n+2} < \frac{5}{N+2} = \frac{5}{\left(\frac{5}{\varepsilon} - 2\right) + 2} \\
&= \frac{5}{5/\varepsilon} = \varepsilon.
\end{aligned}$$

That is, $|a_n - a| < \varepsilon$. So by Definition 5.18 we have shown that $\dfrac{3n+1}{n+2} \to 3$. $\square$

— Chapter 5 Pro-Tips —

- Thinking this formally, especially in terms of truth tables, is something you may not see in any later course, and even if you spent a career doing math research you may never feel the need to whip out a truth table. They really are not used much.[55] That said, this chapter does have its uses. The ability to parse a technical statement, work with quantifiers, negate statements and find the contrapositive of a statement are skills that you should ideally be able to do effortlessly. Working through all this is primarily to help you get to that point.

- All that said, one of the most difficult concepts for students (or anyone) to wrap their heads around is the idea that if P is false, then the implication $P \Rightarrow Q$ is considered true. Fortunately, this particular idea is one that very rarely comes up in higher-level mathematics. In math we almost always are dealing with P's which we are assuming to be true, or know to be true. This weird case where P is known to be false, and yet we are still interested in knowing the truth value of some implication $P \Rightarrow Q$, will likely never play a significant role in your future work. But at this point in your math journey, there is still a benefit to having crossed all of our antecedent t's and dotted all of our consequent i's.[56]

- If you really like logic and are considering studying it further, I thought I'd give you a quick heads up: If you are male and you pursue a PhD in mathematical logic, you will be required to grow a beard and wear some weird type of shoes. I wish there were a way around it, but from my experience I can only assume that is the law.

- Grammar Pro-Tip: In English, every "if, then" sentence has a comma separating the two clauses. Examples:
  - If n is odd, then n^2 is odd.
  - If p and $p + 2$ are both prime, then p is called a *twin prime*.
  - If p and $p + 4$ are both prime, then p is called a *cousin prime*.
  - If p and $p + 6$ are both prime, then p is called a *sexy prime*.[57]
  - If a sexy prime could also be a twin prime and a cousin prime, then a lot of college kids would be very amused.[58]

- One peculiar thing about math is that we typically use "if" in our definitions when we really mean "if and only if." We say "n is even if $n = 2k$ for some $k \in \mathbb{Z}$," even though we really mean that those two are the same thing. With

[55] Note: There is a less cumbersome system called *Boolean algebra*, which sets up the rules to do arithmetic on variables which only take the values "true" or "false." This algebra is used lots in computer science, as well as in logic and statistics.

[56] Plus, when else can you talk about unicorns in math class?

[57] Yes, this is a real term, and yes this is its actual definition.

[58] Sorry college kids, but this is impossible. Try to prove this on your own by thinking about p, $p + 2$, $p + 4$ and $p + 6$, each modulo 3.

an "if" we seem to be leaving open the possibility that $n = 2k$, for $k \in \mathbb{Z}$, could be true without n being even. But this is never intended when stated as a definition. Importantly, though: This is the *only* time in math where we conflate these two.

- De Morgan's logic law said $\sim(P \wedge Q) \Leftrightarrow \sim P \vee \sim Q$. There are similar distributive laws for $\wedge$ and $\vee$, which can also be demonstrated via truth tables. They are:

$$P \wedge (Q \vee R) = (P \wedge Q) \vee (P \wedge R)$$
$$P \vee (Q \wedge R) = (P \vee Q) \wedge (P \vee R)$$

When you do not mix-and-match $\wedge$s and $\vee$s, things associate nicely:

$$P \wedge (Q \wedge R) = (P \wedge Q) \wedge R$$
$$P \vee (Q \vee R) = (P \vee Q) \vee R$$

Notice that, just like with De Morgan's laws, logic rules mirror set theory rules. If you replace statements P, Q and R with sets A, B and C, and you replace logic operators $\wedge$ and $\vee$ with set operations $\cap$ and $\cup$, then the four rules above still hold.

- In Section 5.4 we discussed how to prove the existence of something. In many areas of math, such as differential equations, an important class of problems is to determine not only the *existence* of something, but the *uniqueness* of that thing as well. That is, the aim is to show that there is one and only one of that thing.

- Hopefully I bugged you enough that you have created a study group. Here is your reminder to keep meeting with them; the best team sport is math![59] In fact, a 2002 study found that in the 1990s, the average math paper had 1.63 authors on it, and 81% of all publishing mathematicians in this decade wrote at least one coauthored paper. These rates have been increasing every decade.

- Logicians use slightly different language than mathematicians. As a small example, what mathematicians call a "direct proof" a logician would call a "conditional proof." To them, a "direct proof" would be one in which you establish a proposition without an assumption.

- It should be crystal clear in your mind when an example is sufficient as a proof, and when it is not. If the theorem says "<u>there exists</u> a perfect covering of the 8×8 chessboard," then an example is enough. It alone proves the theorem, because the theorem was an "existential" statement. However, if the theorem is a "universal" statement, like "<u>for all</u> even integers n, the number n^2 is also even," then an example is not enough. A single example will only cover a single n-value, and thus can not prove this universal statement. Getting these confused is a common mistake for budding mathematicians.

[59]Shout-out to my own undergrad study group of Zach, Adam, Hotovy, Corey and Laila, without whom I wouldn't be here today.

- As we have discussed, if you prove that $P \Rightarrow Q$ and $Q \Rightarrow P$, then P and Q are equivalent — if one is true, then so is the other. We wrote this as $P \Leftrightarrow Q$. Suppose you wanted to show that three statements — P, Q and R — are all equivalent to each other, meaning that $P \Leftrightarrow Q$, and $P \Leftrightarrow R$, and $R \Leftrightarrow Q$. Must you show all six of the following?

$$\begin{array}{ccc} P \Rightarrow Q & P \Rightarrow R & R \Rightarrow Q \\ Q \Rightarrow P & R \Rightarrow P & Q \Rightarrow R \end{array}$$

Answer: no.

In fact, if you show $P \Rightarrow Q$, and $Q \Rightarrow R$, and $R \Rightarrow P$, then that's enough. Why is this enough? Why does this, for example, imply that $Q \Rightarrow P$ (that is, if Q is true then P is also true)? Well, if Q is true, then R is true (since $Q \Rightarrow R$), which in turn means P is true (since R is true and $R \Rightarrow P$).

The fact that these three implications are enough can be visualized in the triangular diagram on the right. If any one of these three is true, then by moving around the triangle of implications, you can arrive at any other being true.

$$\begin{array}{ccc} & Q & \\ \nearrow & & \searrow \\ P & \Leftarrow & R \end{array}$$

If you had six statements, you would have to prove six implications to show that any two are equivalent. And note that you have options about which six implications you show! You get to choose! For example, here are two options:

$$\begin{array}{ccccc} P & \Rightarrow & Q & \Rightarrow & R \\ \Uparrow & & & & \Downarrow \\ U & \Leftarrow & T & \Leftarrow & S \end{array} \qquad \begin{array}{ccccc} P & \Rightarrow & U & \Rightarrow & T \\ \Uparrow & & & & \Downarrow \\ R & \Leftarrow & S & \Leftarrow & Q \end{array}$$

Note that showing some pairs are equivalent gives you additional routes.

$$\begin{array}{ccccc} P & \Rightarrow & Q & \Leftrightarrow & R \\ \Uparrow & & \Downarrow & & \\ U & \Leftarrow & T & \Leftrightarrow & S \end{array} \qquad \begin{array}{ccccc} P & \Leftrightarrow & U & \Leftrightarrow & T \\ & & \Updownarrow & & \\ R & \Leftrightarrow & S & \Leftrightarrow & Q \end{array}$$

- One final example of a logical deduction: If interest and talent in mathematics does not favor any demographics criteria, and if it is a righteous goal to allow everyone with an educational passion the opportunity to pursue their goals, then math educators should work diligently to show more young people the beauty of mathematics, and we should all support them in this endeavor.[60]

---

[60] M∀TH.

— Exercises —

Exercise 5.1. Explain why the following is logically correct.

1. Everyone loves my baby;

2. My baby loves only me;

3. Therefore, I am my own baby.

Exercise 5.2. You are investigating *muuuuurrderr*. The crime was committed by either Miss Scarlet, Colonel Mustard or Professor Plum. The murder weapon was either a wrench, a knife or a candlestick. The crime occurred at either noon or midnight. And it occurred in the billiard room, the kitchen or the library. The following additional facts have been established at the scene of the crime.

1. Either the weapon was the wrench or the crime took place in the billiard room (but not both).

2. If the crime took place at midnight, then Colonel Mustard is guilty.

3. Professor Plum is innocent if and only if the weapon was not the wrench.

4. If the weapon was the candlestick, then Miss Scarlet is guilty.

5. If the crime occurred at noon, then it was not in the library and the weapon was not a knife.

Now, as a side hustle, you are also a mathematician. As such, you start wondering what further piece of evidence would conclusively determine the killer. For each of the following, explain how that piece of evidence — if established — would determine the killer.

(a) The crime did not take place in the billiard room.

(b) The murder was not at noon.

(c) The murderer was not Miss Scarlet and the weapon was not the wrench.

(d) The crime took place at noon in the billiard room.

Hint: Remember what the contrapositive tells us.

Exercise 5.3. Determine which of the following are statements. Among those that are, determine whether it is true or false.

(a) $2 + 3 = 5$.

(b) The sets $\mathbb{Z}$ and $\mathbb{Q}$.

(c) The sets $\mathbb{Z}$ and $\mathbb{Q}$ both contain $\sqrt{2}$.

(d) Every real number is an integer.

(e) Every integer is a real number.

(f) $\mathbb{N} \in \mathcal{P}(\mathbb{N})$.

(g) The integer n is a multiple of 5.

(h) $\sin(x) = 1$.

(i) Either $5 \mid n$ or $5 \nmid n$.

(j) 8765309 is a prime number.

(k) 0 is not positive or negative.

(l) Proofs are fun.[61]

Exercise 5.4. Each of the following statements can be written in the form $P \wedge Q$, $P \vee Q$ or $\sim P$. Determine which of these forms each statement takes; write down explicitly what P and Q stand for in your framing.

(a) $2 \mid 8$ and $4 \mid 8$

(b) $x \neq y$

(c) $x < y$

(d) $x \leq y$

(e) n is even while m is not

(f) $x \in A \setminus B$

(g) Either x or y is zero.

(h) 27 is both odd and divisible by 3.

Exercise 5.5. Consider this open sentence:

$$\frac{2n^2 + 5 + (-1)^n}{2} \text{ is prime.}$$

Give an n-value for which this becomes a true statement, and an n-value for which this becomes a false statement.

Exercise 5.6. Give an example of an open sentence. Also, write down an input value that causes your open sentence to be a true statement, and a second input value that causes your open sentence to be a false statement.

Exercise 5.7. Each of the below includes a hidden quantifier. Rewrite each of these sentences in such a way that includes either "for all" or "there exists."

(a) If f is an odd function, then $f(0) = 0$.

(b) The equation $x^3 + x = 0$ has a solution.

[61] ♫♫ I like to prove it prove it. I like to prove it prove it. We like to — Prove It! ♫♫

Exercise 5.8. Rewrite each of the following sentences to be of the form "If P, then Q." Make sure your new wording does not change its meaning. (You don't need to understand what all these are saying to answer this question. However, you may learn all of these before you earn your degree!)

(a) A group is cyclic whenever it is of prime order.

(b) Two graphs have identical degree sequences whenever they are isomorphic.

(c) Being differentiable is a sufficient criterion for a function to be continuous.

(d) In order for f to be continuous, it is necessary that it is integrable.

(e) A set A has infinitely many elements only if $|A| \geq |\mathbb{N}|$.

(f) Whenever a tree has m edges, it has $m + 1$ vertices.

(g) An integer is even provided it is not odd.

(h) A geometric series with ratio r diverges whenever $|r| \geq 1$.

Exercise 5.9. Rewrite each of the following sentences to be of the form "P if and only if Q." Make sure your new wording does not change its meaning.

(a) If $n \in \mathbb{Z}$ then $(n + 1) \in \mathbb{Z}$, and if $(n + 1) \in \mathbb{Z}$ then $n \in \mathbb{Z}$.

(b) For a rectangle to be a square, it is necessary and sufficient that its sides all be the same length.

(c) A matrix A being invertible is equivalent to $\det(A) \neq 0$.

(d) If N is a normal subgroup of G, then $Ng = gN$ for all $g \in G$, and conversely.

Exercise 5.10. Negate the following sentences.

(a) For every prime p, there exists a prime q for which $q > p$.

(b) Every polynomial is differentiable.

(c) If $xy = 0$, then $x = 0$ or $y = 0$.

(d) If mn is odd, then m is odd and n is odd.

(e) If p is prime, then $\sqrt{p} \notin \mathbb{Q}$.

(f) For every $\varepsilon > 0$ there exists an N such that $n > N$ implies $|a_n - a| < \varepsilon$.

(g) For all $\varepsilon > 0$ there exists some $\delta > 0$ such that $|x - a| < \delta$ implies $|f(x) - f(a)| < \varepsilon$.

(h) If I pass Algebra I and Analysis I this semester, then I will take Algebra II or Analysis II next semester.

Exercise 5.11. Give two examples of an implication $(P \Rightarrow Q)$ which are true, but whose converses $(Q \Rightarrow P)$ are not true. One example should be a real-world example, while the other should be an example from math involving the integers (and perhaps even numbers, odd numbers, divisibility, sets, or anything else you wish).

Exercise 5.12. Consider the following triangle, with angles a, b and c.

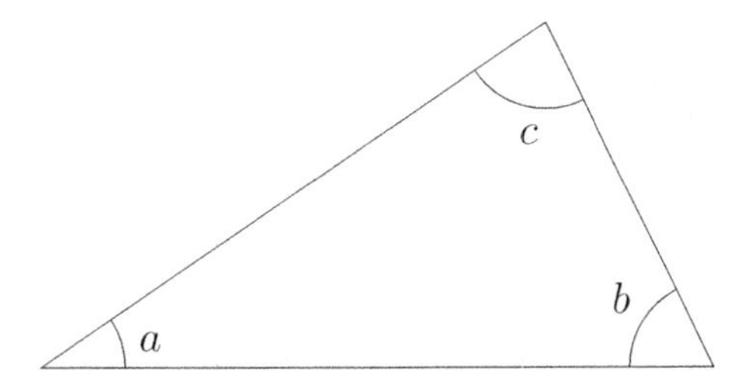

Consider the statement "If $a = 30$ and $b = 50$, then $c = 100$." Is this statement true? Is its converse true? Why or why not?

Exercise 5.13. Prove that for all $x, y \in \mathbb{Q}$, there exists some $z \in \mathbb{Q}$ such that $x < z < y$.

Exercise 5.14. Look up *Fermat's last theorem* and *Goldbach's conjecture.* Write both down on your homework. Is Fermat's last theorem a statement? Is Goldbach's conjecture a statement?

Exercise 5.15. Let P and Q be statements. Determine two statements, each of which is a combination of P and Q, which are logically equivalent (that is, they have the same truth tables). There are many correct answers to this problem.

Exercise 5.16. In the dystopian novel *1984*, the official motto of Oceania is

War is Peace
Freedom is Slavery
Ignorance is Strength

By thinking about each of these as a conjunction of two statements, analyze their logical merit.

Exercise 5.17. True or false: The flying panda in this room is riding a centaur.

Exercise 5.18. Given statements P, Q and R, write the truth tables for the following.

(a) $(\sim P \vee \sim Q) \wedge Q$

(b) $\sim (\sim P \wedge Q)$

(c) $\sim(P \vee Q) \vee (\sim P)$

(d) $\sim(\sim P \vee \sim Q)$

(e) $(P \vee Q) \vee (\sim P \wedge \sim Q)$

(f) $(P \wedge Q) \vee \sim R$

(g) $(P \wedge Q) \vee (P \wedge R)$

(h) $\sim(P \Rightarrow Q)$

(i) $P \vee (Q \Rightarrow R)$

Exercise 5.19. For each of the below, find a compound statement involving P and Q that you could put above the final column to make the truth table make sense.

(a)

P	Q	$\sim P$	$P \vee Q$	$\sim(P \vee Q)$	
True	True	False	True	False	False
True	False	False	True	False	False
False	True	True	True	False	False
False	False	True	False	True	True

(b)

P	Q	$P \wedge Q$	$P \vee Q$	
True	True	True	True	True
True	False	False	True	False
False	True	False	True	False
False	False	False	False	True

Exercise 5.20. Consider the expression $(AxB)y(CzD)$. Using the Internet, pick a random number between 1 and 6. And:

If you get	Replace A with
1	P
2	$\sim P$
3	Q
4	$\sim Q$
5	R
6	$\sim R$

Repeat this procedure for B, C and D.

Next, use the Internet to pick a random number between 1 and 2. And:

If you get	Replace x with
1	$\wedge$
2	$\vee$

Repeat this procedure for y and z.

You have now produced your own logical expression where P, Q and R are statements. Write down which expression you produced and make a truth table for your expression.

Exercise 5.21. Give two examples of statements which are true and whose converse is also true; have one be a real-world example and one a math example. Then, give two examples of statements which are true but whose converse is false; have one be a real-world example and one a math example.

Exercise 5.22. Construct truth tables to prove the following logical equivalences, for propositions P, Q and R.

(a) $P \Leftrightarrow \sim(\sim P)$

(b) $\sim(P \vee Q) \Leftrightarrow \sim P \wedge \sim Q$

(c) $P \wedge (Q \wedge R) \Leftrightarrow (P \wedge Q) \wedge R$

(d) $P \vee (Q \vee R) \Leftrightarrow (P \vee Q) \vee R$

(e) $P \wedge (Q \vee R) \Leftrightarrow (P \wedge Q) \vee (P \wedge R)$

(f) $P \vee (Q \wedge R) \Leftrightarrow (P \vee Q) \wedge (P \vee R)$

(g) $(P \Rightarrow Q) \Leftrightarrow \sim P \vee Q$

(h) $P \Rightarrow (Q \Rightarrow R) \Leftrightarrow (P \wedge Q) \Rightarrow R$

(i) $P \Rightarrow (Q \wedge R) \Leftrightarrow (P \Rightarrow Q) \wedge (P \Rightarrow R)$

(j) $P \Rightarrow (Q \vee R) \Leftrightarrow (P \wedge \sim R) \Rightarrow Q$

(k) $(P \vee Q) \Rightarrow R \Leftrightarrow (P \Rightarrow R) \wedge (Q \Rightarrow R)$

(l) $(P \Rightarrow Q) \Leftrightarrow (P \wedge \sim Q) \Rightarrow (Q \wedge \sim Q)$

(m) $(P \Leftrightarrow Q) \Leftrightarrow (\sim P \vee Q) \wedge (\sim Q \vee P)$

(n) $\sim(P \Leftrightarrow Q) \Leftrightarrow (P \wedge \sim Q) \vee (Q \wedge \sim P)$

Exercise 5.23. In the last exercise you proved the following. For each, describe in your own words why it makes sense that they are equivalent.

(a) $P \Leftrightarrow \sim(\sim P)$

(b) $(P \Rightarrow Q) \Leftrightarrow \sim P \vee Q$

(c) $P \Rightarrow (Q \wedge R) \Leftrightarrow (P \Rightarrow Q) \wedge (P \Rightarrow R)$

(d) $(P \vee Q) \Rightarrow R \Leftrightarrow (P \Rightarrow R) \wedge (Q \Rightarrow R)$

(e) $(P \Rightarrow Q) \Leftrightarrow (P \wedge \sim Q) \Rightarrow (Q \wedge \sim Q)$

(f) $(P \Leftrightarrow Q) \Leftrightarrow (\sim P \vee Q) \wedge (\sim Q \vee P)$

Exercise 5.24. Translate each of the following English sentences to a symbolic sentence with quantifiers.

(a) Every natural number, when squared, remains a natural number.

(b) Every real number has a cube root in the reals.

(c) Not every integer has a square root in the reals.

(d) There exists a smallest natural number.

(e) There exists a largest negative integer.

(f) Every real number, when multiplied by zero, equals 0.

Exercise 5.25. Translate each of the following to plain English, and then write down whether each statement is true or false.

(a) $\exists x \in \mathbb{N}$ such that $3x + 4 = 6x + 13$.

(b) $\exists x \in \mathbb{N}$ such that $\forall y \in \mathbb{N}$, we have $x \leq y$.

(c) $\forall x \in \mathbb{Q}$, $\exists y \in \mathbb{Q}$ such that $x = -y$.

(d) $\forall x \in \mathbb{N}$, $\exists y \in \mathbb{N}$ such that $x = -y$.

(e) $\exists a, b \in \mathbb{N}$ such that $a \neq b$ and $a^b = b^a$.

(f) $\forall x \in \mathbb{R}$, $\exists y \in \mathbb{N}$ such that $x^y \geq 0$.

(g) $\forall x, y \in \mathbb{R}$, $\exists z \in \mathbb{Q}$ such that $x < z < y$.

(h) $\exists x \in \mathbb{N}$ such that $\forall y \in \mathbb{Z}$, we have $x \leq y^2$.

(i) $\forall x \in \mathbb{N}$, $\forall y \in \mathbb{R}$, we have $x \leq y$.

(j) $\forall x \in \mathbb{N}$, $x^2 - x + 41$ is prime.

Exercise 5.26. A *tautology* is a statement which is guaranteed to be true. By finding their truth tables, determine which of the following are tautologies.

(a) $\sim(P \wedge \sim P)$

(b) $(P \wedge Q) \vee (\sim P \wedge \sim Q)$

(c) $(P \wedge Q) \vee (\sim P \vee \sim Q)$

(d) $(P \vee Q) \vee (\sim P \wedge \sim Q)$

(e) $(\sim P \wedge Q) \wedge \sim(P \wedge R)$

(f) $P \Rightarrow (P \vee Q)$

Exercise 5.27. Determine which of the following are true. If it is true, just say so. If it is false, give a counterexample.

(a) There exists some $n \in \mathbb{N}$ such that $\sqrt{n} \in \mathbb{N}$.

(b) There exists some $n \in \mathbb{N}$ such that $\sqrt{n} \notin \mathbb{N}$.

(c) For all $n \in \mathbb{N}$, we have $(20 - n^2) \in \mathbb{N}$.

(d) For all $n \in \mathbb{N}$ there exists some $m \in \mathbb{N}$ such that $(m+1) \mid n$.

(e) For all $x \in \mathbb{R}$ there exists some $y \in \mathbb{R}$ such that $x^2 = y$.

(f) There exists some $x \in \mathbb{R}$ such that for all $y \in \mathbb{R}$, we have $x^2 = y$.

(g) For all $x \in \mathbb{R}$ there exists some $y \in \mathbb{R}$ such that $y^2 = x$.

(h) For all $x \in \mathbb{R}$ there exists some $y \in \mathbb{R}$ such that $y^3 = x$.

Exercise 5.28. Negate the following sentences.

(a) The number n is even.

(b) If $a \mid b$, then $a \mid c$.

(c) The number n is even, but the number m is not.

(d) For every $\varepsilon > 0$ there exists some N such that $|a_n - a| < \varepsilon$ for all $n > N$.

Exercise 5.29. Read the *Introduction to Real Analysis* following this chapter. Then, complete the following.

(a) In Footnote 65, a proof idea was presented. Explain each step of the calculation.

(b) Using blocks or books or bricks or anything else that works, try stacking your objects as demonstrated in this section. Take a picture of the maximum overhang that you obtained.

— Open Question —

A sum of infinitely many numbers is called a *series*. For example, back in Calc II you might have seen the series $\sum_{k=1}^{\infty} \frac{1}{2^k}$. If you were going to try to figure out what this series equals, you would probably start adding up the *partial sums*. That is:

$$\begin{aligned} \frac{1}{2} &= 0.5 \\ \frac{1}{2} + \frac{1}{4} &= 0.75 \\ \frac{1}{2} + \frac{1}{4} + \frac{1}{8} &= 0.875 \\ \frac{1}{2} + \frac{1}{4} + \frac{1}{8} + \frac{1}{16} &= 0.9375 \end{aligned}$$

and if you continued this to, say, 10 terms, you would get this:

$$\frac{1}{2} + \frac{1}{4} + \cdots + \frac{1}{2^{10}} = 0.9990234375.$$

And if you continued this to 100 terms, you would get this:

$$\frac{1}{2} + \frac{1}{4} + \cdots + \frac{1}{2^{100}} = 0.99999999999999999999999999999992111390.$$

As you might guess (or perhaps you already knew), the series $\sum_{k=1}^{\infty} \frac{1}{2^k} = 1$.

Indeed, not only does the sequence of partial sums provide us intuition for what a series equals, it is actually how a series is *defined*. In the Bonus Examples we defined what it means for a sequence to converge. Thus, given a series $\sum_{k=1}^{\infty} a_k$, we can look at the sequence of partial sums:

$$\begin{aligned} s_1 &= a_1 \\ s_2 &= a_1 + a_2 \\ s_3 &= a_1 + a_2 + a_3 \\ s_4 &= a_1 + a_2 + a_3 + a_4, \end{aligned}$$

and so on. This gives us a sequence $s_1, s_2, s_3, s_4, \ldots$, and if this sequence converges to a (as we discussed in the Bonus Examples portion), then we say that $\sum_{k=1}^{\infty} a_k = a$.

That intro is simply to say that when we discuss what a *series* equals in math, what we are really discussing is what a *sequence* is converging to.[62] And while that might sound confusing, it is actually what most people do naturally: To determine an infinite sum, you add up terms as you go along, and see what it is getting close to.

[62] But Lemony Snicket's tale still should have been called *A Sequence of Unfortunate Events*...

In Calc II, you probably learned that an infinite sum can either equal a number (like how $\sum_{k=1}^{\infty} \frac{1}{2^k} = 1$), or can equal ∞, $-\infty$ or can not exist at all. If it equals a number, it *converges*; if it does not equal a number, then it *diverges*.

You also probably learned some important rules for determining whether a series converges or diverges. An important one was the p-test, which said this:

> The series $\sum_{k=1}^{\infty} \frac{1}{k^p}$ converges if and only if $p > 1$.

For example, $\sum_{k=1}^{\infty} \frac{1}{k^3}$ converges.[63] What if, for every k, we scale the "k^3" part by some number between 0 and 1? For instance ... notice that $0 < \sin^2(k) < 1$ for every $k \in \mathbb{N}$. What do you think about the series $\sum_{k=1}^{\infty} \frac{1}{k^3 \sin^2(k)}$? Converge or diverge?

Its first term is

$$1/(1^3 \sin^2(1)) \approx \frac{1}{1 \cdot 0.71} \approx 1.41.$$

Here are some random terms:

k	$\frac{1}{k^3}$	$\sin^2(k)$	$\frac{1}{\sin^2(k)}$	$\frac{1}{k^3 \sin^2(k)}$
2	0.125	0.827	1.209	0.151181
5	0.008	0.920	1.088	0.008700
10	0.001	0.296	3.379	0.003379
50	0.000008	0.069	14.526	0.000116
100	0.000001	0.256	3.900	0.000004

But check out the 355th term:

k	$\frac{1}{k^3}$	$\sin^2(k)$	$\frac{1}{\sin^2(k)}$	$\frac{1}{k^3 \sin^2(k)}$
355	0.00000006	0.0000000009	1100494954.9	24.5982

For most k-values, the $1/k^3$ part is small enough that the $\sin^2(k)$ part doesn't make much of a difference. But for some terms, like the 355th, since $\sin^2(k)$ is soooooo small, sticking it in the denominator makes the whole fraction large.[64] Are there enough terms like this for the series to diverge? That's the open question.

Open Question. Does $\sum_{k=1}^{\infty} \frac{1}{k^3 \sin^2(k)}$ converge or diverge?

[63] This sum is an important number called *Apéry's constant*.

[64] If k is a multiple of π, then $\sin^2(k) = 0$. For $k \in \mathbb{N}$, this will never happen, but it just so happens that 355 is about 113.00001π, so $\sin^2(k)$ is very close to 0.

Introduction to Real Analysis

In the Chapter 5 bonus examples we discussed how to prove that a sequence converges. In fact, that was the first part of your introduction to real analysis! In that section, we gave the following definition for a sequence to converge to a point.

Definition.

Definition 5.18. A sequence $a_1, a_2, a_3, \ldots$ *converges* to $a \in \mathbb{R}$ if for all $\varepsilon > 0$ there exists some N such that $|a_n - a| < \varepsilon$ for all $n > N$.

Definitions in real analysis often use quantifiers like "for all" and "there exists." Indeed, a major idea in real analysis is to ask how something behaves as some aspect of it gets smaller and smaller, or bigger and bigger. For sequences, we were narrowing the window around a, and asking whether the sequence, at some point N, would enter that tight window, and from that point on remain inside of that tight window.

In Chapter 8 we will study functions in detail, but for now rely on your general understanding of what a function is, and what it means for a function to be *continuous*. Got it? Good. Is the definition below what you had in mind?

Definition.

Definition 5.23. Let $f : \mathbb{R} \to \mathbb{R}$. We say that f is continuous at c if for all $\varepsilon > 0$ there exists some $\delta > 0$ such that for every $x \in \mathbb{R}$ for which $0 < |x - c| < \delta$, we have

$$|f(x) - f(c)| < \varepsilon.$$

Probably not! Although I imagine that whatever image you had in mind, you wouldn't have been able to write down what "continuous" means. It is a hard thing to define precisely. And it is even harder when you allow a function to be continuous at one point but not at another, meaning that we must define what it means to be continuous at a point.

Nevertheless, the above definition is still a good one. Indeed, what this definition is saying is that in order for f to be continuous at the x-value of c, it cannot have any jumps right at c. More precisely: If x is some point really close to c, then we demand that $f(x)$ be really close to $f(c)$. Even more precisely: For every $\varepsilon > 0$, we demand that there not be any jump of size larger than ε right around $f(c)$; that is, for any x that is within some δ of c, we demand that $f(x)$ be within ε of $f(c)$. Here's the picture:

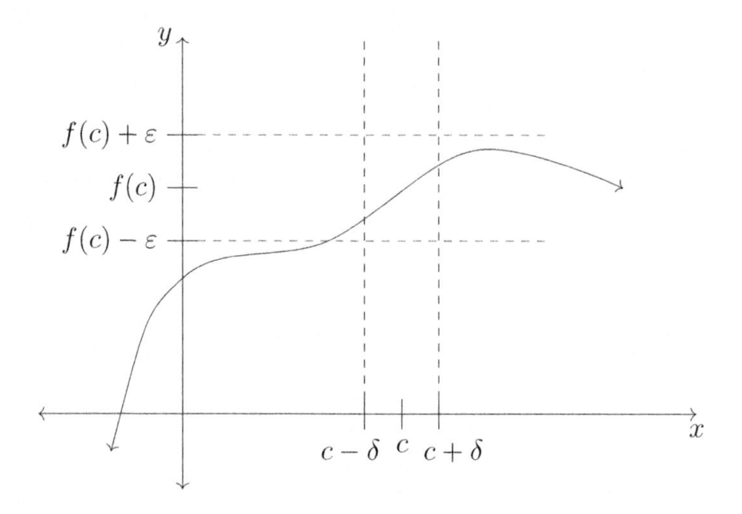

When you studied calculus, you probably learned a definition of the derivative using limits (of a difference quotient), and a definition of the integral using limits (of Riemann sums). When you break both of those down, there are again many hidden quantifiers; you will learn all about these subtleties when you take real analysis.

Other important topics in real analysis are sequences of functions, infinite sums of numbers, and infinite sums of functions. And a final important characteristic of real analysis is how weird and surprising the examples can be. The rest of this introduction aims to give an example combining both of these.

Assume you have a bunch of ordinary blocks ('ordinary' means that each is a rectangular cuboid, has uniform density, and the blocks are identical). If you hang one block off a table, up to half of it can be hanging off the edge without it falling:

For convenience, let's say that each block is 1 foot long, meaning that we achieved $\frac{1}{2}$ foot of overhang. Here's the big question we are interested in asking: Using more blocks, can we achieve more overhang? If so, how much overhang can we achieve?

If you have blocks or books nearby, even if they aren't completely uniform, go ahead and try this out. Come back and continue reading once you've experimented.

If you don't have any blocks or books, call up everyone you know and invite them to a pizza party at your house in 45 minutes. Then order 31 pizzas. Once all the pizza has been eaten you can use the pizza boxes for the experiment.

⋮

What did you find out? First, you probably found that with just two blocks you can achieve more than $\frac{1}{2}$ foot of overhang. Indeed, if a second block is placed on the edge of the table and with just a couple inches of overhang, you can now place the first block on top of this block so that half of it is overhanging the block below it.

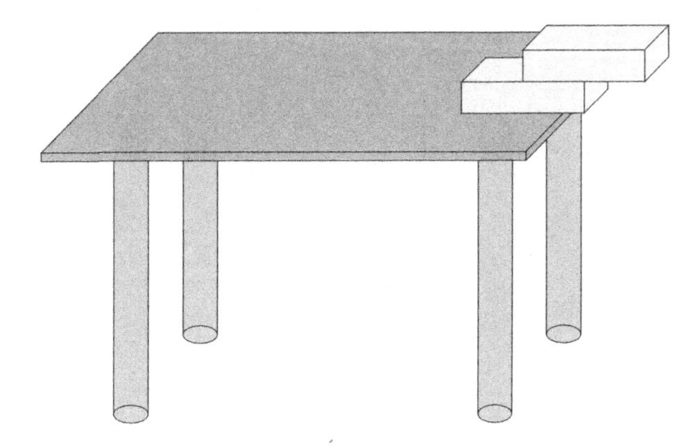

The bottom block "extends the table" in some sense. Here is a simple zoomed-in profile picture of this successful situation:

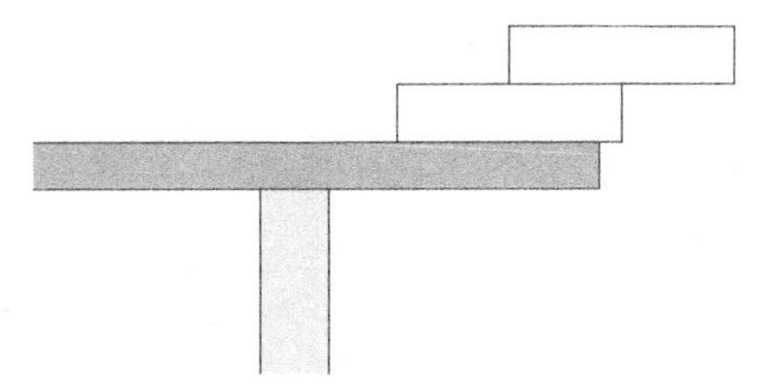

Of course, there is a limit to how much overhang the bottom block can have. For example, if the bottom block had $\frac{1}{2}$ foot of overhang (from the table), and if the top block had an additional $\frac{1}{2}$ foot of overhang (from the bottom block), then there's no way it will stay up. Here is a profile picture of this bad situation:

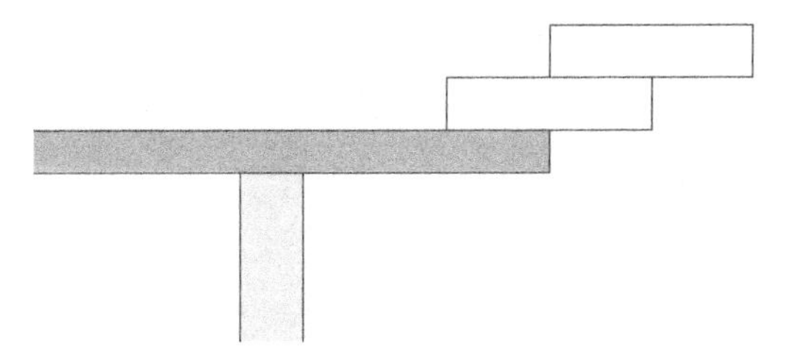

So if the bottom can have some overhang, but not as much as in the above, where is the tipping point? It all comes down to the *center of mass*. With a single block, you can have up to half of it hanging off the edge because the center of mass of the block is exactly in the middle, and provided the center of mass is above the table, it will not topple. If you stack two blocks and neither is sticking out more than halfway from what is below it, then the only way it will topple is if their *collective* center of mass is not above the table.

To proceed, it'll be useful to recall two basic facts about centers of mass. First, to find the center of mass of a collection of objects one may assume that each object's entire mass occurs at its center of mass — this is called a *point mass*. Second, to find the (horizontal-coordinate of the) center of mass of two objects, one can use a *weighted average*, which in the figure below is $\frac{m_1 d_1 + m_2 d_2}{m_1 + m_2}$, where the two point masses (represented with circles) have masses m_1 and m_2, and d_1 and d_2 are included in the figure.

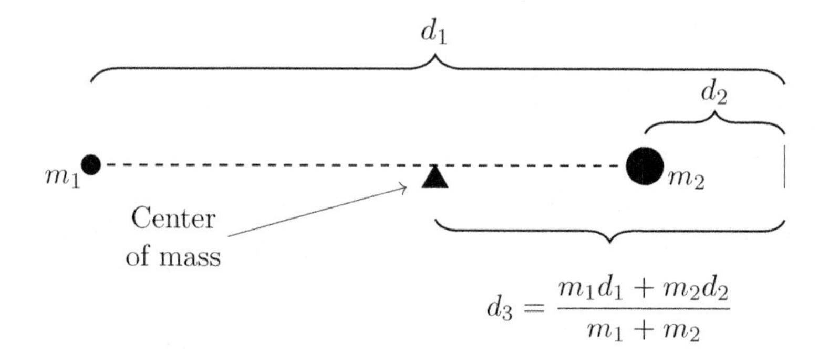

In our problem, all the blocks are identical and therefore have the same weight. For simplicity, let's assume each weighs 1 pound. Then, we can compute the location of the center of mass of two blocks; the optimal placement of the blocks will be such that their collective center of mass is right at the table's edge.

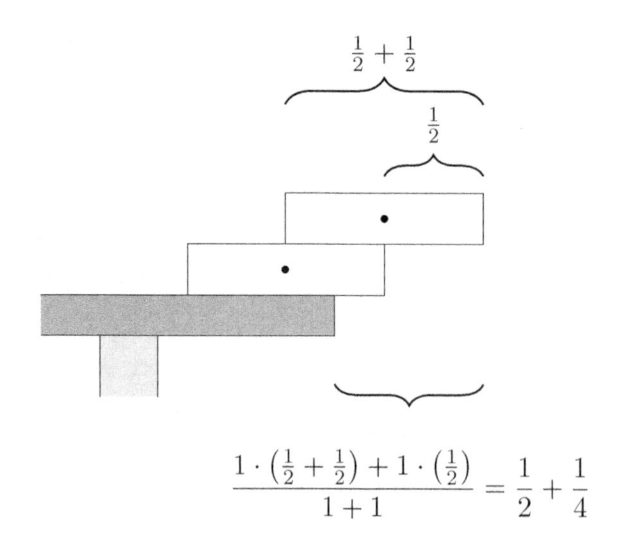

So if the top block hangs out $\frac{1}{2}$ feet from the bottom block, up to $\frac{1}{4}$ feet of the bottom block can hang off the table, giving a combined $\frac{1}{2}+\frac{1}{4}$ feet of overhang. (You'll soon see why I'm writing it as a summation, rather than just $\frac{3}{4}$ feet.)

With three blocks, where the top two use the above (optimal) arrangement, you can view the top two blocks as being a single point mass, with a mass of 2 pounds and a center of mass located $\frac{1}{2}+\frac{1}{4}$ feet from the right side of this two-block, which is optimally placed directly above the edge of the bottom-most block.

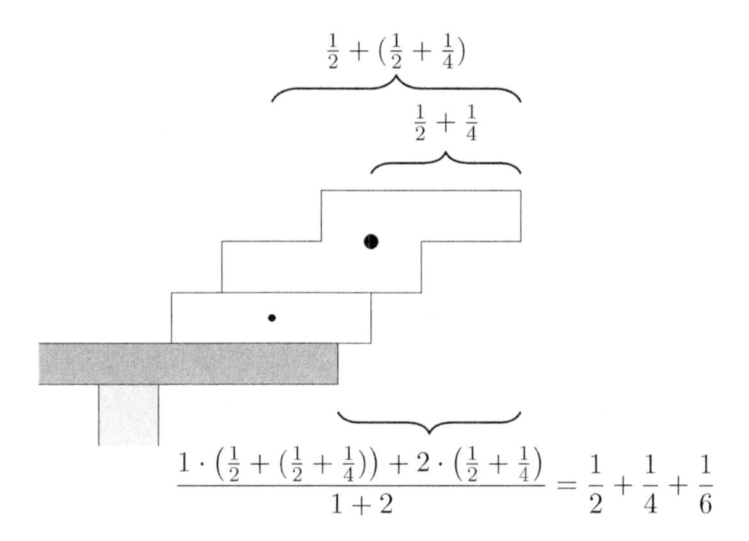

$$\frac{1\cdot(\frac{1}{2}+(\frac{1}{2}+\frac{1}{4}))+2\cdot(\frac{1}{2}+\frac{1}{4})}{1+2}=\frac{1}{2}+\frac{1}{4}+\frac{1}{6}$$

So a third block can safely overhang the table by $\frac{1}{6}$ feet, if one stacks the 2-block solution from before on top of it. In general, the nth block can overhand the table by $\frac{1}{2n}$ feet. Indeed, one can prove this by induction. If n blocks can overhang

$$\frac{1}{2}+\frac{1}{4}+\frac{1}{6}+\cdots+\frac{1}{2n}$$

feet, then using the above center-of-mass argument one can show that $n+1$ blocks, in feet, can overhang

$$\frac{1\cdot(\frac{1}{2}+(\frac{1}{2}+\cdots+\frac{1}{2n}))+n\cdot(\frac{1}{2}+\cdots+\frac{1}{2n})}{1+n}=\frac{1}{2}+\frac{1}{4}+\cdots\frac{1}{2n}+\frac{1}{2(n+1)}.$$

And now, if you've read this far, it's time for a huge payoff. We asked how far this stack can hang out from the table. For example, can you stack enough blocks so that no part of the top block is above the table? Amazingly, yes! In fact, just four blocks is enough, because with four blocks the total overhang, in feet, is

$$\frac{1}{2}+\frac{1}{4}+\frac{1}{6}+\frac{1}{8}\approx 1.04>1.$$

Since there is more than a 1 foot overhang, the top block is entirely to the right of the table!

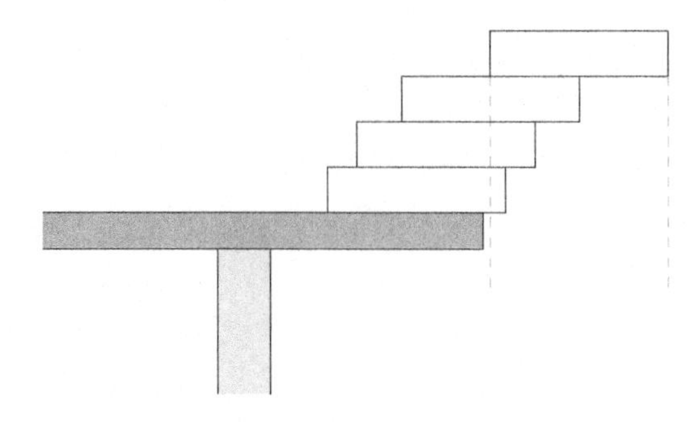

Moreover, because

$$\frac{1}{2}+\frac{1}{4}+\frac{1}{6}+\cdots+\frac{1}{2(31)} \approx 2.01 > 2,$$

with 31 blocks there can be an entire block-length separating the top block and the table! In fact, there is no limit to how big the overhang can be, since[65]

$$\frac{1}{2}+\frac{1}{4}+\frac{1}{6}+\frac{1}{8}+\cdots=\frac{1}{2}\cdot\sum_{k=1}^{\infty}\frac{1}{k}=\infty.$$

As this shows, the limit could even be ∞, provided you used infinitely many blocks! And with some (very large, but finite) number of blocks, you can have a mile of overhang, or a million miles, or anything else. It'll take a lot of blocks, since the harmonic series diverges suuuuuppper slowly, but because it does diverge to ∞, any overhang distance is possible.

---

[65]Note: The series $\sum_{k=1}^{\infty}\frac{1}{k}$ is called the *harmonic series* and does indeed equal ∞, a fact that you probably learned in Calculus II while discussing the series p-test. And here is the proof idea:

$$\begin{aligned}\sum_{k=1}^{\infty}\frac{1}{k} &= 1+\frac{1}{2}+\frac{1}{3}+\frac{1}{4}+\frac{1}{5}+\frac{1}{6}+\frac{1}{7}+\frac{1}{8}+\ \dots\\ &= 1+\left(\frac{1}{2}\right)+\left(\frac{1}{3}+\frac{1}{4}\right)+\left(\frac{1}{5}+\frac{1}{6}+\frac{1}{7}+\frac{1}{8}\right)+\ \dots\\ &\geq 1+\left(\frac{1}{2}\right)+\left(\frac{1}{4}+\frac{1}{4}\right)+\left(\frac{1}{8}+\frac{1}{8}+\frac{1}{8}+\frac{1}{8}\right)+\ \dots\\ &= 1+\left(\frac{1}{2}\right)+\left(\frac{1}{2}\right)+\left(\frac{1}{2}\right)+\ \dots\\ &= \infty.\end{aligned}$$

I will leave you with the image of 31 blocks stacked on top of each other, which I mentioned is the first case in which the top block is separated from the table by an entire block-length.

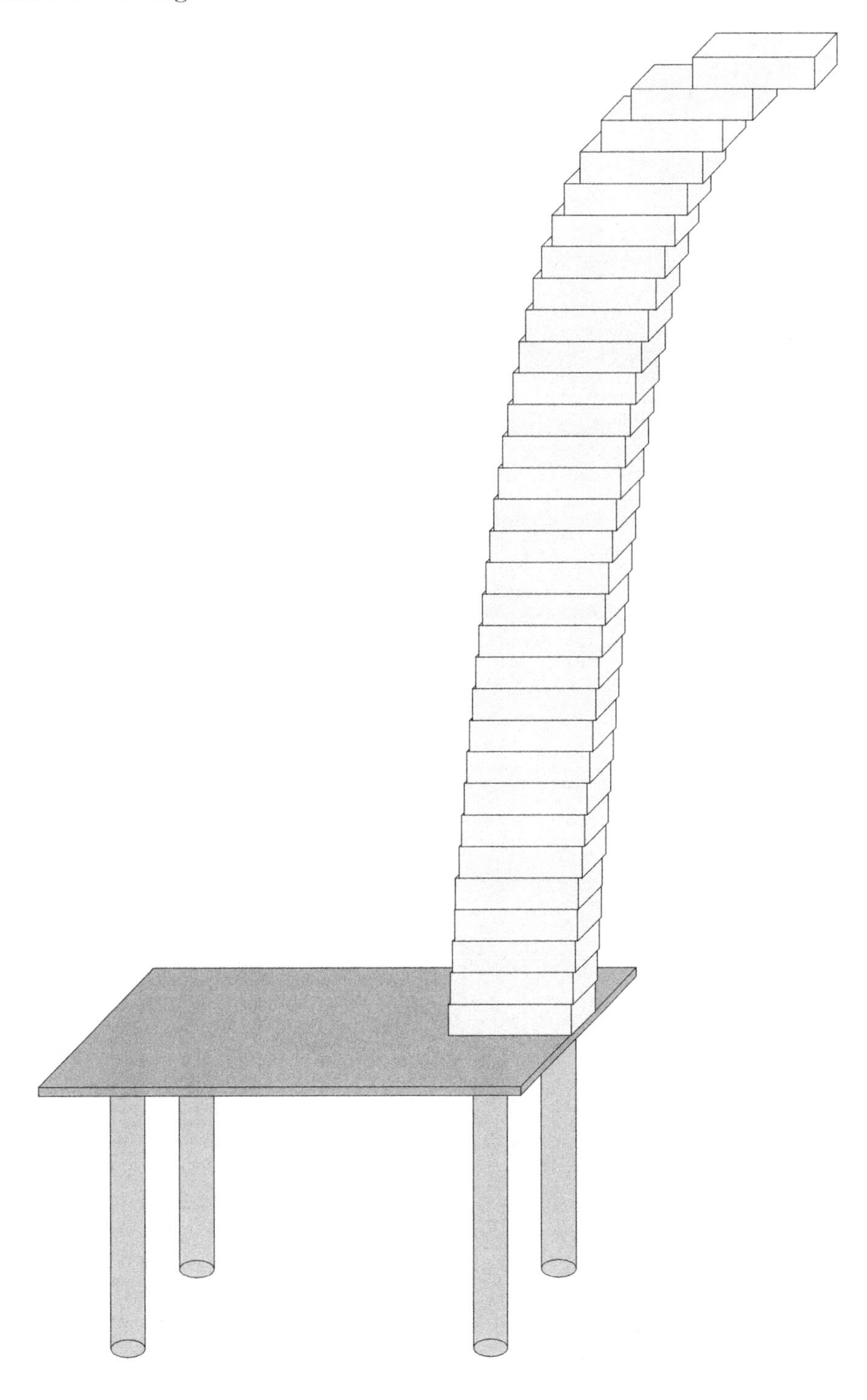

Chapter 6: The Contrapositive

In 1966, cognitive psychologist Peter Wason devised a logic puzzle which is now famous in the biz. Here is an equivalent form of the question:

> You are shown a set of four cards placed on a table (pictured below), each of which has a number on one side and a letter on the other side. Which card, or cards, must you turn over in order to determine whether the following is true or false: If a card shows an even number on one face, then its opposite face is an H?

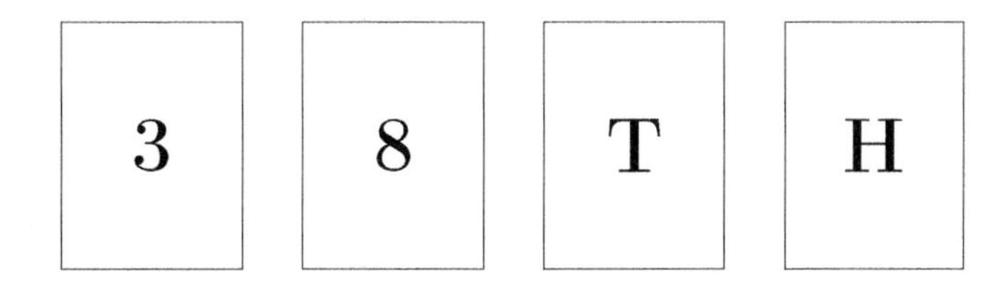

Think about this on your own right now. Seriously, give it a shot. It's easy to keep reading on, but don't! Try it! ...Ok hopefully you did. This is a famous puzzle because it tricked so many people. In Wason's study, fewer than 10% of the people answered it correctly.

I won't tell you the answer immediately, because I really do want you to stop and think about it first. So yeah, go do that. Have a guess in mind.

My next stalling tactic will be to rephrase the question slightly. Suppose four people are each holding a drink (and each is drinking something different), and you're trying to determine whether it is true that "If a person is drinking alcohol, then they are over 21 years old." Observe that the only way this statement could be false is if (1) a person is younger than 21 and (2) their drink is alcoholic. And to compare this to Wason's cards where you only see one side, let's contemplate what this looks like if you only know half this information.

- If you only know that the person is under 21, then the statement will become false if they are drinking alcohol.
- If you only know that alcohol is being drank, then the statement will become false if they are under 21.

With that in mind, let's try Wason's riddle again, but with new cards.

> You are shown a set of four cards placed on a table, each representing one of the four people, each of which is drinking something. Each card contains their age on one side, and what they are drinking on the other. If presented with the four cards below, which must you turn over in order to determine whether the following is true or false: If a person is drinking alcohol, then they are over 21 years old.

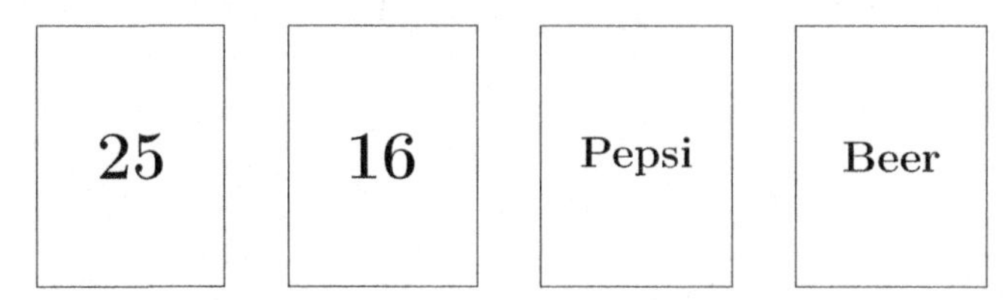

You must turn over the 16 card and the Beer card, right? Because by the bullet points above, the only way that statement could be false is if the 16 year old is drinking alcohol or the beer drinker is under 21. Does that make sense?

And notice what happened: In order to check the validity of "Alcohol $\Rightarrow$ over 21" we turned over the alcohol card and the under 21 card. Does that give you a hint as to how to solve Wason's puzzle?

Indeed, the above argues that the two statements below are logically equivalent.[1]

- If someone is drinking alcohol, then they are over 21.
- If someone is *not* over 21, then they are *not* drinking alcohol.

So what's the answer to the original puzzle? Well, I asked you to try it on your own, and if you still haven't... I still won't tell you—ha ha ha! You still have to figure it out![2] But it's very similar to the alcohol version, so you're almost there. (But if your original guess was to turn over the 8 and the H, then you made the most common mistake, which you may now be able to fix.)

The goal of this puzzle is to provide motivation for the fact that $P \Rightarrow Q$ (e.g., Alcohol $\Rightarrow$ Over 21) being true is logically equivalent to not-$Q \Rightarrow$ not-P (e.g., Under 21 $\Rightarrow$ Non-Alcohol). In fact, this is something we showed at the end of the last chapter in our discussion of the contrapositive.[3] Here is that result:

Theorem.

Theorem 5.16. An implication is logically equivalent to its contrapositive. That is,

$$(P \Rightarrow Q) \quad \Leftrightarrow \quad (\sim Q \;\Rightarrow\; \sim P).$$

[1] Recall what logical equivalence means: if one is true, the other is true too; if one is false, the other is false too.

[2] ...or you can Google it, I guess...it is pretty famous...

[3] Quick review of that: The *contrapositive* of $P \Rightarrow Q$ is $\sim Q \Rightarrow \sim P$, and one can see that these two are logically equivalent ($P \Rightarrow Q$ is true if and only if $\sim Q \Rightarrow \sim P$ is true). This fact was demonstrated by showing that their truth tables align, and is stated formally in Theorem 5.16.

6.1 Finding the Contrapositive of a Statement

Here are examples of taking the contrapositive of a statement.

Example 6.1.

1. $P \Rightarrow Q$: If $n = 6$, then n is even.

   $\sim Q \Rightarrow \sim P$: If n is not even, then $n \neq 6$.

2. $P \Rightarrow Q$: If I just dumped water on you, then you're wet.

   $\sim Q \Rightarrow \sim P$: If you're not wet, then I didn't just dump water on you.

3. $P \Rightarrow Q$: If Shaq is the tallest player on his team, then Shaq will play center.

   $\sim Q \Rightarrow \sim P$: If Shaq is not playing center, then Shaq is not the tallest player on his team.

4. $P \Rightarrow Q$: If you're happy and you know it, then you're clapping your hands.

   $\sim Q \Rightarrow \sim P$: If you're not clapping your hands, then you're either not happy or you don't know it.

5. $P \Rightarrow Q$: If $p \mid ab$, then $p \mid a$ or $p \mid b$.

   $\sim Q \Rightarrow \sim P$: If $p \nmid a$ and $p \nmid b$, then $p \nmid ab$.

For each of these, $P \Rightarrow Q$ and $\sim Q \Rightarrow \sim P$ will have the same truth value. Consider the Shaq example: If the $P \Rightarrow Q$ rule is true, then the $\sim Q \Rightarrow \sim P$ rule is also true. But if, say, their team signs a taller player but they still play Shaq at center, then both statements are false. A common mistake is to think the contrapositive is always true, but all that is being asserted is that the contrapositive is *logically equivalent* to the original implication. So yes, $\sim Q \Rightarrow \sim P$ could be false—but if so, then the original implication will be false as well. Their truth values will always match. Here is a final example where both are false (such as if $n = 9$):

6. $P \Rightarrow Q$: If $3 \mid n$, then $6 \mid n$.

   $\sim Q \Rightarrow \sim P$: If $6 \nmid n$, then $3 \nmid n$.

By the way, since $P \Rightarrow Q$ is logically equivalent to its contrapositive, which is $\sim Q \Rightarrow \sim P$, this new implication must also be logically equivalent to *its* contrapositive. What does this give us? The contrapositive of $\sim Q \Rightarrow \sim P$ is $\sim\sim P \Rightarrow \sim\sim Q$. But since applying "$\sim$" twice gets you back to where you started, this is the same as $P \Rightarrow Q$. So yes, applying a contrapositive a second time gets you a logically equivalent statement—it just happens to be the one we started with.[4]

[4] *Meme that I can't print without paying The Office an annoyingly large amount of money:*
Pam to Michael: Corporate needs you to find the differences between this picture ($P \Rightarrow Q$) and this picture ($\sim Q \Rightarrow \sim P$).
Pam to the camera: They're the same picture.

6.2 Proofs Using the Contrapositive

Let's prove some things. As you learn more proof techniques, you'll discover that often a proposition can be proved in many different ways, using different proof techniques. For example, some of the propositions below could be proven by using either a direct proof or the contrapositive. This is good! It means you have a choice. Also, learning the proof to a proposition helps you understand it; and learning a *second* proof of that proposition helps you understand it even more.[5] Below is the general structure of a proof by contraposition.

Proposition. $P \Rightarrow Q$.

Proof. We will use the contrapositive. Assume not-Q.

≪An explanation of what not-Q means≫ ⟵ Apply definitions and/or other results.

⋮ apply algebra, logic, techniques

≪Hey look, that's what not-P means≫

Therefore not-P.

Since not-$Q \Rightarrow$ not-P, by the contrapositive $P \Rightarrow Q$. □

We begin by proving the converse of Proposition 2.6.

(Amazingly, by pure coincidence, Proposition 2.6's converse is Proposition 6.2. And while I did literally nothing to facilitate it, I am feel unreasonably proud that it occurred.)

Proposition.

Proposition 6.2. Suppose $n \in \mathbb{Z}$. If n^2 is odd, then n is odd.

[5] And, in general, if you know k proofs of a proposition, learning a $(k+1)^{\text{st}}$ proof of it will further deepen your understanding. And so, by induction, you should never stop learning new proofs of a proposition. □

Proof Idea. If we tried to prove this directly, then we would start by applying the definition of oddness to say that $n^2 = 2a + 1$ for some $a \in \mathbb{Z}$. But then what? How do we get to n being odd? The path seems unclear. But the contrapositive of

"If n^2 is odd, then n is odd"

is

"If n is not odd, then n^2 is not odd."

And because $n \in \mathbb{Z}$, this is the same as

"If n is even, then n^2 is even."

And this seems like a much clearer path! In fact, we have already proved a number of results just like this in Chapter 2. So simply by taking the contrapositive we turn a problem with no clear way forward into a routine problem from Chapter 2.

Proof. Suppose $n \in \mathbb{Z}$. We will use the contrapositive. Assume that n is not odd, which means that n is even,[6] by Fact 2.1.[7] By the definition of an even integer (Definition 2.2), this means that $n = 2a$ for some integer a. Then,

$$n^2 = (2a)^2 = 4a^2 = 2(2a^2).$$

And note that since a is an integer, $2a^2$ is an integer too. Therefore, by Definition 2.2, this means that n^2 is even. And since $n \in \mathbb{Z}$, this is equivalent to concluding that n^2 is not odd.

We have shown that if n is not odd, then n^2 is not odd. Thus, by the contrapositive, if n^2 is odd, then n is odd. □

A common question at this point is "how do I know which proof method to use?" Induction is the easiest to spot, because the proposition is asserting something (often a simple equality or single property, rather than an "if, then" condition) holds for every $n \in \mathbb{N}$, or similar. For direct proof versus contrapositive. . . it is trickier, and experience certainly helps.

The last proposition worked well by contrapositive because the contrapositive "flips" P and Q, and if n comes first and n^2 comes second, that's helpful. The next proposition will be done in two parts, and one of those parts works best as a direct proof, while the other works best as a contrapositive. However, to emphasize that you do have choice, I will show how the contrapositive proof could have been done directly.

[6](foot)Note: It is important that we said $n \in \mathbb{Z}$. If $n \in \mathbb{R}$, for example, then n being "not odd" no longer implies that n is even. Because maybe $n = 2.2$ or $n = \pi$. These are not odd numbers since they are not $2k + 1$ for some $k \in \mathbb{Z}$—but that does not imply that they are even.

[7]As I mentioned before, as we go along I will very slightly begin to pull back on the meticulousness that we have been exercising. You are now a much stronger mathematician than you were four chapters ago, and this is the last time that I will formally cite Fact 2.1. It has served us well!

Proposition.

Proposition 6.3. Suppose $n \in \mathbb{N}$. Then, n is odd if and only if $3n + 5$ is even.

Proof Sketch. Remember that to prove an "if and only if" proposition, you need to prove it in "both directions." Indeed, we will prove this proposition by proving

1. If n is odd, then $3n + 5$ is even; and
2. If $3n + 5$ is even, then n is odd.

The first of these is a classic problem that you could have solved in Chapter 2. The second of these can be turned into a classic problem once we apply the contrapositive.

Proof. We will prove this in two parts.

Part 1: If n is odd, then $3n + 5$ is even. Suppose that $n \in \mathbb{N}$ and n is odd. Then, by the definition of an odd number (Definition 2.2), $n = 2a + 1$ for some $a \in \mathbb{Z}$. So,

$$\begin{aligned} 3n + 5 &= 3(2a + 1) + 5 \\ &= 6a + 8 \\ &= 2(3a + 4). \end{aligned}$$

Since $a \in \mathbb{Z}$, also $(3a+4) \in \mathbb{Z}$. And so, by the definition of an even number (Definition 2.2), $3n + 5$ is even. This completes Part 1.

Part 2: If $3n + 5$ is even, then n is odd. Suppose $n \in \mathbb{N}$. We will use the contrapositive. Assume that n is not odd. Since $n \in \mathbb{N}$, this means that n is even, and so by the definition of an even number (Definition 2.2), $n = 2a$ for some $a \in \mathbb{Z}$. Then,

$$\begin{aligned} 3n + 5 &= 3(2a) + 5 \\ &= 6a + 4 + 1 \\ &= 2(3a + 2) + 1. \end{aligned}$$

Since $a \in \mathbb{Z}$, also $(3a + 2) \in \mathbb{Z}$. So, by the definition of an odd number (Definition 2.2), $3n + 5$ is odd. And since $3n + 5$ is an integer, this means that $3n + 5$ is not even.

We have shown that n being not odd implies that $3n + 5$ is not even. Thus, by the contrapositive, if $3n + 5$ is even, then n is odd.[8] This completes Part 2.

We have proven that n being odd implies $3n + 5$ is even, and that $3n + 5$ being even implies n is odd. Combined, these show that n is odd if and only if $3n + 5$ is even, completing the proof. □

[8] For many proofs, having a summarizing sentence is nice but not required. If you recall, it *was* required for induction proofs, and it is again required for contrapositive proofs. Taken at face value, the contrapositive is a different statement than the proposition you are trying to prove. When we write "by the contrapositive" at the end, we are saying that "we just proved that something else is true, but it being true implies that our proposition is true." (In fact, it is *equivalent* to our proposition being true.) Without formally invoking the contrapositive, you have not connected your proof to the proposition you are trying to prove.

In Part 2, we once again benefited from the contrapositive's ability to "flip" the order of P and Q. Starting with n being even made our approach much cleaner.

That said, I'd like to mention that there is actually a way to prove Part 2 as a direct proof. If you assume that $3n+5$ is even, then you can write $3n+5 = 2a$ where $a \in \mathbb{Z}$. The goal now is to show that n is odd by writing n as $2b+1$ for some $b \in \mathbb{Z}$. How do we do it? Notice that $3n+5 = 2a$ implies that $3n = 2a - 5$, but should we now divide both sides by 3? How would that give us $2b+1$?

The trick is to think about $3n$ as $n + 2n$, and then move the $2n$ to the right:

$$\begin{aligned} 3n &= 2a - 5 \\ n &= 2a - 2n - 5 \\ n &= 2a - 2n - 6 + 1 \\ &= 2(a - n - 3) + 1. \end{aligned}$$

And since $a, n \in \mathbb{Z}$, also $(a - n - 3) \in \mathbb{Z}$. Thus, $n = 2b + 1$ where $b = a - n - 3$ is an integer. So, n is odd.

For the next result, recall that in Lemma 2.17 part (iii) we proved that if $p \mid bc$, then $p \mid b$ or $p \mid c$. Let's now prove the indivisibility[9] version of this result.

Proposition.

Proposition 6.4. Let $a, b \in \mathbb{Z}$, and let p be a prime. If $p \nmid ab$, then $p \nmid a$ and $p \nmid b$.

You might hope that this proposition is precisely the contrapositive of Lemma 2.17 part (iii); if it were, then to prove this proposition we could simply apply the contrapositive to Lemma 2.17, and the proof would be done! However, they are not contrapositives of each other. If Lemma 2.17 is $P \Rightarrow Q$, then Proposition 6.4 is $\sim P \Rightarrow \sim Q$, whereas the contrapositive of $P \Rightarrow Q$ is $\sim Q \Rightarrow \sim P$.[10]

We will still use the contrapositive to prove this, but unfortunately we are unable to make use of Lemma 2.17 in our proof. We will have to work a little harder.

***Proof*.** Suppose $a, b \in \mathbb{Z}$ and p is a prime. We will use the contrapositive. Suppose that it is not true that $p \nmid a$ and $p \nmid b$. By the logic form of De Morgan's law (Theorem 5.9), this is equivalent to saying it is not true that $p \nmid a$ *or* it is not true that $p \nmid b$. That is, $p \mid a$ or $p \mid b$. Let's consider these two cases separately.

Case 1. Suppose $p \mid a$, which by the definition of divisibility (Definition 2.8) means that $a = pk$ for some $k \in \mathbb{Z}$. Thus,

$$ab = (pk)b = p(kb).$$

[9]Fun fact: "indivisibility" has more copies of 'i' in it than any other English word. Other winners: "abracadabra" has the most copies of a, knickknack has the most ks, whippersnapper the most ps, bowwow the most ws, and pizzazz has the most zs.

[10]In fact, $\sim P \Rightarrow \sim Q$ is the converse of the contrapositive. (Or, the contrapositive of the converse.)

Since $k, b \in \mathbb{Z}$, also $(kb) \in \mathbb{Z}$. And so, by the definition of divisibility (Definition 2.8), $p \mid ab$.

Case 2. Suppose $p \mid b$, which by the definition of divisibility (Definition 2.8) means that $b = p\ell$ for some $\ell \in \mathbb{Z}$. Thus,

$$ab = a(p\ell) = p(a\ell).$$

Since $a, \ell \in \mathbb{Z}$, also $(a\ell) \in \mathbb{Z}$. And so, by the definition of divisibility (Definition 2.8), $p \mid ab$.

In either case, we concluded that $p \mid ab$, which is equivalent to saying that it is not true that $p \nmid ab$.

We proved that if it is not true that $p \nmid a$ and $p \nmid b$, then it is not true that $p \nmid ab$. Hence, by the contrapositive, this implies that if $p \nmid ab$, then $p \nmid a$ and $p \nmid b$. □

Note that the two cases above were the exact same. Sure, in Case 1 we were focused on a while on Case 2 we were focused on b, but their variable names are literally the only difference between a and b. And sure, we chose different variable names for k and ℓ, but that's it; every mathematical characteristic about these two cases is exactly the same. In fact, when I typed this up, I literally copied Case 1 and pasted it as Case 2. I switched out a few variable names, and was done. Seems a little silly to be proving two things when they are essentially the same, doesn't it?

Mathematicians have agreed that we should be allowed to skip essentially-identical cases. With all those skipped Case-2s, just think of all the trees that can be saved! And the time! We could all knock off work 10 minutes early and spend more time with kids. It seemed like such a no-brainer that the Elders of Math were unanimous on the motion's first ballot. From this, "without loss of generality" was approved.

If you have two cases, like $p \mid a$ and $p \mid b$, and there is literally no mathematical distinction between them, then you are allowed to say "without loss of generality, assume $p \mid a$." This allows you to skip the "$p \mid b$" case entirely. For example, the above proof is rewritten in this condensed way below.

Condensed, Elder-Approved Proof. Suppose $a, b \in \mathbb{Z}$ and p is a prime. We will use the contrapositive. Suppose that it is not true that $p \nmid a$ and $p \nmid b$. By the logic form of De Morgan's law (Theorem 5.9), this is equivalent to saying it is not true that $p \nmid a$ *or* it is not true that $p \nmid b$. That is, $p \mid a$ or $p \mid b$. Without loss of generality, assume $p \mid a$.

By the definition of divisibility (Definition 2.8), this means that $a = pk$ for some $k \in \mathbb{Z}$. Thus,

$$ab = (pk)b = p(kb).$$

Since $k, b \in \mathbb{Z}$, also $(kb) \in \mathbb{Z}$. And so, by the definition of divisibility (Definition 2.8), $p \mid ab$.

We proved that if it is not true that $p \nmid a$ and $p \nmid b$, then it is not true that $p \nmid ab$. Hence, by the contrapositive, this implies that if $p \nmid ab$, then $p \nmid a$ and $p \nmid b$. □

As we close out the main content of this chapter, I wanted to comment again on the fact that as we are learning more sophisticated proof techniques, and as our proofs themselves become more complicated, it is increasingly important to proceed with caution when writing out your own proofs.

The writer Joan Didion once noted that the process of writing is not only to share what you think is true, but to *discover* what you think is true. This is insightful, although it also comes with risk — are you actually discovering what you think is true, or do you risk slowly convincing yourself of falsities, while all of your blind spots remain?

If this is cause for concern with everyday writing, then proof writing demands even greater caution. When writing about politics, people tend to be their easiest market. When writing a proof, though, you must insist on being a (nice) critic of yourself. Constantly test your intuition, probe your ideas, and break things down until they are of their simplest form. It is healthy and productive to approach the first draft of your proof with doubt. And if you can find a friend to read through your proofs in a critical (and nice) way, then all the better.

6.3 Counterexamples

The contrapositive is naturally a shorter topic than our other proof methods, so I am going to steal a few pages here to discuss counterexamples. We first mentioned counterexamples way back on Page 11 as a way to disprove a conjecture. The idea is this: In order to disprove a "universal" statement (like one using the words "for all" or "for every"), it suffices to find one example against it. If I said "every NBA player can dunk," to disprove it you would have to find a single NBA player that cannot dunk. If you did, then that single NBA player is a *counterexample* to the claim "every NBA player can dunk." Indeed, to disprove a statement, you must prove its negation. And the negation of a "for all" statment is always a "there exists" statement. We can do this with a single counterexample.

Likewise, if I said "every prime number is odd," you could disprove that by presenting the counterexample of 2. Of course, sometimes it is more difficult to find a counterexample. Suppose, for instance, you saw this conjecture:

Conjecture 1. If p is prime, then $2^p - 1$ is prime.

Disproof Idea. Notice that even though we did not use language like "for all," this is still a universal statement, as it is asserting something about every prime p. As it turns out, this conjecture is false. Since it is a false universal statement, there must be a counterexample. You could try the first few cases:

$$\begin{aligned}
2^2 - 1 &= 3, \text{ which is prime}\\
2^3 - 1 &= 7, \text{ which is prime}\\
2^5 - 1 &= 31, \text{ which is prime}\\
2^7 - 1 &= 127, \text{ which is prime}
\end{aligned}$$

It wouldn't be until $2^{11} - 1 = 2047 = 23 \cdot 89$ that you would find a counterexample. With this single example, though, we can disprove Conjecture 1.

Disproof. Note that $p = 11$ is prime and that $2^p - 1 = 2^{11} - 1 = 2047 = 23 \cdot 89$, which we demonstrated is composite and hence is not prime. This counterexample disproves Conjecture 1. □

In the last example we had to check five cases to find a counterexample. If you think that is a lot of work, suppose you saw this conjecture:

Conjecture 2. For every $n \in \mathbb{N}$,

$$1 + \frac{1}{2} + \frac{1}{3} + \cdots + \frac{1}{n} < 100.$$

Disproof Idea. If you check the first few n-values, you won't find a counterexample. Even at $n = 100$, you'd find

$$1 + \frac{1}{2} + \frac{1}{3} + \cdots + \frac{1}{100} \approx 5.187,$$

which easily satisfies Conjecture 2. If you then checked $n = 101$, and then $n = 102$, and so on, you'll realize that you are making little progress to finding a counterexample. And yet, as you might remember from your calculus class, Conjecture 2 is false and so a counterexample *does* exist. In fact, the first counterexample is

$$n = 15,092,688,622,113,788,323,693,563,264,538,101,449,859,497.$$

That's right, for $n = 15,092,688,622,113,788,323,693,563,264,538,101,449,859,496$, the conjecture is still satisfied:

$$1+\frac{1}{2}+\frac{1}{3}+\cdots+\frac{1}{15,092,688,622,113,788,323,693,563,264,538,101,449,859,496} < 100.$$

But including just one more term is a counterexample to the conjecture:

$$1+\frac{1}{2}+\frac{1}{3}+\cdots+\frac{1}{15,092,688,622,113,788,323,693,563,264,538,101,449,859,497} > 100.$$

It would require a computer to verify that this is correct (WolframAlpha can do it!), so you will have to take my word on it for now, but this is indeed a counterexample which disproves Conjecture 2. □

Similarly, suppose someone conjectured that $n^{19}+6$ and $(n+1)^{16}+9$ are relatively prime (their gcd equals 1) for every $n \in \mathbb{N}$. You could plug in the first billion natural numbers, and each time you would produce two relatively prime numbers. In fact, a billion would not even be close. The first counterexample is when

$$n = 1,578,270,389,554,680,057,141,787,800,241,971,645,032,008,710,129,107,338,825,798.$$

That number has 61 digits! And while every n before this produces numbers with a gcd of 1, this n value produces a gcd of

$$5,299,875,888,670,549,565,548,724,808,121,659,894,902,032,916,925,752,559,262,837.$$

So yes, sometimes counterexamples come late.

Let's now turn to something a little more manageable: Exponents![11] How do you evaluate a^{b^c}? For example, what does 2^{3^2} equal? Is it

$$2^{(3^2)} \quad \text{or} \quad \left(2^3\right)^2 ?$$

Or does it matter? Consider this conjecture:

Conjecture 3. For every $a, b, c \in \mathbb{N}$,

$$a^{(b^c)} = \left(a^b\right)^c.$$

Disproof Idea. Not only is this false, the example we discussed above is a counterexample, as we show below.

Disproof. Note that $a = 2$, $b = 3$ and $c = 2$ are all natural numbers, and

$$2^{(3^2)} = 2^9 = 512 \neq 64 = 8^2 = \left(2^3\right)^2.$$

This counterexample disproves Conjecture 3. □

(And since you might be wondering, the convention that mathematicians have agreed on is to start from the top and work down. So, $2^{3^2} = 2^{(3^2)} = 2^9 = 512$. Likewise, $2^{3^{2^3}} = 2^{3^8} = 2^{6561}$.)

Let's do one more example. Just like the last three conjectures, the next conjecture will be false (otherwise it would be impossible to find a counterexample). This example comes from set theory.

Conjecture 4. If A, B and C are sets and $A = B \setminus C$, then $B = A \cup C$.

Disproof Idea. If you try out a few examples, you might find cases where this conjecture is satisfied. For example, if $B = \{1,2,3,4\}$ and $C = \{1,2\}$, then $A = B \setminus C = \{1,2,3,4\} \setminus \{1,2\} = \{3,4\}$. That's the setup. And the conclusion is satisfied, as $A \cup C = \{1,2,3,4\}$, which is indeed B. This shows that at least *some* sets satisfy

[11] I know a terrible exponent joke, but I'm 2^2 to tell it.

this conjecture. But in order for this conjecture to be true, it would have to hold for *every* triple of sets with the given conditions. Let's try to find a triple which does not follow the conjecture.

One way to find a counterexample for a sets problems is to draw up a Venn diagram. This might show you what you need in your counterexample. Here is $B \setminus C$, shaded in a Venn diagram:

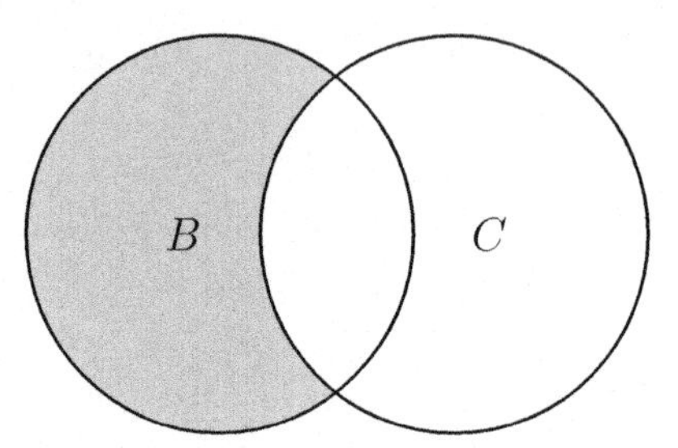

According to this problem, $A = B \setminus C$, which means that A is also the shaded region. In order for the conclusion to hold — that is, for $B = A \cup C$ — we would need the B circle to equal the union of the shaded region (which is A) and the intersection region (which is $B \cap C$). Thus, we need there to be no points here

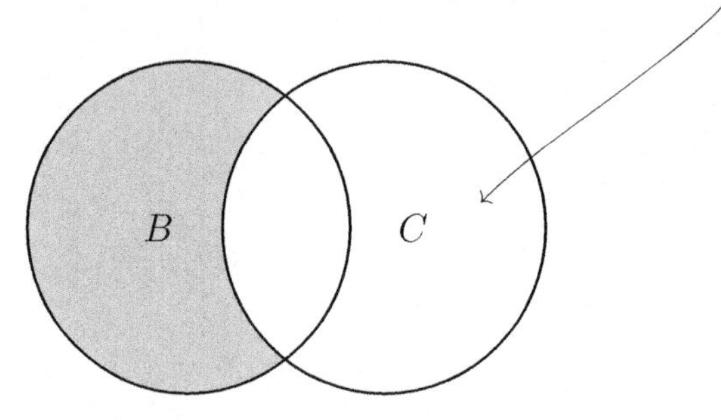

So all we need to do to find a counterexample is to pick a set C which contains at least one point not in B. For example, if $B = \{1, 2, 3, 4\}$, then choosing $C = \{4, 5\}$ should work; it contains the point 5, which is outside of B.

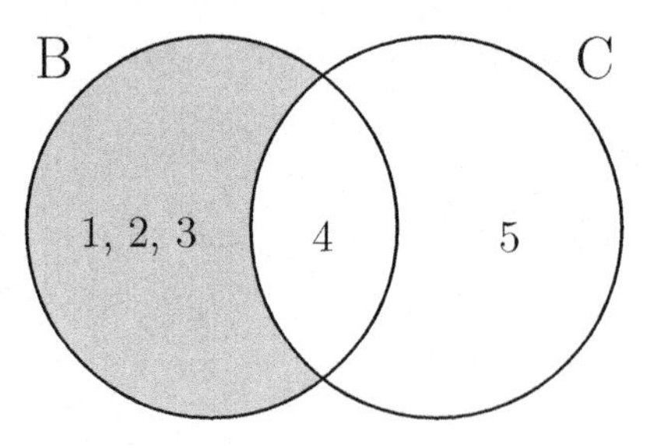

Disproof. Define the sets $A = \{1, 2, 3\}$, $B = \{1, 2, 3, 4\}$ and $C = \{4, 5\}$. Then, notice that $A = B \setminus C$, as $\{1, 2, 3\} = \{1, 2, 3, 4\} \setminus \{4, 5\}$. Does $B = A \cup C$? Nope: $B = \{1, 2, 3, 4\}$, but $A \cup C = \{1, 2, 3, 4, 5\}$. Thus, these three sets form a counterexample to Conjecture 4. □

In the exercises you will have more practice finding counterexamples to statements. For our Bonus Examples section, we turn back to the contrapositive.

6.4 Bonus Examples

As we have discussed, when presented with a problem it is often tough to know which proof technique to try first. This can be a hard question, and the more practice you have the better you'll get. But even with lots of practice, you'll often have to try a couple proof techniques before you can make one work. That all said, in the Proof Idea to the following proposition, we will discuss another way to spot a contrapositive.

Proposition.

Proposition 6.5. Suppose $a, b, n \in \mathbb{N}$. If $36a \not\equiv 36b \pmod{n}$, then $n \nmid 36$.

Proof Idea. The fact that this proposition says a lot of things are *not* happening is one indication that the contrapositive could be worthwhile. The contrapositive adds a "not" to both P and Q, so it will turn "$n \nmid 36$" into "$n \mid 36$," and "$36a \not\equiv 36b \pmod{n}$" into "$36a \equiv 36b \pmod{n}$." In both cases, this looks better. By definition, $n \mid 36$ means that $36 = nk_1$ for some $k_1 \in \mathbb{Z}$, so if that is our assumption, we have a clear forward direction. And if our goal is to show that $36a \equiv 36b \pmod{n}$, which means that $n \mid (36a - 36b)$, which in turn means that $(36a - 36b) = nk_2$ for some $k_2 \in \mathbb{Z}$, then this is a clearer target to aim for.

Proof. Suppose $a, b, n \in \mathbb{N}$. We will use the contrapositive. Assume that $n \mid 36$. By Definition 2.8,

$$36 = nk$$

for some $k \in \mathbb{Z}$. We will now prove that $36a \equiv 36b \pmod{n}$ by showing that $n \mid (36a - 36b)$. Since $36 = nk$,

$$\begin{aligned} 36a - 36b &= nka - nkb \\ &= n(ka - kb). \end{aligned}$$

And since $k, a, b \in \mathbb{Z}$, also $(ka - kb) \in \mathbb{Z}$. By Definition 2.8, this means that $n \mid (36a - 36b)$ and hence, by Definition 2.14, that $36a \equiv 36b \pmod{n}$.

We have shown that $n \mid 36$ implies that $36a \equiv 36b \pmod{n}$. Thus, by the contrapositive, $36a \not\equiv 36b \pmod{n}$ implies that $n \nmid 36$. □

In the next bonus example, we will assume that the quadratic formula that you learned in high school is true. (SPOILER ALERT: it is.) We will also make use of a couple of this book's exercises, which we state now as a lemma.

Lemma.

Lemma 6.6. This lemma has two parts.

(i) If $m \in \mathbb{Z}$, then $m^2 + m$ is even.

(ii) If $a \in \mathbb{Z}$ and a^2 is even, then a is even.

The proof of this lemma is asked of you in the exercises. Part (i) is Exercise 2.8,[12] and part (ii) is Exercise 6.5 part (a).[13] We now use this result to prove the following proposition.

Proposition.

Proposition 6.7. Suppose a is an integer. If a is odd, then $x^2 + x - a^2 = 0$ has no integer solution.

Proof Idea. Again, since the conclusion "$x^2 + x - a^2 = 0$ has no integer solution" is saying something can *not* happen, it makes sense to try the contrapositive. Assuming that there *is* an integer solution seems like a good place to start a proof, especially since the quadratic formula tells us exactly what such a solution can look like.

Proof. Suppose that a is an integer. We will use the contrapositive. Assume that it is false that $x^2 + x - a^2 = 0$ has no integer solutions; that is, assume that there is some integer m such that

$$m^2 + m - a^2 = 0.$$

By the quadratic formula[14] and then some algebra,

$$\begin{aligned} m &= \frac{-1 \pm \sqrt{1^2 - 4(1)(-a^2)}}{2(1)} \\ 2m &= -1 \pm \sqrt{1 + 4a^2} \\ 2m + 1 &= \pm\sqrt{1 + 4a^2} \\ 4m^2 + 4m + 1 &= 1 + 4a^2 \\ m^2 + m &= a^2. \end{aligned}$$

Next, observe that $m^2 + m$ is guaranteed to be even by Lemma 6.6 part (i).[15] Thus, since we just deduced that $m^2 + m = a^2$, this means that a^2 must be even. And

[12]Main idea: $m^2 + m$ is either a sum of two even numbers (if m is even) or the sum of two odds (if m is odd). In either case, this sum will be even.

[13]Hint: It is very similar to Proposition 6.2.

[14]Did your high school teacher make you sing the formula? Are you singing it in your head right now??

[15]*"Yo, lemma help you prove that proposition."*

since a is an integer, a^2 being even implies that a is even, by Lemma 6.6 part (ii). In particular, this means that a is not odd.

We have shown that if it is false that $x^2 + x - a^2 = 0$ has no integer solutions, then it is also false that a is an odd integer. By the contrapositive, if a is an odd integer, then $x^2 + x - a^2 = 0$ has no integer solution. □

There is also a nice *proof by contradiction* of this proposition, which is another proof technique, and is the topic of the next chapter.

— Chapter 6 Pro-Tips —

- We have discussed how $P \Rightarrow Q$ is logically equivalent to $\sim Q \Rightarrow \sim P$. Therefore, when someone discovers a theorem, they have a choice of how to express it. If they write it as $P \Rightarrow Q$, but their proof flips it around and proves the contrapositive, then why not just write the theorem as $\sim Q \Rightarrow \sim P$ to begin with, so the proof can be a direct proof? Sometimes these decisions are made behind the scenes.[16]

- As we saw in Proposition 6.3, when proving an if-and-only-if statement, say $P \Leftrightarrow Q$, then it is quite common to break the proof into two parts: a proof that $P \Rightarrow Q$, followed by a proof that $Q \Rightarrow P$. And it is often the case that in this situation, one of these two proofs is best done as a direct proof while the other is best done using the contrapositive. This is a good trick to keep in mind.

- Unrelated to the contrapositive, but a general Pro-Tip: When reading math books, pay extra close attention to theorems/propositions/lemmas/corollaries that have names. Typically something gets a name because it is used a lot or is a deep and important result. These will likely play an outsized role in your homework, exams, future courses, and beyond.

  Most theorems have boring names, like being named after the first person to prove it (or the first person to prove it after Euler proved it[17]). Others are named in a way that summarizes what they say, like the *orbit-stabilizer theorem*. But some others have fun names. Below are the actual names of some theorems, for your enjoyment. And as of early 2021, all but one of them has a Wikipedia page if you'd like to learn more!

  - The Ham Sandwich Theorem
  - The Ugly Duckling Theorem
  - The No Free Lunch Theorem
  - The Chicken McNugget Theorem
  - The Art Gallery Theorem
  - The BEST Theorem
  - The British Flag Theorem
  - The Butterfly Theorem
  - The Edge-of-the-Wedge Theorem
  - The Hairy Ball Theorem[18]
  - The Envelope Theorem
  - The Infinite Monkey Theorem[19]
  - The Pizza Theorem
  - The Star of David Theorem
  - The Omniscient Rabbit Theorem.[20]

- This is your periodic reminder that struggling through tough ideas and proofs is a really important skill in math. After awhile, you may even start to enjoy it.

[16] *Shhhh... don't tell our secrets!*

[17] Or, sadly, the first European to prove it. See: Stigler's Law.

[18] This joke is trivial, and is left as an exercise to the reader.

[19] If do you look this up on Wikipedia, make sure you read the "Actual monkeys" section. It's hilarious.

[20] This one I proved! (But, sadly, this is the one that does not (yet?) have a Wikipedia page.)

It is also one of the most important skills that one can gain from a challenging math class like an intro to proofs class. The classic high school math student question is, "When are we going to use this material in the real world?" There are many good answers to this question, but one is that the soft skills like grit and mental tenacity are some of the most important skills in order to succeed "in the real world," and few other classes instill these skills better than a math class. As I wrote in the Chapter 1 Pro-Tips, no teacher can download the Math Castle into your brain.[21] This forces you to struggle and get better at dealing with new and uncomfortable ideas, which is good.

- When you take real analysis, counterexamples will become very important. In math, it is important to know how far an idea can take you, and a good way to know just how far is too far — is with a counterexample. By having a large collection of strange mathematical objects at your disposal, you will improve your sense of where the boundaries are in math, and you'll become better at spotting when a conjecture seems plausible or implausible.

  It is also a good idea to collect some favorite mathematical objects: a favorite polynomial, discontinuous function, integrable function, group, ring, field, graph, etc. And if these objects are a little unusual in some way, then all the better. Then, whenever you have a conjecture, you can check whether it might be true by testing it on your favorite applicable object to see how it does. This will also help you get a deeper understanding of those objects, which is a great way to anchor your mathematical knowledge.

- When testing a conjecture or coming up with your own conjecture, it is also wise to be cognizant of the "boundary case" or any special values. What happens at 0? What happens at 1? What happens as it grows large? As it goes off to negative infinity? Indeed, mathematicians often immediately jump to the extreme cases when checking something in their minds. This not only helps them check plausibility, but it can help them get a sense of what is happening once the messiness of the "small cases" get averaged out.

- Are you a slow thinker? Does this concern you? I'll let Laurent Schwartz[22] respond. In his autobiography, he wrote this: "I was, and still am, rather slow. I need time to seize things because I always need to understand them fully... If a new question arose, [other students] answered before me. Towards the end of the eleventh grade, I secretly thought of myself as stupid. I worried about this for a long time... I never talked about this to anyone, but I always felt convinced that my imposture would someday be revealed: the whole world and myself would finally see that what looked like intelligence was really just an illusion... At the end of the eleventh grade, I took the measure of the situation, and came to the conclusion that rapidity doesn't have a precise relation to intelligence. What is important is to deeply understand things and their relations to each other. This is where intelligence lies. The fact of being quick or slow isn't really relevant."

[21] Although Elon Musk seems to be trying.

[22] Schwartz won the Fields Medal — the most prestigious prize in mathematics. He was brilliant.

— Exercises —

Exercise 6.1. Explain in your own words the difference between the contrapositive, the converse and a counterexample.

Exercise 6.2. Give 4 examples of implications, and for each write down their contrapositive. Have two be real-world examples, and two be math examples.

Exercise 6.3.

(a) What is the contrapositive of "If $n^2 - 4n + 7$ is even, then n is odd"?

(b) Suppose that $n \in \mathbb{Z}$. Prove that if $n^2 - 4n + 7$ is even, then n is odd.

Exercise 6.4.

(a) What is the contrapositive of "If mn is odd, then m is odd and n is odd"?

(b) Suppose that $m, n \in \mathbb{Z}$. Prove that if mn is odd, then m is odd and n is odd.

Exercise 6.5. Suppose $n \in \mathbb{Z}$. Prove the following.

(a) If n^2 is even, then n is even.

(b) If $5n^2 + 3$ is even, then n is odd.

(c) If $n^2 + 4n + 5$ is even, then n is odd.

Exercise 6.6. Suppose $n \in \mathbb{Z}$. Prove the following.

(a) If $4 \nmid n^2$, then n is odd.

(b) If $8 \nmid (n^2 - 1)$, then n is even.

(c) If $3 \nmid n^2$, then $3 \nmid n$.

(d) If $3 \nmid (n^2 - 1)$, then $3 \mid n$.

Exercise 6.7. Suppose $m, n, t \in \mathbb{Z}$. Prove the following.

(a) If $m^2(n^2 + 5)$ is even, then m is even or n is odd.

(b) If $(m^2 + 4)(n^2 - 2mn)$ is odd, then m and n are odd.

(c) If $m \nmid nt$, then $m \nmid n$ and $m \nmid t$.

Exercise 6.8. Suppose $x, y \in \mathbb{R}$. Prove the following.

(a) If $x + y \geq 2$, then $x \geq 1$ or $y \geq 1$.

(b) If $x^3 + xy^2 \leq y^3 + yx^2$, then $x \leq y$.

(c) If $x^2 + 2x + \frac{1}{2} < 0$, then $x < 0$.

(d) If $x^2 - 5x + 6 < 0$, then $2 < x < 3$.

(e) If $x^3 + x > 0$, then $x > 0$.

Exercise 6.9. Consider the three sets $A = \{1, 2, 3, 4\}$, $B = \{6, 7, 8, 9, 10\}$ and $C = \{1, 2, 3, 4, 5, 6, 7, 8, 9, 10\}$. Suppose that $n \in C$. Prove that if $\dfrac{n(n+1)(2n+1)}{6}$ is divisible by 6, then $n \in A \cup B$.

Exercise 6.10. Define the *Fibonacci sequence* to be the sequence $F_1, F_2, F_3, F_4, \ldots$, where F_1 and F_2 both equal 1, and every term thereafter is the sum of the previous two: $F_n = F_{n-1} + F_{n-2}$ for $n = 3, 4, 5, 6, \ldots$. Thus, the sequence begins

$$1 \ , \ 1 \ , \ 2 \ , \ 3 \ , \ 5 \ , \ 8 \ , \ 13 \ , \ 21 \ , \ 34 \ , \ 55 \ , \ \ldots.$$

Prove that if F_n is not a perfect cube, then $n \notin \{1, 2, 6\}$.

Exercise 6.11. Make up your own problem that is easier to solve with a contrapositive than with a direct proof or a proof by induction.

> **Note.** For the next exercise, recall that one way to prove $P \Leftrightarrow Q$ is to prove $P \Rightarrow Q$ and then $Q \Rightarrow P$. Each direction will be its own proof, and these two proofs may use different methods.

Exercise 6.12. Suppose $n \in \mathbb{Z}$.

(a) Prove that n is even if and only if $n^2 + 1$ is odd.

(b) Prove that n is odd if and only if $n^2 + 2n + 6$ is odd.

(c) Prove that $(n + 1)^2 - 1$ is even if and only if n is even.

Exercise 6.13. What is the contrapositive to the pigeonhole principle? Which do you think seems more clear: The pigeonhole principle or its contrapositive?

Exercise 6.14. Find an implication in a song, and write down the implication as an "if, then" statement. Then, write down its contrapositive. For example, the contrapositive to "If a guy is a scrub, then he can't get no love from me" is "If a guy can get love from me, then he is not a scrub."

Exercise 6.15. Come up with a real-world claim that is false, and prove that it is false by exhibiting a counterexample to the claim.

Exercise 6.16. The following statements are all false. For each, find a counterexample.

(a) If $a, b, c \in \mathbb{N}$, then $a^2 + b^2 \neq c^2$.

(b) If $a, b \in \mathbb{N}$, then $a + b < ab$.

(c) If $x, y \in \mathbb{R}$, then $|x + y| = |x| + |y|$.

(d) If $x, y \in \mathbb{R}$ and $|x + y| = |x - y|$, then $y = 0$.

(e) If $x \in \mathbb{R}$, then $x^2 - x \geq 0$.

(f) If $x \in \mathbb{R}$, then $\dfrac{x^2 + x}{x^2 - x} = \dfrac{x + 1}{x - 1}$.

(g) If $a, b \in \mathbb{N}$ and $a \neq b$, then $a^b \neq b^a$.

(h) If $n \in \mathbb{N}$, then $n^2 + 17n + 17$ is prime.

(i) If $n \in \mathbb{N}$, then $2n^2 - 4n + 31$ is prime.

(j) If $n^2 + 5n$ is even, then n is odd.

(k) If $a, b, c \in \mathbb{N}$ and $a \mid bc$, then $a \mid b$ or $a \mid c$.

(l) If $a, b \in \mathbb{Z}$, $a \mid b$ and $b \mid a$, then $a = b$.

(m) If A, B and C are sets and $A \subseteq B \cup C$, then $A \subseteq B$ or $A \subseteq C$.

(n) If A and B are sets, then $|A \cup B| = |A| + |B|$.

(o) If A, B and C are sets, then $A \setminus (B \cap C) = (A \setminus B) \cap (A \setminus C)$.

(p) If A, B, C and D are sets, then $(A \times B) \cup (C \times D) = (A \cup C) \times (B \cup D)$.

(q) If A, B and C are sets and $A \times C = B \times C$, then $A = B$.

Exercise 6.17. Read the *Introduction to Big Data* following this chapter. Then, draw out your own small internet, including enough links that it is interesting. Following the method of that section, determine the Google PageRank ranking of the websites in your internet.

— Open Question —

In this chapter, we continued our discussion of even, odd and prime numbers, and we discussed counterexamples. Here is a famous problem that combines all of these.

Choose some $n \in \mathbb{N}$, and then perform the following operation to it:

- If n is even, replace it with $n/2$;
- If n is odd, replace it with $3n+1$.

Then, repeat. For example, if $n = 13$, this n turns into $3(13) + 1 = 40$. Then 40 turns into 20, which turns into 10, which turns into 5, which turns into 16, then 8, then 4, then 2, then 1. And technically, you could continue, but 1 turns into 4 which turns into 2 which turns into 1, so we just go in loops at this point. Since it is now boring, we say that as soon as the sequence reaches 1, we stop.

Picking a number n and performing these operations, one produces a sequence of numbers called a *Collatz sequence*. Here is the Collatz sequence starting at $n = 42$, which also terminates at 1:

$$42 \longrightarrow 21 \longrightarrow 64 \longrightarrow 32 \longrightarrow 16 \longrightarrow 8 \longrightarrow 4 \longrightarrow 2 \longrightarrow 1.$$

If n is a power of 2, then its Collatz sequence will terminate at 1 because every step of the way the number will be even so it just keeps getting halved. For example, here is the Collatz sequence starting at $n = 128$ (which is 2^7):

$$128 \longrightarrow 64 \longrightarrow 32 \longrightarrow 16 \longrightarrow 8 \longrightarrow 4 \longrightarrow 2 \longrightarrow 1.$$

Moreover, if any n value ever reaches a power-of-2, then from that point on it will just keep getting halved, and hence will terminate at 1. Here is a way to visualize some more cases:

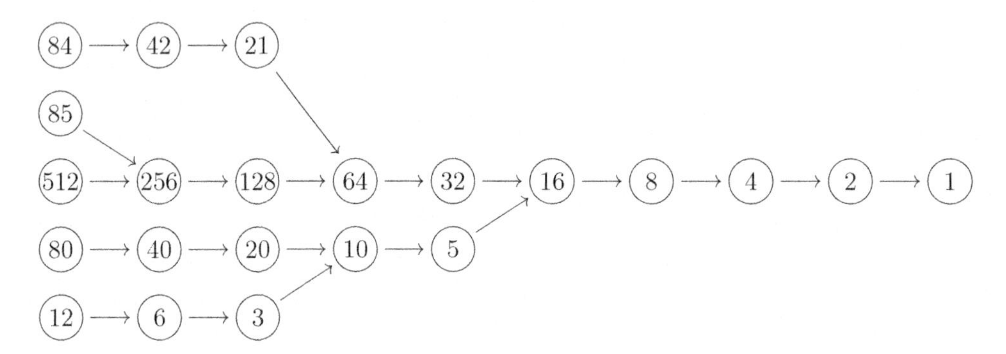

This process uses relatively simple rules and operations, yet here is a surprisingly difficult question: For which numbers n does the Collatz sequence starting at n terminate at 1? Most researchers guess that every natural number will terminate at 1, but so far nobody has been able to prove it. This is called the *Collatz conjecture*.

A computer has checked this conjecture for all n less than 2^{68} (a number with 20 digits), and no counterexample was found. So if you are going to start searching for a counterexample (an n which does not eventually reach 1), then you will have to start your search with *realllyy* big numbers. Nevertheless, I hope I convinced you in Section 6.3 that just because a conjecture holds for a long time, that does not mean it is true. The Collatz conjecture is probably true, but we still need proof.

Why is finding a proof so difficult? One challenge is that n's prime factorization does not seem too helpful. Indeed, the prime structure is destroyed whenever n is odd. Any prime that divides n will not divide $3n + 1$.

How else might we try to prove it? To generate ideas, let's look at a similar problem. Suppose we started with an $n \in \mathbb{N}$, but instead we followed these rules:

- If n is even, replace it with $n/2$;
- If n is odd, replace it with $n + 1$.

Notice that whenever n is even, in the very next step n gets smaller, as it turns into $n/2$. Next, notice that whenever n is odd, in *two* steps it will get smaller, since in the next step it will turn into $n + 1$, which will be even, and hence in the following step it will turn into $\dfrac{n+1}{2}$. And with a little algebra you can show that for any $n \in \mathbb{N}$ (except for $n = 1$, when we would be done anyways),

$$\frac{n+1}{2} < n.$$

Thus, either in the next step or in two steps, the number n is decreasing. This can be seen with an example:

$$19 \longrightarrow 20 \longrightarrow 10 \longrightarrow 5 \longrightarrow 6 \longrightarrow 3 \longrightarrow 4 \longrightarrow 2 \longrightarrow 1.$$

Except for a few moments where the numbers bump up by 1, they're steadily trending downward. And if the sequence keeps going down, then it can only end at 1.

Could a similar argument work for the Collatz conjecture? If n is even, then n decreases to $n/2$. If n is odd, then it increases to $3n + 1$, which is even, and hence turns into $\frac{3n+1}{2}$, which is bigger than n (it's about $1.5n$). So if n is odd, after two steps the number is bigger—it's not just steadily decreasing to 1. If you ask what happens after three steps, and then four steps, and so on, there is an interesting probability argument that can be made to suggest that the Collatz conjecture is true for at least most natural numbers, but it falls short of a proof of all natural numbers.

How could the theorem be false? Perhaps there is a sequence which drifts upward toward infinity. Or perhaps there is a number n_1 that is sent to a number n_2 which is sent to n_3 and so on, but eventually this sequence wraps back around and one of its terms is sent back to n_1. That is, maybe there is a cycle like the $1 \to 4 \to 2 \to 1 \to 4 \ldots$ cycle. Is that possible?

Despite its simple setup, decades ago Paul Erdős said "Mathematics may not be ready for such problems." In 2019, a year before I write this, Terrence Tao made major progress on the problem, getting "about as close as one can get to the Collatz conjecture without actually solving it." Perhaps this will cause new activity between the time I write this and the time you read it! I hope so!

Open Question. Does every Collatz sequence terminate at 1?

Introduction to Big Data[23]

"Big" is a relative term. You put Shaq next to almost anyone, but especially gymnast Simone Biles, and he looks enormous. But if he stands with Yao Ming... suddenly he doesn't look so big.

The same applies to data. Ten lines on a spreadsheet could be a lot of data if it is being processed by the tired eyes of an underpaid intern. Meanwhile, processing the text of every book ever published is small in the face of the NSA's computing capabilities.

The term *big data* refers to the rapidly increasing quantity and variety of data over a wide array of fields, from healthcare to finance to security. We gather data to gain insight, but how do we store, process, analyze and access all of these data?[24] Sophisticated mathematical techniques have been developed which make use of spreadsheets, databases, matrices and graphs, with the general understanding that data should be stored and operated upon as a whole set, rather than thinking about each datum individually. Let's talk about one of these techniques in the context of an application we all make use of: the Internet.

[23]Most of the *Introduction to* sections relate to the chapter they follow. This is an exception. There was not a natural topic to follow the contrapositive. So let's talk about big data!

[24]Pro-Tip: *Data* is the plural form, while a single piece of data is a *datum*. If you're talking about data science, toss "these data are" into your conversation, or a couple uses of "a datum," and you will reap instant credibility.

I remember the days before Google. I remember being in first grade and having to write a short report on dogs. My family didn't have a book to help me, nor did we have a set of the Encyclopedia Britannica—which would be the standard resource for such a project—and so I turned to my Plan B: the Internet. After a long and noisy dial-up process, I was on! But... how do I find websites about dogs? I remember trying `dog.com` and `dogs.com`, and when they weren't helpful, I was out of ideas.

Why? Because this was before Google! There were a few search engines out there, but they were bad and I probably didn't even know about them. These primitive search engines had succeeded in indexing lots and lots of websites, so if you searched for "facts about dogs" it could return websites that used all three words, but they struggled mightily to rank them by importance. The magic behind Google is that the "good stuff" always winds up near the top. For example, "Long Form Math" returns nearly 500 million results—but this book's companion website, `longformmath.com`, is at the top of the list.[25]

So how does Google do it? The one piece of information that they can easily collect about a website is the number of links it has to other websites, and which websites it links to. As a *much* smaller example, consider an internet with five websites, and the following links ($A \to B$ means website A linked to website B).

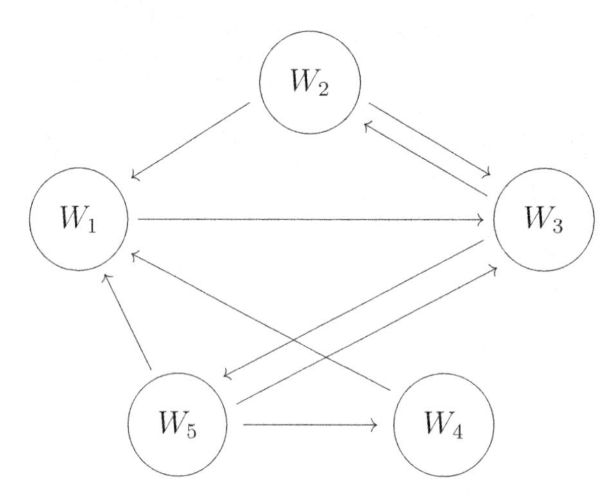

Our goal is to assign to each website a numerical *value*. Let v_1 be the value of website W_1, let v_2 be the value of website W_2, and so on. You can think about a link as an endorsement—if $A \to B$, then in some ways website A has endorsed website B as being important. Now, there are two tensions here:

- It should be more valuable to be endorsed by an important website than a bad website.
- It should be less valuable to be endorsed by a website if that website endorses lots of other websites.

You can think about this like a political endorsement in a presidential primary. It should be considered more valuable to be endorsed by the governor than by a

[25]Please check it out!

city council member. But if someone endorses both you and someone else, then it shouldn't be worth quite as much as if you were the only candidate they supported; it is basically a half endorsement.

With this intuition, look at the internet graph above, and for now let's consider website W_1.

W_1 endorsed by:	W_2	W_4	W_5
Which has value:	v_2	v_4	v_5
Number of websites they endorsed:	2	1	3

Website W_1, in a sense, received half of an endorsement from W_2, because W_2 also endorsed one other website. Website W_1 also received a full endorsement from W_4 and a third of an endorsement from W_5. Therefore, if we wish to assign a numerical value v_1 to the website W_1, and if this value is to be determined by these endorsements, it would make sense to want

$$v_1 = \frac{v_2}{2} + v_4 + \frac{v_5}{3}.$$

Doing this for all other websites, this would produce the following.

$$\begin{aligned} v_1 &= \frac{v_2}{2} + v_4 + \frac{v_5}{3} \\ v_2 &= \frac{v_3}{2} \\ v_3 &= v_1 + \frac{v_2}{2} + \frac{v_5}{3} \\ v_4 &= \frac{v_5}{3} \\ v_5 &= \frac{v_3}{2} \end{aligned}$$

Big question. Is it possible to assign numerical values to v_1, v_2, v_3, v_4 and v_5 so that the above are all satisfied?

Although it might seem silly to put these into vectors, having these all satisfied is indeed the same thing as saying

$$\begin{bmatrix} v_1 \\ v_2 \\ v_3 \\ v_4 \\ v_5 \end{bmatrix} = \begin{bmatrix} \frac{1}{2}v_2 + v_4 + \frac{1}{3}v_5 \\ \frac{1}{2}v_3 \\ v_1 + \frac{1}{2}v_2 + \frac{1}{3}v_5 \\ \frac{1}{3}v_5 \\ \frac{1}{2}v_3 \end{bmatrix}.$$

But why did we do this? Because we can write the right-hand side as a product!

$$\begin{bmatrix} v_1 \\ v_2 \\ v_3 \\ v_4 \\ v_5 \end{bmatrix} = \begin{bmatrix} 0 & \frac{1}{2} & 0 & 1 & \frac{1}{3} \\ 0 & 0 & \frac{1}{2} & 0 & 0 \\ 1 & \frac{1}{2} & 0 & 0 & \frac{1}{3} \\ 0 & 0 & 0 & 0 & \frac{1}{3} \\ 0 & 0 & \frac{1}{2} & 0 & 0 \end{bmatrix} \begin{bmatrix} v_1 \\ v_2 \\ v_3 \\ v_4 \\ v_5 \end{bmatrix}.$$

It may feel like we have only made things more complicated, since we moved from a collection of five equations to a matrix product. However, if we call this matrix A, and we call the vector v, then we are looking for a vector v such that

$$v = Av.$$

As it turns out, once the number of websites grows into the millions or billions, this becomes a *much* easier and more holistic approach. Indeed, finding such a vector not only earned the founders of Google (Larry Page and Sergey Brin) tens of billions of dollars each, but also is really important mathematically.

Eigenvalues and Eigenvectors

Three things: First, a vector v for which $v = Av$ is called an *eigenvector* of A. Second, we typically write it in the other order: $Av = v$. And third, as it turns out, the math is very much the same whether we want Av to be equal to v or a scaled version of v, like $2v$ or $3v$. Mathematicians like to work in as general a setting as possible, so let's also include that scaling constant in our discussion.

I suspect most readers will have taken linear algebra before, and will be familiar with eigen-stuff. But if not — or in case you could use a review — here is one of the most important definitions from that field, formally stated.

Definition.

Definition 6.8. Let A be an $n \times n$ matrix. If $\lambda \in \mathbb{R}$ and

$$Av = \lambda v$$

for some nonzero vector v, then λ is an *eigenvalue* of A. Moreover, the corresponding v are the *eigenvectors* corresponding to λ.

Google's algorithm (the one we just discussed, which by the way is called the *PageRank* algorithm) corresponds to the case where $\lambda = 1$, giving just $Av = v$.

How to find A's eigenvalues.

One of the great things about this approach is that it allows us to use linear algebra to solve a tough problem. This is good, because computers are really good at doing linear algebra. You may remember the procedure to find the eigenvalues and eigenvectors of a matrix, but if not, or you would still like a refresher, below is a discussion and an example.

Given a matrix A, its eigenvalues are the numbers λ for which

$$Av = \lambda v$$

for some *nonzero* vector v. Following your instincts from high school algebra,[26] let's get a 0 on one side of the equation. From there, we will include the identity matrix I (recall that $Iv = v$), and then factor out the v:

$$\begin{aligned} Av - \lambda v &= 0 \\ Av - \lambda Iv &= 0 \\ (A - \lambda I)v &= 0. \end{aligned}$$

In other words, we want all λ such that the matrix $(A - \lambda I)$ has a nontrivial null space. And in linear algebra you learn that this is the same as asking for all λ such that $\det(A - \lambda I) = 0$, which is a straightforward computation.

Example 6.9. Find all eigenvalues of $A = \begin{bmatrix} 0 & 4 \\ 1 & 0 \end{bmatrix}$.

Solution. We are seeking all λ such that

$$\begin{aligned} \det(A - \lambda I) &= 0 \\ \det\left(\begin{bmatrix} 0 & 4 \\ 1 & 0 \end{bmatrix} - \lambda \begin{bmatrix} 1 & 0 \\ 0 & 1 \end{bmatrix}\right) &= 0 \\ \det\left(\begin{bmatrix} -\lambda & 4 \\ 1 & -\lambda \end{bmatrix}\right) &= 0 \\ (-\lambda)(-\lambda) - (1)(4) &= 0 \\ \lambda^2 - 4 &= 0 \\ (\lambda - 2)(\lambda + 2) &= 0 \\ \lambda = 2 \quad \text{or} \quad \lambda &= -2 \end{aligned}$$

And so, the eigenvalues of A are 2 and -2. □

[26]Example: To solve a quadratic equation like $x^2 - 5x = 6$, first you'd move the six to left.

How to find A's eigenvectors.

Finding eigenvectors, you might recall, was an exercise in row reduction. Below is an example.

Example 6.10. Find all eigenvectors of $A = \begin{bmatrix} 0 & 4 \\ 1 & 0 \end{bmatrix}$ corresponding to $\lambda = 2$.

Solution. We want to find all v such that

$$\begin{aligned} Av &= 2v \\ Av - 2Iv &= 0 \\ (A - 2I)v &= 0. \end{aligned}$$

Note that this is the same thing as

$$\begin{bmatrix} -2 & 4 \\ 1 & -2 \end{bmatrix} \begin{bmatrix} v_1 \\ v_2 \end{bmatrix} = \begin{bmatrix} 0 \\ 0 \end{bmatrix},$$

which can be solved via the augmented matrix

$$\left[\begin{array}{cc|c} -2 & 4 & 0 \\ 1 & -2 & 0 \end{array}\right].$$

And now we row reduce:

$$\left[\begin{array}{cc|c} -2 & 4 & 0 \\ 1 & -2 & 0 \end{array}\right] \xrightarrow{2R_2 + R_1} \left[\begin{array}{cc|c} -2 & 4 & 0 \\ 0 & 0 & 0 \end{array}\right] \xrightarrow{-\frac{1}{2}R_1} \left[\begin{array}{cc|c} 1 & -2 & 0 \\ 0 & 0 & 0 \end{array}\right].$$

This implies two equations:

$$\begin{aligned} v_1 - 2v_2 &= 0 \\ 0 &= 0. \end{aligned}$$

There is a degree of freedom in this system of equations. If we treat the second variable as free (so we let it equal some variable t), this gives the following:

$$\begin{aligned} v_2 &= t \\ v_1 &= 2t. \end{aligned}$$

This then gives the solutions

$$\begin{bmatrix} v_1 \\ v_2 \end{bmatrix} = \begin{bmatrix} 2t \\ t \end{bmatrix} = t \begin{bmatrix} 2 \\ 1 \end{bmatrix}.$$

Thus, the set of all eigenvectors is the following:

$$\left\{ \begin{bmatrix} 2t \\ t \end{bmatrix} : t \in \mathbb{R} \right\}.$$

□

In our example of a 5-website internet, we were searching for a solution to this equation:

$$\begin{bmatrix} v_1 \\ v_2 \\ v_3 \\ v_4 \\ v_5 \end{bmatrix} = \begin{bmatrix} 0 & \frac{1}{2} & 0 & 1 & \frac{1}{3} \\ 0 & 0 & \frac{1}{2} & 0 & 0 \\ 1 & \frac{1}{2} & 0 & 0 & \frac{1}{3} \\ 0 & 0 & 0 & 0 & \frac{1}{3} \\ 0 & 0 & \frac{1}{2} & 0 & 0 \end{bmatrix} \begin{bmatrix} v_1 \\ v_2 \\ v_3 \\ v_4 \\ v_5 \end{bmatrix}.$$

Doing so would produce the following eigenvector:

$$\begin{bmatrix} v_1 \\ v_2 \\ v_3 \\ v_4 \\ v_5 \end{bmatrix} = \begin{bmatrix} 4/3 \\ 1 \\ 2 \\ 1/3 \\ 1 \end{bmatrix}.$$

Although websites W_2 and W_5 were assigned the same numerical value, and hence will be tied in importance, Google would rank these websites as:

$$W_3 > W_1 > W_2 = W_5 > W_4.$$

This 5×5 matrix is small enough to explain and it captures all the important points of understanding, but of course the real Internet is not so small. And neither are other real-world matrices for which these techniques are used.[27] When Google computes an eigenvector to rank websites, their matrix could have over a billion rows and columns — one for every website they've indexed. That makes their job hard because, well, that's a really big matrix.

However, they do have some things going for them. First, the matrix is *sparse*, meaning that most of the entries are 0 — because each website links to a very small percentage of the billions of websites out there. Moreover, the matrix is *left stochastic*, meaning the columns necessarily sum to 1. And mathematicians have found sophisticated techniques to significantly improve the computation time needed to analyze sparse, left stochastic matrices.

There is also a lot of theory developed around matrices of this form. They always have an eigenvalue of 1, meaning our Google searches are guaranteed to have a ranking. An $n \times n$ matrix is guaranteed to have n eigenvalues (possibly with some repeated, possibly with some being complex numbers), and if λ is such an eigenvalue, then $|\lambda| \leq 1$.

[27]This exact technique has been found useful in social network analysis, recommender systems, biology, chemistry and neuroscience. There are also many useful variants of it.

These matrices are also intimately related to Markov chains, which are a collection of states, and a system that describes a randomized procedure for moving from one state to another. An example of this is below.

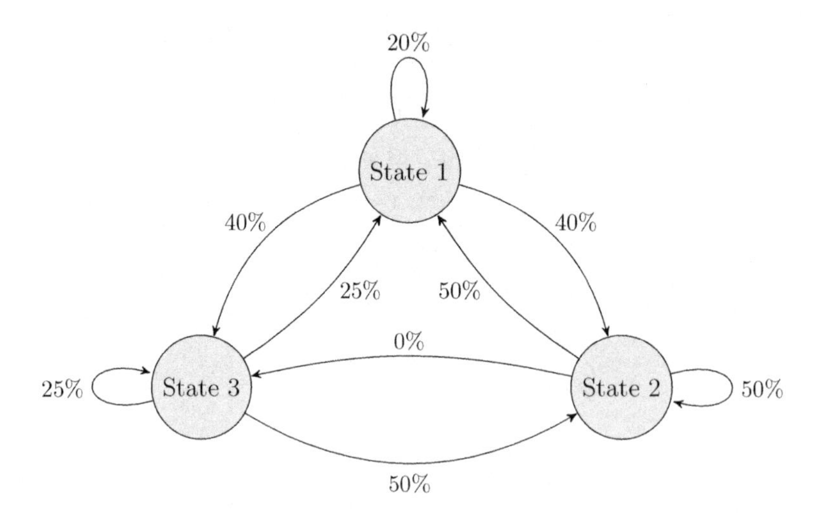

Indeed, you can think about the Internet as a Markov chain, in which you move from website to website by clicking on a random link on each website you visit.

Linear algebraic techniques[28] are one of the most important tools for data[29] science, but much more mathematics plays a significant role. This includes:

- Statistics, probability and measure theory;
- Optimization, especially convex optimization;
- Discrete math, especially graph theory;
- Differential calculus; and
- Functional and numerical analysis.

Data science is a big and important area right now. And if the projections remain true, it will still be growing whenever you're reading this, with many companies eager to pay you a lot of money to do exciting mathematics and live in a fun city. Which sounds like a pretty nice life to me.

[28]By the way, the mathematician in me wants to interrupt this discussion to point out that linear algebra is way more than matrix computations. Matrix algebra provides "easy" ways to solve theoretical questions, but the theory itself is what is really going on. And the theory is beautiful! There is wonderful geometry underlying the theory of systems of linear equations on vector spaces, and if your first course in linear algebra ignored this, I encourage you to take an advanced course. As Fields Medalist Michael Atiyah says,

> "Algebra is the offer made by the devil to the mathematician. The devil says: I will give you this powerful machine, it will answer any question you like. All you need to do is give me your soul: Give up geometry and you will have this marvelous machine."

[29]Oh, and by the way, "data" is pronounced "data," not "data."

Chapter 7: Contradiction

Suppose someone stole Mrs. Figg's purse at 8pm last night; they grabbed it right from her arms. If Carmen is a suspect, but the detective finds conclusive evidence that Carmen was across town at 8pm, then there is no way for her to have grabbed Mrs. Figg's purse. For mathematical reasons, think about it like this: Assume Carmen did steal it. Then Carmen was in two places at once. This is absurd, so she must not have done it.

These arguments are called *reductio ad absurdum* — reduction to absurdity — and are used often in everyday conversation. Suppose your mom asks "so... are you dating anyone right now?" And, annoyed, you respond "no, if I were I would have told you." Then — whether or not you're being truthful — you are making a *reductio ad absurdum* argument. Your argument is essentially: If I were dating, then you would know — but you don't know. Therefore, I must not be dating. Here, you are assuming for the argument that you are dating, and then deducing the absurdity that this would mean you would have told her, which contradicts the reality that she has not been told.

In math one might say, "There is not a largest integer, because if there were and we called it N, then $N + 1$ would be a larger integer." Again, you are assuming what ends up being false — that there exists a largest integer N. And we are showing that such an assumption would imply something absurd — that N is the largest and yet $N + 1$ is larger.

This is the main idea behind our next proof technique: *proof by contradiction.* Recall from our logic chapter that if a statement Q is true, then $\sim Q$ is false. And if Q is false, then $\sim Q$ is true. We described this relationship with the simplest of all truth tables:

Q	$\sim Q$
True	False
False	True

The important observation is that Q and $\sim Q$ cannot both be true, and cannot both be false. In fact, we could even draw out a truth table for that:

Q	$\sim Q$	$Q \wedge \sim Q$
True	False	False
False	True	False

This is analogous to Carmen not being in two different places at once, your mom being told and not being told, or N and $N+1$ both being integers while N is the largest.

The big idea is this: If you start with something true and apply correct logic to it, you will never arrive at something false. So it can't be true that Carmen stole the bag, if that would imply the falsity that she can be in two places at once. Indeed, *if your assumptions imply something false, then something you assumed had to be false as well.* Here's a schematic summary of these thoughts:

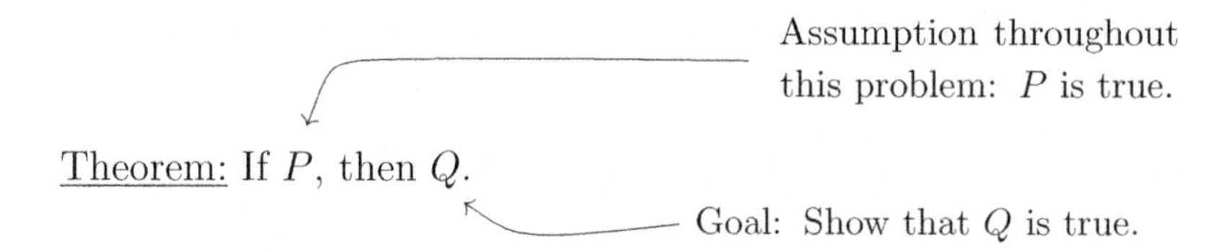

By the truth tables: Either Q is true or $\sim Q$ is true, but not both.
This gives two options:

- P is true and Q is true $(P \wedge Q)$
- P is true and $\sim Q$ is true $(P \wedge \sim Q)$

If $P \wedge \sim Q$ implies anything false, that can't be the correct option.
That is, it must be $P \wedge Q$. Thus, we have shown $P \Rightarrow Q$!

By *reductio ad absurdum*

Another way to see it: The truth table for $P \Rightarrow Q$ can be unintuitive, so there is risk to using it to motivate a new proof method. Nevertheless, it might be enlightening for you. Notice that the only way that $P \Rightarrow Q$ can be false, is if P is true and Q is false.

P	Q	$P \Rightarrow Q$
True	True	True
True	False	False
False	True	True
False	False	True

Thus, this is the only case we have to rule out in order to prove our theorem: that $P \Rightarrow Q$ is true. So, if you assume that P is true and Q is false, and manage to use that to deduce a contradiction, then you will have ruled out the one and only bad case, which in turn means that the theorem must be true! *Voilà!*

One final comment: What if the theorem is not written in the form "If P, then Q"? What if it's just a simple statement "P"? For example, suppose the theorem said "257 is a Fermat prime." This is just "P". Must you first turn it into an implication, like "If $p = 257$, then p is a Fermat prime."? Answer: no. As long as you can negate the theorem, you can use that. So, to prove "257 is a Fermat prime," your proof could begin with "Assume 257 is not a Fermat prime," and then search for a contradiction.

7.1 Two Warm-Up Examples

Our first two propositions are of the form "P" rather than "If P, then Q."

Proposition.

Proposition 7.1. There does not exist a largest natural number.

Proof Idea. Notice that this proposition is of the "P" variety, not the "$P \Rightarrow Q$" variety.[1] Thus, the proof will proceed by assuming $\sim P$ (the negation of P).

The negation of "there does not exist a largest natural number" is "there *does* exist a largest natural number." So our proof strategy will be to assume that there is a largest natural number and then show why that leads to something false (the contradiction). One (of several) ways to do this is to assume that N is the largest natural number and then show that $N+1$ must be larger — if it weren't, we could deduce that $0 \geq 1$, which is a contradiction. Here's that:

Proof. Assume for a contradiction that there is a largest element of $\mathbb{N}$, and call this number N. Being larger than every other natural number, N has the property that $N \geq m$ for all $m \in \mathbb{N}$.

Observe that since $N \in \mathbb{N}$, also $(N+1) \in \mathbb{N}$. And so, by assumption,

$$N \geq N + 1.$$

Subtracting N from both sides,

$$0 \geq 1.$$

This is a contradiction[2] since we know that $0 < 1$, and therefore there must not be a largest element of $\mathbb{N}$. □

Sometimes the contradiction is something that we knew was false before we started the problem, like $0 \geq 1$. But sometimes the contradiction is something within the problem itself: At some point we assume P and then later we deduce $\sim P$. The next proposition is an example of that.

[1]Now, one *could* turn it into a "$P \Rightarrow Q$" if one really wanted. For example, it could instead be stated as: "If $\mathbb{N}$ is the set of natural numbers, then $\mathbb{N}$ does not have a largest element." Or, equivalently: "If N is larger than every natural number, then $N \notin \mathbb{N}$." Or, equivalently: "If N is a natural number, then there exists a natural number larger than N." But these all feel contrived.

[2]
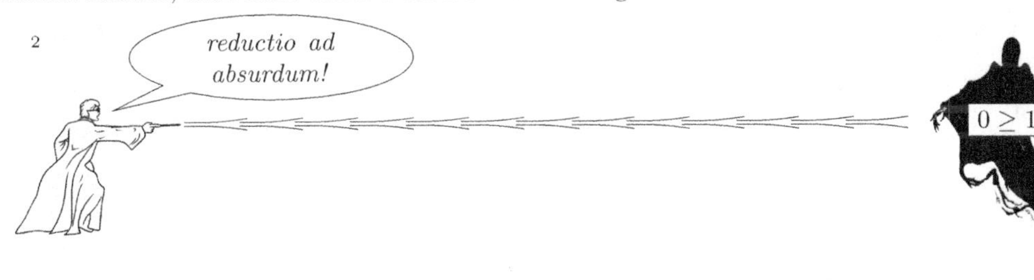

Proposition.

Proposition 7.2. There does not exist a smallest positive rational number.

Scratch Work. In order to use a proof by contradiction, let's suppose that there *does* exist a smallest positive rational number, and let's call such a number q; this means $q = \frac{a}{b}$ for some $a, b \in \mathbb{Z}$ such that $a, b > 0$ (they are in $\mathbb{Z}$ by the definition of a rational number, and they are positive because $q > 0$). Why does this lead to a contradiction? Because we can now find a smaller such number! For example, $\frac{a}{2b}$ will be such a number. If $\frac{a}{b}$ is rational and positive, then $\frac{a}{2b}$ will be, too. Now, how can we prove that $\frac{a}{2b}$ is smaller? Let's do some scratch work:

$$\begin{aligned} \frac{a}{2b} &< \frac{a}{b} \\ a &< 2a && \text{(multiply both sides by } 2b\text{)} \\ 0 &< a. && \text{(subtract } a \text{ from both sides)} \end{aligned}$$

Exclamation, not factorial

Ah ha! We know $a > 0$! So now if we do this same scratch work in reverse, we will be starting with something we know ($0 < a$) and concluding with the statement we want ($\frac{a}{2b} < \frac{a}{b}$)! And this concluding inequality gives us our contradiction.

Proof. Assume for a contradiction that there is a smallest positive rational number, and call this number q. Then, since q is rational,

$$q = \frac{a}{b}$$

for some $a, b \in \mathbb{Z}$. And since q is positive, we may assume that $a, b > 0$. Then, by starting with $0 < a$, adding a to both sides, and then dividing by the positive number $2b$, we get this:

$$\begin{aligned} 0 &< a \\ a &< 2a \\ \frac{a}{2b} &< \frac{a}{b}. \end{aligned}$$

We have shown that $\frac{a}{2b} < \frac{a}{b}$. Moreover, $\frac{a}{2b}$ is a positive rational number, since a and $2b$ are positive integers. This contradicts our assumption that q was the smallest positive rational number,[3] and completes the proof. □

[3]

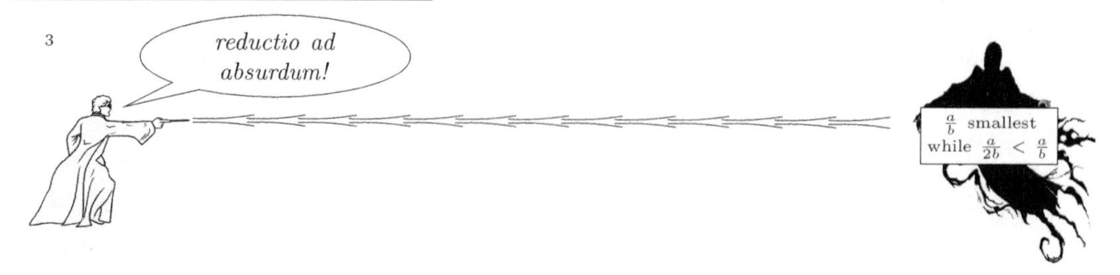

Our first proposition contradicted the fact that $0 < 1$, and our second contradicted $\frac{a}{b}$ being the smallest by finding a smaller. For practice, try to flip these: Try to write a second proof of Proposition 7.1 by contradicting N being the largest by finding a larger, and write a second proof of Proposition 7.2 by contradicting $0 < 1$. Both are similar.[4]

Our proofs by contradiction follow one of two forms. If the theorem is an "if, then" statement (Example: If n is even, then n^2 is even), then here is the general form:

Proposition. If P, then Q.

Proof. Assume for a contradiction P and $\sim Q$.

≪ An explanation of what these mean ≫ ⟵ Apply definitions and/or other results.

⋮ apply algebra, logic, techniques

≪ Hey look, that contradicts something we know to be true. ≫

We obtained a contradiction, therefore $P \Rightarrow Q$. □

If the theorem is a simple statement (Example: There is no largest natural number), then this is the general structure:

Proposition. P.

Proof. Assume for a contradiction $\sim P$.

≪ An explanation of what this means ≫ ⟵ Apply definitions and/or other results.

⋮ apply algebra, logic, techniques

≪ Hey look, that contradicts something we know to be true. ≫

We obtained a contradiction, therefore P. □

[4]Hint: For the first, use $0 < 1$ to show that $N + 1$ is larger. For the second, assume $\frac{a}{2b} > \frac{a}{b}$ and do algebra until you reach the contradiction $0 > 1$.

If you're thinking that proofs by contradiction are a little weird, I agree! You assume something false in order to show that it's true? It's a little strange. In fact, I was inspired enough by this strangeness to write a short poem about it. Enjoy.

The strangest proof method is contradiction.
One chooses to enter a land of fiction:
False things assumed true in a warped depiction,
And *then* one searches for logical friction?!

It's like giving a crook a prison eviction,
With hopes they'll relapse from *your* dereliction.
You set up a sting in a math jurisdiction,
And call your job done — if you get a conviction.

Now that you're warmed up, and I've confessed to my own insecurities, let's do some more examples!

7.2 Examples

Our next example comes from set theory.

Proposition.

Proposition 7.3. Prove that if A and B are sets, then $A \cap (B \setminus A) = \emptyset$.

Scratch Work. This makes sense when you look at the Venn diagram.

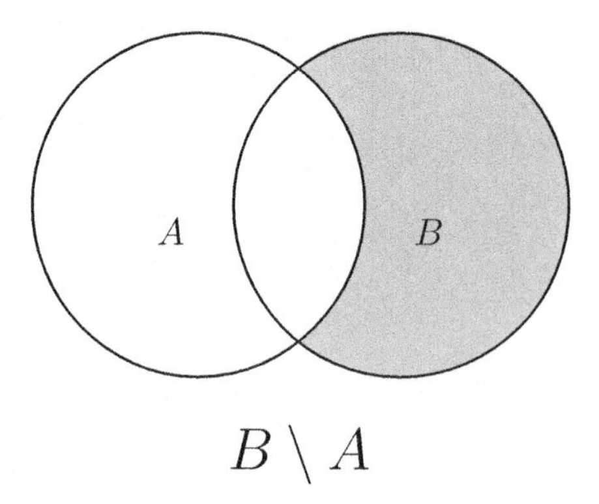

$B \setminus A$

Notice that the shaded portion is $B \setminus A$, and none of that shaded portion falls inside of the A circle. Since there are no elements inside of both A and $B \setminus A$, it makes sense that $A \cap (B \setminus A)$ is the empty set.

Believing the proposition is one thing, but how do we prove it? The conclusion to the proposition is that $A \cap (B \setminus A) = \emptyset$. Being equal to the empty set means that you do not have any elements. Since this is saying that something does *not* happen, you might wonder whether a contrapositive proof would work here. Unfortunately, the contrapositive is "If $A \cap (B \setminus A) \neq \emptyset$, then A or B is not a set." But how would you prove something is not a set? This does not look promising. A proof by contradiction, on the other hand, still allows you to assume that $A \cap (B \setminus A) \neq \emptyset$ but instead of wondering about non-sets you simply have to find some contradiction. This sounds better.

In fact, our contradiction proof is really about following our instincts. If we assume for a contradiction that $A \cap (B \setminus A) \neq \emptyset$, then what does that imply? If you're not the empty set, then you contain an element, so there is some $x \in A \cap (B \setminus A)$. Ok, what does that mean? Well, being in an intersection means that $x \in A$ and $x \in B \setminus A$. Ok, what does that mean? Well, $x \in B \setminus A$ has a definition of its own. And we are nearly at a contradiction. See if you can work it out, and then take a look at the below.

***Proof*.** Assume for a contradiction that $A \cap (B \setminus A) \neq \emptyset$. Then there exists some $x \in A \cap (B \setminus A)$. By the definition of the intersection (Definition 3.8), this implies $x \in A$ and $x \in (B \setminus A)$. By the definition of set subtraction (Definition 3.9), $x \in (B \setminus A)$ means that $x \in B$ and $x \notin A$. Note that in the previous two sentences we deduced that both $x \in A$ and $x \notin A$, which gives the contradiction. □

Our next proposition is one more bread-and-butter example of a proof by contradiction, the type of problem that tests the skill without much hoopla on top.

Proposition.

Proposition 7.4. Prove that there do not exist integers m and n for which $15m + 35n = 1$.

Scratch Work. Assume for a contradiction that there *are* integers m and n where

$$15m + 35n = 1.$$

What can we do now? Well, even though we don't know what m and n are, you could plug in some numbers for m and n to get a feel for what's going on. Doing so might help you notice that $15m + 35n$ is always a multiple of 5.

That sounds really promising, because supposedly this multiple of 5 is equal to 1 (since $15m + 35n = 1$), and 1 is not a multiple of 5. A multiple of 5 is equal to something that is not a multiple of 5? Seems like a contradiction to me! We will have to make that precise in some way, but that should do it.

Proof. Assume for a contradiction that there do exist integers m and n for which $15m + 35n = 1$. Since $m, n \in \mathbb{Z}$, also $(3m + 7n) \in \mathbb{Z}$. Dividing both sides by 5 gives

$$3m + 7n = \frac{1}{5}.$$

This is a contradiction, since we had said that $3m + 7n$ is an integer, and $\frac{1}{5}$ is not an integer. $\square$

There is often more than one way to prove something. For example, the above proof could have begun the same way, by assuming that there are $m, n \in \mathbb{Z}$ for which $15m + 35n = 1$, but then factoring out the 5 to get

$$5(3m + 7n) = 1.$$

Since $(3m + 7n) \in \mathbb{Z}$, this means that $5k = 1$ where $k \in \mathbb{Z}$; by the definition of divisibility (Definition 2.8), this means $5 \mid 1$. However, clearly $5 \nmid 1$, giving the contradiction.

The two proofs relied on similar ideas, even though they diverged at the end. This is common for proofs by contradiction, because once you enter a land of fiction, there are likely contradictions all over the place, and *any* contradiction you find is sufficient to conclude the proof.

And now, ladies and gentlemen, it is time for a real treat:

7.3 The Most Famous Proof in History

This is a book on proofs, so it would be a dereliction of duty to not include the most famous proof in the history of mathematics — Euclid's proof of the infinitude of primes. (Or, in his words, "Prime numbers are more than any assigned multitude of prime numbers.")

In the following proof, recall that if $n \geq 2$ is a natural number, then n is either prime or composite — and being composite means you are a product of primes.[5]

Theorem.

Theorem 7.5. There are infinitely many prime numbers.

Proof Sketch. Since the proof is by contradiction, it will begin by supposing there are only finitely many primes, say $p_1, p_2, \ldots, p_k$. To find a contradiction, our goal will

[5]We defined these terms in Definition 2.16, and in Theorem 4.8, the fundamental theorem of arithmetic, we proved that every such n is prime or a product of primes. This was a proof by strong induction.

be to prove that this list of primes is incomplete; there must be a prime left out. Over two millennia ago, Euclid had the idea to consider what happens when you multiply together this supposed list of all the primes, and then add one: $p_1p_2p_3 \dots p_k + 1$. Why? Consider this for some subsets of the primes:

If the only primes were	Then consider	The Contradiction:
2 and 3	$2 \cdot 3 + 1 = 7$	7 is a new prime!
2, 3 and 5	$2 \cdot 3 \cdot 5 + 1 = 31$	31 is a new prime!
2, 3, 5 and 7	$2 \cdot 3 \cdot 5 \cdot 7 + 1 = 211$	211 is a new prime!
2, 3, 5, 7 and 11	$2 \cdot 3 \cdot 5 \cdot 7 \cdot 11 + 1 = 2311$	2311 is a new prime!
2, 3, 5, 7, 11 and 13	$2 \cdot 3 \cdot 5 \cdot 7 \cdot 11 \cdot 13 + 1$ $= 30031 = 59 \cdot 509$	59 and 509 are both new primes!
2, 3, 5, 7, 11, 13 and 17	$2 \cdot 3 \cdot 5 \cdot 7 \cdot 11 \cdot 13 \cdot 17 + 1$ $= 510511 = 19 \cdot 97 \cdot 277$	19, 97 and 277 are all new primes!

The fourth row, for instance, shows why 2, 3, 5 and 7 can't be the only primes. Since 2, 3, 5 and 7 all divide $2 \cdot 3 \cdot 5 \cdot 7$, there is no way that any of them divide $2 \cdot 3 \cdot 5 \cdot 7 + 1$. If they tried, they would get a remainder of 1! But of course, $2 \cdot 3 \cdot 5 \cdot 7 + 1$ is still a positive integer, and so is either a prime or a product of primes, so there must be *new* primes in there — primes other than 2, 3, 5 or 7.

This was Euclid's big idea. If the only primes are $p_1, p_2, \dots, p_k$, then consider $(p_1 \cdot p_2 \cdot p_3 \cdot \dots \cdot p_k) + 1$. Either this number is prime, in which case it is a *new* prime, since it is bigger than each p_i, or[6] it is composite, in which case it is a product of *new* primes by our reasoning above. In either case, our assumption that $p_1, p_2, \dots, p_k$ was a complete list of all the primes is contradicted. This is how the proof is traditionally presented, although being rigorous in this last step can be a little subtle. One way to make it precise is to use modular arithmetic, which we do in our proof below.

Proof. Suppose for a contradiction that there are only finitely many primes, say k in total. Let $p_1, p_2, p_3, \dots, p_k$ be the complete list of prime numbers, and consider the number $N = p_1 \cdot p_2 \cdot p_3 \cdot \dots \cdot p_k$, which is the product of every prime. Next, consider the number $N + 1$, which is $(p_1 \cdot p_2 \cdot p_3 \cdot \dots \cdot p_k) + 1$. Using $N + 1$, we will find a prime not appearing in the list $p_1, p_2, \dots, p_k$, which will give us our desired contradiction. First note that, being a natural number, $N + 1$ must either be prime or composite, so consider these two cases.

<u>Case 1: $N + 1$ is prime.</u> Since every prime is an integer at least 2, and $N + 1$ is the product of all the primes plus one, $N + 1$ is certainly larger than each p_i. So if $N + 1$ is a prime number, it must be larger than all the primes we had previously considered, and hence is a new prime.

<u>Case 2: $N + 1$ is composite.</u> We begin by showing that no p_i can divide $N + 1$. To do so, remember that by the definition of modular congruence, for any integers a and

[6] By the way, when I write "bigger than each p_i," what I mean is that it is bigger than p_1 and p_2 and p_3 and ... and p_k. In general, when you see a mathematician write "each p_i," what they mean is: look at the context in the problem, and consider all the values of i for which p_i is defined.

b, we have $a \mid b$ precisely when $b \equiv 0 \pmod{a}$. For instance, because $p_i \mid N$, we know

$$N \equiv 0 \pmod{p_i}.$$

Then by applying Proposition 2.15 part (i), we may add 1 to each side to produce

$$N + 1 \equiv 1 \pmod{p_i}.$$

We have shown that $N + 1 \not\equiv 0 \pmod{p_i}$, implying that $p_i \nmid (N + 1)$. And since p_i was arbitrary, this proves that none of our k primes divide $N + 1$.

We assumed that $p_1, p_2, \ldots, p_k$ was the complete list of prime numbers. And recall that $N + 1$ is assumed to be composite, which means it is a product of primes. But since none of the p_i divide $N + 1$, there must be some other prime number, q, which divides $N + 1$. Hence, we have again found a new prime.

In either case we have contradicted the claim that $p_1, p_2, \ldots, p_k$ was an exhaustive list of the prime numbers.[7] Therefore, there must be infinitely many primes. □

Just as atoms are the building blocks of nature, primes are the building blocks of numbers. It takes only 94 elements from the periodic table to construct all of Earth; imagine the intricacies with infinitely many primal building blocks!

This theorem was first proved by Euclid, the great ancient mathematician who is often called "The father of geometry."[8] Euclid wrote the book *Elements*, the most important math book in history. Nowadays, new books are published all the time, and no one book is used everywhere. But *Elements* served as *the* primary math textbook for over *two thousand years*. This is remarkable. And, I can confirm, it's enough to make any author drool.

I also really like the image of Euclid, two and a half millennia ago, setting aside his geometric constructions to puzzle over the primes. In fact, it inspired me to write the following poem. So if you're not sick of my poetry yet...I hope you enjoy!

---

[7] 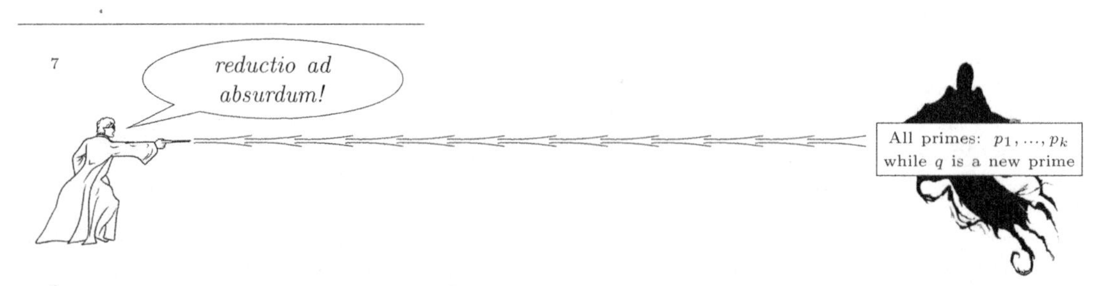

[8] Euclid showed that there are pros and cons to geometry. Pro: tractor. Con: structions.

Civilizations rise and fall;
Their rulers fade away.
It's profound ideas, above all,
That time cannot decay.

Twenty five hundred years ago,
As games of thrones were waged,
An old man sat in candles' glow,
His probing mind engaged.

Parchment scattered, compass askew.
Thoughts turned from lines and arcs
Towards integers — deep questions grew;
His eyes flickered with sparks.

The integers go on and on,
Forever up they climb.
But in this eternal marathon
Will they outlive their prime(s)?

A largest prime? Last of its kind?
What then could we infer?
A story grew within the mind
Of mathematics' Homer.

Soon in his thoughts the truth shone through:
An infinitude of primes.
Below's his proof — and just for you,
This version even rhymes!

Assume for a contradiction
That there're k primes in all.
(For such proofs, the assumed fiction
Will be its own downfall.)

Let's call these primes p_1, p_2,
And so on to p_k;
k could be a zillion and two —
It's *finite*, so that's okay.

Multiply together every prime
And call the answer N.
This integer — far up the climb —
Has *every* prime within.

Since every prime under the Sun
Divides N perfectly,
No p_i divides $N + 1$,
And *that* is this proof's key.

(Followers of the Mod Rabbi
Follow a different path.
N is 0 mod each p_i —
Because of *higher* math.

But then, we see, that $N + 1$,
Is 1 mod each p_i,
So every single prime, bar none,
Ain't dividing that guy.)

We've found a number — N+1 —
That *no* prime can divide?!
Contradiction! So we are done!
The theorem's verified!

This simple N had primed the pump
And powered the proof with ease;
Primal ideas pushed math to jump
To its modern prestige.

The best-known proof in history
Is still taught everywhere;
Profound, simple, and gracefully
It shows us beauty bare.

And yet, even with all this play,
And all the ink we spill,
The proof's still as fresh as the day
It came from Euclid's quill.

"It is impossible to be a mathematician without being a poet in the soul." – Sofia Kovalevskaya. Now, this doesn't mean you have to write poems or like poetry. I think Kovalevskaya's point is that the part of mathematics that so many mathematicians find attractive is when you find a creative connection that you never saw before; it's being able to see something from a new and enlightening perspective; it's when you realize that two seemingly-disparate ideas rhyme in an unexpected way. It's the poetry of ideas that inspires so many mathematicians.

7.4 The Pythagoreans

The Pythagorean theorem says if a and b are the lengths of the legs of a right triangle, and c is the length of the hypotenuse, then $a^2 + b^2 = c^2$. Here's[10] the picture:

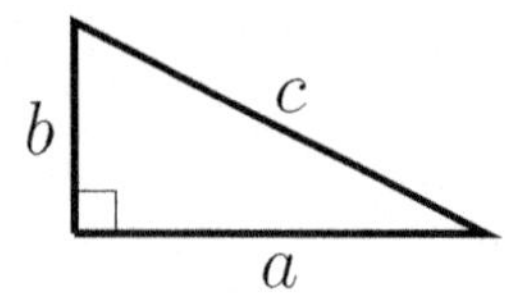

The Pythagoreans proved this by placing four copies of this triangle (non-overlapping) into an $(a+b) \times (a+b)$ square, and asking this question: What's the area of the portion of the square *not* covered by any triangle? Of course, it doesn't matter how we place the triangles into the square — the non-covered area is the same regardless. In fact, that was their key to prove this theorem: If we strategically place them in *two* different ways, we will get *two* different answers to the same question, which will give us exactly what we want.

The Pythagoreans first placed them like below. Doing so, the non-triangle area is the area of one $a \times a$ square and one $b \times b$ square.

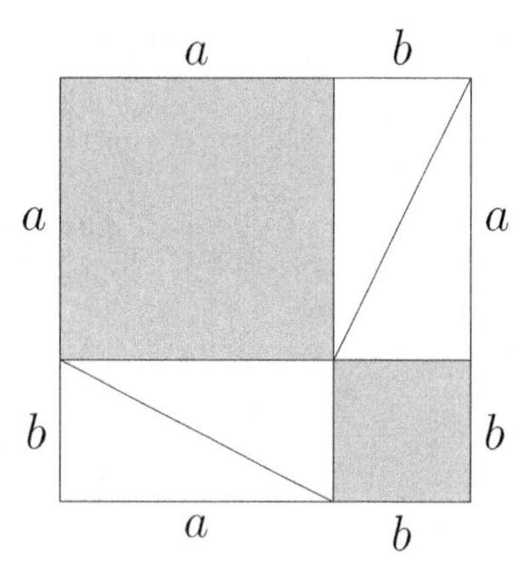

But if we place the triangles differently, we can answer the question a second time! This time, the answer to the question is the area of a $c \times c$ square.

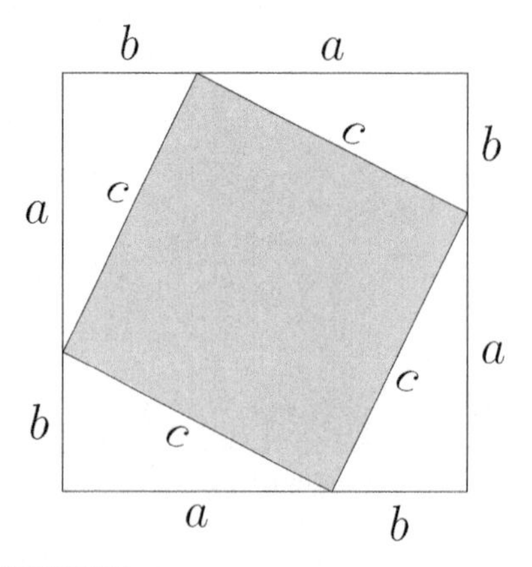

[10]Pythagoras: $a^2 + b^2 = c^2$. Einstein: $E = mc^2$. Corollary: $E = ma^2 + mb^2$?

And since both are answers to the same question, they must equal each other.

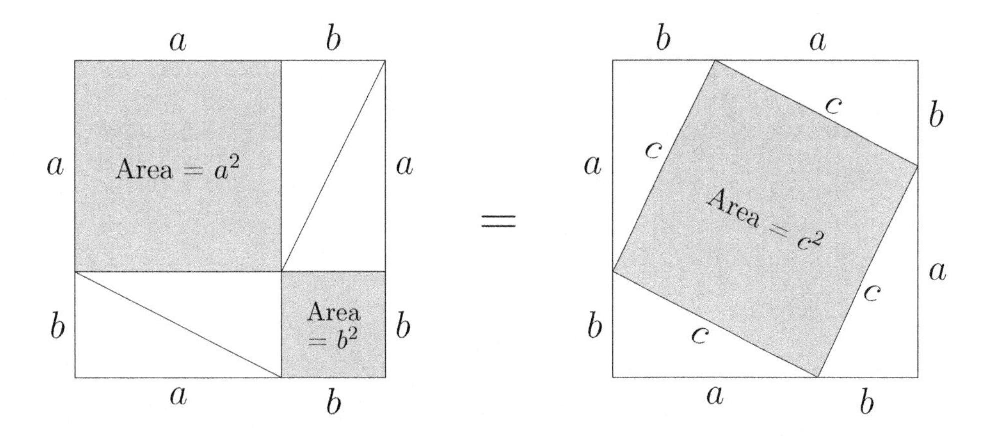

That is,

$$a^2 + b^2 = c^2.$$

This theorem is significant not only for its own merits, but also because it is the key to proving that irrational numbers exist. Sadly, despite having the key, Pythagoras[11] lived and died believing that all numbers were rational. But after his death, his school of thought — Pythagoreanism — lived on.[12] About a century after his death, a Pythagorean named Hippasus proved what is now *the* classic proof of one of *the* classic theorems — that $\sqrt{2}$ is irrational. The existence of irrational numbers was a radical idea, and contradicted some of their fundamental philosophical beliefs.

As the legend goes, the other Pythagoreans were so horrified by this theorem that they took Hippasus out to sea and threw him overboard, killing him. They then made a pact to never tell the world of his discovery. This has got to be one of the worst cover-ups in history, as today his proof is probably the second most known proof in the world, only behind Euclid's proof which we just discussed.[13]

In fact, even if only to stick it to the murderous, anti-intellectual Pythagoreans one last time, let's discuss Hippasus' proof that $\sqrt{2}$ is irrational, which is another proof by contradiction.

[11]Pythagoras' story is cloaked in legend — but fortunately the legends are all highly amusing. Aristotle wrote that Pythagoras had a golden thigh, was born with a golden wreath upon his head, and that after a deadly snake bit him, he bit the snake back, which killed it; he was supposedly the son of Apollo, and it was said that a priest of Apollo gave him a magic arrow that allowed him to fly; the philosophers Porphyry and Iamblichus both reported that Pythagoras once persuaded a bull not to eat beans, and convinced a notoriously violent bear to swear that it would never harm a living thing again — and the bear was true to his word. What is odd is that none of his own writings have survived, and most of the credible writings about him were done long after his death. Some have even suggested that he was not a real person... but this is certainly a minority opinion among historians.

[12]Some may argue that "school of thought" is a bit generous. It was basically a cult.

[13]They tried to stay discrete and discreet, and failed on both counts.

Theorem.

Theorem 7.6. The number $\sqrt{2}$ is irrational.

Proof. Assume for a contradiction that $\sqrt{2}$ is rational. Then, there must be some nonzero integers p and q where

$$\sqrt{2} = \frac{p}{q}.$$

Moreover, we may assume that this fraction is written in *lowest terms*, meaning that p and q have no common divisors. Then,

$$\sqrt{2}q = p.$$

And by squaring both sides,

$$2q^2 = p^2.$$

Since $q^2 \in \mathbb{Z}$, by the definition of divisibility this implies that $2 \mid p^2$, and hence $2 \mid p$ by Lemma 2.17 part (iii).[14] By a second application of the definition of divisibility, this means that $p = 2k$ for some nonzero integer k. Plugging this in,

$$\begin{aligned} 2q^2 &= p^2 \\ 2q^2 &= (2k)^2 \\ 2q^2 &= 4k^2 \\ q^2 &= 2k^2. \end{aligned}$$

Therefore, $2 \mid q^2$, and hence $2 \mid q$, again by Lemma 2.17 part (iii). But this is a contradiction: We had assumed that p and q had no common factors, and yet we proved that 2 divides each.[15] Therefore $\sqrt{2}$ cannot be rational, meaning it is irrational.[16] □

This theorem is really important and fundamental, and as such mathematicians have searched for additional proofs of it. I'd like to share a geometric one that I really enjoy.

[14] *"Yo, lemma help you prove that theorem."*

[15]

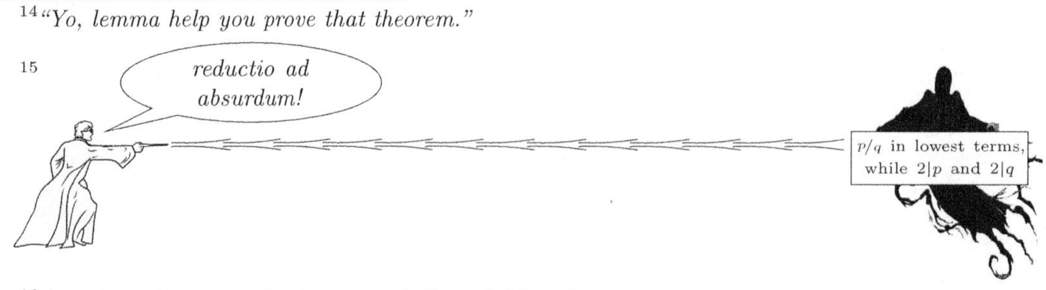

[16] How does that taste, Pythagoreans? Bitter? Mmmhmm.

We begin the same way we began the last proof: Assume for a contradiction that $\sqrt{2} = \dfrac{p}{q}$, where $p, q \in \mathbb{N}$, and assume the fraction is written in lowest terms.[17] This implies that $2q^2 = p^2$, but this time let's think about this as $p^2 = 2q^2$. Or, better yet,

$$p^2 = q^2 + q^2.$$

Since p and q are integers, p^2 represents the area of a square with side length p, and each q^2 represents the area of a square with side length q.

$$p^2 \quad = \quad q^2 \quad + \quad q^2$$

Now, to appreciate the punch line, you have to remember that $\sqrt{2} = \frac{p}{q}$ was written in lowest terms. In particular, this means that there do <u>not</u> exist any smaller a and b for which $\sqrt{2} = \frac{a}{b}$. Our contradiction will be to find such an a and b.

Getting back to the squares above, we are now going to imagine each square is a piece of paper and we are going to place the two q^2 squares on top of the p^2 square. If one q^2 square is placed in the lower-left, and the other is placed in the upper-right, this happens:

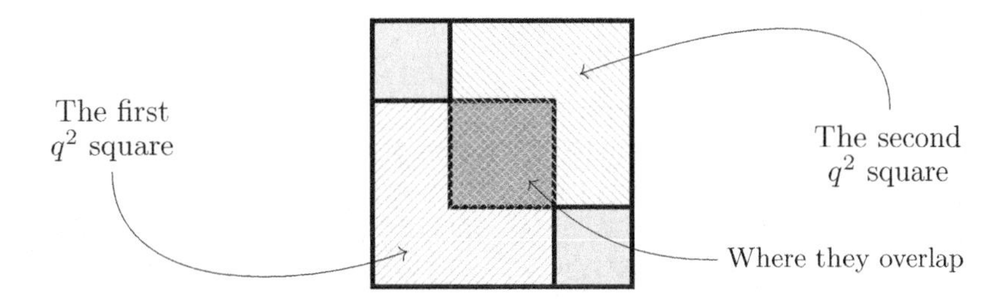

Notice that there is one square region in the middle that was covered twice, and two small squares in the upper-left and lower-right that were not covered at all. And remember: The amount of area in the p^2 square is equal to the amount of area in the two q^2 squares. Therefore, the area that was covered twice must equal the area that was not covered at all! Let's suppose the middle square has dimensions $a \times a$, and the two corner squares have dimensions $b \times b$. Then, this reasoning shows that

$$a^2 \quad = \quad b^2 \quad + \quad b^2$$

[17]Once again, there is fine line between our numerator and denominator.

And those a and b must also be integers, since they are the difference of integers from the overlap picture:

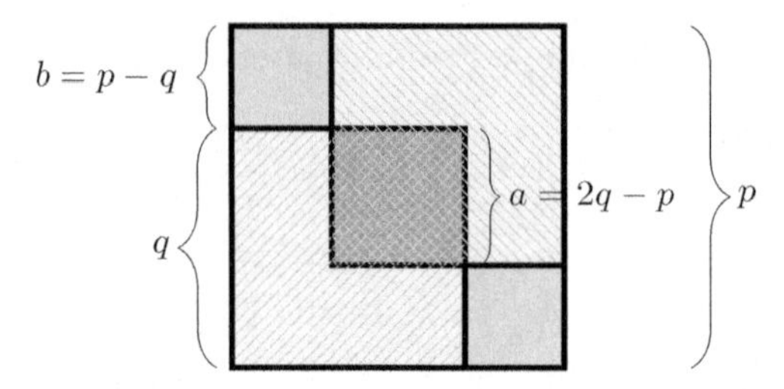

We had assumed that p and q were the smallest integers for which $\sqrt{2} = p/q$, and yet the above image shows that a and b are also integers, and since $a^2 = b^2 + b^2$, which implies $2b^2 = a^2$, we have $2 = a^2/b^2$. And so, finally, by taking the square root of each side, we see that

$$\sqrt{2} = \frac{a}{b}.$$

We have shown that a and b are integers with the above property. The picture above also shows that a is smaller than p, and b is smaller than q. Combined, this contradicts our assumption that p and q are the smallest integers where $\sqrt{2} = p/q$. □

Of historical significance, this theorem shows that the hypotenuse of the triangle below is irrational.

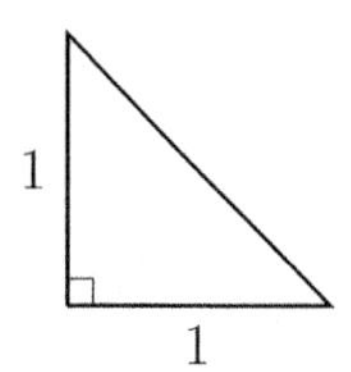

The fact that irrational numbers exist explains why we need the real numbers $\mathbb{R}$—the rational numbers $\mathbb{Q}$ are clearly not enough! Next, note that while $\sqrt{2}$ is not a ratio of integers, it *is* a root of $x^2 - 2 = 0$, which is a polynomial with integer coefficients. Big Question: Is every irrational number a root of a polynomial with integer coefficients? Big Answer: Nope! In 1851, Joseph Liouville proved that

$$\sum_{k=1}^{\infty} \frac{1}{10^{k!}} = 0.110001000000000000000000100\ldots$$

is not the root of any polynomial with integer coefficients.

The irrational numbers were thus partitioned into *algebraic numbers*, which are the roots of such polynomials, and *transcendental numbers*, which are not. Today, π and e are the most famous numbers which have been proved to be transcendental.

Starting with $\mathbb{N}$, our number system was extended to $\mathbb{N}_0$, to $\mathbb{Z}$, to $\mathbb{Q}$, and to $\mathbb{R}$. In fact, it also extends to the *complex numbers*, the *quaternions*, the *hyperreals*, and more. But while it is nice to talk about this progression as orderly and natural, its history is less so. Our progression to today was filled with confusion and misunderstanding. Prominent mathematicians even disagreed about the number zero as late as the 16th century! Nevertheless, progress has marched on.

Now it's time for our next proposition, and I'll let you decide whether it is absurdly deep or deeply absurd.

Proposition.

Proposition 7.7. Every natural number is interesting.

Scratch Work. Let's check the first ten natural numbers:

- 1 is the smallest natural number. Interesting!
- 2 is the only even prime number. Interesting![18]
- 3 equals $\sqrt{1+2\sqrt{1+3\sqrt{1+4\sqrt{1+\sqrt{1+5\sqrt{1+\dots}}}}}}$. Interesting!
- 4 is the largest number of colors needed to color a typical map.[19] Interesting!
- 5 is the smallest degree of a general polynomial that cannot be solved in radicals.[20] Interesting!
- 6 is the sum and product of the same three numbers.[21] Interesting!
- 7 is the smallest n such that the regular n-gon cannot be constructed with a ruler and compass. Interesting!
- 8 is the last Fibonacci number which is a perfect cube. Interesting!
- 9 is how many regular polyhedra there are.[22] Interesting!
- 10 has the property that among any 10 consecutive integers, there is at least one that is relatively prime to all the others. Interesting!

[18] In some ways, 2 is the oddest prime.

[19] This is called *the four color theorem.* You can learn more by turning to page 440!

[20] The general quadratic, $ax^2 + bx + c = 0$ has the quadratic formula as a general solution: $x = \frac{-b\pm\sqrt{b^2-4ac}}{2a}$. The general cubic and quartic polynomials also have formulas for their roots using only arithmetic operations and roots. For 250 years, one of the biggest unsolved problems was to find a formula for the general quintic. Finally, in 1823, a 20 year old Niels Abel proved the remarkable fact that *no such formula is possible* by inventing group theory, a course you will likely take soon. An introduction to group theory follows Chapter 9.

[21] In particular, $6 = 1 \cdot 2 \cdot 3$ and $6 = 1 + 2 + 3$. Want to learn more? Check out page 87!

[22] A regular polyhedron is a 3-dimensional figure where each of its faces is the same regular polygon. For example, a cube is a regular polyhedron because each face is a square. There are 5 convex polyhedra: the classical Platonic solids; and 4 star polyhedra: Kepler-Poinsot stellated polyhedra. And, amazingly, that's it. They look like this:

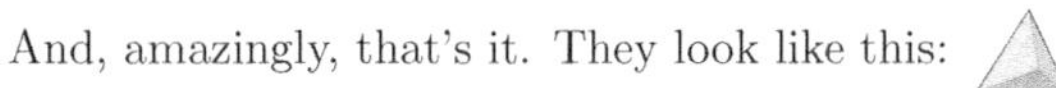

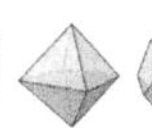
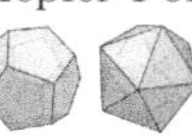

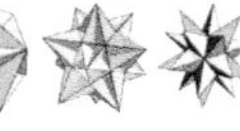

Ok, so far we have established that each of the first 10 natural numbers are interesting. But if we are to prove that *all* of the infinitely many natural numbers are interesting, listing properties one-by-one won't cut it. We will have to be clever.

Asserting something is true for all $n \in \mathbb{N}$ is typically a strong suggestion that we should at least consider using induction, but as you will see, in this case a proof by contradiction works out nicely.

Proof. Assume for a contradiction that not every natural number is interesting. Then, there must be a *smallest* uninteresting number,[23] which we call n. But being *the smallest uninteresting number* is a very interesting property for a number to have! So n is both uninteresting and interesting, which gives the contradiction.[24] Therefore, every natural number must be interesting. □

Sure, this was just a fun example, and "interesting" is impossible to define in the way we mean it.[25] But one of its main ideas was a good one: If we are assuming that there exist uninteresting numbers, let's find a specific one that is in fact interesting. Don't think about all of the cases, focus on a special one; in this case, the special one is the smallest one.[26]

It's like if someone time traveled to today, from 1911 — just before the Titanic was set to embark on its infamous voyage. If this time traveler claimed that the Titanic was unsinkable, and you wanted to prove to them otherwise, then (after telling them to go kill Hitler) what would you say? You wouldn't give them a lecture on the subtle weaknesses in the hull's rivets, you'd just show them the specific day in 1912 on which it sunk. There are often many reasons why something is false; the art is to identify the *simplest* reason why it is false.

[23]This is because of the fact that every nonempty set of natural numbers must contain a smallest element. This is sometimes called the *well-ordering principle*.

[24]
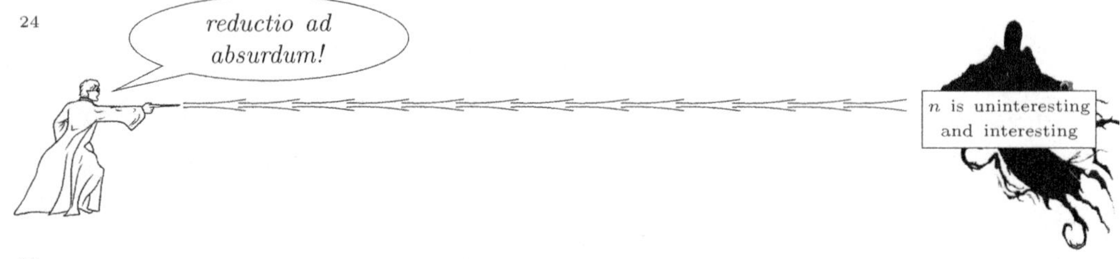

[25]Just for fun, what if we asked about a precise smallest-example? What if we asked, "What is the smallest natural number that no human will ever say out loud?" Now, we have clearly defined our parameters.

How big do you think this number is? Typing this, I just said out loud the number 1,254,231,439. Over a billion — that's a big number in most real-world contexts, and while there are many bigger numbers that have been said aloud (ever tell someone your mail's tracking number?), I doubt *every* number less than 1,254,231,439 will be spoken aloud by a human. But that's just my guess — what do you think? Is the smallest in the ten-thousands? Hundred-thousands? Millions? Ten-millions? Hundred-millions? Billions? Ten-billions?

What if we asked for the smallest number that nobody will ever explicitly think about? Or the smallest number (in base 10) that currently appears on a webpage? How big do you think those numbers are?

[26]We will discuss this further in this chapter's Bonus Examples.

Comparing Proof Methods

"Don't hate the proof, hate the axioms."

– Ice-T in Dimension C-314[27]

There are some mathematical purists who believe that direct proofs, contrapositive proofs, and induction proofs are better than proofs by contradiction. First, in support of this belief, we write proofs to explain our ideas to others, and to convince them that we are correct. But the best proofs not only convince us *that* the result is true, but also *why* it is true. And proofs by contradiction tend to struggle on this score. Indeed, even the act of proving something by contradiction can prohibit your understanding, since you end up spending all your time thinking about the way things *aren't* rather than the way things *are*.

What are reasons to rebel against these purists' beliefs? First, proof by contradiction is a perfectly valid proof method, and so ignoring a tool in your toolbox seems silly. But most importantly for someone at the beginning of their journey with proofs: It's not just any tool, it's a really freaking powerful tool! Contradiction is often the most powerful proof technique you have. Moreover, thinking about the lies can still aid understanding.

British science fiction writer Arthur Clarke formulated three "laws" for writings in his field. The third is the most famous ("Any sufficiently advanced technology is indistinguishable from magic."), but let's talk about his second law: "The only way of discovering the limits of the possible is to venture a little way past them into the impossible." Indeed, to best understand the borders between truths and falsehoods, reality and fantasy, it is useful to probe it from the dark side too. Clarke said it well, but for the sake of balance I'll close the main content of this chapter with some wise words from Darth Vader:

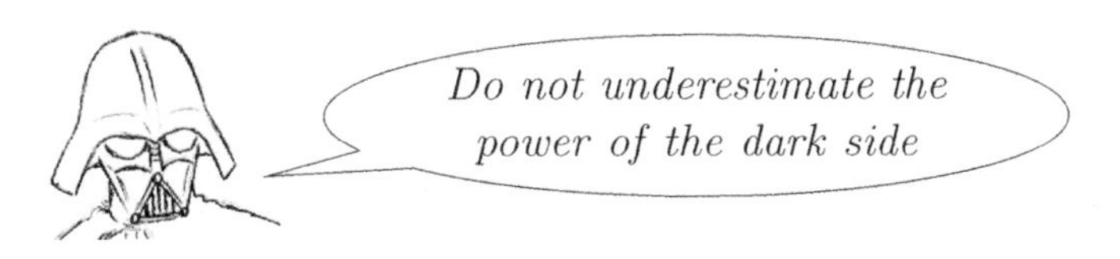

7.5 Bonus Examples

Our first bonus example comes from computer science and is called *the halting problem.* Suppose you write a computer program, you run it, and it seems to be taking too long. Could it be the case that your program is running an *infinite loop*? For example,

[27]The Ice-T in the parallel universe where all his raps were about math. And Jerry's happy.[28]

[28]P.S. If you don't get this Ice-T joke, ask one of your millennial profs.

let's consider a program written in the following pseudocode:

Input: A number N
while $N > 1$ **do**
  if N is even **then**
    $N \to (N + 2)$
  else
    $N \to (N - 2)$
  end if
end while

If $N = 5$ is input, then the program will recognize that $N > 1$ and so will loop through the **if** statement, turning N into $N - 2 = 3$. And since the new N is still larger than 1, it will again loop through the **if** statement, turning N into $N - 2 = 1$. Since N is no longer larger than 1, the program will halt after these two loops. Likewise, if $N = 7$ were input, it would halt after three loops.

What about if $N = 6$ were input? One loop would turn N into 8. A second loop would turn N into 10. A third turns it into 12. And so on. At every step we still have $N > 1$, and since N is only ever climbing higher, this will never stop being true. It's an infinite loop!

Now, with a program this simple, an experienced coder would know to avoid such an infinite loop, but in a really long and complicated program, an infinite loop may be introduced without its coder realizing it. How do we tell whether there is a bug like this in our code?

If code is the problem, could code also be the solution? Is it possible to write a program which can tell whether our code has an infinite loop? If such a program existed, we could plug *an entire program* into it, and its output would either be "This program has an infinite loop" or "This program does not have an infinite loop"? That would be quite a neat program! Does it exist?

This question is known as the halting problem, since the goal is to determine whether a program exists that can always determine whether or not other programs will eventually halt.

So, does a halt-detecting program exist? Sadly, the answer is no. We have to sniff out our own infinite loops, because no program exists that can always do it for us. And I'm not simply saying that the code monkeys have all tried their best and so far they've failed but, who knows, there's always tomorrow. No, I would never disparage computer scientists like that. All I'm saying is that the mathematicians swooped in to save the day and proved that such a program is impossible. Check it out:

Theorem.

Theorem 7.8. Assume that P is an arbitrary program and i is a possible input of P; we write $P(i)$ to be the result of plugging input i into the program P. There does not exist a program $H(P(i))$ which determines whether $P(i)$ will eventually halt.

Proof. Assume for a contradiction that such a program H did exist. Create a new program $T(x)$; its input, x, is itself a program with some input. Now, we define the program $T(x)$ as follows:

Input: A program x, with its own input
Run $H(x)$
if $H(x)$ answers "Program x will halt" **then**
  begin an infinite loop
else
  halt
end if

The program T is designed to run counter to x: If the input program x was going to halt, then T begins an infinite loop. And if the input program was going to run forever, then T says to halt.

The program T accepts as input any program. And since T is itself a program, we are allowed to *plug T into itself!* What is the result? Well, since $T(T)$ is a program, like any program either $T(T)$ contains an infinite loop or it does not. Let's consider each of these two cases.

Case 1. Observe that if $T(T)$ has an infinite loop, then like all programs with infinite loops, it will not halt—but by looking at the above pseudocode for T, we see that if $T(T)$ has an infinite loop, then it *will* halt! This is a contradiction.

Case 2. Conversely, if $T(T)$ does *not* have an infinite loop, then like all programs without an infinite loop it must eventually halt—but by looking at the above pseudocode for T, we see that if $T(T)$ will eventually halt, then it will begin an infinite loop which will prevent it from halting! This is again a contradiction.

Whether T does or does not have an infinite loop, we have reached a contradiction.[29] And since T was built from H, our assumption that there exists a halting program H must have been incorrect. This concludes the proof. □

The person who discovered this was Alan Turing, who formulated a mathematical definition of a computer program, allowing him to prove interesting results like this one, studying the limitations of computers. In particular, the study of *undecidability*.[30]

29
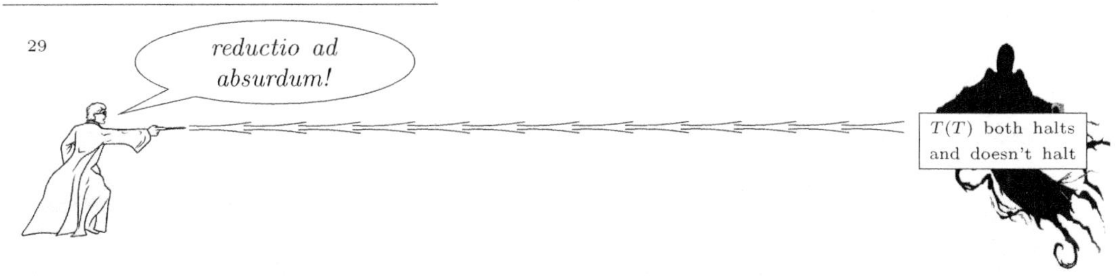

30 Five years earlier, a deep, mathematical result was published that decidedly halted the mathematical world. To learn about the extent of mathematical undecidability, check out Kurt Gödel's *incompleteness theorems*.

Proof by Minimal Counterexample

Earlier in this chapter, we proved that every natural number is interesting. The way we did this was by assuming for a contradiction that not every number is interesting. Under this assumption, there exist uninteresting natural numbers, and so there must exist a *smallest* uninteresting natural number.

Despite it being a silly example, there is an important idea behind it which is sometimes called *proof by minimal counterexample.* Consider a theorem which asserts something is true for every natural number, and you are attempting to prove it by contradiction. Then, you would assume for a contradiction that not every natural number satisfies the result — that is, you're assuming there is at least one counterexample. Well, among all of the counterexamples, one of them must be the *smallest.*[31] And thinking about that smallest counterexample — such as the smallest uninteresting number — can at times be a powerful variant of proof by contradiction.

In Chapter 4, we used strong induction to prove the fundamental theorem of arithmetic. There's another slick proof of this theorem that uses a proof by minimal counterexample, which I would like to show you now.

Theorem.

Theorem 4.8 (*Fundamental theorem of arithmetic*). Every integer $n \geq 2$ is either prime or a product of primes.

Proof. Assume for a contradiction that this is not true. Then there must be a minimal counterexample; let's say N is the smallest natural number greater than or equal to 2 which is neither prime nor the product of primes. Since it is not prime, by definition this means that it is composite: $N = ab$ for some $a, b \in \{2, 3, \ldots, N-1\}$.

We now make use of the fact that N is assumed to be the *minimal* counterexample to this result — which means that everything smaller than N must satisfy the result. In particular, since a and b are smaller than this smallest counterexample, a and b must each be prime or a product of primes.

This gives us a contradiction: Since $N = ab$, if a and b are each prime or a product of primes, then their product — which equals N — must be as well. This contradicts our assumption that N was a counterexample,[32] completing the proof. $\square$

[31] Again, this is because of the fact that every nonempty set of natural numbers must contain a smallest element, which is sometimes called the *well-ordering principle.*

[32]

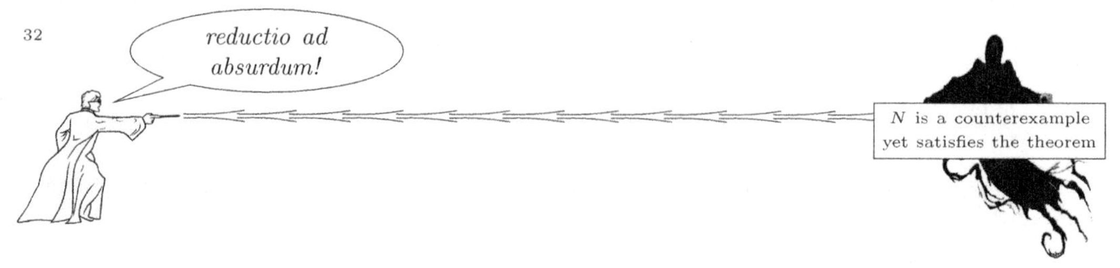

Another way to think about this proof is that it argues that if N were a counterexample, then since $N = ab$, it can't possibly be that both a and b are primes or a product of primes, since as we just saw, that would produce a contradiction. And therefore it must be the case that either a or b is also a counterexample. This implies that every counterexample produces a smaller counterexample — every N produces an a or a b. But this is a contradiction, since you cannot repeatedly find smaller and smaller natural numbers — at some point you reach the bottom.

Proof of the Division Algorithm

Way back in Chapter 2, I promised to show you a proof of the Division Algorithm (Theorem 2.11), and the time has arrived. As a reminder, here is what that theorem said:

Theorem.

Theorem 2.11 (*The division algorithm*). For all integers a and m with $m > 0$, there exist unique integers q and r such that

$$a = mq + r$$

where $0 \leq r < m$.

***Proof*.** Fix any two integers a and m for which $m > 0$. The theorem asserts two things:

1. That there *exist* integers q and r for which $a = mq + r$ and $0 \leq r < m$, and

2. That those q and r are *unique*.

We will prove existence and uniqueness separately, beginning with existence.

Existence. First, note that if $a = 0$, then by simply choosing $q = 0$ and $r = 0$, the theorem follows. Thus, we may assume that $a \neq 0$.

Next, we will argue that if the theorem holds for all positive a, then it also holds for all negative a. Indeed, assume that $a > 0$, and suppose a and m can be expressed as

$$a = mq + r$$

where $0 \leq r < m$. Then, $-a$ has an expression as well. In particular, if we let $q' = -q - 1$ and $r' = m - r$, then[33]

$$mq' + r' = m(-q-1) + (m - r) = -mq - m + m - r = -(mq + r) = -a.$$

[33]Example: If $a = 13$ and $m = 3$, then $a = m \cdot 4 + 1$, whereas $-a = m \cdot (-5) + 2$.

Therefore, for these integers q' and r',

$$-a = mq' + r',$$

where $0 \leq r' < m$. Because of this, any expression for $a > 0$ immediately produces one for $-a$. Thus, we need only prove the case where a is a *positive* integer. We will do this with a proof by minimal counterexample.

Fix any $m > 0$, and assume for a contradiction that not every $a \in \mathbb{N}$ satisfies the theorem, which in turn means that there is a smallest a for which the theorem fails. Consider three cases.

Case 1: $a < m$. In this case, we can simply let $q = 0$ and $r = a$, and we have obtained

$$a = m \cdot q + r$$

with $0 \leq r < m$, and the theorem is satisfied.

Case 2: $a = m$. In this case, we can simply let $q = 1$ and $r = 0$, and we have obtained

$$a = m \cdot q + r$$

with $0 \leq r < m$, and the theorem is satisfied.

Case 3: $a > m$. Recall that the theorem assumes that $m > 0$, and so in this case we have $a > m > 0$. In particular, note that $a > a - m$ and also $a - m > 0$.

Since a is the *smallest* positive counterexample to this theorem, and $a - m$ is both positive and less than a, the integer $a' = a - m$ must satisfy this theorem! That is, there must exist integers d and s for which

$$(a - m) = m \cdot d + s$$

with $0 \leq s < m$. By moving the m on the left side over, $a = m \cdot d + s + m$. By factoring,

$$a = m \cdot (d + 1) + s.$$

Thus, by letting $q = d+1$ and $r = s$, we have shown that our smallest counterexample is not a counterexample at all:

$$a = mq + r$$

with $0 \leq r < m$. Since there cannot exist a smallest counterexample, there cannot exist any counterexample. Thus, for each a and m, there must exist numbers q and r as the theorem asserts.

Uniqueness. Assume for a contradiction that for our fixed a and m, that the q and r are not unique. That is, assume there exist two different representations of a,

$$a = mq + r \qquad \text{and} \qquad a = mq' + r',$$

where $q, r, q', r' \in \mathbb{Z}$ and $0 \leq r, r' < m$. Then,

$$mq + r = mq' + r'.$$

By some algebra, $r - r' = mq' - mq$, which means

$$r - r' = m(q - q').$$

Since q and q' are integers, $q - q'$ is also an integer, which means the above expression matches the definition of divisibility (Definition 2.8)! That is, $m \mid (r - r')$. Notice that since $0 \leq r, r' < m$, the difference $r - r'$ would have these restrictions: $-m < r - r' < m$. And the only number in this range which is divisible by m is zero. That is, $r - r' = 0$, or $r = r'$.

Next, since $r = r'$, the fact that $r - r' = m(q - q')$ implies that

$$0 = m(q - q').$$

Since $m > 0$, we may divide both sides by m, which means $0 = q - q'$, or $q = q'$.

We assumed that

$$a = mq + r \qquad \text{and} \qquad a = mq' + r'$$

were two different representations of a and m, but we have proven that $q = q'$ and $r = r'$, proving that they are in fact the same representation, giving the contradiction and concluding the proof. □

— Chapter 7 Pro-Tips —

- One mistake that students sometimes make is to automatically use a proof by contradiction when the proof they have in mind is in fact a direct proof or a contrapositive proof. Indeed, if you are proving $P \Rightarrow Q$ and your proof goes "Assume P and $\sim Q$. <math math math> We have now shown that Q is true, which contradicts our assumption that $\sim Q$ is true, and therefore $P \Rightarrow Q$ must be true," then you shouldn't be using a proof by contradiction; if you prove Q as a part of your proof, then you should be using a direct proof.

  Likewise, if you are proving $P \Rightarrow Q$ and your proof goes "Assume P and $\sim Q$. <math math math> We have now shown that not-P is true, which contradicts our assumption that P is true, and therefore $P \Rightarrow Q$ must be true," then you shouldn't be using a proof by contradiction; if you prove not-P as a part of your proof, then you should be using a contrapositive.

  It is not uncommon for students to instinctively pursue a proof by contradiction on each problem, when doing so only adds an unnecessary layer of complication.[34]

- When using a direct proof, you usually know whether or not you have successfully reached your conclusion. If you prove $P \Rightarrow Q$ by a direct proof and you make a mistake on your journey from P to Q, you will likely not arrive at Q, and so you will know that you still have work to do. Indeed, a mistake will typically throw you off course and it would take a second equal-but-opposite mistake to arrive at Q. There could certainly be steps that require more justification than you gave, but you were at least on a legitimate path.

  And since a proof by contrapositive is essentially a direct proof from not-Q to not-P, the same lesson holds there.

  Likewise, if within a proof by induction you make a mistake within the induction step, then you will rarely reach where you need to get to. And so, once again, it will be clear to you that a mistake was made and must be hunted down.

  For a proof by contradiction, though, this is not so simple. Suppose you are trying to prove $P \Rightarrow Q$, and you begin your proof by assuming for a contradiction that P is true and Q is false. The goal now is to identify *anything else* that is false, which will give us the contradiction. Consequently, there's no longer a single target that is either hit or not — in a sense, there are innumerable potential targets, since *any* contradiction counts. Thus, if you make a mistake along the way, who is to say that the "contradiction" you find is the result of the not-Q assumption, rather than a byproduct of your mistake?

  Unfortunately, this Pro-Tip does not contain a silver bullet. There's no foolproof way to patch this vulnerability. All I can say is that a proof by contradiction requires more care than other proofs, for this reason. It is easier to deceive yourself into accepting a flawed proof. Proving by contradiction is a

[34] You're a math major, not a proof-by-contradiction major. Remember that!

powerful tool, but heightened risks call for a heightened attention to detail.[35]

- It is always difficult to know when one proof technique is better than another, but I will continue my efforts to give you some general advice. One instance in which a proof by contradiction is particularly useful is for universal statements, like when the theorem asserts that something is true "for every" or "for all" such-and-such. For example, "every natural number is interesting" and "Every integer $n \geq 2$ is prime or a product or primes." The reason is that when you assume for a contradiction that the theorem is false, what that tells you is that there is a *particular* element which does *not* satisfy the theorem.

  This can be quite powerful, as it allows you to focus on a single element, and what its existence implies — what does it mean if there is an uninteresting number, or a number that is not a product of primes? Also, when you assume such an element exists, what you're really assuming is that there is at least one element with those properties. But if there are possibly more such elements, then maybe there is a smallest one, or a largest one, or one which is special in some other way. If so, then this allows you to work with more: Rather than supposing you have a randomly selected counterexample to the theorem, you suppose that you have a very special counterexample (one with extra assumptions piled on top). And then you're really cookin'.

- A proof by minimal counterexample is deeply connected to a proof by induction. Typically, a minimal counterexample proof works by showing that if S_k is a counterexample, then S_{k-1} (or some other previous term) must be a counterexample too, which prohibits there being a *minimal* counterexample. Whereas a typical proof by induction works by showing that if S_k satisfies the theorem, then S_{k+1} does too.

  In essence, an induction proof shows that S_k being true implies S_{k+1} is also true, while a minimal counterexample proof show that S_k being false implies S_{k-1} is also false. There are many variations to these "typical" proof formats, but all of the variations operate on the same core ideas (which are related to the "well-ordering principle," which we have mentioned many times).

- Math is always typed up using a program called LaTeX. This is software that you can download to your computer for free, or you can use it at websites like `Overleaf.com`. LaTeX was used to typeset this book, and probably every other math book you have ever read. It allows you to make math symbols, Greek letters, and beautiful graphics. There are many packages that you can load which supply you with a wealth of shortcuts to create symbols and graphics.

  It really is wonderful, and if you pursue mathematics further (or even if you don't), it would be worthwhile to learn some LaTeX. Get good at it, and you'll never look back; I haven't written a document in Word in a decade. And it has been a good decade.

---

[35] *With great power comes great responsibility*

— Exercises —

Exercise 7.1. Consider the four similar-sounding words: contrapositive, contradiction, converse and counterexample. Explain the similarities and differences between these.

Exercise 7.2. Suppose $a, b \in \mathbb{R}$.

(a) Prove that if a is rational and b is irrational, then $a + b$ is irrational.

(b) Prove that if a is rational and ab is irrational, then b is irrational.

(c) Give an example of two irrational numbers whose sum is rational.

(d) Give an example of two irrational numbers whose product is rational.

(e) Give an example of a rational number and an irrational number whose product is rational.

Exercise 7.3. Prove that there do not exist integers m and n where $12m+21n = 1$.

Exercise 7.4. Prove that no odd integer can be expressed as the sum of three even integers.

Exercise 7.5. Prove that if A, B and C are sets, $A \cap C \subseteq B$, and $a \in C$, then $a \notin A \setminus B$.

Exercise 7.6. Suppose that A and B are sets inside a universal set U. Prove the following.

(a) $A \cap A^c = \emptyset$

(b) $A \times \emptyset = \emptyset$

(c) $A^c \cap (B \cap A) = \emptyset$

(d) $B \cap (A \setminus B) = \emptyset$

Exercise 7.7. Let A, B and C be sets. Prove that $A \setminus B$, $B \setminus A$ and $A \cap B$ are all pairwise disjoint.

Exercise 7.8. Let $\mathbb{Q}^+$ be the set of positive rational numbers. Prove that if $x \in \mathbb{Q}^+$, then there is some $y \in \mathbb{Q}^+$ such that $y < x$. Provide two proofs of that fact — one using a direct proof and one using a proof by contradiction.

Exercise 7.9. Suppose p is a prime number. Also, suppose $m \in \mathbb{N}$, with $m \geq 2$.

(a) Prove that $\sqrt{5}$ is irrational.

(b) Prove that $\sqrt{20}$ is irrational.

(c) Prove that $\sqrt{p}$ is irrational.

(d) Prove that $\sqrt{10}$ is irrational.

(e) Prove that $\sqrt{2} + \sqrt{5}$ is irrational.

(f) Prove that $\sqrt[m]{2}$ is irrational.

Exercise 7.10. Look back at the geometric proof that $\sqrt{2}$ is irrational that we discussed beginning on Page 307. Note that the side lengths of the smaller squares that arose in our proof were $(p-q) \times (p-q)$ and $(2q-p) \times (2q-p)$. Using these expressions and some algebra, provide a second algebraic proof that if $\sqrt{2} = p/q$ and these p and q are the smallest natural numbers giving this, then this leads to a contradiction.

Exercise 7.11.

(a) Come up with your own conjecture as to which $n \in \mathbb{N}$ have the property that $\sqrt{n}$ is irrational.

(b) Come up with your own conjecture as to which $n, k \in \mathbb{N}$ have the property that $\sqrt[k]{n}$ is irrational.

Exercise 7.12. Are there infinitely many composite numbers? Prove your answer.

Exercise 7.13. Suppose $a, b, c \in \mathbb{Z}$. Prove that if $a^2 + b^2 = c^2$, then a or b is even.

Exercise 7.14. Suppose $n \in \mathbb{Z}$. Prove that if $n \nmid m$ for every $m \in \mathbb{N}$, then $n = 0$.

Exercise 7.15. Prove that if x and y are positive real numbers, then $x+y \geq 2\sqrt{xy}$.

Exercise 7.16. Assume that x and y are positive real numbers such that $x - 4y < y - 3x$. Prove that if $3x > 2y$, then $12x^2 + 10y^2 < 24xy$. In fact, prove it three times, once with each of our main proof methods: a direct proof, a contrapositive proof, and a proof by contradiction.

Exercise 7.17. Prove that if $n \in \mathbb{Z}$, then $4 \nmid (n^2 + 2)$.

Exercise 7.18. Suppose $m, n \in \mathbb{Z}$. Prove that if $4 \mid (m^2 + n^2)$, then m and n are not both odd.

Exercise 7.19. Prove that there do not exist three distinct numbers a, b and c for which $a + b + c$, ab, ac, bc and abc are all equal.

Exercise 7.20. Prove that there does not exist $n \in \mathbb{N}$ for which $n^2 + n + 1$ is a perfect square.

Exercise 7.21. Prove that there do not exist $m, n \in \mathbb{N}$ for which $m^2 - n^2 = 1$.

Exercise 7.22. Assume that $a, b > 0$. Prove that $\sqrt{a+b} \neq \sqrt{a} + \sqrt{b}$.

Exercise 7.23. Prove that the graphs of $y = x^2 + x + 5$ and $y = x + 1$ do not intersect.

Exercise 7.24. Provide a second proof of Exercise 2.14 using a proof by contradiction.

Exercise 7.25. Prove that $x \in \mathbb{R}$ is irrational if and only if it has a different distance from each rational number.

Exercise 7.26. A *magic square* is an $n \times n$ matrix where the sum of the entries in each row, column and diagonal equal the same value. For example,

8	1	6
3	5	7
4	9	2

is a 3×3 matrix whose three rows, three columns, and two diagonals each sum to 15. Thus, this is a magic square.

Prove that the following cannot be completed to form a magic square.

1	2	3	
	4	5	6
7		8	
	9		10

Exercise 7.27.

(a) Prove that there are arbitrarily long arithmetic progressions consisting of relatively prime numbers.

(b) Prove that there does not exist an infinitely long arithmetic progression consisting of relatively prime numbers.

(c) Prove that there are arbitrarily long intervals of natural numbers containing no prime numbers.

— Open Question —

In this chapter we discussed the Pythagorean theorem, which said that if a and b are the lengths of the legs of a right triangle, and c is the length of the hypotenuse, then $a^2 + b^2 = c^2$. If a, b and c are all natural numbers, then they are called a *Pythagorean triple*. For example, $3, 4, 5$ and $5, 12, 13$ are Pythagorean triples.

Another way to picture it: c is the length of the diagonal of an $a \times b$ rectangle. Thus, $c = \sqrt{a^2 + b^2}$, giving this picture:

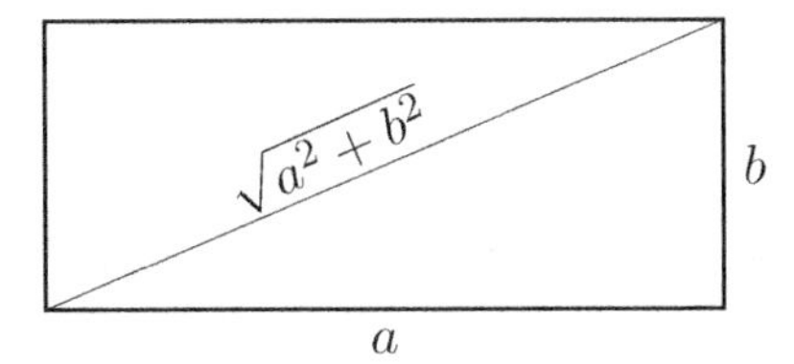

The three-dimensional version of an $a \times b$ rectangle is an $a \times b \times c$ rectangular box.

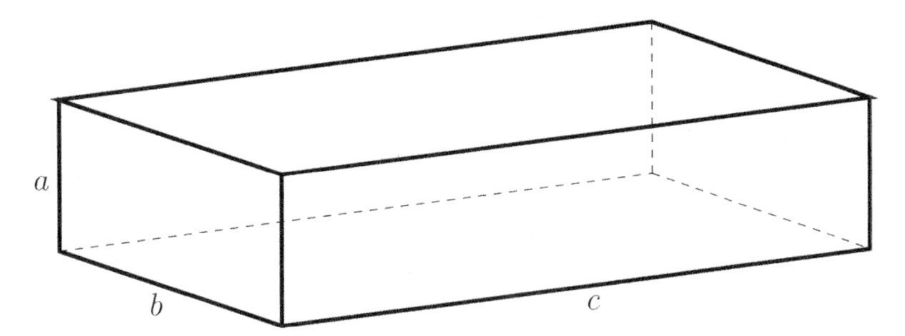

Now there are many diagonals! The left face is an $a \times b$ rectangle and has a diagonal of length $\sqrt{a^2 + b^2}$. The front face is an $a \times c$ rectangle and has a diagonal of length $\sqrt{a^2 + c^2}$. The top face is similar. The long diagonal from one corner to the opposite corner requires a little more thought, but it has length $\sqrt{a^2 + b^2 + c^2}$.

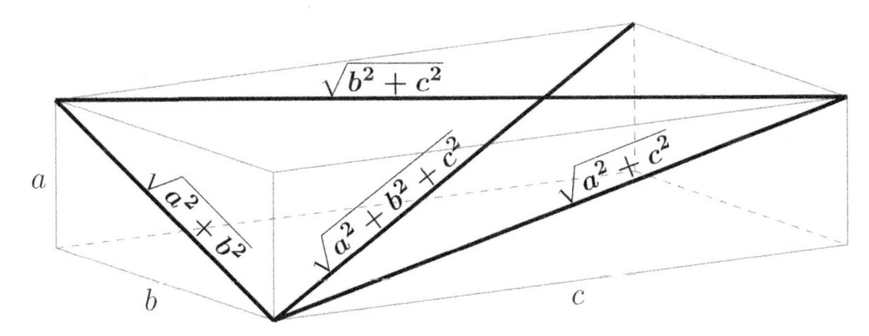

Is it possible to choose natural numbers a, b and c to that all four of these diagonals are also natural numbers? This is the three-dimensional version of a Pythagorean "triple." No one knows if such numbers exist, but if they do, the resulting box is called an *Euler brick*. Thus, this is the open question:

Open Question. Does there exist an Euler brick?

Introduction to Game Theory

You and a classmate Tom are under investigation. The two of you submitted nearly-identical essays for your assignment on the Banach-Tarski paradox. There are two options: You two worked together when you weren't supposed to, or one of you cheated off the other. Your professor calls you two into her office one at a time, and you two have no chance to discuss anything. She tells you that if you two simply worked together, then that is bad, but not worthy of being reported to the university, who would expel a proven cheater. She therefore lays out the possible outcomes:

- If you two both say you worked together, you will fail that assignment, but that's it.
- If you both accuse the other of cheating off them, then the university will decide neither to be credible and so neither will be expelled. But the professor says she would act harshly, and you would both fail her class.
- If you accuse Tom, and Tom says you worked together, then the university will expel Tom and you will get no punishment.
- If Tom accuses you and you say you worked together, then the university will expel you and Tom will get no punishment.

You can visualize this with a matrix:

		Tom's Answer	
		Worked Together	Blame You
Your Answer	Worked Together	You: Fail Assignment Tom: Fail Assignment	You: McDonald's Hiring? Tom: No Punishment
	Blame Tom	You: No Punishment Tom: McDonald's Hiring?	You: Fail Class Tom: Fail Class

You don't know Tom too well, so you don't know what he will do. And for the sake of this problem, the truth about who cheated is not important. We simply ask: If you and Tom are "rational actors" (i.e., selfish logicians), and will therefore choose the option which minimizes your own penalty, what will you and Tom do?

Game Theory

This is what *game theory* is all about. It seeks to answer the question of what rational actors will do when presented with such situations, which are called *games*. Game theory deals with recreational games, like chess, poker and Call of Duty,[36] as well as real-world situations like the above in which the set up and rules are clear, and the actions of one player affects the outcome of another player. And although the cheating example, like many math applications, may feel contrived, game theory presents itself naturally and often in the real world. Two notable examples are military strategy and economics, which are filled with game theory.

There are five binary criteria which help to classify games. They are:

1. **Cooperative vs. Non-Cooperative.** A cooperative game allows players to collaborate and negotiate, while in non-cooperative games this is forbidden.
2. **Normal Form vs. Extensive Form.** A normal form game is one with a matrix representation like the above, while an extensive form game is more complicated and is modeled by a tree.
3. **Sequential vs. Simultaneous.** In a sequential game, players take turns with their moves or actions, allowing the players to respond to each other, while in a simultaneous game the moves happen at the same time.
4. **Zero-Sum vs. Nonzero-Sum.** Zero-sum games are ones in which the advantages gained by one player produce an equal loss by the other player(s). A nonzero-sum game does not have such a property.
5. **Symmetric vs. Asymmetric.** In a symmetric game, each player's optimal strategy is identical, while in an asymmetric game the best strategies will not be identical between players.

Our cheating example is:

1. Non-cooperative, since you and Tom do not have the chance to speak;
2. Normal, since the situational outcomes could be displayed with a matrix;
3. Simultaneous, since by keeping you and Tom separate, it is equivalent to you two answering simultaneously;
4. Nonzero-sum, since your possible outcomes (even though we didn't quantify them) are not in balance and sum to a negative loss; and
5. Symmetric, since you and Tom are in identical situations with identical consequences.

[36] There is also some game theory if you and Tom complete your proofs homework together:

	You: Play Call of Duty	You: Do Proofs Homework
Tom: Play Call of Duty	0% grade	70% grade
Tom: Do Proofs Homework	70% grade	90% grade

Nash Equilibria

Sheldon and Leonard are playing a non-cooperative game, and each has a strategy. Suppose each strategy is optimal in the sense that if Sheldon changes his strategy (but Leonard doesn't change his), then Sheldon only hurts himself. And, symmetrically, Leonard can also not improve his outcome, given Sheldon's strategy. Each is using the best possible reply to the other's strategy, and so, since they are not collaborating, neither player will deviate from their strategy.

If such strategies exist for a game, then such strategies are said to be in *Nash equilibrium*. This is one of the most important ideas in game theory, and is named after its pioneer, John Nash.[37] This suggests another question: Does every game have a Nash equilibrium? John Nash proved one great theorem about this: Every finite non-cooperative game contains a Nash equilibrium.

If a game has a unique Nash equilibrium, then two rational players will play that Nash equilibrium strategy, since any other pair of strategies can be improved upon by at least one player. Our cheating example is a non-cooperative, finite[38] game, and so it must have a Nash equilibrium, which will tell us how you should react when cornered by your professor. Below is the matrix again, for reference.

		Tom's Answer	
		Worked Together	Blame You
Your Answer	Worked Together	You: Fail Assignment Tom: Fail Assignment	You: McDonald's Hiring? Tom: No Punishment
	Blame Tom	You: No Punishment Tom: McDonald's Hiring?	You: Fail Class Tom: Fail Class

I claim that the Nash equilibrium is that both you and Tom will blame the other, which will result in you both failing the class. Let's check if this is true.

- If Tom plans on blaming you, can you improve your position by changing your strategy? Nope! You will either fail the class with your current strategy or get expelled from the university by changing your strategy, so certainly you will not change.

- If you plan on blaming Tom, can Tom improve his position by changing his strategy? Nope! He will either fail the class with his current strategy or get expelled by changing his strategy, so certainly he will not change.

[37]The book and movie *A Beautiful Mind* are great, and mostly accurate, portrayals of Nash's life.

[38]By the way, if a game gives the players infinitely many possible moves, then it is not guaranteed to have a Nash equilibrium. Example: the game in which two players each shout out a number at the same time, and the person who shouted out the larger number wins. There is certainly no Nash equilibrium for this game, since any strategy can be improved because you can always choose a larger number.

In both cases, "changing strategy" was straightforward to think about, since there was literally only one other strategy possible.[39] In a game of chess, if you changed strategies you have an enormous number of other strategies to pursue. And so the analysis of that game is... not so easy.

So there you have it! Both you and Tom, being rational actors, would choose to blame the other. It is true that if you were able to collude, and you both trusted the other to follow through with the plan, then you could work together to get a lighter sentence. But since you cannot talk to Tom, and if you could you two still couldn't trust each other... it is optimal to just blame the other.

This cheating scenario is commonly phrased in terms of two prisoners in separate cells, each asked whether they robbed the bank. If they accuse each other, they split the jail sentence. If one accuses the other while the other stays silent, the accused gets the full sentence while the other walks free. If neither confesses, they both get slapped with tax evasion, which includes a small jail sentence (and conviction is certain). Because of this setup, the problem is famously called the *prisoner's dilemma.*[40]

The Minimax Theorem

In a penalty kick in soccer, the kicker aims left or right, and the goalkeeper decides which side to jump towards. This decision has to happen before the kick, because by the time the goalkeeper sees which side the ball is going towards, it would be too late to get there. The sports analytics revolution is changing much, but even before pro sports teams put mathematicians on their payroll, the best athletes often behaved very close to the Nash equilibrium, learning simply by trial and error.

In this simple model of penalty kicks, the game's Nash equilibrium will be a *mixed strategy.* That is, it will be randomized—you certainly wouldn't choose left every time or right every time, you would choose each with a certain probability.

John von Neumann considered finite, zero-sum, non-cooperative games like this one, and considered all mixed strategies for such a game. To do this, he first assumed that whatever strategy you choose, your opponent will respond optimally; that is, among all possible response strategies that your opponent could choose, they choose the one that maximizes your loss. Then he thought, your goal should be to select whatever strategy *minimizes* this *maximum loss*? Whatever strategy does this is called the *minimax strategy.*

Von Neumann proved that a minimax strategy always exists. Moreover, he showed that if Player 1 plays his minimax strategy, and Player 2 plays her minimax strategy, then the Player 1's expected losses will equal Player 2's expected gains. This implies that when two rational players play a game, they will both utilize the minimax strategies, and the expected outcome of the game is fully determined by the game itself. This is called the *value* of the game, and can often be computed directly. This is sometimes called the fundamental theorem of game theory. Von Neumann certainly thought so, saying "there could be no theory of games... without that theorem."

[39]Well, sort of. There are what are called *mixed strategies* in which you make your choice based on some randomized procedure. Those are not important now, but we will talk about them soon.

[40]Career goal: Prove something in my research using the prisoner's dilemma, and them name my result *the prisoner's ditheorem.*

Strategy-Stealing Arguments

One particularly fun and delicious game is called *chomp*. Chomp is played between two people and uses a chocolate bar[41] — a rectangular array of chocolate squares. Here are the rules:

- The upper-left square is called the *poison square*, and whoever eats that square loses.
- Player 1 goes first, after which the players take turns. On each turn, a player selects one of the remaining squares, and then eats that square and all of the squares below and to the right of the chosen square.

For instance, below is an example of a mid-game selection, and its consequence:

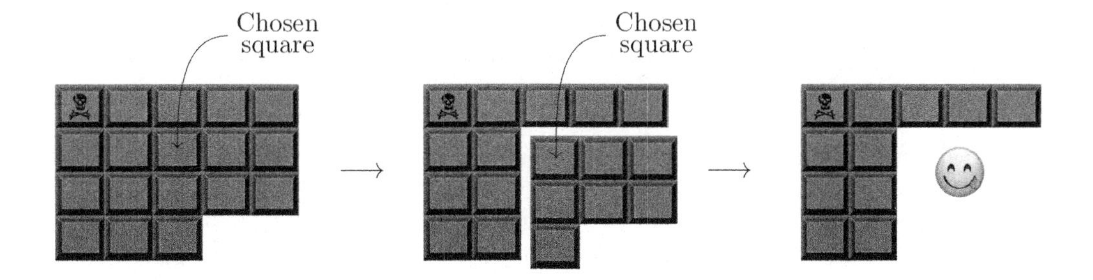

Below is an example game using a 4×7 bar. This game lasted for six moves, and if you count it out you'll find that Player 2 was forced to eat the poison square at the end, which means that Player 1 won this game.

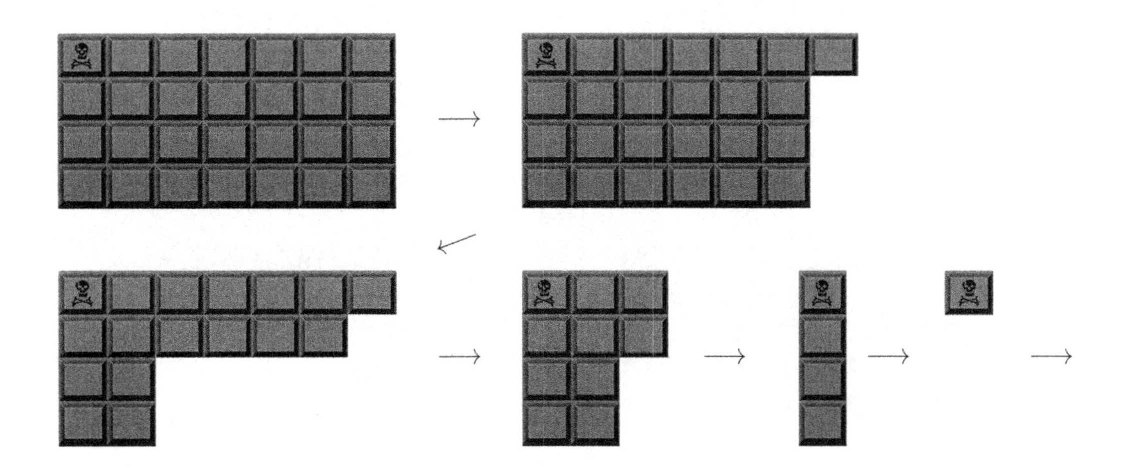

[41] Pro-Tip: When you spend hours making a virtual chocolate bar for one problem, it pays to use it again for another. Likewise, if you write a math talk, go ahead and spend twice as long on it to make it really good, and then find several more venues to give the talk. Your time-spent-per-delivery will go down, and you'll get to share the talk with more people.

Before reading on, find someone to play a few rounds of chomp with, and see if you can develop a strategy. Or at least play a few games against yourself![42]

A common question for games like this is: Does there exist a winning strategy for Player 1? That is, is there a strategy which would guarantee a win for Player 1? If not, does there exist a winning strategy for Player 2? Because there are only finitely many ways this game can play out and, unlike a game like tic-tac-toe, each game must end with someone winning, one of these two people must have a winning strategy. So, is it Player 1 or Player 2? And what is that strategy?

Proposition.

Proposition 7.9. If two players play chomp on a 1×1 chocolate bar, Player 2 will win. If they play on any other $m \times n$ chocolate bar, then Player 1 has a winning strategy.

Proof. If the chocolate bar is 1×1, then Player 1 is forced to eat the poison square on the first move, and so Player 2 wins.

As for a larger chocolate bar, assume for a contradiction that Player 2 has a winning strategy. Then Player 1 can execute what is called a *strategy-stealing argument.* Given any such bar, have Player 1 select the bottom-right square on their first move; this removes only that square.

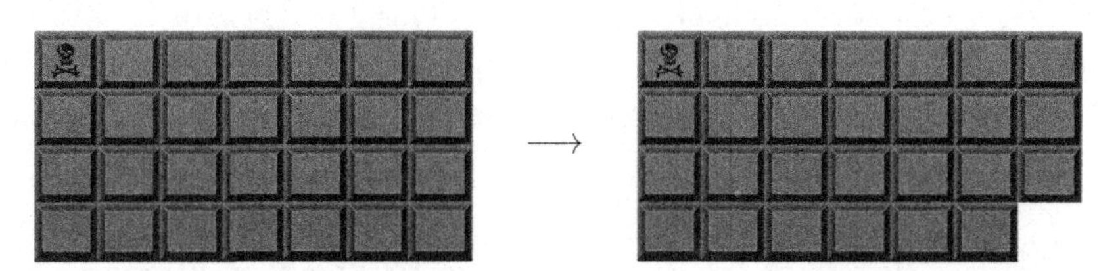

Because we are working under the assumption that Player 2 has a winning strategy, there must be a move that Player 2 can make which will eventually lead to victory. For example, perhaps this is the move:

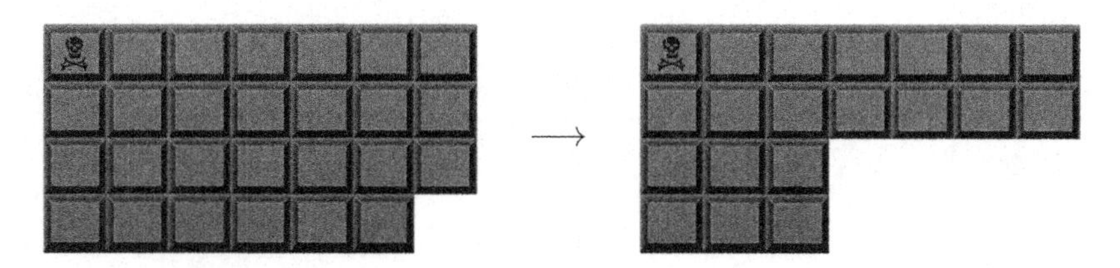

But notice that Player 1 could have made that move as their first move! And since, as we said, this move will eventually lead to victory, this shows that Player 1 in fact had a winning strategy (even though we can't say what it is). This contradicts our assumption that Player 2 had a winning strategy, and completes the proof. □

[42]Even when you lose, you win!

Chapter 8: Functions

8.1 Approaching Functions

I made a meme describing your journey towards understanding functions.[1]

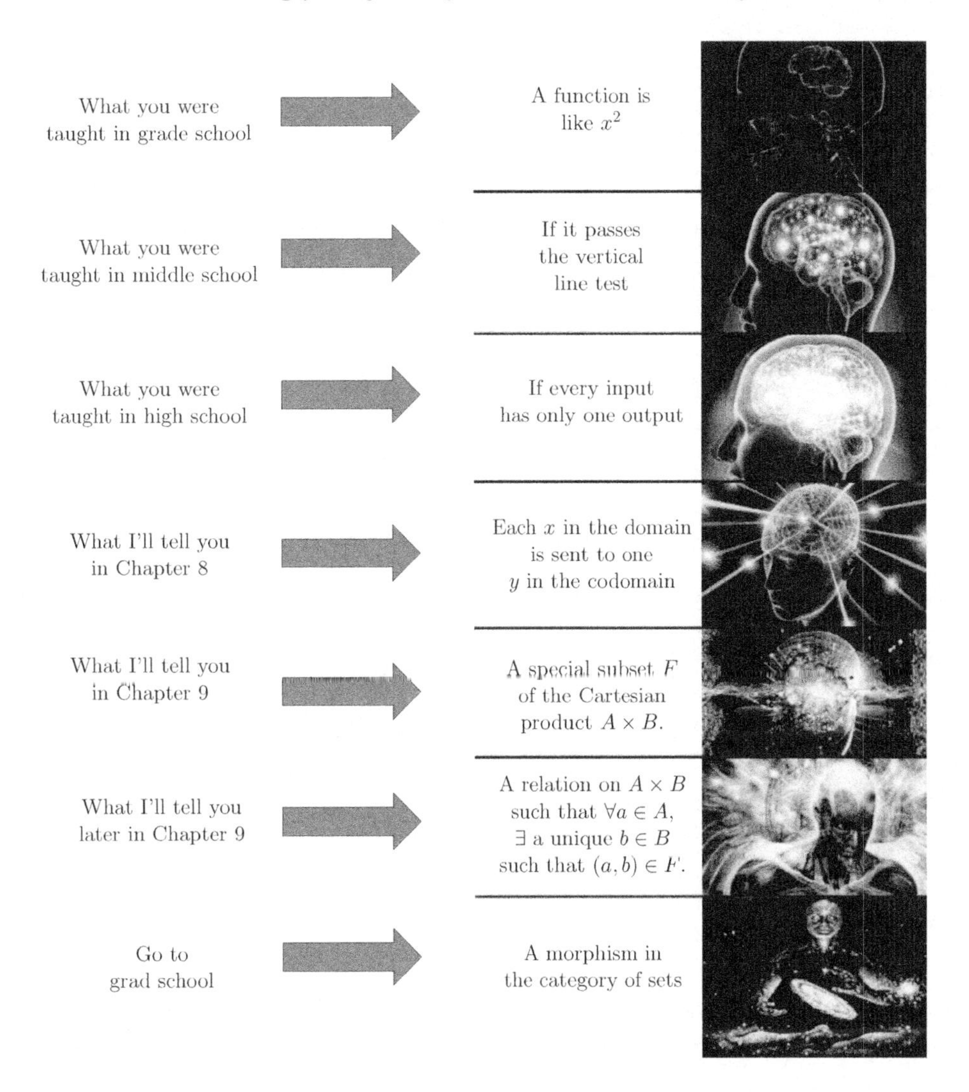

[1]Here's to hoping that memes have a longer half-life than I fear!

Sets are fundamental to almost everything we do in mathematics, for two big reasons. First, they are "simple," and foundations should be simple. And second, because functions are based on sets, and functions are *eevverryyywhheeerrree* in mathematics. The reason you have been told what a function is so many different times in your life is that you began studying functions way back in elementary school, when you learned how to find the area of basic shapes; in middle school, when you took an entire class studying functions and equations; in high school, when you took more algebra, pre-calc, and maybe some calculus, too. The more you look in math, the more functions you will find. So yeah, they're kind of a big deal.

In Chapter 3 we studied the static properties of sets, but things get more interesting and dynamic when we start applying functions to those sets.

I will continue the confusing practice of your foreteachers by giving you several definitions of a function, as your brain gradually expands.

Definition.

Definition 8.1. Given a pair of sets A and B, suppose that each element $x \in A$ is associated, in some way, to a unique element of B, which we denote $f(x)$. Then f is said to be a *function* from A to B. This is often denoted[2]

$$f : A \to B.$$

Furthermore, A is called the *domain* of f, and B is called the *codomain* of f. The set $\{f(x) : x \in A\}$ is called the *range* of f.

Intuitively, you can think of the domain as the inputs of f, the range as the outputs of f, and the codomain as a possibly-larger set in which all the outputs live. But at the end of the day, all three are just sets. They correspond to each other via f, but they are just sets.

When you were young(er), the domain and codomain were usually the set $\mathbb{R}$. The range, though, varied a lot. The range consists only of the elements in the codomain which get hit — that is, y is in the range if there is an x in the domain that maps to it: $f(x) = y$. For example, if $f : \mathbb{R} \to \mathbb{R}$ is given by $f(x) = 2x$, then the range is $\mathbb{R}$. But if $f : \mathbb{R} \to \mathbb{R}$ is given by $f(x) = x^2$, then the range is the set of nonnegative real numbers: the interval $[0, \infty)$; 4 and 9 are in the range because $f(2) = 4$ and $f(-3) = 9$, but -1 is not in the range, because no $x \in \mathbb{R}$ has the property that $f(x) = -1$. Before showing you some diagrams, we have a Recurring Theme Alert.

Recurring Theme Alert. When discussing functions, the ideas of *existence* and *uniqueness* will come up repeatedly. In fact, this began with Definition 8.1. We defined a function $f : A \to B$ to be a rule which sends each $x \in A$ to some $f(x) \in B$. What this means is that $f(x)$ must *exist* (it must be equal to some $b \in B$), and it must be *unique* (it must be equal to only *one* $b \in B$).

[2]Note: "$f : A \to B$" is typically read "f from A to B."

A function's domain and codomain can each be *any* set. For example, here's a graphical way to write some function f:

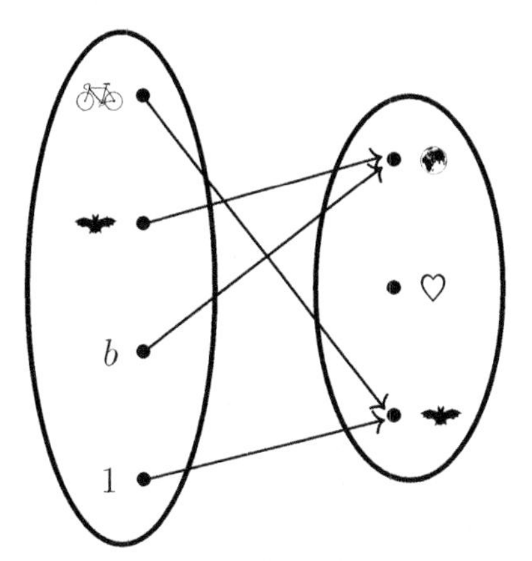

This is a function with domain $\{1, b,$ 🦇, 🚲$\}$, codomain $\{$🦇, ♡, 🌍$\}$, and range $\{$🦇, 🌍$\}$. For example, $f(1) =$ 🦇, so 🦇 is in the range. However, there does not exist any $x \in \{1, b,$ 🦇, 🚲$\}$ such that $f(x) =$ ♡, which is why ♡ is not in the range.[3]

For a diagram like this to *not* represent a function, it would have to fail either the existence or the uniqueness part of being a function, as discussed in the Recurring Theme Alert. Below are two examples.

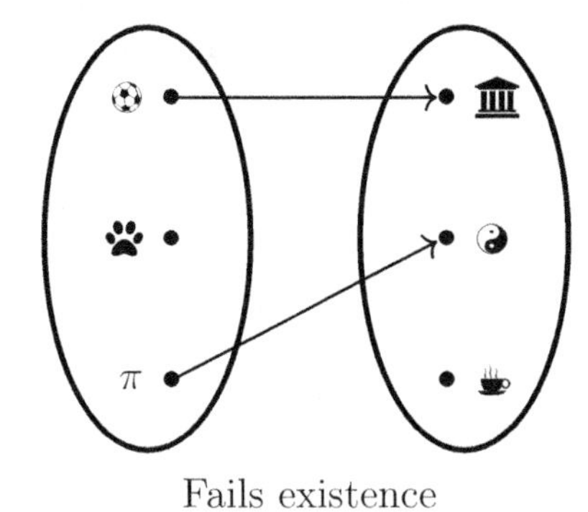

Fails existence

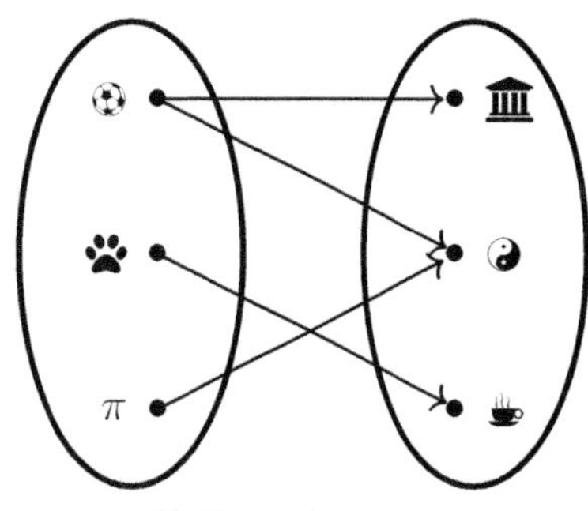

Fails uniqueness

It is perfectly ok to have two arrows pointing at the same dot in the codomain (or zero arrows, or more than two arrows), but for the domain the rules are rigid: one and only one line must emanate from each dot. So the two diagrams above would *not* be functions;[4] the first because 🐾 is being sent to nowhere, and the second because ⚽ is being sent to two places.

[3]Another metaphor: Suppose Cupid is shooting arrows at a target; he hits different spots on the target, but never misses the target completely. Then, Cupid is like a function. The codomain is the target, since those are the possible points that can get hit. The range is the set of points on the target that actually do get hit by an arrow. So, for the function above, since ♡ was not in the range, Cupid missed love. :-(

[4]We use similar language in the real world. If you dial someone's number, but the call goes nowhere, then you would say your phone isn't functioning. Or if you dialed a single phone number, but half the time the call went to Mikaela and half the time the call went to Brandon, then you would again say that your phone is not functioning. If something is properly functioning, it always responds to an action with a single, anticipated reaction.

In high school you were probably taught the *vertical line test* to check whether a graph corresponds to a function. Below are two examples of that.

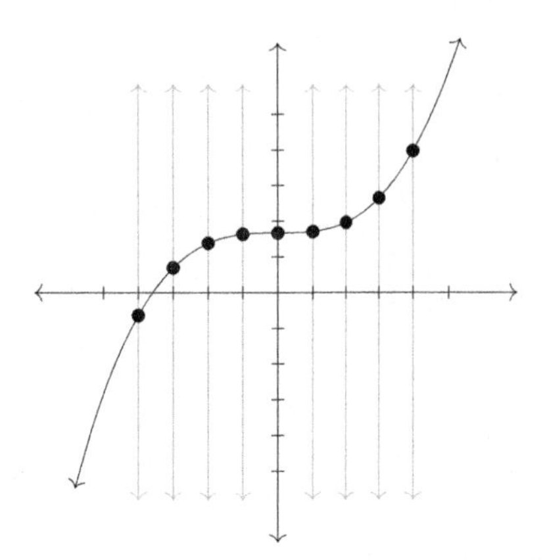

Passes vertical line test

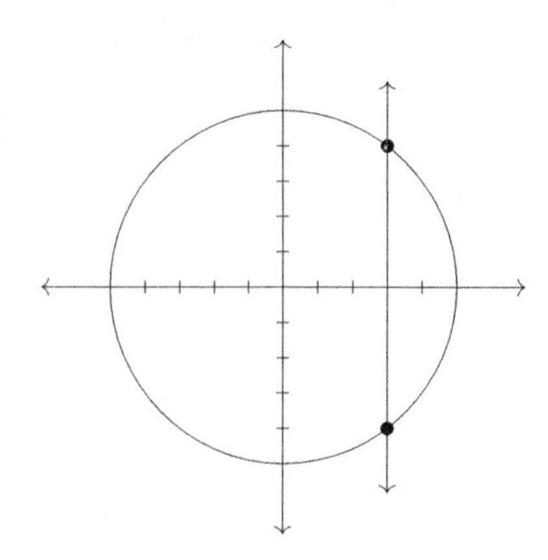

Fails vertical line test

The vertical line test says that if every vertical line hits the graph in one (existence) and only one (uniqueness) spot, then the graph corresponds to a function. So the left example would be a function, while the right example would not be, according to this test.

Does the vertical line test ever fail? Could f be a function and yet have a graph which fails the vertical line test? This answer to this question actually comes down to what you define a *graph* to be. If graphing a function on the xy-plane means that it is a function from $\mathbb{R}$ to $\mathbb{R}$ (where the x-axis is the domain), then the vertical line test will never fail. But if you do not insist on this, then the vertical line test *could* fail. Take a moment to try to think up an example, and then check out the one in the footnote.[5] But let's now look at more examples of functions.

Example 8.2.

- $f : \mathbb{R} \to \mathbb{Z}$ where $f(x) = \lfloor x \rfloor$.
  - This is the floor function, where you just round down. E.g., $\lfloor 3.2 \rfloor = 3$, and $\lfloor -3.2 \rfloor = -4$.
- $g : \{1, 2, 3, 4, 5\} \to \{1, 4, 9, 16, 25, 36, 49\}$ where $g(x) = x^2$.
  - This is the usual square function, except that you cannot plug anything besides $1, 2, 3, 4$ or 5 into g. For example, $g(-2)$ and $g(6)$ do not mean anything since g is only defined to accept elements from its domain. Notice that g's range is $\{1, 4, 9, 16, 25\}$.
- Recall that $\mathbb{N}_0 = \{0, 1, 2, 3, 4, \dots\}$, let $\mathcal{P}$ be the set of subsets of $\mathbb{Z}$ which contain finitely many elements. Then, consider the function $|\cdot| : \mathcal{P} \to \mathbb{N}_0$ where $|S|$ is the cardinality of set S.

[5]An example: $f : \mathbb{R} \to (\mathbb{R} \times \mathbb{R})$ where $f(x) = (5\cos(x), 5\sin(x))$. Here, the codomain is the Cartesian product $\mathbb{R} \times \mathbb{R}$. That is, it is the set of ordered pairs (a, b) where $a, b \in \mathbb{R}$.

  - For example, $|\{-1, 5, 12\}| = 3$.
  - Also, note how this isn't written as most functions are, like $f(x)$ or $g(t)$. Here, the bars go around the input. To make this clearer, we add a little dot between the bars as a "placeholder," to show where the element will go: $|\cdot|$.

- Let S be the set containing all students in your Intro to Proofs class. And let $G : S \to \{\text{A,B,C,D,F}\}$ where $G(s)$ equals the letter grade that student s received on their last homework assignment.
  - For this function, we hope that the range does not equal the codomain. Hopefully everybody passed!

8.2 Injections, Surjections and Bijections

There are three important classes of functions to discuss next. First up are injections.

Definition.

Definition 8.3. A function $f : A \to B$ is *injective* (or *one-to-one*) if $f(a_1) = f(a_2)$ implies that $a_1 = a_2$.

Example 8.4. Here is an example and a non-example:

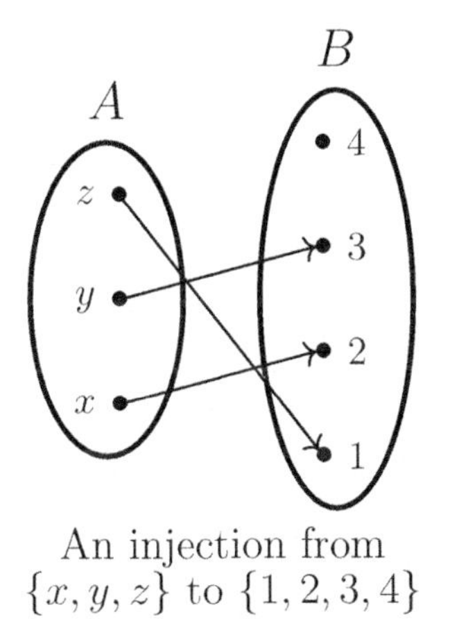

An injection from $\{x, y, z\}$ to $\{1, 2, 3, 4\}$

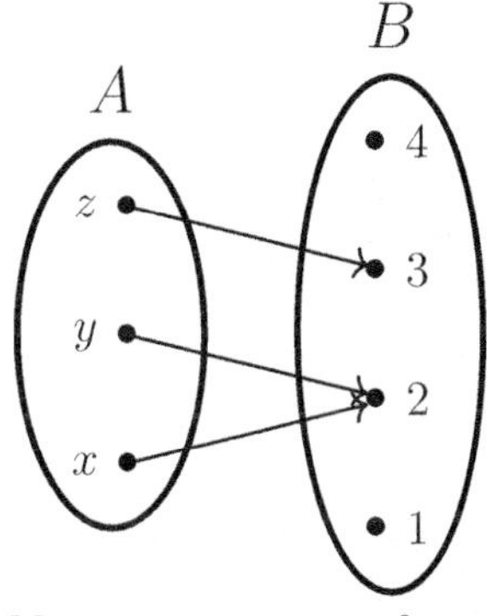

Not an injection from $\{x, y, z\}$ to $\{1, 2, 3, 4\}$

While both of these are functions (verify this on your own), does it make sense why the first satisfies the definition to be injective, but the second does not? The second example is not injective because $f(x) = 2$ and $f(y) = 2$. So we have $f(x) = f(y)$ while $x \neq y$, as these are distinct elements in the domain. Basically, to be injective means that you do not have two arrows pointing at the same point.

Interestingly, the contrapositive provides another way to think about an injection. Recall that the contrapositive turns an implication like "$f(a_1) = f(a_2)$ implies that

$a_1 = a_2$" into a logically equivalent implication, and even for definitions this can at times be useful. Applying the contrapositive to (the second half of) the injection definition gives this:

Equivalent to Definition 8.3. A function $f : A \to B$ is *injective* (or *one-to-one*) if $a_1 \neq a_2$ implies that $f(a_1) \neq f(a_2)$.

So a function is injective if different points in the domain are sent to different points in the codomain. No two arrowheads collide.

Definition.

Definition 8.5. A function $f : A \to B$ is *surjective* (or *onto*) if, for every $b \in B$, there exists some $a \in A$ such that $f(a) = b$.

Example 8.6. Here is an example and a non-example:

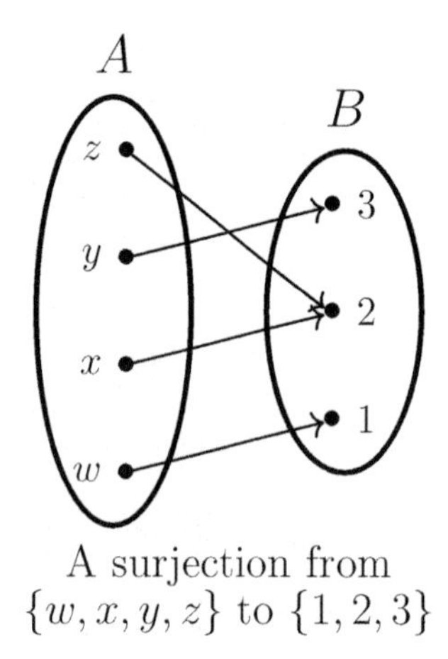

A surjection from $\{w, x, y, z\}$ to $\{1, 2, 3\}$

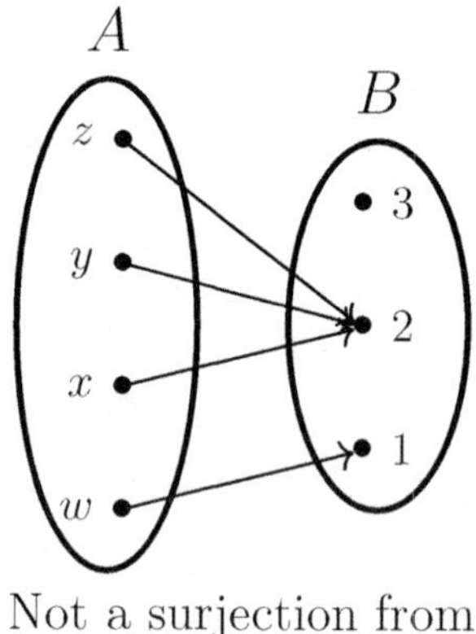

Not a surjection from $\{w, x, y, z\}$ to $\{1, 2, 3\}$

Again, read through the definition and convince yourself that the first example satisfies it while the second does not. For the second, it is not true that for *every* $b \in \{1, 2, 3\}$ there exists some $a \in \{w, x, y, z\}$ such that $f(a) = b$. Why? Because $b = 3$ does not have this property! In terms of arrows, this means every dot in B has at least one arrow pointing at it.

Let's take a look at another way to define this same idea, by again applying the contrapositive (and doing a little rearranging).

Equivalent to Definition 8.5. A function $f : A \to B$ is *surjective* (or *onto*) if there does not exist any $b \in B$ for which $f(a) \neq b$ for all $a \in A$.

Let's check back in with our recurring theme.

> **Recurring Theme Alert.** When defining a function $f : A \to B$, the ideas of *existence* and *uniqueness* were focused on A—for every $x \in A$, we demanded that $f(x)$ exist and be unique. To be injective and surjective, the attention shifts to B. To be surjective means that B has an existence criterion (for every $b \in B$, there *exists* some $a \in A$ that maps to it). And to be injective means that B has a uniqueness-type criterion (for every $b \in B$, there is *at most one* $a \in A$ that maps to it).

Finally, we arrive at bijections. The easiest way to define a bijection would be through a Venn diagram:

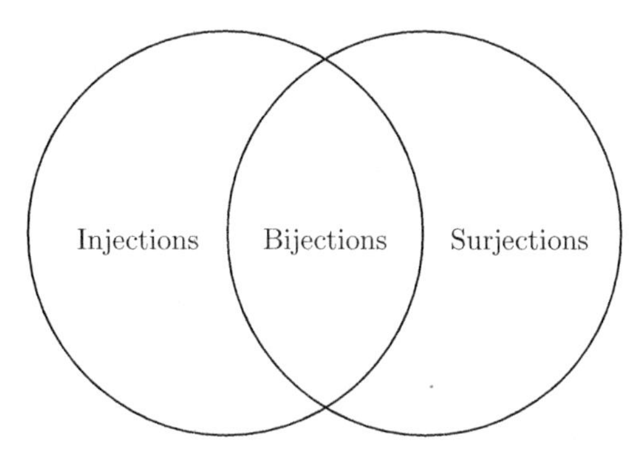

But, if you want to be all formal about it, here ya go:

Definition.

> **Definition 8.7.** A function $f : A \to B$ is *bijective* if it is both injective and surjective.

Let's look at some examples and non-examples. For starters, if you pick a random function, then it is certainly possible—in fact, likely—that it is neither injective nor surjective. Here is such a case:

	Injective	Surjective	Bijective
$z \mapsto 3$, $y \mapsto 2$, $x \mapsto 2$ (codomain $1, 2, 3, 4$)	X	X	X

What about functions that are injective or surjective (or both)? The next table covers these three cases.

	Injective	Surjective	Bijective
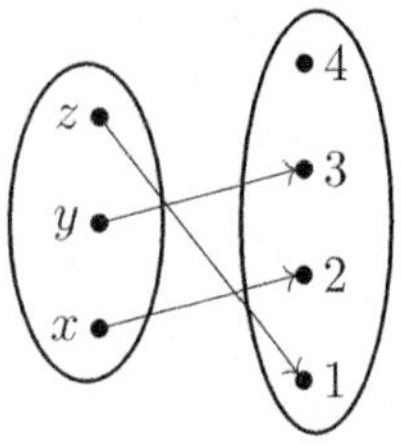	✓	X	X
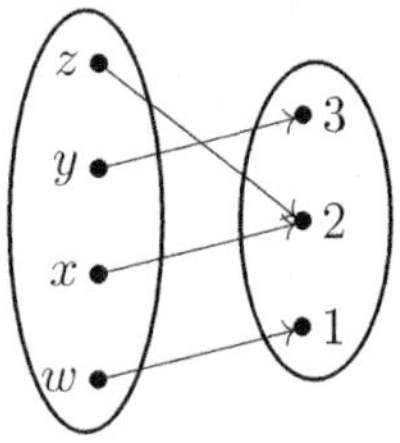	X	✓	X
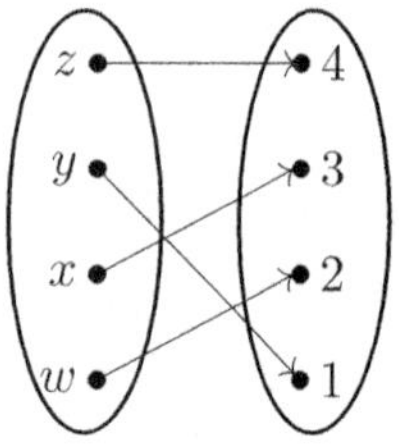	✓	✓	✓

Being bijective means that every element in A is paired up with precisely one element in B. As an analogy, you could think about f as putting elements in A into relationships with elements in B. Being injective means all the relationships are monogamous, while being not injective means there is at least one polygamous person. Being surjective means that everyone has found love,[6] while being not surjective means at least one person (in B) is left out. And being bijective therefore means everyone has found love in a monogamous relationship.

In terms of arrows, being a bijection means that every dot on the left has precisely one arrow emanating from it, and every dot on the right has precisely one arrow entering it. (And yes, that sentence is screaming for another Recurring Theme Alert.)

Recurring Theme Alert. Defining a function $f : A \to B$ placed existence and uniqueness criteria on A. If f is both injective and surjective, then this adds existence and uniqueness criteria to B. Thus, if f is a bijection, then it has these criteria on both sides: Every $a \in A$ is mapped to precisely one $b \in B$, and every $b \in B$ is mapped to by precisely one $a \in A$. In effect, this pairs up each element of A with an element of B; namely, a is paired with $f(a)$ in this way.[7]

[6] Simply being a function means that everyone in A has found love. The surjectivity guarantees that everyone in B has also found love.

[7] Foreshadowing Alert: For f to be a function, we demanded existence and uniqueness criteria on A. If $f : A \to B$ is a bijection, then we demand those same criteria *[footnote continues on next page]*

Proving $x$jectiveness, for $x \in \{\text{in}, \text{sur}, \text{bi}\}$

Based on its definition, this is the outline to prove a function is injective.

Proposition. $f : A \to B$ is an injection.

Proof. Assume $x, y \in A$ and $f(x) = f(y)$.

⋮ apply algebra,
logic, techniques

Therefore, $x = y$.

Since $f(x) = f(y)$ implies $x = y$, f is injective. □

Alternatively, one could use the contrapositive, which would mean one starts by assuming $x \neq y$, and then concludes that $f(x) \neq f(y)$.

Next, here's the outline for a surjective proof.

Proposition. $f : A \to B$ is a surjection.

Proof. Assume $b \in B$.

⋮ Find an $a \in A$ where $f(a) = b$,
and show it works

Since every $b \in B$ has an $a \in A$ where $f(a) = b$, f is surjective. □

It is important to remember that when you choose $b \in B$ (at the start of your proof), it must be completely arbitrary. If $B = \mathbb{R}$, make sure you are never assuming that b is positive or negative or nonzero or an integer or anything like that. Your work must be valid regardless what the b is. Recall that this is what we mean when we say that we chose an *arbitrary* $b \in B$. The only exception to this is if you divide up your work into cases where you have, say, the negative case and the nonnegative case, or the zero and the nonzero case. But if you do that, then within *each case* you

---

on B. Thus, if we decided to switch things up and use B as a domain, A as a codomain, and have f map things "in reverse," then that is perfectly fine as far as the definition of a function is concerned. Indeed, by discussing bijections now, we are allowing ourselves to discuss a function's *inverse* later.

must choose an arbitrary b, and collectively the cases must cover all the options in B.

Again, one could instead proceed via the contrapositive, although that tends to be less common for surjection arguments.

To prove a function is a bijection, one way is to prove both of the above. A separate way (using inverses) will be discussed later. Let's do some examples.

Example 8.8. Let $\mathbb{R}^+$ denote the nonnegative real numbers. Prove the following.

(a) $f : \mathbb{R} \to \mathbb{R}$ where $f(x) = x^2$ is not injective, surjective or bijective.

(b) $g : \mathbb{R}^+ \to \mathbb{R}$ where $g(x) = x^2$ is injective, but not surjective or bijective.

(c) $h : \mathbb{R} \to \mathbb{R}^+$ where $h(x) = x^2$ is surjective, but not injective or bijective.

(d) $k : \mathbb{R}^+ \to \mathbb{R}^+$ where $k(x) = x^2$ is injective, surjective and bijective.

Scratch Work.[8] Here we have four different functions. Each squares its input, but a function is not only the operation, but the domain and codomain as well. And since their domains/codomains do not match, they are all different functions. This allows them to have different properties, as we are proving here.

From $\mathbb{R} \to \mathbb{R}$, or subsets thereof, the injective property is essentially a "horizontal line test." If every horizontal line hits the function in only one place, then the function is injective. But if any horizontal line hits the function in more than one place, then the function is not injective.[9]

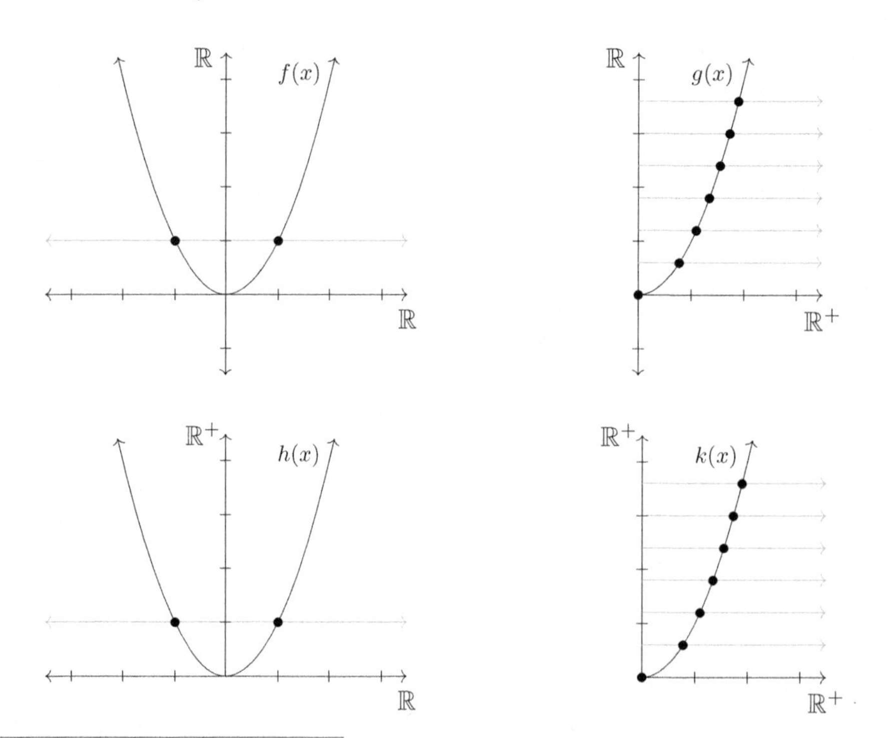

[8]This is your periodic encouragement to try out problems on your own before reading the solution.

[9]That is, a horizontal line through the *range*. If $y = b$ is a horizontal line and b is in the range, then it must only hit the function once if it is to be injective. (If b is in the codomain but not in the range, then it will necessarily hit the function zero times.)

Thus, it makes sense that f and h are not injective while g and k are. To show that f and h are *not* injective it will suffice to find any two points from the domain which map to the same value in the codomain; I believe 1 and -1 should work (or 2 and -2, or any other such pair).[10]

In order to prove that g and k *are* injective, we will assume that, say, $g(x) = g(y)$, and we will try to prove that $x = y$. This would be some natural scratch work:

$$\begin{aligned} g(x) &= g(y) \\ x^2 &= y^2 \\ \sqrt{x^2} &= \sqrt{y^2} \\ x &= y. \end{aligned}$$

Is every step legit? The only question mark would be the final step. What if the third equation is $\sqrt{2^2} = \sqrt{(-2)^2}$? This is true, but if you "cancel" the square root and the square, then you would get $-2 = 2$, which is false. In general, if $\sqrt{x^2} = \sqrt{y^2}$, there are two options: $x = y$ or $x = -y$. This is sometimes written succinctly as $x = \pm y$. And while this fact prohibits f and h from being injective, the same is not true for g and k, since their domains do not include negative numbers. So for g and k, the above scratch work will prove that $x = y$, which proves these functions are injective.

What about surjectivity? To show that f and g are not surjective we simply have to find some y in the codomain which nothing maps to — any negative number should work. But h and k are different. Since their codomains do not include negative numbers, they will be surjective. To show this, we will pick any y in the codomain, and find the specific x in the domain such that $h(x) = y$ (in part (c)) and $k(x) = y$ (in part (d)); the value $x = \sqrt{y}$ should work in both cases.

***Proof*. Part (a).** Observe that $f(-2) = f(2)$ while $-2 \neq 2$, showing that f is not injective. Next, observe that $f(x) = x^2 \geq 0$ for all $x \in \mathbb{R}$, showing that there does not exist an $x \in \mathbb{R}$ for which $f(x) = -4$. And since -4 is in f's codomain, this proves that f is not surjective. Since f is neither injective nor surjective, it is also not bijective.

Part (b). Similar to part (a), because $g(x) = x^2 \geq 0$ for all $x \in \mathbb{R}^+$, there does not exist an $x \in \mathbb{R}$ for which $g(x) = -4$. And since -4 is in g's codomain, this proves that g is not surjective, which also proves that g is not bijective.

To see that g is injective, assume $x, y \in \mathbb{R}^+$ and $g(x) = g(y)$. Then,

$$\begin{aligned} g(x) &= g(y) \\ x^2 &= y^2 \\ \sqrt{x^2} &= \sqrt{y^2}. \end{aligned}$$

In the reals, this gives two possibilities: $x = y$ or $x = -y$. However, since we know $x, y \in \mathbb{R}^+$, the only option is that they are both positive, and so $x = y$. We have

[10] You can think about this as finding a *counterexample* to the claim that f is injective.

shown that $g(x) = g(y)$ implies $x = y$, thus g is an injection.

Part (c). The fact that h is not injective is just like with f: Note that $h(-2) = h(2)$ while $-2 \neq 2$, showing that h is not injective. To show that h is a surjection, pick any b in its codomain, $\mathbb{R}^+$. Since $b \geq 0$, its positive square root exists. Let's call this square root x; that is, $x = \sqrt{b}$. Since $x \in \mathbb{R}^+$ as well and

$$h(x) = x^2 = (\sqrt{b})^2 = b,$$

we have shown that for every $b \in \mathbb{R}^+$ there exists an $x \in \mathbb{R}^+$ such that $h(x) = b$. This proves that h is a surjection.

Part (d). The fact that k is an injection follows the same exact reasoning as with g, and the fact that k is a surjection follows the exact same reasoning as with h. And because k is both an injection and a surjection, it is also a bijection. □

As you can see, proofs that a function is *not* an injection or *not* a surjection are shorter than proofs that a function *is* an injection or surjection. For example, to prove that $f : \mathbb{R} \to \mathbb{R}$ where $f(x) = \sin(x)$ is not injective you simply have to note that $\sin(0) = \sin(\pi)$, and to prove it is not surjective you simply have to note that $-1 \leq \sin(x) \leq 1$ for all x, and so there does not exist any x for which $\sin(x) = 17$. Again, this should feel like you are searching for a counterexample to a claim that they are injective and surjective.

Let's do another example where the function is a bijection.

Example 8.9. The function $f : (\mathbb{Z} \times \mathbb{Z}) \to (\mathbb{Z} \times \mathbb{Z})$ where $f(x, y) = (x + 2y, 2x + 3y)$ is a bijection.

Scratch Work. If we prove that f is an injection and a surjection, then that will prove that f is a bijection. For the injection proof, the scratch work is basically just the proof, so let's instead focus on proving that f is a surjection. To do so, we must show that for an arbitrary $(a, b) \in \mathbb{Z} \times \mathbb{Z}$, there exists some $(x, y) \in \mathbb{Z} \times \mathbb{Z}$ such that $f(x, y) = (a, b)$; that is, every element in the codomain gets hit. How do we find such an (x, y)? Scratch work! We want[11]

$$f(x, y) = (a, b)$$
$$(x + 2y, 2x + 3y) = (a, b).$$

When are two ordered pairs equal? Well, certainly $(2, 3) \neq (4, 5)$, but also $(2, 3) \neq (2, 5)$ and $(2, 3) \neq (4, 3)$. In order to have two ordered pairs equal, they must be equal in *both* coordinates! Thus, in order to have $(x + 2y, 2x + 3y) = (a, b)$, we must have

$$x + 2y = a \qquad \text{and} \qquad 2x + 3y = b.$$

[11]Note that if f takes in a single number x as input, then we write $f(x)$; we put parentheses around the input. So, since in this case we are inputting an ordered pair like (x, y) into f, should we write it as $f\big((x, y)\big)$? Meaning, shouldn't we put another set of parentheses around the (x, y) input? Sometimes this is done, but it is in fact more common to drop the outer parentheses and just write $f(x, y)$. In most cases it looks nicer and does not add any confusion. (Also, frankly, it is hard to remember to put the second set there!)

We want to figure out which x and y make this work, so we do algebra to solve for x and y. And if you've taken linear algebra, you probably have lots of experience with these sorts of calculations.[12]

$$\begin{aligned} x + 2y &= a \\ 2x + 3y &= b \end{aligned} \xrightarrow{\text{First equation x2}} \begin{aligned} 2x + 4y &= 2a \\ 2x + 3y &= b \\ \hline y &= 2a - b \end{aligned} \;\text{(subtract)} \xrightarrow{\text{Plug in this } y} \begin{aligned} x + 2y &= a \\ x + 2(2a - b) &= a \\ \Rightarrow \quad x &= -3a + 2b \end{aligned}$$

According to this scratch work, in order for $f(x, y) = (a, b)$, we would need $(x, y) = (-3a + 2b, 2a - b)$. We will use this in the surjective half of our proof below.

Proof. We will prove that f is injective and surjective.

Injective. Assume $f(x_1, y_1) = f(x_2, y_2)$. We aim to show $(x_1, y_1) = (x_2, y_2)$, which is true provided $x_1 = x_2$ and $y_1 = y_2$. Notice that

$$\begin{aligned} f(x_1, y_1) &= f(x_2, y_2) \\ (x_1 + 2y_1, 2x_1 + 3y_1) &= (x_2 + 2y_2, 2x_2 + 3y_2). \end{aligned}$$

Two ordered pairs are equal provided their first coordinates are the same and their second coordinates are the same. Thus, the above tells us that

$$\begin{aligned} x_1 + 2y_1 &= x_2 + 2y_2 \\ 2x_1 + 3y_1 &= 2x_2 + 3y_2. \end{aligned} \qquad (⚽)$$

Multiplying the top equation by 2 gives

$$\begin{aligned} 2x_1 + 4y_1 &= 2x_2 + 4y_2 \\ 2x_1 + 3y_1 &= 2x_2 + 3y_2. \end{aligned}$$

Subtracting the bottom from the top leaves

$$y_1 = y_2. \qquad (\heartsuit)$$

To conclude that $x_1 = x_2$, we plug y_2 in for y_1 (by equation ($\heartsuit$)) into equation (⚽):

$$x_1 + 2y_2 = x_2 + 2y_2.$$

Canceling the $2y_2$ from both sides, we see that

$$x_1 = x_2.$$

Combined with equation ($\heartsuit$), we have at last deduced that $(x_1, y_1) = (x_2, y_2)$.

[12]A pair of equations like $\begin{aligned} x + 2y &= a \\ 2x + 3y &= b \end{aligned}$, where none of the terms are raised to any power, is called a *system of linear equations*, which may be familiar if you have taken a course in linear algebra.

We have shown that if $f(x_1, y_1) = f(x_2, y_2)$, then $(x_1, y_1) = (x_2, y_2)$, proving that f is an injection.

Surjective. Pick any $(a, b) \in \mathbb{Z} \times \mathbb{Z}$. We wish to find some $(x, y) \in \mathbb{Z} \times \mathbb{Z}$ such that $f(x, y) = (a, b)$, and (based on our scratch work) we claim that $(x, y) = (-3a + 2b, 2a - b)$ works. First, note that since $a, b \in \mathbb{Z}$, also $(-3a + 2b), (2a - b) \in \mathbb{Z}$. This implies that $(x, y) \in \mathbb{Z} \times \mathbb{Z}$, as required.

Second, note that

$$\begin{aligned} f(x, y) &= f(-3a + 2b, 2a - b) \\ &= \Big((-3a + 2b) + 2(2a - b)\ ,\ 2(-3a + 2b) + 3(2a - b)\Big) \\ &= \Big(-3a + 2b + 4a - 2b\ ,\ -6a + 4b + 6a - 3b\Big) \\ &= (a, b). \end{aligned}$$

We showed that for any (a, b) from the codomain, there exists some (x, y) from the domain such that $f(x, y) = (a, b)$. Thus, f is surjective.

Since f is both an injection and a surjection, it is a bijection. □

Preventing (in/sur)jectiveness by Set Comparison

You might have noticed that in our early examples of injectivity and surjectivity, where we were looking at domains and codomains of finite size, if $|A| > |B|$ then it is impossible for a function $f : A \to B$ to be injective. Likewise, if $|A| < |B|$, then it is impossible for a function $f : A \to B$ to be surjective.

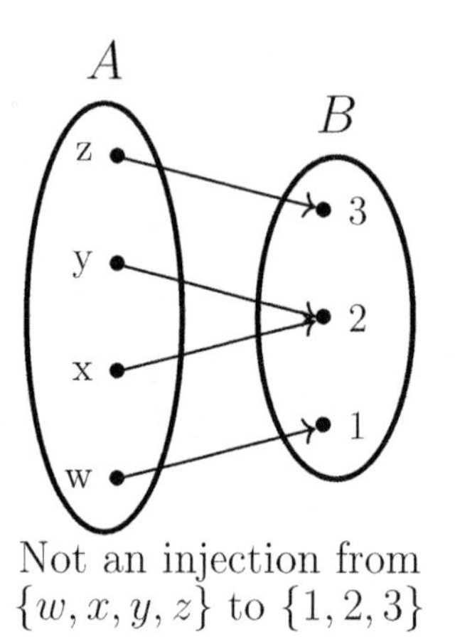

Not an injection from $\{w, x, y, z\}$ to $\{1, 2, 3\}$

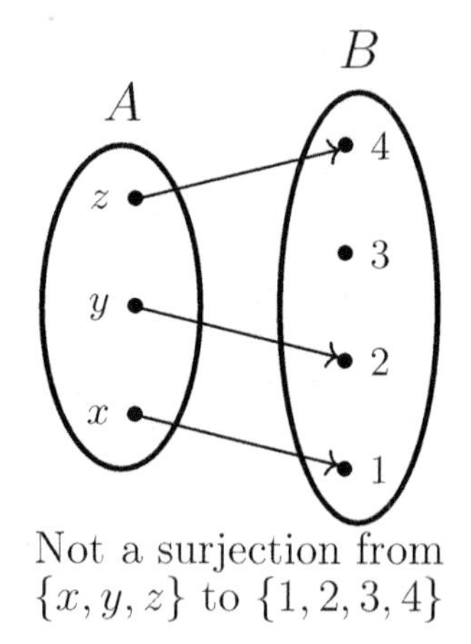

Not a surjection from $\{x, y, z\}$ to $\{1, 2, 3, 4\}$

If this reminds you of the pigeonhole principle, then great job! You're exactly right! If this does not remind you of the pigeonhole principle, well, I think you're pretty great anyways. Below is the function version of the pigeonhole principle.

Theorem.

Theorem 8.10 (*The func-y pigeonhole principle*)**.** Suppose A and B are finite sets and $f : A \to B$ is any function.

(a) If $|A| > |B|$, then f is not injective.

(b) If $|A| < |B|$, then f is not surjective.

Proof Idea. For part (a), you can think about A as a set of pigeons and B as a set of pigeonholes. The function $f : A \to B$ is the rule which tells each pigeon which pigeonhole to go into. For example, consider the function

$$f : \{-2, -1, 0, 1, 2, 3, 4\} \to \{0, 1, 2, 3, 4\} \quad \text{where} \quad f(x) = |x|.$$

This can be pictured like this:

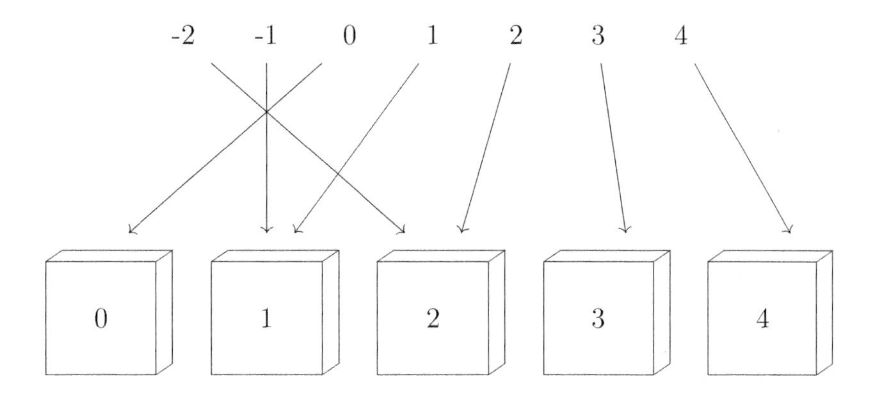

Or, perhaps the function is

$$f : \{-2, -1, 0, 1, 2, 3, 4\} \to \{0, 1, 2, 3, 4\} \quad \text{where} \quad f(x) = (-1)^x + 2.$$

Then, the picture would be this:

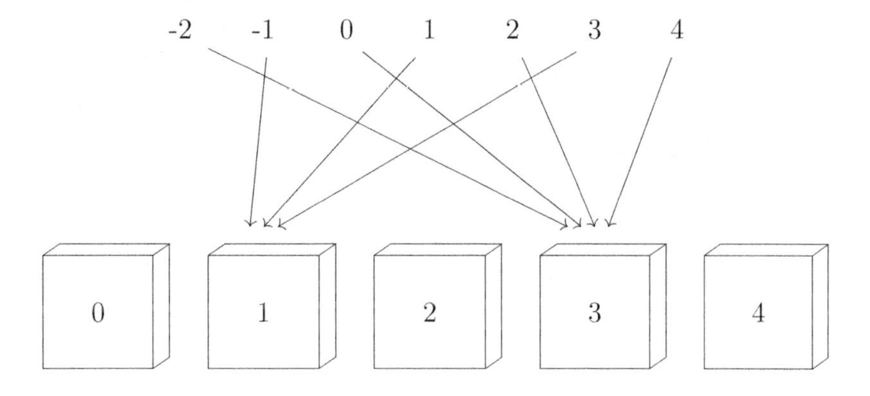

No matter what the function is, if it is a function from a domain of size $|A| = 7$ to a codomain of size $|B| = 5$, then the pigeonhole principle tells us that there must be two things in the same box. The first example has 2 and -2 in the same box (among other pairs), and the second example has -1 and 3 in the same box (among other pairs).

In general, if $f : A \to B$ where $|A| > |B|$, then at least one of the boxes will have had two elements placed into it by f. And two things (say, a and b) in the same box means that $f(a) = f(b)$, which is precisely what it means to say that f is not injective, which will prove (a).

Part (b) is slightly different—it comes down to the fact that a function sends a point in the domain to just one point in the codomain. Thus, the $|A|$ points in the domain can hit at most $|A|$ points in the codomain. So if $|A| < |B|$, it is certainly not hitting them all—meaning f is not surjective. This is like saying: If 7 pigeons go to live in 10 pigeonholes, there will be at least one pigeonhole with no pigeons living in it.[13]

Proof. Part (a). Consider each element in A to be an object and each element of B to be a box. Given an $a \in A$, place object a into box b if $f(a) = b$. Since there are more objects than boxes, by the pigeonhole principle (Principle 1.5) at least one box has at least two objects in it. That is, $f(a_1) = f(a_2)$ for some distinct a_1 and a_2, implying that f is not injective.

Part (b). Since f is a function, each $a \in A$ is mapped to only one $b \in B$. Thus, k elements in A can map to at most k elements of B. And so the $|A|$ elements in A can map to at most $|A|$ elements in B. However, since $|A| < |B|$, there must be some elements in B that are not hit. This means that f is not surjective. □

It is again useful to think about what the contrapositive tells us. For finite sets A and B, applying the contrapositive to the func-y pigeonhole principle gives this:

(a) If $f : A \to B$ is injective, then $|A| \leq |B|$.

(b) If $f : A \to B$ is surjective, then $|A| \geq |B|$.

Viewing the statements this way is beneficial for another reason: By combining the above two statements, we can see that in order for f to be a bijection—meaning an injection and a surjection—we would need $|A| = |B|$.

It is also worth mentioning that this theorem still holds true in the case that $|A|$ and/or $|B|$ are/is infinite.[14] But proving this to be the case would take us too far afield.

[13]It's easy to get the terms "injective" and "surjective" mixed up, but perhaps this pigeon metaphor is a way to remember which is which. A function is INjective if every pigeon goes IN their own box (no sharing; they each get their own). A function that is SURjective allows for the possibility that a pigeon will be SURprised when they enter their box to find another pigeon already there (they are forced to share boxes because all the other boxes were full).

[14]When both are infinite, some particularly exciting stuff happens! Stay tuned for an Introduction to Cardinality following this chapter, in which these exciting implications are explored.

8.3 The Composition

Suppose you have a function $g : A \to B$ and a function $f : B \to C$. Then the outputs of g can be used as inputs of f![15] For example, if $a \in A$, then $g(a)$ is in B, which is the domain of f, and so you can plug it into f to get $f(g(a))$, which is something in C.

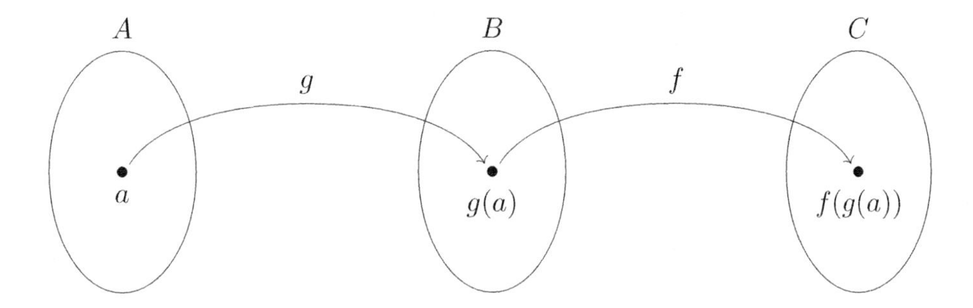

By applying g and then f, we in effect create a single function from A to C. This function we denote $f \circ g$. Below is the picture.[16]

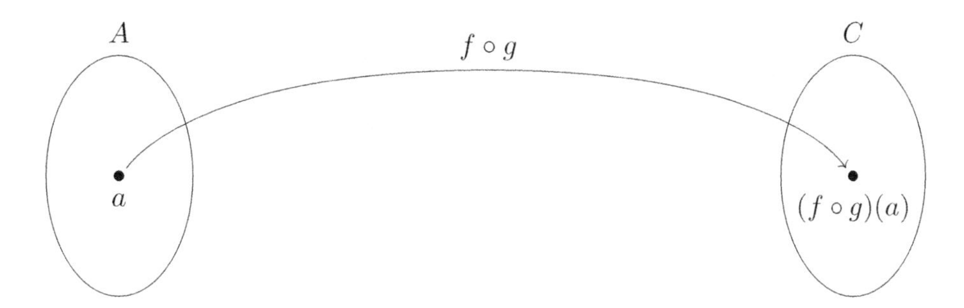

We also give this function a special name: We call it the *composition* of f with g.

Definition.

Definition 8.11. Let A, B and C be sets, $g : A \to B$, and $f : B \to C$. Then the *composition* function is denoted $f \circ g$ and is defined as thus:

$$(f \circ g) : A \to C \qquad \text{where} \qquad (f \circ g)(a) = f(g(a)).$$

[15] Free short story idea: g is a mild-mannered function, living a happy little life. She is particularly proud of her codomain on which she has imprinted her whole image. Then, in the distance, riding across the range, is f. A smooth-talking, mean-valued jerk, f has worked his way into the local government's higher powers. And under the authority of the Composition Committee, f obtains a function injunction to seize g's codomain to be his own domain! Drama ensues. Suggested title: *Eminent Domain*.

[16] Note: The notation "$f \circ g$" is properly read "f composed with g" or simply "f of g." It should *not* be read "fog." But. . . I suspect that won't stop many of you.

That's the main idea: It's a function inside of a function.[17]

Example 8.12. Recall that $\mathbb{R}^+$ is the set of nonnegative reals, and $\mathbb{N}_0$ is the set $\mathbb{N} \cup \{0\}$. Below are three examples of function composition. As you read through them, remember that $f \circ g$ is a function from g's domain to f's codomain.

- Suppose

$$g : \mathbb{R} \to \mathbb{R} \qquad \text{where} \qquad g(x) = x + 1$$
$$f : \mathbb{R} \to \mathbb{R}^+ \qquad \text{where} \qquad f(x) = x^2.$$

Then,

$$(f \circ g) : \mathbb{R} \to \mathbb{R}^+ \qquad \text{where} \qquad (f \circ g)(x) = (x+1)^2.$$

- Suppose

$$g : \mathbb{R} \to \mathbb{R}^+ \qquad \text{where} \qquad g(x) = x^2$$
$$f : \mathbb{R}^+ \to \mathbb{R}^+ \qquad \text{where} \qquad f(x) = x + 1.$$

Then,

$$(f \circ g) : \mathbb{R} \to \mathbb{R}^+ \qquad \text{where} \qquad (f \circ g)(x) = x^2 + 1.$$

- Recall that the *floor function* rounds integers down; e.g., $\lfloor 5.7 \rfloor = 5$. Suppose[18]

$$g : \mathbb{R} \to \mathbb{Z} \qquad \text{where} \qquad g(x) = \lfloor x \rfloor$$
$$f : \mathbb{Z} \to \mathbb{N}_0 \qquad \text{where} \qquad f(x) = |x|.$$

Then,

$$(f \circ g) : \mathbb{R} \to \mathbb{N}_0 \qquad \text{where} \qquad (f \circ g)(x) = \big|\lfloor x \rfloor\big|.$$

For instance, $g(3.2) = 3$, and $f(3) = 3$, implying that $(f \circ g)(3.2) = 3$. And $g(-3.2) = -4$, and $f(-4) = 4$, implying that $(f \circ g)(-3.2) = 4$.

Ok, let's prove some things.

[17]SPOILER ALERT: What follows is the plot to the movie *Inception*. Opening plot: Let $d(t)$ be the dream function. *Audience: *Very interested.** Plot development: What if $d(d(t))$? *Audience: *Whoa!!** Plot twist: What if $d(d(d(t)))$? *Audience: *Loses their minds.**

[18]Some say that $e^{i\pi} + 1 = 0$ is the most beautiful equation in math. But have you seen $\lfloor \pi \rfloor - \lceil e \rceil = 0$? Stunning.

Theorem.

Theorem 8.13. Suppose A, B and C are sets, $g : A \to B$ is injective, and $f : B \to C$ is injective. Then, $f \circ g$ is injective.

Proof Sketch. We want to show that the function $f \circ g : A \to C$ is injective. That is, given any $a_1, a_2 \in A$, we want to show that $(f \circ g)(a_1) = (f \circ g)(a_2)$ implies $a_1 = a_2$. Written differently, if $f(g(a_1)) = f(g(a_2))$, then $a_1 = a_2$. Basically, the proof consists of two copies of the standard injective proof, with the definition of a composition thrown in (which is what makes it challenging). Here is the proof overview:

$$f(g(a_1)) = f(g(a_2)) \quad \overset{f \text{ is injective}}{\Longrightarrow} \quad g(a_1) = g(a_2) \quad \overset{g \text{ is injective}}{\Longrightarrow} \quad a_1 = a_2.$$

To do this, we will first use the fact that $f : B \to C$ is injective, which tells us that for any $b_1, b_2 \in B$, if $f(b_1) = f(b_2)$, then $b_1 = b_2$. And that works for *any* two elements in B. In particular, note that $g(a_1)$ and $g(a_2)$ are in B!

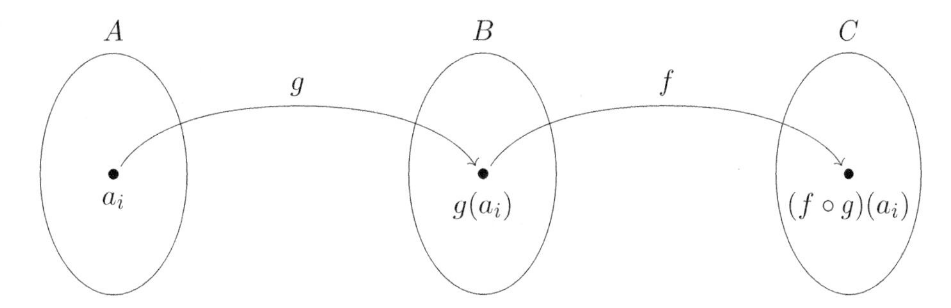

Since $g(a_1)$ and $g(a_2)$ are in B and f is injective, this tells us that $f(g(a_1)) = f(g(a_2))$ implies $g(a_1) = g(a_2)$.

Next is a direct application of $g : A \to B$ being injective. We have $a_1, a_2 \in A$ and $g(a_1) = g(a_2)$, which by injectivity means $a_1 = a_2$. Boom!

***Proof*.** Since $(f \circ g) : A \to C$, to show that $f \circ g$ is injective we must show that, for any $a_1, a_2 \in A$, if $(f \circ g)(a_1) = (f \circ g)(a_2)$, then $a_1 = a_2$. To this end, assume $a_1, a_2 \in A$ and $(f \circ g)(a_1) = (f \circ g)(a_2)$. Applying the definition of the composition,

$$f(g(a_1)) = f(g(a_2)).$$

Since $f : B \to C$ is an injection, if $f(b_1) = f(b_2)$ for any $b_1, b_2 \in B$, then $b_1 = b_2$. In particular, observe that $g(a_1), g(a_2) \in B$ and $f(g(a_1)) = f(g(a_2))$, and so $g(a_1) = g(a_2)$.

Likewise, $g : A \to B$ is injective and we just showed that $g(a_1) = g(a_2)$ where $a_1, a_2 \in A$. This implies that $a_1 = a_2$.

We have shown that for $a_1, a_2 \in A$, if $(f \circ g)(a_1) = (f \circ g)(a_2)$, then $a_1 = a_2$. Thus, $(f \circ g)$ is an injection. $\square$

That was cool! No wonder they're called ***fun**ctions*![19] Let's prove a similar result for surjections!

Theorem.

Theorem 8.14. Suppose A, B and C are sets, $g : A \to B$ is surjective, and $f : B \to C$ is surjective. Then $f \circ g$ is surjective.

Proof Idea. We want to repeat the main idea from the last proof, in which we first apply the property to one function, and then to the other. In this case, to prove $f \circ g : A \to C$ is surjective, we want to prove that for any $c \in C$, there exists some $a \in A$ such that $(f \circ g)(a) = c$. To get from C back to A, the trick is to stop off in B along the way. Here's the three-step overview, beginning on the right:

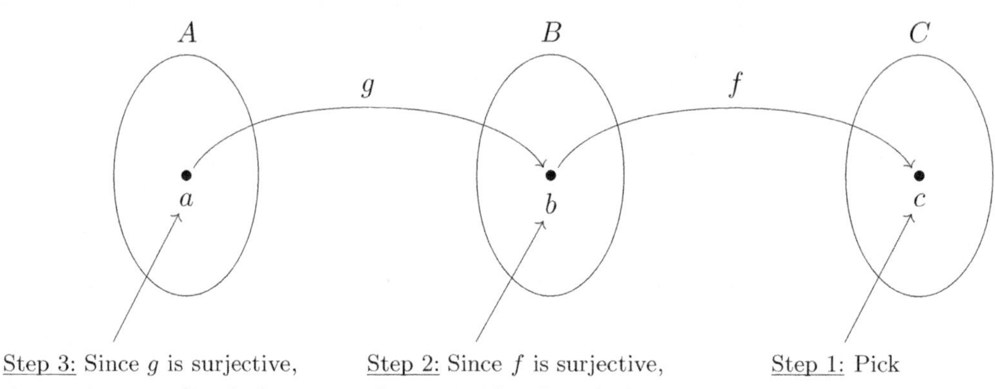

Combined, we have found an $a \in A$ such that $f(g(a)) = c$, proving that $f \circ g$ is surjective.

Proof. Since $(f \circ g) : A \to C$, to show that $f \circ g$ is surjective we must show that, for any $c \in C$, there exists some $a \in A$ such that $(f \circ g)(a) = c$. To this end, pick any $c \in C$.

Since $f : B \to C$ is surjective and $c \in C$, there must be some $b \in B$ such that $f(b) = c$. Next, since $b \in B$ and $g : A \to B$ is surjective, there must be some $a \in A$ such that $g(a) = b$.

For an arbitrary $c \in C$, we have found an $a \in A$ such that

$$(f \circ g)(a) = f(g(a)) = f(b) = c,$$

completing the proof. □

[19]Ok, that one was terrrible.

Next, recall that a *corollary* is a result that follows quickly from previous results. Our previous two theorems quickly give the following corollary.

Corollary.

Corollary 8.15. Suppose A, B and C are sets, $g : A \to B$ is bijective, and $f : B \to C$ is bijective. Then $f \circ g$ is bijective.

***Proof*.** By Theorem 8.13, $f \circ g$ is an injection. By Theorem 8.14, $f \circ g$ is a surjection. Thus, by the definition of a bijection (Definition 8.7), $f \circ g$ is a bijection. □

Airport Metaphor

As we close out this section, here is one final note. Notice that in our definition of function composition (Definition 8.11) we had functions g and f where $g : A \to B$, and $f : B \to C$. Notice that B is used twice—that is, the codomain of g is the domain of f. This makes sense: In order for $f \circ g : A \to C$ to be a function, every a that we plug in must be sent to some $c \in C$. And since $f \circ g$ is the function $f(g(x))$, if we plug a into this function, g sends that a to something in B (since $g : A \to B$), and f can handle something in B since $f : B \to C$; that is, $f(g(a)) = f(b) = c$, for some $b \in B$ and $c \in C$. Thus, $f(g(x))$ succeeds in taking something in A and sending it to something in C.

It's like if you were flying from Los Angeles to New York City, but you had a layover in Chicago. If Delta Airlines can fly you from Los Angeles to Chicago, and United Airlines can fly you from Chicago to New York City, then you can successfully piece them together to form a travel plan from LA to NYC.

Here, Delta : $\{\text{LA}\} \to \{\text{Chicago}\}$; and United : $\{\text{Chicago}\} \to \{\text{NYC}\}$; and so, United $\circ$ Delta : $\{\text{LA}\} \to \{\text{NYC}\}$. However, if instead Delta : $\{\text{LA}\} \to \{\text{Chicago}\}$ and United : $\{\text{Detroit}\} \to \{\text{NYC}\}$, then you can *not* piece them together. The codomain of g has to match the domain of f for it to work out.

Notice, though, that we don't need the codomain of g to *equal* the domain of f. If we had $g : A \to B$ and $f : D \to C$ where $B \subseteq D$, that would be enough (for the definition, and for these last two theorems). As long as $g(a)$ is a part of f's domain, then $f(g(a))$ will make sense, which is all we need.

Continuing the above metaphor, it is in fact the case that Chicago has two different airports: O'Hare Airport and Midway Airport. If it were the case that Delta has flights from LA to O'Hare, and United has flights from both O'Hare to NYC *and* has flights from Midway to NYC, then as a traveler you are still happy, right? You will just use the O'Hare layover! United has flights from Midway that you won't use, but it's perfectly fine that they are there, right? This shows why the flight functions Delta : $\{\text{LA}\} \to \{\text{O'Hare}\}$ and United : $\{\text{O'Hare, Midway}\} \to \{\text{NYC}\}$ is an acceptable setup to discuss the composition function United$\circ$Delta : $\{\text{LA}\} \to \{\text{NYC}\}$.

8.4 Invertibility

In the reals, the *multiplicative identity* is 1, because $a \cdot 1 = a$ for all $a \in \mathbb{R}$. Every nonzero number has a *multiplicative inverse*. For example, the multiplicative inverse of 4 is $\frac{1}{4}$, because $4 \cdot \frac{1}{4} = 1$. And the multiplicative inverse of $\frac{1}{3}$ is 3, because $\frac{1}{3} \cdot 3 = 1$. The multiplicative inverse is whatever you have to multiply by to get the multiplicative identity element of 1.

Likewise, in the reals the *additive identity* is 0, because $a + 0 = a$ for all $a \in \mathbb{R}$. Every real number has an *additive inverse*. For example, the additive inverse of 6 is -6, because $6 + (-6) = 0$. And the additive inverse of -2 is 2, because $(-2) + 2 = 0$. The additive identity is whatever you have to add to get the additive identity element of 0.

There is also an *identity function*, which is analogous to 1 and 0 above in that when you apply it to any function, the function is unchanged. Except instead of multiplication and addition, the operation is function composition. In this way, many functions will also have inverses.

Definition.

Definition 8.16. For a set A, the *identity function* on A is the function

$$i_A : A \to A \qquad \text{where} \qquad i_A(x) = x \text{ for every } x \in A.$$

The *inverse* of a function $f : A \to B$, if it exists, is the function $f^{-1} : B \to A$ such that $f^{-1} \circ f = i_A$ and $f \circ f^{-1} = i_B$.

An inverse undoes the function. If x was sent to y by f (that is, $f(x) = y$), then y will be sent to x by f^{-1} (that is, $f^{-1}(y) = x$). Here's a small example.

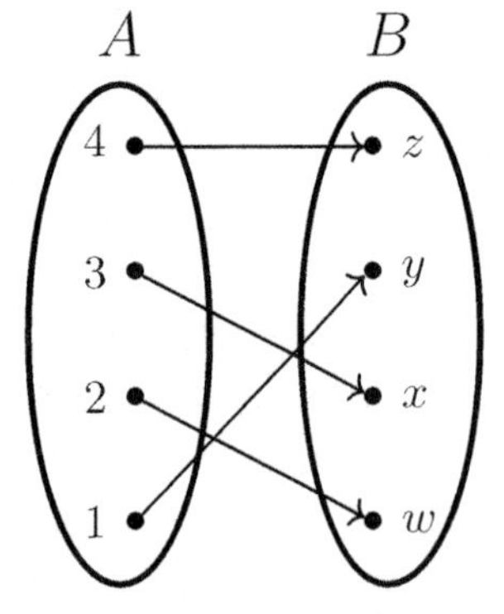

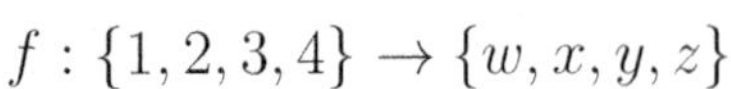

$f : \{1, 2, 3, 4\} \to \{w, x, y, z\}$

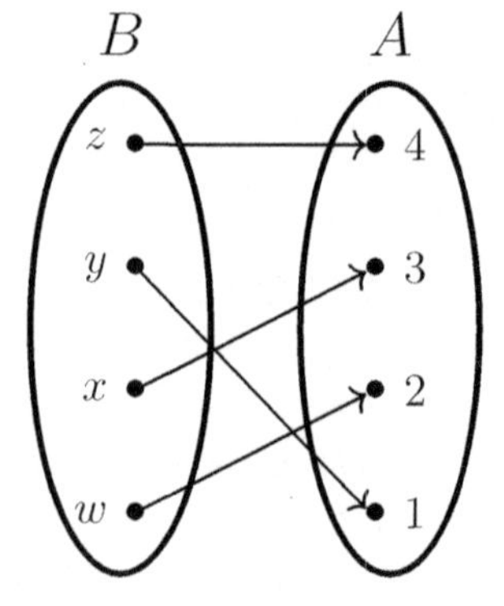

$f^{-1} : \{w, x, y, z\} \to \{1, 2, 3, 4\}$

Combined, we get the identity function. Here is $f^{-1} \circ f$, which equals i_A:

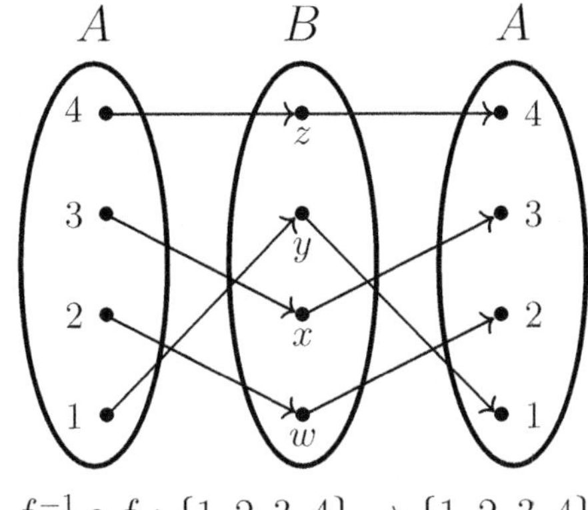

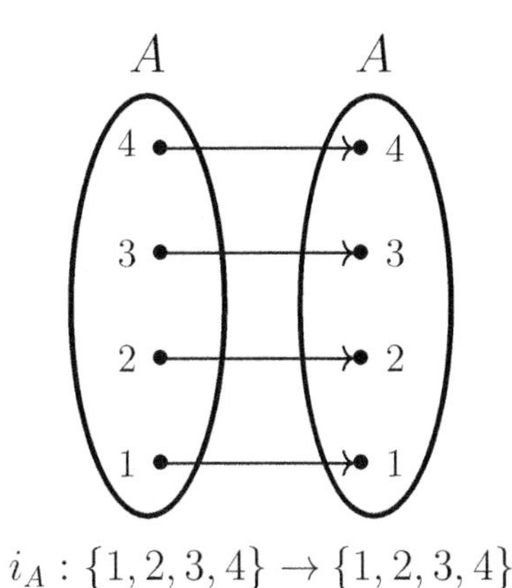

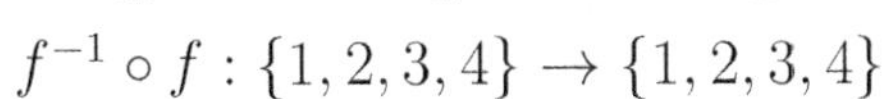

$f^{-1} \circ f : \{1, 2, 3, 4\} \to \{1, 2, 3, 4\}$ $\qquad$ $i_A : \{1, 2, 3, 4\} \to \{1, 2, 3, 4\}$

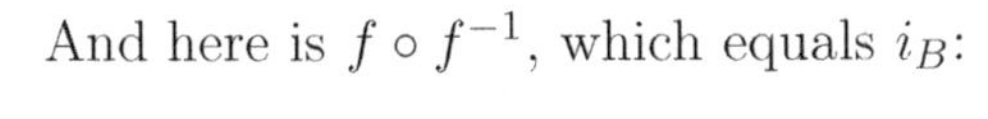

And here is $f \circ f^{-1}$, which equals i_B:

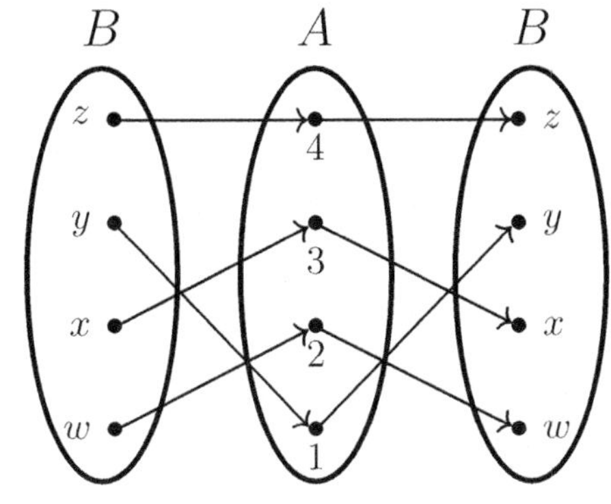

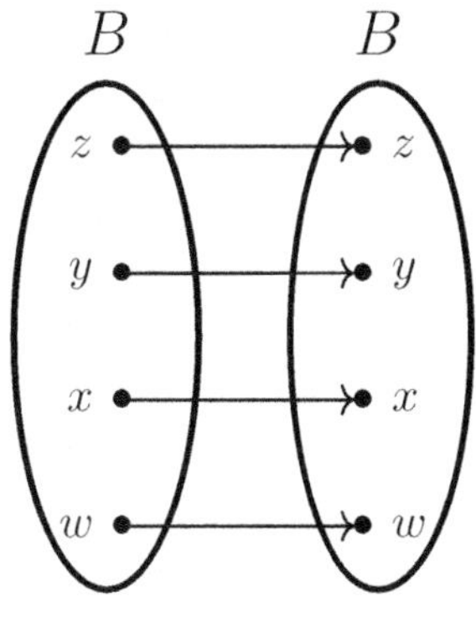

$f \circ f^{-1} : \{w, x, y, z\} \to \{w, x, y, z\}$ $\qquad$ $i_B : \{w, x, y, z\} \to \{w, x, y, z\}$

For example, if $f : \mathbb{R} \to \mathbb{R}$ where $f(x) = x + 1$, then $f^{-1} : \mathbb{R} \to \mathbb{R}$ is the function $f^{-1}(x) = x - 1$. To see this, simply note that

$$(f \circ f^{-1})(x) = f(f^{-1}(x)) = f(x - 1) = (x - 1) + 1 = x,$$

showing that $(f \circ f^{-1})(x) = x$, meaning that it is equal to the identity function ($i_{\mathbb{R}}(x) = x$) on $\mathbb{R}$. Likewise,

$$(f^{-1} \circ f)(x) = f^{-1}(f(x)) = f^{-1}(x + 1) = (x + 1) - 1 = x.$$

Here are a couple more examples.

- If $f : \mathbb{R}^+ \to \mathbb{R}^+$ where $f(x) = x^2$, then $f^{-1} : \mathbb{R}^+ \to \mathbb{R}^+$ is the function $f^{-1}(x) = \sqrt{x}$.

- If $f : [0, 1] \to [0, 2]$ where $f(x) = 2x$, then $f^{-1} : [0, 2] \to [0, 1]$ is the function $f^{-1}(x) = \frac{1}{2}x$.

And this is a great opportunity to mention a couple important functions — $\arctan(x)$ and $\ln(x)$ — which are *defined* as the inverses to other important functions.

- If $\tan : (-\pi/2, \pi/2) \to \mathbb{R}$ is the tangent function, then its inverse is defined to be $\arctan : \mathbb{R} \to (-\pi/2, \pi/2)$, and is called the arctangent function.[20]
- If $\exp : \mathbb{R} \to \mathbb{R}^+$ is the exponential function (that is, $\exp(x) = e^x$), then its inverse is defined to be $\ln : \mathbb{R}^+ \to \mathbb{R}$, and is called the natural logarithm.

Returning now to our small example from $A = \{1, 2, 3, 4\}$ to $B = \{w, x, y, z\}$, notice that the f we chose in our example was a bijection. This was no accident. In fact, *only* bijections have inverses. The reason for this can be seen in two parts: (1) why f must be an injection, and (2) why f must be a surjection. Let's start with the latter. Suppose that f were not a surjection; e.g., like this:

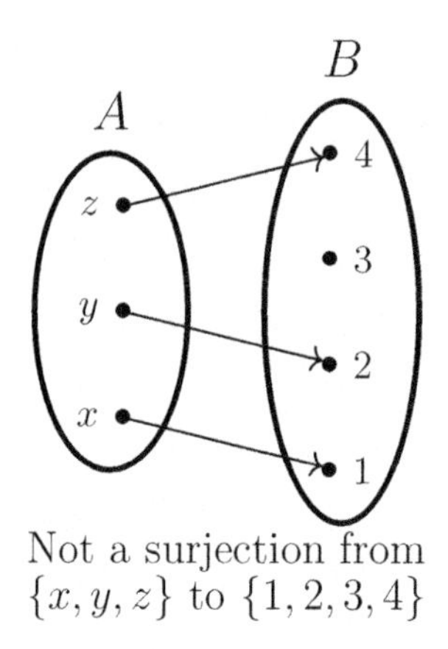

Not a surjection from $\{x, y, z\}$ to $\{1, 2, 3, 4\}$

Why does this function not have an inverse? Because where would f^{-1} send 3? A function has to send everything somewhere, so if the function f^{-1} did exist, it would need to know what to do with 3. But f is silent on this point. This is why f must be surjective in order for f^{-1} to exist.

Next, suppose that f were not an injection; e.g., like this:

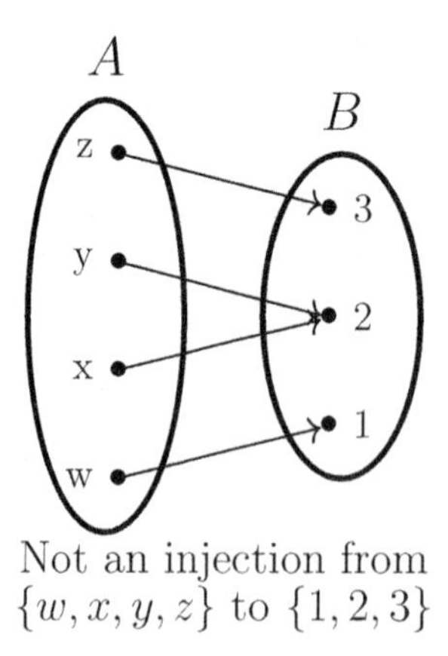

Not an injection from $\{w, x, y, z\}$ to $\{1, 2, 3\}$

[20]Note: Some people use $\tan^{-1}(x)$ instead of $\arctan(x)$ for the arctangent. This is probably due to some dope at Texas Instruments in the '70s who decided that it would be easier to squeeze $\tan^{-1}$ onto a calculator key, and consequently a million students get confused thinking $\tan^{-1}(x) = \frac{1}{\tan(x)}$.

Why does this function not have an inverse? Because where would f^{-1} send 2? A function has to send everything to only one destination, so if the function f^{-1} did exist, it would need to know what to do with 2. But f has two conflicting demands on this point. This is why f must be injective in order for f^{-1} to exist.

We have argued that f must be injective and must be surjective; that is, if f is invertible, then f is a bijection. Is the converse true? Is it true that *every* bijection is invertible? The answer is yes, by similar reasoning to the above. Bijections are functions which pair up each element in the domain with an element in the codomain, in such a way that every element in both sets is in a single pair. And what determines the pair? The function f does: a is paired with b if $f(a) = b$. This allows the inverse to pair up b with a, by having $f^{-1}(b) = a$. Perfect! This discussion is the idea behind the proof of the following theorem.

Theorem.

Theorem 8.17. A function $f : A \to B$ is invertible if and only if f is a bijection.

Proof. First, suppose that $f : A \to B$ is invertible. We will prove that f is both an injection and a surjection, which will prove that f is a bijection. To see that f is a surjection, choose any $b \in B$. We aim to find an $a \in A$ such that $f(a) = b$. To this end, let $a = f^{-1}(b)$, which exists and is in A because $f^{-1} : B \to A$. Now, simply observe that the definition of an invertible function (Definition 8.16) implies

$$f(a) = f(f^{-1}(b)) = b.$$

This proves that f is a surjection.

To see that f is an injection, let $a_1, a_2 \in A$ and assume $f(a_1) = f(a_2)$. Note that $f(a_1)$ (and hence $f(a_2)$, since they're equal) is an element of B due to the fact that $f : A \to B$. And so, since $f^{-1} : B \to A$, we may apply f^{-1} to both sides:

$$\begin{aligned} f(a_1) &= f(a_2) \\ f^{-1}(f(a_1)) &= f^{-1}(f(a_2)) \\ a_1 &= a_2, \end{aligned}$$

by the definition of the inverse. Thus, f is an injection. And since we already showed that f is a surjection, it must be a bijection. This concludes the forward direction of the theorem.

As for the backwards direction, assume that f is a bijection. For $b \in B$, we will now define $f^{-1}(b)$ like this:

$$f^{-1}(b) = a \quad \text{if} \quad f(a) = b.$$

That is, we are defining f^{-1} to act as we would want an inverse from B to A to act, without yet claiming that f^{-1} is a function. Our goal now is to demonstrate that this definition of f^{-1} satisfies the conditions to be a function, which would prove that f

is invertible. To do so, recall that to be a function there is an existence condition ($f^{-1}(b)$ must be equal to some $a \in A$) and a uniqueness condition ($f^{-1}(b)$ must be equal to only one $a \in A$). We will check these separately.

Existence: Let $b \in B$. Since f is surjective, there must be some $a \in A$ such that $f(a) = b$. Hence, by our definition of f^{-1}, we have $f^{-1}(b) = a$. We have shown that for every $b \in B$ there exists at least one $a \in A$ for which $f^{-1}(b) = a$, which concludes the existence portion of this argument.

Uniqueness: Suppose $f^{-1}(b) = a_1$ and $f^{-1}(b) = a_2$, for some $b \in B$ and $a_1, a_2 \in A$. By the definition of f^{-1}, this means that $f(a_1) = b$ and $f(a_2) = b$. But since f is injective, this means that $a_1 = a_2$. We have shown that $f^{-1}(b)$ cannot be equal to two different elements of A, which concludes the uniqueness portion of this argument.

Combined, these two parts show that $f^{-1} : B \to A$ is a function. Moreover, by the way we defined f^{-1}, we see that $f^{-1}(f(a)) = a$ for all $a \in A$, and $f(f^{-1}(b)) = b$ for all $b \in B$. Thus, by the definition of the inverse (Definition 8.16), f^{-1} is the inverse function of f, proving that f is invertible.

We have proved the forwards and backwards directions of Theorem 8.17, which completes its proof. □

8.5 Bonus Examples

Your first bonus example might be of particular interest to the computer scientists in the audience. There are some algorithms whose goal is to take in a file and compress it (i.e., reduce the number of bits needed to represent it, so that you can store it or send it more efficiently). The ideal such algorithms are called *lossless compression algorithms*; the "lossless" refers to the fact that once you shrink a file, you want to be able to invert this process later, to get back the original file without losing data. Furthermore, a *universal lossless compression algorithm* is one which can take in any file and compress it. This would of course be the gold standard of compression algorithms, but sadly it does not exist. Many types of files can certainly be compressed and then later inverted without losing data, but there is no *universal* lossless compression algorithm.

Proposition.

Proposition 8.18. There does not exist a universal lossless compression algorithm.

Proof. Assume for a contradiction that there does exist a universal lossless compression algorithm. Messages are sequences of bits (0 or 1), and a compression algorithm takes in a message of length n and outputs a message of a smaller length—length at most $n - 1$. Let A be the set of all messages of length at most n, and let B be the set of all messages of length at most $n - 1$. Then, by applying the lossless compression algorithm to every string in A, the algorithm can be viewed as a function $f : A \to B$.

Observe that $|A| > |B|$, as there are certainly more strings of length at most n than strings of length at most $n-1$. So, by the func-y pigeonhole principle (Theorem 8.10), f is not an injection. Then, by the definition of a bijection (Definition 8.7), this also means that f is not a bijection. And by Theorem 8.17, this means that f is not invertible. This is a contradiction: Since f is our compression function, being able to retrieve a compressed file x would be equivalent to asking for $f^{-1}(x)$. And so if f^{-1} does not exist, then such an algorithm is impossible. $\square$

Another way to think about this is that, because f is not injective, there must exist two different files, a and b, for which $f(a) = f(b)$. That is, two files which were compressed to the same smaller file. And it makes sense that it is impossible to undo this compression, because when presented with the file $f(a)$, it is impossible to tell if it was a or b before the compression.

If a compression algorithm only accepts files of a certain type—say, iPhone pictures—then there very well may be a way to compress them in an invertible way. What Proposition 8.18 shows is that compression algorithms have to be selective about what they accept, since we have proven they cannot be universal.

Practice With Specific Functions

Let's now do a couple concrete problems which practice working with specific functions.

Example 8.19. Prove that the function $f : (0, \infty) \to (0, 1)$ where $f(x) = \dfrac{1}{x+1}$ is a bijection.

Scratch Work. Being a bijection means you are an injection and a surjection. So a standard way to approach a problem like this is to demonstrate those two separately. How do we do that? Usually, we just use their definitions and the structural overview on Page 339. To show f is an injection, we assume that $f(x) = f(y)$ and do algebra to show that $x = y$. To show that f is a surjection, we pick an element $b \in (0, 1)$ and show that there is some $x \in (0, \infty)$ such that $f(x) = b$. Let's do some scratch work to determine which x we should use in our proof.

$$\begin{aligned}
f(x) &= b \\
\frac{1}{x+1} &= b \\
\frac{1}{b} &= x + 1 \\
\frac{1}{b} - 1 &= x
\end{aligned}$$

Looks like if we pick $x = \frac{1}{b} - 1$, then $f(x) = b$. The x we choose must be from our domain; is this x? Since $b \in (0, 1)$, notice that $\frac{1}{b} > 1$. So $\frac{1}{b} - 1 > 0$. So yes, everything seems ok there. Let's write out a proof! Starting with the injective part.

Proof. We will prove that f is injective and surjective.

Injective. Suppose $x, y \in (0, \infty)$ and $f(x) = f(y)$. That is,

$$\frac{1}{x+1} = \frac{1}{1+y}.$$

Simplifying,

$$\begin{aligned} y+1 &= x+1 \\ y &= x \\ x &= y. \end{aligned}$$

We have shown that if $f(x) = f(y)$, then $x = y$. Thus, f is injective.

Surjective. Suppose that $b \in (0, 1)$. We wish to find an x from the domain for which $f(x) = b$.

Let $x = \frac{1}{b} - 1$. Notice that $b \in (0, 1)$ implies the following: Since $b < 1$ and b is positive, we can divide both sides by b to get $1 < \frac{1}{b}$. Hence, $0 < \frac{1}{b} - 1$, which means that $0 < x$. Thus, we have demonstrated that $x \in (0, \infty)$, our function's domain.

Next, observe that

$$f(x) = f\left(\frac{1}{b} - 1\right) = \frac{1}{\left(\frac{1}{b} - 1\right) + 1} = \frac{1}{1/b} = b.$$

That is, for any $b \in (0, 1)$ we found an $x \in (0, \infty)$ for which $f(x) = b$. Thus, f is surjective.

Since f is both injective and surjective, f is bijective. □

Since f is a bijection, it is invertible. How do we find its inverse? Back in pre-calc you might have learned that to find the inverse of $y = \dfrac{1}{x+1}$, you should "switch x and y and solve." That is,

$$x = \frac{1}{y+1} \quad \Rightarrow \quad y + 1 = \frac{1}{x} \quad \Rightarrow \quad y = \frac{1}{x} - 1 \quad \Rightarrow \quad f^{-1}(x) = \frac{1}{x} - 1.$$

First, this probably looks familiar: It is the x we found in the surjective portion of the last proof! And if you spend 20 seconds staring at the above computation, and then another 15 seconds comparing it to the computation in last example's scratch work, you'll realize that this is no coincidence. But does the pre-calc strategy work in general? And if so, why?

Answer: If your function $f(x)$ has a simple-enough formula that you can set $y = f(x)$, switch the variables to get $x = f(y)$, and solve this equation for x, then yes, you will have found a formula for the inverse.[21]

[21] And if they don't have a simple formula, then at times we just give the inverse a name and roll with it, like with $\log(x)$ and $\arctan(x)$.

The inverse (when it exists) of a function f is the function that undoes it. If f sends x to y (meaning, $y = f(x)$), then f^{-1} sends that same y back to that same x. This suggests a slightly simpler way to find inverses, avoiding all the "switching x and y" business: if we started with $y = f(x)$ and simply solved for x, this would gives us the answer for f^{-1}, only it would be expressed as a function of y instead of a function of x. For example, the above computation would instead look like this:

$$y = \frac{1}{x+1} \quad \Rightarrow \quad x + 1 = \frac{1}{y} \quad \Rightarrow \quad x = \frac{1}{y} - 1 \quad \Rightarrow \quad f^{-1}(y) = \frac{1}{y} - 1.$$

We get the same inverse as before, just written in terms of a different variable.

Now, it seems like teachers decades ago decided that to have a function written in terms of y would be confusing to students, so they told students to first switch the 'x' and the 'y' so that in the end you get a function in terms of x. Perhaps this was an attempt to lower the blood pressures of a million anxious students. While I understand this, to me it seems silly, since (1) it is an extra and unnecessary step, and (2) writing the inverse in terms of y would help emphasize that f and f^{-1} have their domains and codomains switched. But alas, I don't have the power to change such things.[22]

Now, this is just discussion; we haven't proven any theorems about this. So if you use either of these approaches to find an inverse, consider it just scratch work. At the end you should still verify that your inverse is correct by checking it against the definition of an inverse. Below is an example where we do this.

Example 8.20. Find the inverse of $f : \mathbb{R} \to \mathbb{R}$ where $f(x) = 2x + 5$.

Scratch Work. Let $y = 2x + 5$. Solving for x,

$$y = 2x + 5 \quad \Rightarrow \quad y - 5 = 2x \quad \Rightarrow \quad x = \frac{y-5}{2}.$$

So $f^{-1}(y) = \dfrac{y-5}{2}$ should be the inverse (written in terms of the variable y).

Proof. We claim that $f^{-1} : \mathbb{R} \to \mathbb{R}$ where $f^{-1}(x) = \dfrac{x-5}{2}$ is the inverse function of f. First, note that $f^{-1}(x) = \dfrac{x-5}{2}$ is a function. To prove that it is the inverse of f, we will check whether it satisfies the inverse definition (Definition 8.16). Note that

$$f(f^{-1}(x)) = f\left(\frac{x-5}{2}\right) = 2\left(\frac{x-5}{2}\right) + 5 = (x-5) + 5 = x$$

and

$$f^{-1}(f(x)) = f^{-1}(2x+5) = \frac{(2x+5)-5}{2} = \frac{2x}{2} = x,$$

for any $x \in \mathbb{R}$. Thus, we have shown that $f \circ f^{-1}$ and $f^{-1} \circ f$ are both the identity function on $\mathbb{R}$. Therefore, by the definition of the inverse (Definition 8.16), we have proven that $f^{-1}(x) = \dfrac{x-5}{2}$ is the inverse of f. □

[22] Future Pre-Calc Teachers of the World: You *do* have the power!

— Chapter 8 Pro-Tips —

- Injections, surjections and bijections are worth spending extra time to learn well. They will be important in your later courses.

- When working with function compositions, it can be especially useful to draw pictures. Even just a few blobs and arrows, like we did in this chapter, can help. (More broadly, if you can draw any picture in your scratch work, do it!)

- To reiterate a point that we discussed earlier, in the airport metaphor: Notice that in our definition of function composition (Definition 8.11) we had functions g and f where $g : A \to B$, and $f : B \to C$. Notice that we don't really *need* the codomain of g to equal the domain of f. If we had $g : A \to B$ and $f : D \to C$ where $B \subseteq D$, that would be enough. As long as $g(a)$ is a part of f's domain, then $f(g(a))$ will make sense. Sometimes the definition of a function composition is written in this other way, since it is more general.

- In math, it doesn't *really* matter what you choose as a variable name. You can say "let $x \in \mathbb{N}$" or "$m : \mathbb{R} \to \mathbb{R}$ is a function given by $m(f) = f^2$." You could say, "let a be a set and $A \in a$." You *could*. A computer wouldn't care. But despite mathematicians working in the realm of pure truths and deep ideas... we are humans too, darn it, and we like what we like.

  Mathematicians have come to prefer using x and y to represent a real number. We like to use k, m and n as representing integers. We like to use p and q as prime numbers. We like z for a complex number. We like ε for a small positive number. We use capital letters for our sets and lower case for our elements, and while we sometimes dress ourselves from head to toe in drab, mismatched clothes,[23] we would never dare mismatch our elements and our sets: we always let $a \in A$ and $b \in B$. This is who we are.

- In this chapter, we emphasized the importance of *existence* and *uniqueness* when discussing functions. These ideas appear in tandem a lot in mathematics. If you have taken a class in differential equations, you may have seen a theorem that guarantees the existence and uniqueness of a solution to an initial value problem. If you have taken linear algebra, you may have seen a theorem that specifies when a system of linear equations has a unique solution (which is not just a uniqueness condition but an existence condition, too, as it also guarantees

[23]First day of work at my tenure-track job was a simple pre-semester department meeting. I remember wondering how nicely I should dress. A button-up rather than a t-shirt — that seemed obvious... It's nearly 100° out, but shorts seemed unprofessional... should the pants be nice or are jeans ok? What about sneakers? I erred on the side of caution, drove to work, showed up to the meeting... and there were shorts, t-shirts and Hawaiian shirts everywhere. One prof wasn't even wearing *shoes*. And when all of these attires continued into the semester, I realized that as a mathematician there are some things you can get away with.

Once you start giving talks at various departments, especially ones whose colloquia are open to the public, you will start to wonder whether the ragged fella in the back is a disheveled emeritus professor, a homeless person, or the guy who will be taking you out to Thai food later that night. It can, on occasion, be genuinely hard to tell.

that the solution exists). In many cases, the number of solutions was either 0, 1 or ∞, and the theorem specified when the answer was 1.

As you take more math, you may come across many more theorems of this type. Now, uniqueness can be thought of as meaning, "if a and b are both instances of what we are discussing, then $a = b$." In some areas of math, the "=" sign is generalized. In the next chapter, for example, we will discuss what are called *equivalence relations*, in which we do just that. You may discover versions of "existence and uniqueness" where 'existence' has the same meaning as always, but 'uniqueness' is generalized to allow a collection of different solutions to be regarded as the same, and hence satisfy the uniqueness criterion.

- Suppose you are trying to prove a result, or you are toying around with some concept. If you're able to apply some idea or process, it is worth checking what happens when you apply that idea again. And again, and again. The derivative, for example, is useful, but when we apply it additional times we attain the second and third derivatives, which provide even more information. In the *Introduction to Cardinality*, following this chapter, we will use an idea to find a subset of $\mathbb{N}$ which has the same size as $\mathbb{N}$. But we won't stop there, by applying similar ideas again and again we will discover much more. These are higher-level concepts, but often even a simple idea can be really powerful when iterated.

  Another version of this idea is one applies some idea to *itself.* We saw an example of this in this chapter's Bonus Examples, with the halting problem. Applying our halting function to itself was the key.

- If you have proved a theorem, it is worth taking some time to see if you can find a better proof of that same theorem. It's like the standard advice when playing chess: "Once you've found a good move, start looking for a better one."

- The details are really important to ensure a proof is correct, but when you are reading proofs it is wise to be primarily interested in the big picture. What is the main idea of this proof? What is the general strategy? What previous theorem is used to make it work? When you walk away from a proof, if you can identify the big picture idea, then that typically means that you understood the heart of the proof. Also, it is much easier to remember a main idea than some small details, so focusing on the big picture is also a good way to retain the math that you learn.

- You have now completed the chapter on functions, which contained some of the most abstract ideas in all 8 chapters thus far. That said, the next—and final—chapter of this book is going to be even more abstract. It is on a topic called *relations*. As always in math, it is a good idea to take things slow and read over paragraphs many times. You should also be doing plenty of exercises, challenging yourself to try to figure out parts of a proof before you finish reading it, discussing the math with others, etc. Those practices will be even more important when reading abstract math, and I would encourage you to take that active-learning mindset with you as you begin the final chapter.

— Exercises —

Exercise 8.1. For each of the diagrams below, determine whether the diagram represents a function. If it does, determine whether the function is injective, surjective, bijective, or none of these.

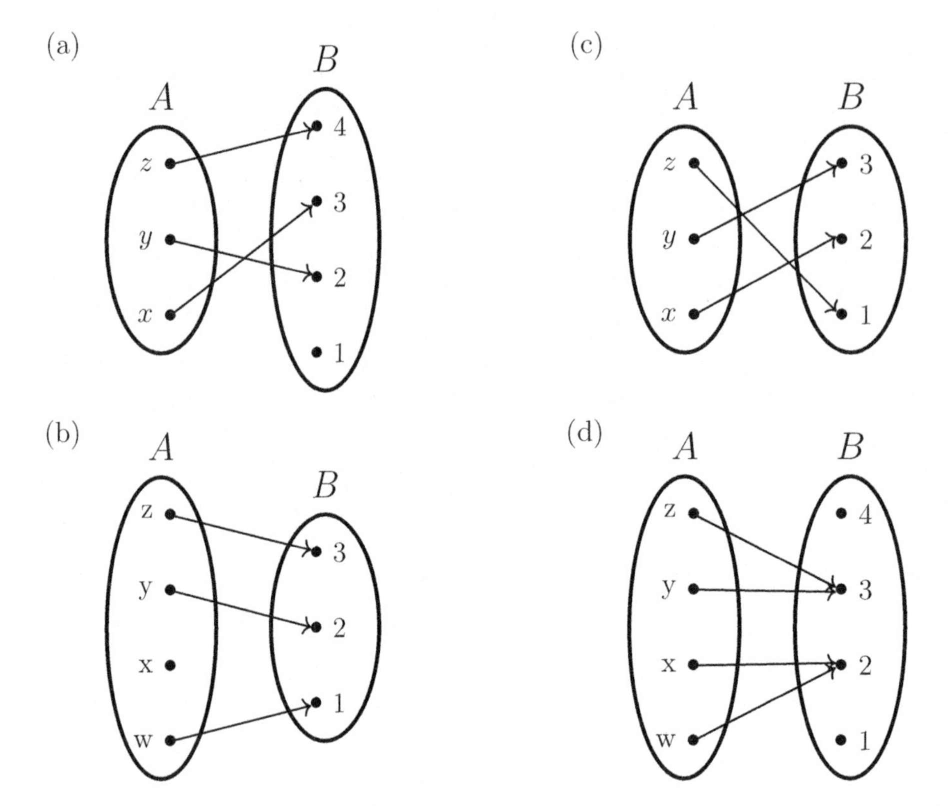

Exercise 8.2. Give two reasons why "$f : \mathbb{R} \to \mathbb{R}$ where $f(x) = \pm\sqrt{x}$" is not a function.

Exercise 8.3. In pre-calculus you may have written things like this:

$$\frac{x^2+x-6}{x+3} = \frac{(x+3)(x-2)}{x+3} = \frac{\cancel{(x+3)}(x-2)}{\cancel{x+3}} = x-2.$$

This seems to suggest that $f(x) = \frac{x^2+x-6}{x+3}$ and $g(x) = x - 2$ are the same function. Explain why they are not.

Exercise 8.4.

(a) Define $f : \mathbb{N} \to \mathbb{Z}$ where $f(n) = n - 5$. Determine the range of f.

(b) Define $g : \mathbb{R} \to \mathbb{R}$ where $g(x) = \lfloor x \rfloor$; this is the floor function from Example 8.2, just with a new codomain. Determine the range of g.

(c) Define $h : \mathbb{R} \setminus \{0\} \to \mathbb{R}$ where $h(x) = \frac{1}{x^2}$. Determine the range of h.

Exercise 8.5. In words, describe the range of the function $f : \mathbb{N} \times \mathbb{N} \to \mathbb{N}$ where $f(m, n) = 2^m 3^n$.

Exercise 8.6. Consider "$f : \mathbb{Z} \to \mathbb{Z}$ where $f(x) = y$ if $x \equiv y \pmod 6$." Is f a function?

Exercise 8.7. Determine whether each of the following is an injection, surjection, bijection or none of these. Prove your answers.

(a) $f : \mathbb{R} \to \mathbb{R}$ where $f(x) = 2x + 7$

(b) $g : \mathbb{R} \to \mathbb{Z}$ where $g(x) = \lfloor x \rfloor$

(c) $h : \mathbb{R} \to \mathbb{R}$ where $h(x) = \frac{1}{x^2+1}$

(d) $j : \mathbb{R} \to \mathbb{R}$ where $j(x) = x^2$

(e) $k : \mathbb{N} \to \mathbb{N}$ where $k(x) = x^2$

(f) $m : \mathbb{R} \setminus \{-1\} \to \mathbb{R}$ where $m(x) = \frac{2x}{x+1}$

(g) $n : \mathbb{Z} \to \mathbb{N}$ where $n(x) = x^2 - 2x + 2$

(h) $p : \mathbb{N} \to \mathbb{N}$ where $p(x) = |x|$

(i) $q : (-\infty, -10) \to (-\infty, 0)$ where $q(x) = -|x + 4|$

(j) $r : (-\infty, 0) \to (-\infty, 0)$ where $r(x) = -|x + 4|$

(k) $s : \mathbb{N} \to \mathbb{N} \times \mathbb{N}$ where $s(x) = (x, x)$

(l) $t : \mathbb{Z} \times \mathbb{Z} \to \mathbb{Q}$ where $t(m, n) = \frac{m}{|n|+1}$

(m) $u : \mathbb{Z} \times \mathbb{Z} \to \mathbb{Z} \times \mathbb{Z}$ where $u(m, n) = (m + n, 2m + n)$

(n) $v : \mathbb{Z} \times \mathbb{Z} \to \mathbb{Z}$ where $v(m, n) = 3m - 4n$

Exercise 8.8. Let A be a set. Consider the function $f : A \to \mathbb{R}$ where $f(x) = 7$. Determine conditions on A that cause f to be injective.

Exercise 8.9. Let $A = \mathcal{P}(\mathcal{P}(\mathbb{N}))$ and $B = \mathbb{N}$. Define $f : A \to B$ to be the function where $f(S) = \displaystyle\bigcup_{x \in S} x$.

(a) Find $f(\{\{1, 2\}, \{3, 4\}\})$.

(b) Is f injective? Surjective? Bijective?

Exercise 8.10. Let A and B be finite sets for which $|A| = |B|$, and suppose $f : A \to B$. Prove that f is injective if and only if f is surjective.

Exercise 8.11. To convert from F degrees Fahrenheit to C degrees Celsius, one can use the formula

$$F = \frac{9}{5}C + 32.$$

Determine a formula to convert from Celsius to Fahrenheit, and show that these two formulas are inverse functions of each other. (You may assume the domain and codomain are both $\mathbb{R}$.)

Exercise 8.12. Determine whether each of the following is invertible. If it is, write down its inverse and show that it satisfies Definition 8.16. If it is not, then that means it is either not injective, not surjective, or both. If it is not injective, give two distinct values x and y from the function's domain for which $f(x) = f(y)$. If it is not surjective, give an element of the codomain which is not hit by the function. You do not need to prove that your answers are correct.

(a) $k : \mathbb{R} \to \mathbb{R}$ where $k(x) = \frac{x}{5} + 2$

(b) $f : \mathbb{R} \to \mathbb{R}$ where $f(x) = \sqrt[3]{x}$

(c) $g : \mathbb{R}^+ \to \mathbb{R}$ where $g(x) = x^4$, where $\mathbb{R}^+$ is the set of nonnegative real numbers

(d) $h : [0, \pi] \to [-1, 1]$ where $h(x) = \cos(x)$

(e) $j : [0, \pi] \to [-2, 2]$ where $j(x) = \cos(x)$

(f) $k : \mathbb{N} \to \mathbb{Z}$ where $k(x) = \dfrac{(-1)^n(2n-1)+1}{4}$

Exercise 8.13. Let $A = \mathbb{R} \setminus \{2\}$ and let $f(x) = \dfrac{3x}{x-2}$.

(a) Determine a set B for which $f : A \to B$ is a bijective function.

(b) For the set B from part (a), find the inverse of $f : A \to B$.

Exercise 8.14. Consider the function $f : \mathbb{R} \setminus \{1\} \to \mathbb{R} \setminus \{3\}$ where $f(x) = \dfrac{3x}{x-1}$. Prove that $f^{-1} : \mathbb{R} \setminus \{3\} \to \mathbb{R} \setminus \{1\}$ where $f^{-1}(x) = \dfrac{x}{x-3}$ is indeed the inverse of f by showing that it satisfies the definition of an inverse.

Exercise 8.15. Consider the functions $f : \mathbb{Z} \times \mathbb{Z} \to \mathbb{Z} \times \mathbb{Z}$ where $f(m, n) = (5m - 3n, 2n)$, and $g : \mathbb{Z} \times \mathbb{Z} \to \mathbb{Z} \times \mathbb{Z}$ where $g(m, n) = (3m + 2n, 4n - m)$. Find formulas for $f \circ g$ and $g \circ f$.

Exercise 8.16. Let $f : \mathbb{Z} \to \mathbb{Z}$ where $f(x) = x^2 + x$, and let $g : \mathbb{Z} \to \mathbb{Z}$ where $g(x) = 4x + 3$.

(a) Write down $(f \circ g)(x)$ and $(g \circ f)(x)$ as expanded polynomials.

(b) Determine the ranges of $(f \circ g)(x)$ and $(g \circ f)(x)$.

Exercise 8.17. Give an example of functions f and g such that $f \circ g$ is injective and g is injective, but f is not injective. Write down f, g and $f \circ g$, but you do not need to prove that your example works.

Exercise 8.18. Give an example of functions f and g such that $f \circ g$ is surjective and f is surjective, but g is not surjective. Write down f, g and $f \circ g$, but you do not need to prove that your example works.

Exercise 8.19. Consider the functions $f : \mathbb{R} \to \mathbb{R}$ where $f(x) = 2x + 1$ and $g : \mathbb{R} \to \mathbb{R}$ where $g(x) = 3x - 2$.

(a) Find $(f \circ g)(x)$.

(b) Find $(f \circ g)^{-1}(x)$.

(c) Find $f^{-1}(x)$.

(d) Find $g^{-1}(x)$.

(e) Find $(g^{-1} \circ f^{-1})(x)$.

(f) What do you notice? Prove that your observation always holds.

Exercise 8.20. For functions $f : \mathbb{R} \to \mathbb{R}$, $g : \mathbb{R} \to \mathbb{R}$ and $h : \mathbb{R} \to \mathbb{R}$, prove or disprove each of the following conjectures.

(a) **Conjecture 1:** $(f + g) \circ h = (f \circ h) + (g \circ h)$

(b) **Conjecture 2:** $h \circ (f + g) = (h \circ f) + (h \circ g)$

For this problem, recall that "$f + g$" is the function $(f + g)(x) = f(x) + g(x)$.

Exercise 8.21. Write down your own definition for what you think it should mean to say a function $f : \mathbb{R} \to \mathbb{R}$ is *continuous*.

Exercise 8.22. Assume $f : \mathbb{R} \to \mathbb{R}$ is a function. We say that f is *increasing* if $x > y$ implies that $f(x) > f(y)$, for any $x, y \in \mathbb{R}$. Below are two conjectures. For each, either prove they are true or find a counterexample.

(a) **Conjecture 1:** If $f : \mathbb{R} \to \mathbb{R}$ is an increasing function, then f is injective.

(b) **Conjecture 2:** If $f : \mathbb{R} \to \mathbb{R}$ is an increasing function, then f is surjective.

Exercise 8.23. How many functions are there from $\{1, 2, 3\}$ to $\{1, 2, 3\}$? For $n \in \mathbb{N}$, how many functions are there from $\{1, 2, \dots, n\}$ to $\{1, 2, \dots, n\}$? How many bijections are there from $\{1, 2, \dots, n\}$ to $\{1, 2, \dots, n\}$?

Definition. Let A be a set. A *permutation* of A is a bijection $f : A \to A$.

This definition will be used in the next exercise.

Exercise 8.24. In a previous math class you may have learned that a permutation is a rearrangement of a collection of objects. For example, "1 a 🦇 ☺" is a permutation of "☺ 🦇 1 a". Explain why the definition in the box above jives with your rearrangement intuition.

Definition. Let $f : A \to B$ be a function, and assume $X \subseteq A$ and $Y \subseteq B$. The *image* of X is

$$f(X) = \{y \in B : y = f(x) \text{ for some } x \in X\},$$

and the *inverse image* of Y is

$$f^{-1}(Y) = \{x \in A : f(x) \in Y\}.$$

This definition will be used in the next four exercises.

Exercise 8.25. Let $A = \{-4, -3, -2, -1, 0, 1, 2, 3, 4\}$, $B = \{0, 1, 4, 6, 9, 12, 16, 25\}$ and $f : A \to B$ be the function $f(x) = x^2$. Determine the following.

(a) $f(\{1, 2, 3\})$

(b) $f(\{-4, 4\})$

(c) $f(\{-3, -1, 0, 2, 4\})$

(d) $f^{-1}(\{0, 9\})$

(e) $f^{-1}(\{6\})$

(f) $f^{-1}(B)$

Exercise 8.26. Let $f : \mathbb{R} \setminus \{0\} \to \mathbb{R}$ be the function $f(x) = \frac{1}{x}$. Determine the following.

(a) $f((0, 3])$

(b) $f((-2, -1) \cup [3, 4])$

(c) $f^{-1}(\{2\})$

(d) $f^{-1}([-1, 1])$

(e) $f(f^{-1}(\mathbb{R}))$

(f) $f^{-1}(f(\mathbb{R} \setminus \{0\}))$

Exercise 8.27. Let $f : A \to B$ be a function, and assume $C, D \subseteq A$ and $E, F \subseteq B$. In parts (a) to (d), prove the following relationships between sets.

(a) $f(C \cap D) \subseteq f(C) \cap f(D)$

(b) $f(C \cup D) = f(C) \cup f(D)$

(c) $f^{-1}(E \cap F) = f^{-1}(E) \cap f^{-1}(E)$

(d) $f^{-1}(E \cup F) = f^{-1}(F) \cup f^{-1}(F)$

(e) Prove that if f is injective, then $f(C) \cap f(D) = f(C \cap D)$.

Exercise 8.28. Let $f : A \to B$ be a function, and assume $C, D \subseteq A$. Provide an example showing that

$$f(C \cap D) = f(C) \cap f(D)$$

might be false.

Exercise 8.29. Suppose A is a set and $f : A \to \emptyset$ is a function. Explain why f is a bijection.

Exercise 8.30. Let $f : A \to B$ be a function. For each of the following conjectures, prove it or provide a counterexample.

(a) **Conjecture 1:** If $X \subseteq A$, then $X = f^{-1}(f(X))$.

(b) **Conjecture 2:** If $X \subseteq A$, then $X \subseteq f^{-1}(f(X))$.

(c) **Conjecture 3:** If $Y \subseteq B$, then $Y = f(f^{-1}(Y))$.

(d) **Conjecture 4:** If $Y \subseteq B$, then $Y \subseteq f(f^{-1}(Y))$.

Exercise 8.31. Read the *Introduction to Cardinality* following this chapter. Then, complete the following.

(a) Prove that there are uncountably many irrational numbers.

(b) Prove that $|\mathbb{N}|$ is the smallest infinity. That is, prove that if $A \subseteq \mathbb{N}$, then either A has finitely many elements or $|A| = |\mathbb{N}|$.

— Open Questions —

If a and b are rational numbers, do you think $a+b$ must be a rational number? What about ab? What about a^b? Think about each before reading on...

...If a and b are rational numbers, then $a+b$ and ab are both rational numbers. Indeed, if $a = \frac{m}{n}$ and $b = \frac{s}{t}$, where $m, n, s, t \in \mathbb{Z}$, then

$$a + b = \frac{m}{n} + \frac{s}{t} = \frac{mt}{nt} + \frac{sn}{tn} = \frac{mt + sn}{nt} \qquad \text{and} \qquad ab = \frac{m}{n}\frac{s}{t} = \frac{ms}{nt}$$

are both rational numbers. However, it *is* possible that a and b are rational, but a^b is irrational. For example, if $a = 2$ and $b = 1/2$, then $a^b = \sqrt{2}$, which we showed in Theorem 7.6 to be irrational.

Next, what if a and b are *irrational* numbers? Do you think $a+b$ must be irrational? What about ab? What about a^b? Think about each before reading on...

...Once again, our counterexamples rely on the fact that $\sqrt{2}$ is irrational. First, note that this implies that $-\sqrt{2}$ is irrational. Next, observe that

$$\sqrt{2} + (-\sqrt{2}) = 0 \qquad \text{and} \qquad \sqrt{2} \cdot \sqrt{2} = 2$$

are both rational. This shows that if a and b are irrational, it is possible that $a+b$ is rational, and it is possible that ab is rational.

What about a^b? Can it be rational? It can! And we can again use $\sqrt{2}$ to give an example. This time, though, it's a bit more complicated. Consider the number

$$\sqrt{2}^{\sqrt{2}}.$$

Is this number rational or irrational? Let's consider these two possibilities separately.

Case 1: Suppose $\sqrt{2}^{\sqrt{2}}$ is rational. Thus, we have discovered two irrational numbers, $a = \sqrt{2}$ and $b = \sqrt{2}$, for which a^b is rational. So we have an example!

Case 2: Suppose $\sqrt{2}^{\sqrt{2}}$ is irrational. Thus, if we let $a = \sqrt{2}^{\sqrt{2}}$ and $b = \sqrt{2}$, then both of these are irrational numbers and using rules of exponents,

$$a^b = \left(\sqrt{2}^{\sqrt{2}}\right)^{\sqrt{2}} = \sqrt{2}^{\sqrt{2}\cdot\sqrt{2}} = \sqrt{2}^2 = 2,$$

which is rational. Thus, we have again found an example![24]

The next most famous irrational numbers are π and e. Are combinations of these numbers rational or irrational? Here are some open questions:

Open Questions. Are $\pi + e$, $\pi - e$, πe, π^e, $\pi^{\sqrt{2}}$, e^e and π^π rational or irrational?[25]

[24]It turns out that $\sqrt{2}^{\sqrt{2}}$ is irrational, so it's Case 2 that gives us our irrational^{irrational} = rational example, but that's much harder to prove, and our proof goes through just fine without caring whether it's Case 1 or Case 2 that does it.

[25]Fun fact: It has been proven that $\pi + e$ and πe cannot both be rational. But it is unknown whether they are both irrational, or just one of them is.

Introduction to Cardinality

Bijections and Cardinality

My research is in a field of math called *combinatorics.* One central problem in combinatorics is to count sets of things. The scientific philosopher Ernst Mach went as far as to say "Mathematics may be defined as the economy of counting. There is no problem in the whole of mathematics which cannot be solved by direct counting." It's a beautiful thought,[26] but even I would not go *quite* that far; nevertheless, I do think that the best solutions in math are those that use counting.

Recall that the number of elements in a set A is called the *cardinality* of that set. For example,

$$|\{a, b, c\}| = 3,$$

and

$$|\{1, 4, 9, 16, 25, \dots, 100\}| = 10,$$

and

$$|\mathbb{Z}| = \infty.$$

While the cardinality of finite sets seems simple enough, you might agree that when discussing sets of infinite size, the idea of size seems somewhat murky. Are we ok saying that $\mathbb{N}$ and $\mathbb{Z}$ and $\mathbb{R}$ all have the same size, even though $\mathbb{Z}$ contains all of $\mathbb{N}$ and more, and $\mathbb{R}$ seems vastly larger than both? We need a precise characterization of what we mean when we say that two sets have the same size.

What mathematicians settled on is this: two sets have the same size if there is a way to pair up the elements between the two sets. And if this language of "pairing up" elements sounds familiar, it is because in Chapter 8 we drew lots of diagrams arguing that a *bijection* does precisely that! So what does it mean to say that two sets have the same size? It means there is a bijection between them. This is known as *the bijection principle.*

[26] And it's at least semi-faintly-plausible. If he had said, "There is no problem in the whole of mathematics which cannot be solved by the quadratic formula," now *that* would have been quite the hot take.

Principle.

Principle 8.21 (*The bijection principle*). Two sets have the same size if and only if there is a bijection between them.

Example 8.22. One reason that the sets $\{1, 2, 3\}$ and $\{a, b, c\}$ have the same size is that the elements can be paired up like this:

$$1 \leftrightarrow b \qquad 2 \leftrightarrow a \qquad 3 \leftrightarrow c$$

And one reason that $\{x, y, z\}$ and $\{m, a, t, h\}$ do not have the same size is that the elements cannot be paired up. Whenever you try, one element from the second set won't get a pair. In particular, by Theorem 8.10, there do not exist any surjections $f : \{x, y, z\} \to \{m, a, t, h\}$, and hence there cannot exist a bijection between these sets. □

The really cool thing about the bijection principle is that this definition of the size of a set even applies to infinite sets. And that implies some truly fascinating things.

Counting Infinities

The ability to "pair up" elements between two sets is what it means for them to have the same size — this is perfectly intuitive for finite sets with nothing too counterintuitive resulting, but with infinite sets... well, some pretty neat stuff pops out. Indeed, the pluralization in this section's title was your first sign of the miracles to come.

Hilbert's Hotel

> "No other question has ever moved so profoundly the spirit of man; no other idea has so fruitfully stimulated his intellect; yet no other concept stands in greater need of clarification, than that of the infinite."
>
> – David Hilbert

We begin by talking about the set of problems related to the so-called Hilbert's Hotel. Assume that there is a hotel, called Hilbert's Hotel, which has infinitely many rooms in a row.

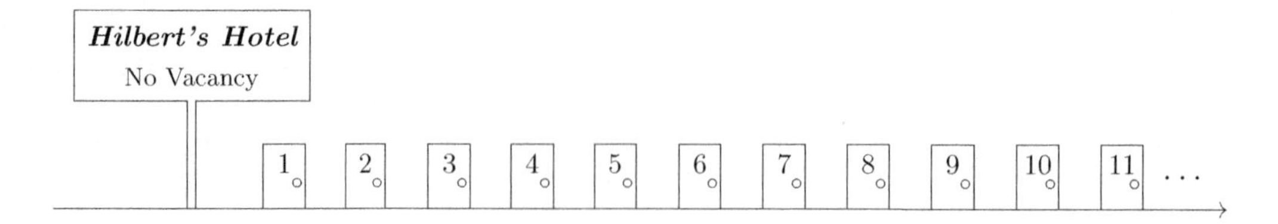

- Assume every room has someone in it, and so the "No Vacancy" sign has been turned on. With most hotels, this would mean that if someone else arrives at the hotel, they will not be given a room. But this isn't the case with Hibert's Hotel. If, for $n \in \mathbb{N}$, the patron in room n moves to room $n+1$, then nobody is left without a room and suddenly room 1 is completely open! So the new customer can go to room 1. We created a room out of nothing![27]

- Now imagine 2 people arrived to the hotel. Can we accommodate them? Certainly! This time, just have everyone move from room n to room $n+2$. This leaves rooms 1 and 2 open to the newcomers, and we are again good-to-go.

- What if, however, we have infinitely many people lined up wanting a room. Can we accommodate *all* of them? Yes! We still can! Just have the person in room n move to room $2n$. Then all of the odd-numbered rooms are vacant and the infinite line of people can take these rooms.[28]

The first point of this exercise is to simply realize that weird stuff can happen when dealing with the infinite. The second point, though, is to realize that each time the people switched rooms, those same exact people got new rooms. So in the first example when they each just moved one room down, that should mean that there are just as many rooms from 1 to ∞ as there are from 2 to ∞... And likewise for the others.

Indeed, with this in mind, let's talk about sizes of specific sets. But first, a ditty:

∞ bottles of beer on the wall,
∞ bottles of beer.
Take one down, pass it around,
∞ bottles of beer on the wall.

(repeat)

Specific Sets

Example 8.23. There are the same number of natural numbers as there are natural numbers larger than 1 (that is, $|\mathbb{N}| = |\{2,3,4,\dots\}|$). What's the bijection that shows this? Let

$$f : \mathbb{N} \to \{2,3,4,\dots\} \qquad \text{where} \qquad f(n) = n+1.$$

In other (non-)words, this is the pairing

$$1 \leftrightarrow 2 \qquad 2 \leftrightarrow 3 \qquad 3 \leftrightarrow 4 \qquad 4 \leftrightarrow 5 \; \dots$$ □

[27]Make sure you take a moment to appreciate how remarkably, wonderfully weird this is.
[28]Make sure you take a moment to appreciate how remarkably, wonderfully weird this is.

The Moral. Two sets can have the same size even though one is a *proper* subset of the other.[29]

Example 8.24. There are the same number of natural numbers as even natural numbers (that is, $|\mathbb{N}| = |\{2, 4, 6, 8, \dots\}|$). What's the bijection that shows this? Let

$$f : \mathbb{N} \to \{2, 4, 6, 8, \dots\} \qquad \text{where} \qquad f(n) = 2n.$$

In other (non-)words, this is the pairing

$$1 \leftrightarrow 2 \qquad 2 \leftrightarrow 4 \qquad 3 \leftrightarrow 6 \qquad 4 \leftrightarrow 8 \ \dots$$ □

The Moral. Two sets can have the same size even though one is a proper subset of the other and the larger one has *infinitely many more elements* than the smaller one.[30]

In a similar way, one can prove that $|\mathbb{N}| = |\mathbb{Z}|$. Indeed, a bijection $f : \mathbb{N} \to \mathbb{Z}$ can be given by following this pattern:

$$\begin{aligned} f(1) &= 0 \\ f(2) &= 1 \\ f(3) &= -1 \\ f(4) &= 2 \\ f(5) &= -2 \\ f(6) &= 3 \\ &\vdots \end{aligned}$$

Intuitively, this is what our bijection is doing:

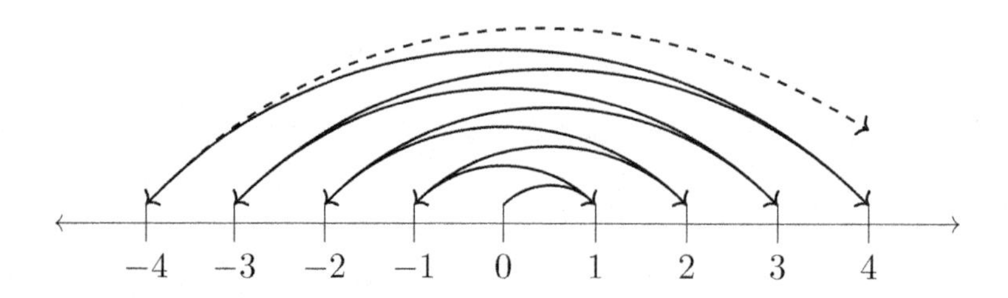

One way to write such a function is this:

$$f : \mathbb{N} \to \mathbb{Z} \qquad \text{where} \qquad f(n) = \begin{cases} n/2 & \text{if } n \text{ is even;} \\ -(n-1)/2 & \text{if } n \text{ is odd.} \end{cases}$$

[29]Make sure you take a moment to appreciate how remarkably, wonderfully weird this is.
[30]Make sure you take a moment to appreciate how remarkably, wonderfully weird this is.

In fact, one can even prove the remarkable fact that $|\mathbb{Z}| = |\mathbb{Q}|$. We won't discuss the details, but just as we winded our way through the integers in the previous diagram, you can do likewise with the rational numbers. Here is the diagram which accompanies (the positive portion of) that argument:[31]

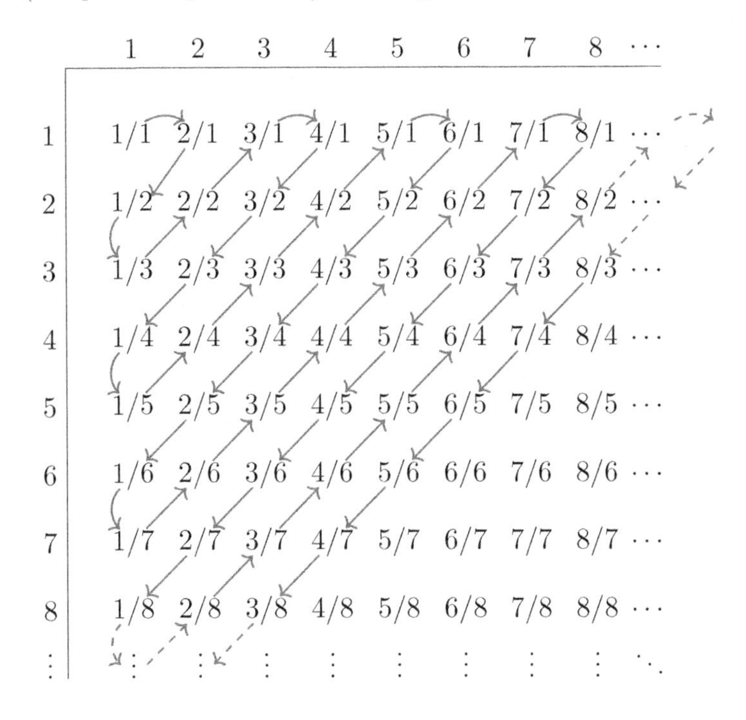

With this, it is the case that $|\mathbb{N}| = |\mathbb{Z}| = |\mathbb{Q}|$. Now, at this point you might be tempted to predict that the reason all these sets have the same size is that they all have infinitely many elements, and maybe all infinities are the same and that's all there is to it... But amazingly that's not actually the case, as the next result states.

Theorem.

Theorem 8.25. There are more real numbers than natural numbers.

This implies that some infinities are bigger than others.[32]

Proof. Since $\mathbb{N} \subseteq \mathbb{R}$, clearly $|\mathbb{N}| \leq |\mathbb{R}|$. To show that they are not equal, we must prove that there is no bijection between $\mathbb{R}$ and $\mathbb{N}$. Let's again use the "pairing up" idea. We will prove it by contradiction. In fact, we will prove the stronger statement that there are more real numbers in $(0, 1)$ than there are natural numbers. (This of course would prove the larger statement since then we could say $|\mathbb{R}| \geq |(0, 1)| > |\mathbb{N}|$.)

[31] An explicit bijection $f : \mathbb{N} \times \mathbb{N} \to \mathbb{N}$ is given by $f(m, n) = \dfrac{(m + n - 1)(m + n - 2)}{2} + m$. This can also be written using a binomial coefficient: $f(m, n) = \binom{m+n-1}{2} + m$.

[32] An infinite set of size $|\mathbb{N}|$ is said to be *countable*. Infinite sets of size larger than $|\mathbb{N}|$ are said to be *uncountable*. Thus, $\mathbb{N}$, $\mathbb{Z}$ and $\mathbb{Q}$ are countable, while $\mathbb{R}$ is uncountable.

Assume for a contradiction that there does exist some way to pair up the naturals with the reals in $(0, 1)$. Writing these reals in decimal notation, assume the pairing is this:

$$\begin{array}{l}
1 \leftrightarrow 0.a_{11}\ a_{12}\ a_{13}\ a_{14}\ a_{15}\ a_{16}\ a_{17}\ a_{18}\ \dots \\
2 \leftrightarrow 0.a_{21}\ a_{22}\ a_{23}\ a_{24}\ a_{25}\ a_{26}\ a_{27}\ a_{28}\ \dots \\
3 \leftrightarrow 0.a_{31}\ a_{32}\ a_{33}\ a_{34}\ a_{35}\ a_{36}\ a_{37}\ a_{38}\ \dots \\
4 \leftrightarrow 0.a_{41}\ a_{42}\ a_{43}\ a_{44}\ a_{45}\ a_{46}\ a_{47}\ a_{48}\ \dots \\
5 \leftrightarrow 0.a_{51}\ a_{52}\ a_{53}\ a_{54}\ a_{55}\ a_{56}\ a_{57}\ a_{58}\ \dots \\
6 \leftrightarrow 0.a_{61}\ a_{62}\ a_{63}\ a_{64}\ a_{65}\ a_{66}\ a_{67}\ a_{68}\ \dots \\
7 \leftrightarrow 0.a_{71}\ a_{72}\ a_{73}\ a_{74}\ a_{75}\ a_{76}\ a_{77}\ a_{78}\ \dots \\
8 \leftrightarrow 0.a_{81}\ a_{82}\ a_{83}\ a_{84}\ a_{85}\ a_{86}\ a_{87}\ a_{88}\ \dots \\
\vdots
\end{array}$$

So we are assuming that on the left of the arrows is every natural number, and on the right of the arrows is every number in the interval $(0, 1)$, and they are paired up in some way. (And note that each a_{ij} is some digit, from 0 to 9.) This proof is due to Georg Cantor and his next idea is quite brilliant. He said, focus now on the "diagonal" of the above. That is, focus on the numbers of the form a_{ii}.

$$\begin{array}{l}
1 \leftrightarrow 0.\textcircled{a_{11}}\ a_{12}\ a_{13}\ a_{14}\ a_{15}\ a_{16}\ a_{17}\ a_{18}\ \dots \\
2 \leftrightarrow 0.a_{21}\ \textcircled{a_{22}}\ a_{23}\ a_{24}\ a_{25}\ a_{26}\ a_{27}\ a_{28}\ \dots \\
3 \leftrightarrow 0.a_{31}\ a_{32}\ \textcircled{a_{33}}\ a_{34}\ a_{35}\ a_{36}\ a_{37}\ a_{38}\ \dots \\
4 \leftrightarrow 0.a_{41}\ a_{42}\ a_{43}\ \textcircled{a_{44}}\ a_{45}\ a_{46}\ a_{47}\ a_{48}\ \dots \\
5 \leftrightarrow 0.a_{51}\ a_{52}\ a_{53}\ a_{54}\ \textcircled{a_{55}}\ a_{56}\ a_{57}\ a_{58}\ \dots \\
6 \leftrightarrow 0.a_{61}\ a_{62}\ a_{63}\ a_{64}\ a_{65}\ \textcircled{a_{66}}\ a_{67}\ a_{68}\ \dots \\
7 \leftrightarrow 0.a_{71}\ a_{72}\ a_{73}\ a_{74}\ a_{75}\ a_{76}\ \textcircled{a_{77}}\ a_{78}\ \dots \\
8 \leftrightarrow 0.a_{81}\ a_{82}\ a_{83}\ a_{84}\ a_{85}\ a_{86}\ a_{87}\ \textcircled{a_{88}}\ \dots \\
\vdots
\end{array}$$

All real numbers in $(0, 1)$ were supposed to be paired up, but we are now going to create a real number in $(0, 1)$ that was not in that above list. The new real number will be different than the first number in its 1st position, different than the second number in its 2nd position, different that the third number in the 3rd position, and so on. The number will have decimal expansion

$$b = 0.b_1\ b_2\ b_3\ b_4\ b_5\ b_6\ b_7\ b_8\ \dots$$

where $b_i \neq a_{ii}$ for all i. To keep it simple, let's just choose

$$b_i := \begin{cases} 1 & \text{if } a_{ii} \neq 1 \\ 2 & \text{if } a_{ii} = 1 \end{cases}.$$

Then notice that, despite $b \in (0, 1)$, b is nowhere in our list! We know b is not the number paired up with 1 because b and that number are different in the first position ($b_1 \neq a_{11}$). We know b is not paired up with 2 because b and that number are different in the second position ($b_2 \neq a_{22}$). In general, we know b is not paired up with k because b and that number are different in the kth position ($b_k \neq a_{kk}$). So this real number b is not anywhere to be found! Thus we have reached a contradiction; clearly we were unable to pair up all the reals, if b got left out. □

All of our work has led up to this remarkable fact:

Theorem.

Theorem 8.26. There are different sizes of infinity. Moreover, $\mathbb{N}$, $\mathbb{Z}$ and $\mathbb{Q}$ are all the same size, while $\mathbb{R}$ is larger.[33]

Do you think there is a smaller infinity than $|\mathbb{N}|$? Or is $|\mathbb{N}|$ the smallest? Do you think that there are any sizes of infinity between $|\mathbb{N}|$ and $|\mathbb{R}|$? Or not? What size of infinity do you think $|\mathbb{R}^2|$ and $|\mathcal{P}(\mathbb{R})|$ are? The same size as $|\mathbb{R}|$? Bigger? How many sizes of infinity are there? Two? Three? Infinitely many? More?

If you find this material a little disquieting, you are not alone. When Cantor's theorems were first published a century and a half ago, many of the great mathematicians of the day responded with disgust. Henri Poincaré called it a "grave disease" infecting mathematics, Leopold Kronecker accused Cantor of being a "corrupter of youth" (a charge that Socrates was sentenced to death for!), and many Christian theologians thought his work against the notion of a unique infinite was an affront to "God's exclusive claim to supreme infinity."[34] Cantor struggled with this for decades.

On the other hand, if this material interests you, then I applaud you and encourage you to read up on *Russell's paradox* and *Cantor's theorem*, as they are the next steps down a fascinating rabbit hole (which, quite literally, is a bottomless pit of mystery). And you should know that Cantor's legacy has been fully restored. The criticisms of the past have been replaced ten times over with praise and accolades. One early defender was the great David Hilbert. Towards the end of Cantor's career, Cantor was awarded the highly-prestigious Sylvester Medal by the Royal Society for his mathematical research. Some criticized this move, but Hilbert — characteristically ahead of his time — recognized the brilliance and importance of Cantor's work, saying:

"No one shall expel us from the paradise that Cantor has created."

[33] Make sure you take a moment to appreciate how remarkably, wonderfully weird this is.

[34] Or, less melodramatically:

> There was a young fellow from Trinity,
> Who took $\sqrt{\infty}$.
> But the number of digits
> Gave him the fidgets;
> He dropped Math and took up Divinity.
>
> —George Gamow

Chapter 9: Relations

For millennia, "math" basically meant geometry or primitive number theory. The Pythagorean theorem, for example, was phrased geometrically; it wasn't the algebraic equation $a^2 + b^2 = c^2$ that we teach our kids to ramble off today. Indeed, the use of variables to describe unknowns in any sort of algebraic equations didn't make their first appearance *until after Christopher Columbus' famous voyage to not-America*, and even then it was far from our modern notation. It is a surprisingly recent innovation! And abstract algebra, for its part, was still centuries away.

The 1600s saw the likes of Issac Newton, who led a charge to use mathematics to understand the physical world. The 1700s saw the likes of Leonhard Euler, who ushered in a purer form of mathematics which required no physical application or real world connection. These ideas led to a movement of abstraction and generalization which flourished in the 1800s, and is the topic of this chapter.

9.1 Equivalence Relations

Consider modular arithmetic. We studied how

$$\begin{aligned} -2 &\equiv 3 \pmod 5 \\ 3 &\equiv 3 \pmod 5 \\ 8 &\equiv 3 \pmod 5 \\ 13 &\equiv 3 \pmod 5. \end{aligned}$$

Since $-2, 3, 8$ and 13 are all congruent to $3 \pmod 5$, there is an equivalence of sorts between these numbers. This is perhaps emphasized by the fact that you flip the order and what you get is still true:

$$\begin{aligned} 3 &\equiv -2 \pmod 5 \\ 3 &\equiv 3 \pmod 5 \\ 3 &\equiv 8 \pmod 5 \\ 3 &\equiv 13 \pmod 5. \end{aligned}$$

Furthermore, $-2 \equiv 3 \pmod 5$ and $3 \equiv 13 \pmod 5$ show that both -2 and 13 are congruent to $3 \pmod 5$, but more than that, they are also congruent to each other, as $-2 \equiv 13 \pmod 5$. Here are more:

$$3 \equiv 13 \pmod 5$$
$$13 \equiv 8 \pmod 5$$
$$-2 \equiv 8 \pmod 5$$
$$13 \equiv 3 \pmod 5.$$

For each of these, if you check the definition of modular congruence (Definition 2.14) you will find that they hold. For example, $3 \equiv 13 \pmod 5$ because $5 \mid (3 - 13)$, because $5(-2) = 3 - 13$.

So in this mod-5 way, every number in $\{-2, 3, 8, 13\}$ is equivalent to every other number in this set, including to itself. And this will extend:

> **Mod-5 Property.** If you pick any two numbers in the set
>
> - $\{\dots, -10, -5, 0, 5, 10, 15, 20, 25, \dots\}$,
>
> then they will be mod-5 equivalent to each other. Moreover, each number in this set is also mod-5 equivalent to itself. This property also holds for each of the following sets:
>
> - $\{\dots, -9, -4, 1, 6, 11, 16, 21, 26, \dots\}$
> - $\{\dots, -8, -3, 2, 7, 12, 17, 22, 27, \dots\}$
> - $\{\dots, -7, -2, 3, 8, 13, 18, 23, 28 \dots\}$,
> - $\{\dots, -6, -1, 4, 9, 14, 19, 24, 29, \dots\}$
>
> These five sets are called the *equivalence classes* of the mod-5 *equivalence relation*. They also have this important property: They completely partition $\mathbb{Z}$; that is, every integer is in precisely one of these five sets.

Quick break to talk about partitions: A partition is simply any way to break up a set into a collection of subsets. For example, a partition of $\{1, 2, 3, 4, 5\}$ is the collection $\{1, 2\}$, $\{3, 5\}$ and $\{4\}$. Another partition is $\{1, 3, 4, 5\}$ and $\{2\}$. What's important here is that each of the five elements went into one and only one of the parts. As long as the entire set is divided up, and you didn't allow any element to go into more than one of the parts, then you have a partition.

Definition.

Definition 9.1. A *partition* of a set A is a collection of nonempty subsets of A for which each element of A is in one and only one[1] of the subsets.[2]

[1] **Recurring Theme Alert.** "one and only one" is an *existence and uniqueness* condition.

[2] For the curious, there is a more formal way to define a partition, and it is this: A *partition* is a collection of nonempty sets $\{P_i\}_{i \in S}$ such that (1) $P_i \subseteq A$ for all i, (2) $\bigcup_{i \in S} P_i = A$, and (3) $P_i \cap P_j = \emptyset$ for all $i \neq j$. See if you can convince yourself that the two definitions are equivalent.

Here are five more examples: A partition of $\mathbb{Z}$ is the set of evens and the set of odds. Another partition of $\mathbb{Z}$ is the positive integers, the negative integers and $\{0\}$. Another is the non-17 integers and $\{17\}$. Another is the five sets in the Mod-5 Property section on the previous page. And the simplest partition of $\mathbb{Z}$ is simply $\mathbb{Z}$ — a partition with only one part.

Ok, let's now get back to equivalence relations. We were just looking at (maximally-sized) sets of numbers which are all mod-5 equivalent to each other, and we found that there are five such sets. Moreover, these sets comprised a partition of $\mathbb{Z}$:

- $\{\dots, -10, -5, 0, 5, 10, 15, 20, 25, \dots\}$
- $\{\dots, -9, -4, 1, 6, 11, 16, 21, 26, \dots\}$
- $\{\dots, -8, -3, 2, 7, 12, 17, 22, 27, \dots\}$
- $\{\dots, -7, -2, 3, 8, 13, 18, 23, 28 \dots\}$
- $\{\dots, -6, -1, 4, 9, 14, 19, 24, 29, \dots\}$

The deep insight from modern algebraists is to ask what properties are required to give us this partition property? Is it necessary that 5 is prime? Or, if we switched mod 5 to mod 6 in that example, would a partition of $\mathbb{Z}$ still be produced?[3] Or, even better, can you describe the properties that produce a partition without any mention of mods?

To start our journey of abstraction, just for a moment let's use this notation:[4]

$$a \sim b \qquad \text{if} \qquad a \equiv b \pmod 5.$$

Given[5] this definition of $\sim$, see if you can prove each of the three properties in the following box.

- $a \sim a$ for all $a \in \mathbb{Z}$;
- If $a \sim b$, then $b \sim a$ for all $a, b \in \mathbb{Z}$; and
- If $a \sim b$ and $b \sim c$, then $a \sim c$ for all $a, b, c \in \mathbb{Z}$.

Not only does the mod-5 property satisfy these three important properties, but it turns out that these three are *precisely* what is required to produce this equivalence/partition property. Here's what I mean: Suppose "$\sim$" no longer means mod-5 equivalence on $\mathbb{Z}$. Instead, suppose you were only told that A is some set,

[3]Spoiler: Yes, you still get a partition. However, the mod-6 equivalence relation will produce six equivalence classes instead of five.

[4]Note: In later examples, $a \sim b$ will mean something else. So we are not saying $a \sim b$ means $a \equiv b \pmod 5$ forever. In each problem, it takes on a new meaning, but all the meanings are going to be connected.

[5]Pro-Tip: "$\sim$" is typically read as "tilde," which is pronounced "till-duh."

and for some pairs of elements from A, you are told that a is "related" to b (denoted $a \sim b$)[6] while for others pairs you are told that a is "not related" to b (denoted $a \not\sim b$). But you are told nothing else about A and you have no idea what rule is determining which pairs have $a \sim b$ and which have $a \not\sim b$.

Given that setup, here's the miracle: If $\sim$ satisfies the three bullet-point properties in the box above, then the set will naturally partition itself into equivalence classes. And if $\sim$ does not satisfy one or more of these three properties, then the set will not partition itself into equivalence classes (meaning that at least one element will either be in no equivalence class, or in more than one equivalence class). Here's a quick example of the latter:

Example 9.2. Suppose $A = \mathbb{N}$ and we say that

$$a \sim b \qquad \text{if} \qquad a \geq b.$$

You may notice that $a \sim a$ for all $a \in \mathbb{N}$; for example, $3 \geq 3$ and $15 \geq 15$. You may also notice that if $a \sim b$ and $b \sim c$, then $a \sim c$; for example, since $10 \geq 6$ and $6 \geq 3$, it is also true that $10 \geq 3$. So the first and third bullet points are satisfied. But what about the second? If $a \sim b$ is it true that $b \sim a$? Nope! For example, $5 \geq 3$, but $3 \not\geq 5$.

And so our next theorem says that this relation will not partition the numbers from A into "equivalence classes." To demonstrate this through a particular case, think about all the numbers a for which $6 \sim a$. If you put all of them into a set it would be this:

$$\{1, 2, 3, 4, 5, 6\}.$$

What about the set of all numbers a for which $4 \sim a$? Or $8 \sim a$? Those are these sets:

$$\{1, 2, 3, 4\} \qquad \text{and} \qquad \{1, 2, 3, 4, 5, 6, 7, 8\}.$$

You see? It's not working! There's no grand partition happening. Just take a look at the three sets above: In the set that 6 generated, we included 4 (because $6 \sim 4$), however in the set that 4 generated, we did not include 6 (because $4 \not\sim 6$). And 2 is in all three of these sets! A partition has to have every number in exactly one set. So this $\sim$ relation on $\mathbb{N}$ is not producing the "equivalence classes" and partition properties that the mod-5 $\sim$ relationship on $\mathbb{Z}$ produced. □

Returning now to the three bullet points, these three properties are indeed *precisely* what's important in order to produce this equivalence class/partition property. Mod-5 equivalence has all sorts of properties related to divisibility and prime factorizations and the division algorithm and remainders. There is a lot that can be said about

[6]To be clear, this $\sim$ is not the same as the "not" symbol from when we studied logic. In math, symbols are often reused, and as you will see, $\sim$'s exact meaning in this chapter will change from problem to problem. In stats, they also use this symbol to write things like $X \sim N(0, 1)$, to assert the distribution of a random variable. In asymptotics, they use this symbol to say that two functions are growing at basically the same rate, like $\pi(x) \sim \frac{x}{\ln(x)}$; in fact, this final use of $\sim$ is also an equivalence relation, which we will be discussing in a moment. (Note: $\pi(x)$ is an important function which we defined in the *Introduction to Number Theory*.)

mod-5 equivalence, but for equivalence classes and partitions, all that matters is that they satisfy those three properties — the rest is fluff. Likewise, the equivalence classes in the last example failed to produce a partition, and as we will soon prove, this was solely because $\sim$ was not "symmetric," which we will define next. This is the art of discovering what *really* matters to obtain a result. Let's now formally record these definitions and results.

Definition.

Definition 9.3. An *equivalence relation* on a set A is an ordered relationship between pairs of elements of A for which the pair is either *related* or is *not related.* If $a, b \in A$, we denote $a \sim b$ if a is related to b, and $a \not\sim b$ if a is not related to b.

For $\sim$ to be an equivalence relation, it also must satisfy the following three properties.

- Reflexive: $a \sim a$ for all $a \in A$;
- Symmetric: If $a \sim b$, then $b \sim a$ for all $a, b \in A$; and[7]
- Transitive: If $a \sim b$ and $b \sim c$, then $a \sim c$ for all $a, b, c \in A$.

Lastly, if $\sim$ is an equivalence relation and $a \in A$, define the *equivalence class generated by* a to be the set

$$\{b \in A : a \sim b\}.$$

We have already discussed how mod-5 congruence is an equivalence relation, and we mentioned that mod-6 congruence is as well. We will soon see several more examples. But as we mentioned at the start, this chapter is focused on abstraction and generalization, and while the idea of an equivalence relation is quite general, we can make it even more general by not demanding that it satisfy the reflexive, symmetric and transitive properties. This is the idea of a *relation.*

Definition.

Definition 9.4. A *relation* on a set A is any ordered relationship between pairs of elements of A for which the pair is either *related* or is *not related.* If $a, b \in A$, we denote $a \sim b$ if a is related to b, and $a \not\sim b$ if a is not related to b.

Lastly, if $\sim$ is a relation and $a \in A$, define the *class generated by* a to be the set

$$\{b \in A : a \sim b\}.$$

[7]The symmetric property would be read either like "If a is related to b, then b is related to a" or "If a till-duhs b, then b till-duhs a."

Since mod-5 congruence is an equivalence relation, and a relation generalizes the idea of an equivalence relation, mod-5 congruence is also an example of a relation. Likewise, mod-6 congruence is a relation. We also have seen an example of a relation that is not an equivalence relation: In Example 9.2, where $a \sim b$ if $a \geq b$.

Equivalence relations generalize the idea of equality. For numbers a, b and c, we certainly know that $a = a$ (like $4 = 4$). We also know that if $a = b$, then certainly $b = a$ (like if $a = 5$, then saying $a = b$ and $b = a$ both just mean b is also 5). Finally, if $a = b$ and $b = c$, then of course $a = c$ (again, that's just saying that if a and b are the same thing, and b and c are the same thing, then of course a and c are the same, too).

As it turns out, the equal sign is far from the only symbol which shares these three properties. If we use $\equiv_5$ to mean congruence modulo 5, then we have already discussed how $\equiv_5$ has these three properties,[8] and we will see many more examples to come. And now, here's the main result of the chapter.

Theorem.

Theorem 9.5. Assume $\sim$ is a relation on A. The relation $\sim$ partitions the elements of A into classes[9] if and only if $\sim$ is an equivalence relation.

If $\sim$ is a relation, then it might be an equivalence relation, and it might not be. What this theorem is saying is that if $\sim$ *is* an equivalence relation, then the equivalence classes will form a partition of A; just like in the mod-5 example, how the five sets on page 381 formed a partition of $\mathbb{Z}$. And if the relation $\sim$ is *not* an equivalence relation, then the classes do *not* form a partition of A; just like in Example 9.2, how the classes overlapped and thus were not a partition of $\mathbb{N}$. Mod-5 congruence was an equivalence relation and produced a partition, while greater-than-or-equal-to was not an equivalence relation and did not produce a partition.

Theorem 9.5 is hard to understand, so before we prove it, let's look at some more examples.

Example 9.6. Recall that the *floor function* is the function that rounds down. For example, $\lfloor 2.6 \rfloor = 2$. Now, let $\sim$ be the relation on $\mathbb{R}$ where

$$a \sim b \qquad \text{if} \qquad \lfloor a \rfloor = \lfloor b \rfloor.$$

For positive values, this would mean they have the same integer part; for example, $12.4 \sim 12.85$ since $\lfloor 12.4 \rfloor = \lfloor 12.85 \rfloor = 12$.

[8] For example, "if $a \equiv_5 b$, then $b \equiv_5 a$" is true, since this is just notation for "if $a \equiv b \pmod 5$, then $b \equiv a \pmod 5$." And that can be quickly proved by the definition of modular congruence.

[9] Subtle note: The theorem refers to partitioning into *classes*, rather than into *equivalence classes*, even though the theorem itself tells us that $\sim$ will be an equivalence relation and hence they will be equivalence classes. However, we use the relation term of "class" because we state the theorem before the proof, and so we can't use the theorem to refer to them as equivalence classes, since doing so is only possible after the proof! Have I confused you yet?

We can verify that $\sim$ is an equivalence relation[10] by checking that it satisfies Definition 9.3: it is reflexive because certainly $\lfloor a \rfloor = \lfloor a \rfloor$ for any $a \in \mathbb{R}$; it is symmetric because if $\lfloor a \rfloor = \lfloor b \rfloor$, then certainly $\lfloor b \rfloor = \lfloor a \rfloor$; and it is transitive because if $\lfloor a \rfloor = \lfloor b \rfloor$ and $\lfloor b \rfloor = \lfloor c \rfloor$, then $\lfloor a \rfloor = \lfloor c \rfloor$. Each of these is immediate because the equal sign already has these properties; e.g., if I told you $x = y$ you would immediately know that $y = x$.

By Theorem 9.5, this means that the equivalence classes must then partition all of $\mathbb{R}$, and indeed they do. The class of all numbers that are equivalent to 12.4 is the set of numbers in the interval $[12, 13)$; that is, all numbers x such that $12 \leq x < 13$. Indeed, the equivalence classes for $\sim$ are all intervals of the form $[n, n+1)$ for $n \in \mathbb{Z}$. Moreover, by Theorem 9.5 this means that the equivalence classes must then partition all of $\mathbb{R}$, and they do: every $x \in \mathbb{R}$ is in precisely one of these intervals:

$$\dots, [2,3), [3,4), [4,5), [5,6), [6,7), \dots. \qquad \square$$

Example 9.7. Let $\sim$ be the relation on $\mathbb{Z}$ where

$$a \sim b \qquad \text{if} \qquad a + b \text{ is even.}$$

For example, $2 \sim 4$ and $2 \sim -14$ since $2 + 4 = 6$ and $2 + (-14) = -12$ are both even, while $2 \not\sim 3$ since $2 + 3 = 5$ is not even. Let's check whether $\sim$ is an equivalence relation.[11]

Reflexive: To see that $a \sim a$ for all $a \in \mathbb{Z}$, simply note that $a + a = 2a$ is even by the definition of an even number. Therefore, $a \sim a$. This proves that $\sim$ is reflexive.

Symmetric: Assume that $a \sim b$ for some $a, b \in \mathbb{Z}$. This means that $a + b$ is even, which of course also means that $b + a$ is even, which implies that $b \sim a$, proving that $\sim$ is symmetric.

Transitive: Assume that $a \sim b$ and $b \sim c$, for some $a, b, c \in \mathbb{Z}$. Then $a + b$ is even and $b + c$ is even. By the definition of evenness, $a + b = 2k$ and $b + c = 2\ell$ for some $k, \ell \in \mathbb{Z}$. Adding these equalities together and then doing some algebra,

$$\begin{aligned}
(a+b) + (b+c) &= 2k + 2\ell \\
a + 2b + c &= 2k + 2\ell \\
a + c &= 2k + 2\ell - 2b \\
a + c &= 2(k + \ell - b).
\end{aligned}$$

[10]But first, especially for these relation problems, do lots of examples! Make sure you fully understand the relation. Here, $12.4 \sim 12.85$ and $12.85 \sim 12.554$ and $12.54 \sim 12$ and $12.4 \not\sim 13.4$, $12.4 \not\sim 11.9$, and $12.67 \not\sim -2.24$. Looks like all the numbers between 12 and 13 are related to each other, but none of them are related to anything else.

[11]But first, do some examples again to make sure you understand the relation! Here, $2 \sim 8$ since $2 + 8 = 10$, which is even; meanwhile, $2 \not\sim 7$ since $2 + 7 = 9$, which is not even. So 2 and 8 will end up in the same equivalence class, while 2 and 7 will end up in different equivalence classes. Do more on your own to get a feel for the relation before we try to prove anything about it. And try to guess how many equivalence classes there will be, and what they will look like!

And because $k + \ell - b$ is an integer, this shows that $a + c$ is even, and so $a \sim c$, proving that $\sim$ is transitive.

Combined, this shows that $\sim$ is an equivalence relation on $\mathbb{Z}$. Moreover, by Theorem 9.5 this means that the equivalence classes must then partition all of $\mathbb{Z}$, and indeed they do. Do you see the equivalence classes? There are only two of them... One is the set of even integers $\{\dots, -4, -2, 0, 2, 4, \dots\}$, and the other is the set of odd integers $\{\dots, -5, -3, -1, 1, 3, 5, \dots\}$. Any two elements from the same set, including an element with itself, will have an even sum (because the sum of two evens is even, and the sum of two odds is even). However, any two elements from different sets will not have an even sum (because an even plus an odd is not even). Therefore, if a and b are from the same set, then $a \sim b$, but if a and b are from different sets, then $a \not\sim b$. Lastly, note that these two equivalence classes do indeed partition $\mathbb{Z}$, since every integer is either even or odd, but none are both. $\square$

We won't go through the details, but when you took geometry you probably wrote things like $\triangle ABC \sim \triangle DEF$ to indicate that these two triangles were *similar* to each other (meaning, each triangle is just a scaled version of the other). In fact, this was the first interesting equivalence relation you ever learned! You even used the correct symbol: $\sim$. It's reflexive: $\triangle ABC \sim \triangle ABC$; symmetric: if $\triangle ABC \sim \triangle DEF$, then $\triangle DEF \sim \triangle ABC$; and transitive: if $\triangle ABC \sim \triangle DEF$ and $\triangle DEF \sim \triangle XYZ$, then $\triangle ABC \sim \triangle XYZ$. It checks out!

Our final example is one I particularly enjoy.

Example 9.8. Let $\mathcal{D}$ be the set of words in the English dictionary and $\sim$ be the relation for which

$$a \sim b \qquad \text{if} \qquad \text{the word } a \text{ rhymes with the word } b.$$

Then, $\sim$ is an equivalence relation. For example, think of the word "math."

- Reflexive ($a \sim a$ for all $a \in \mathcal{D}$)
    - Example: "math" rhymes with "math."
- Symmetric (If $a \sim b$, then $b \sim a$ for all $a, b \in \mathcal{D}$)
    - Example: If "math" rhymes with "path," then also "path" rhymes with "math."
- Transitive (If $a \sim b$ and $b \sim c$, then $a \sim c$ for all $a, b, c \in \mathcal{D}$)
    - Example: If "math" rhymes with "path" and "path" rhymes with "bath," then also "math" rhymes with "bath."

In fact, the rhyming poets and singer-songwriters in the audience will be well-aware of websites like `rhymezone.com` where you type in a word and it tells you all other words which rhyme with that word. Said differently, you give the website a word and the website gives you back that word's equivalence class in $\mathcal{D}$! $\square$

The main point of this example is to drive intuition. In the mod-5 sense, we imagine that 4 and 9 and 14 all rhyme, while 4 and 6 do not rhyme. In the floor-function sense, we imagine that 3.4 and 3.9 and π all rhyme, while π and e do not. In each problem, we used an equivalence relation $\sim$ to define a new mathematical rhyme scheme on a set A, and then we stood back and watched as each new math-rhyming property partitions the set.[12]

There are many more real-world examples of equivalence relations. These include "has the same birthday as" and "is the same height as." In Exercise 9.20, you are asked to come up with more real-world examples.

Now that you have seen some concrete examples and have begun to build a little intuition, let's prove Theorem 9.5.

Proof of Theorem 9.5

Theorem 9.5's proof will be aided by some notation and a lemma. First, the notation.

Notation.

Notation 9.9. Given a set A and an equivalence relation $\sim$ on A, recall that the equivalence class of an element $a \in A$ is the set

$$\{x \in A : a \sim x\}.$$

We denote this set by $[a]$.

As an example, let $\sim$ be the mod-5 equivalence relation on $\mathbb{Z}$. Then,

$$\begin{aligned}
[0] &= \{\dots, -10, -5, 0, 5, 10, 15, 20, 25, \dots\} \\
[1] &= \{\dots, -9, -4, 1, 6, 11, 16, 21, 26, \dots\} \\
[2] &= \{\dots, -8, -3, 2, 7, 12, 17, 22, 27, \dots\} \\
[3] &= \{\dots, -7, -2, 3, 8, 13, 18, 23, 28 \dots\} \\
[4] &= \{\dots, -6, -1, 4, 9, 14, 19, 24, 29, \dots\} \\
[5] &= \{\dots, -10, -5, 0, 5, 10, 15, 20, 25, \dots\} \\
[6] &= \{\dots, -9, -4, 1, 6, 11, 16, 21, 26, \dots\} \\
[7] &= \{\dots, -8, -3, 2, 7, 12, 17, 22, 27, \dots\} \\
&\vdots
\end{aligned}$$

[12] *"It is impossible to be a mathematician without being a poet in the soul."* – Sofia Kovalevskaya.

Also note that $[2] = [7] = [12]$, and $[-4] = [1] = [6]$, and so on. Next, we will need the following lemma in the proof of the Theorem 9.5.

Lemma.

Lemma 9.10. Suppose $\sim$ is an equivalence relation on a set A, and let $a, b \in A$. Then,

$$[a] = [b] \qquad \text{if and only if} \qquad a \sim b.$$

Proof Idea. The forward direction will be, charmingly enough, straightforward. As for the backward direction, we will assume that $a \sim b$ and try to prove that $[a] = [b]$. In it, don't forget that according to Notation 9.9, $[a]$ and $[b]$ are sets! And in Section 3.3 we outlined one way to prove that two sets are equal: We will prove that $[a] \subseteq [b]$, and $[b] \subseteq [a]$. The proof itself will not be terribly interesting, it will simply require some careful applications of the fact that $\sim$ is an equivalence relation, and hence is reflexive, symmetric and transitive.

***Proof*.** For the (straight)forward direction, assume that $[a] = [b]$. Observe that since $\sim$ is reflexive, $b \sim b$ and so $b \in [b]$. And since $[a] = [b]$, this in turn means that $b \in [a]$, which by Notation 9.9 implies $a \sim b$. This concludes the forward direction.

As for the backward direction, we begin by assuming $a \sim b$, and we aim to prove that $[a] = [b]$. This will be accomplished by demonstrating that $[a] \subseteq [b]$ and $[b] \subseteq [a]$. To prove the former, choose any $x \in [a]$; we will show that $x \in [b]$. By assumption we have $a \sim b$, and because $x \in [a]$ we have $a \sim x$. That is,

$$a \sim b \qquad \text{and} \qquad a \sim x.$$

By the symmetry property of $\sim$,

$$b \sim a \qquad \text{and} \qquad a \sim x.$$

By the transitivity property of $\sim$,

$$b \sim x.$$

And so, by Notation 9.9,

$$x \in [b].$$

We have shown that $x \in [a]$ implies $x \in [b]$, and hence $[a] \subseteq [b]$.

The reverse direction is nearly the same. Let $x \in [b]$, which means $b \sim x$. Combining this, the transitivity of $\sim$, and our assumption that $a \sim b$, and we get $a \sim x$, which means $x \in [a]$. And since $x \in [b]$ implies $x \in [a]$, we have $[b] \subseteq [a]$.

We have shown that $[a] \subseteq [b]$ and $[b] \subseteq [a]$, which proves that $[a] = [b]$. This concludes the backward direction, and hence the proof. □

Let's now prove Theorem 9.5, which for your reference is the following.

Theorem.

Theorem 9.5. Assume $\sim$ is a relation on a set A. The relation $\sim$ partitions the elements of A into classes if and only if $\sim$ is an equivalence relation.

Proof. We will first prove the forward direction, and then the backward direction.

– Forward Direction –

Assume that $\sim$ partitions the elements of A into classes, say $\{P_i\}_{i \in \mathcal{S}}$, where $\mathcal{S}$ is some indexing set.[13] And recall that by the definition of a class, that if $x, y \in P_i$, then y is in the same class as x, meaning that $x \sim y$. We aim to prove that $\sim$ is reflexive, symmetric and transitive.

First, we prove that $\sim$ is reflexive. Pick any $a \in A$. Recall that being a partition means that each $a \in A$ is in precisely one class; let's say $a \in P_i$. This will look like a strange statement, but is of course true that if $a \in P_i$, then $a \in P_i$. And this is in fact all we need to prove $\sim$ is reflexive, as this shows that a is in the same class as itself. And since being in the same class means that they are related, we conclude that $a \sim a$, meaning $\sim$ is reflexive.

Next, we prove that $\sim$ is symmetric. Choose any $a, b \in A$ such that $a \sim b$; we aim to prove that $b \sim a$. By the definition of a class, $a \sim b$ means that b is in the class generated by a. Now, the classes are simply the sets $\{P_i\}_{i \in \mathcal{S}}$, so this is simply saying that the set that a is in (say, P_i), is also the set that b is in. That is, $a, b \in P_i$. Moreover, since $\{P_i\}_{i \in \mathcal{S}}$ is a partition, this is the only set that a and b are in. And since P_i is just a set which contains a and b, this also means that a is in the class generated by b. That is, $b \sim a$. Since $a \sim b$ implied $b \sim a$, we have shown that $\sim$ is symmetric.

Finally, we prove that $\sim$ is transitive. Choose any $a, b, c \in A$ such that $a \sim b$ and $b \sim c$; we aim to prove that $a \sim c$. Once again, since b is in the class generated by a, and c is in the class generated by b, this simply means $a, b \in P_i$ and $b, c \in P_j$ for some parts P_i and P_j. And recall that $\{P_i\}_{i \in \mathcal{S}}$ formed a *partition* of A, which by definition means that each element of A is in precisely one of these sets. And so, since b can only be in one class while $b \in P_i$ and $b \in P_j$, it must be the case that $P_i = P_j$. Hence, a and c are in fact in the same class. Thus, c is indeed in the class generated by a, giving $a \sim c$. This proves that $\sim$ is transitive.

– Backward Direction –

Next, we prove the backward direction. Assume that $\sim$ is an equivalence relation; we aim to prove that its equivalence classes partition A. First, recall that to be a partition means that every element is in one and only one class. To see that every element is in *at least* one equivalence class, simply note that each $a \in A$ is in its own

[13]For example, if there happens to be 8 partition sets, then $\mathcal{S} = \{1, 2, 3, 4, 5, 6, 7, 8\}$. If there happens to be $|\mathbb{N}|$ partition sets, then $\mathcal{S} = \{1, 2, 3, 4, \dots\}$.

equivalence class, since $\sim$ being reflexive implies that $a \sim a$. Said differently, each $a \in A$ is certainly in at least one equivalence class, because $a \in [a]$!

We have shown that every element is in at least one equivalence class. The final condition to be a partition (Definition 9.1) is that each element is in *only one* equivalence class. That is, we wish to show:

- No element is in two or more distinct equivalence classes.

Now, observe that there exists an element in two distinct classes if and only if there are two distinct equivalence classes that overlap. So the above is equivalent to this:

- If any two classes overlap, then they cannot be distinct.

To turn this into symbols, note that equivalence classes like $[a]$ and $[b]$ being distinct simply means $[a] \neq [b]$.[14] And for there to be overlap between them means $[a] \cap [b] \neq \emptyset$. Thus, the above is equivalent to saying this:

- For any $a, b \in A$, if $[a] \cap [b] \neq \emptyset$, then $[a] = [b]$.

This is what we will prove. To this end, assume that $a, b \in A$ and $[a] \cap [b] \neq \emptyset$. Then there exists some $c \in A$ such that

$$c \in [a] \qquad \text{and} \qquad c \in [b].$$

By Notation 9.9,

$$a \sim c \qquad \text{and} \qquad b \sim c.$$

By the symmetry property of $\sim$,

$$a \sim c \qquad \text{and} \qquad c \sim b.$$

By the transitive property of $\sim$,

$$a \sim b.$$

By Lemma 9.10, we have[15]

$$[a] = [b].$$

This completes the backward direction and hence the proof. □

This theorem tells us that a relation produces a partition precisely when that relation is an equivalence relation. But it doesn't tell us what that partition looks like. It also does not tell us which partitions correspond to some equivalence relation and which ones do not. As it turns out, for *every* partition of A there is an equivalence relation which produces precisely that partition.

[14] For example, with mod-5 equivalence, $[1] = [6]$. So even if $a \neq b$, we could still have $[a] = [b]$. But two sets are distinct provided they are not exactly equal, like how $\{1, 2, 3\}$ and $\{2, 3, 4\}$ are distinct sets. So to determine distinctness, one must consider them as sets.

[15] *"Yo, lemma help you prove that theorem."*

Proposition.

Proposition 9.11. Let A be a set. Given any partition of A into sets P_i, there is an equivalence relation whose equivalence classes are precisely these sets P_i.

Proof. Define a relation $\sim$ on A such that $a \sim b$ if a and b belong to the same part (that is, $a, b \in P_i$ for some i), and $a \not\sim b$ if a and b do not belong to the same part (that is, a and b are not both in P_i for any i).

This rule is, by its very definition, producing the partition into the given P_i. Moreover, we can see that $\sim$ is an equivalence relation by checking the three properties, as required by Definition 9.3. First, $\sim$ is reflexive because $a \sim a$ is simply saying that a is in the same partition set as itself, which is certainly true for every $a \in A$. Next, $\sim$ is symmetric because if $a \sim b$, then this means a and b are in the same partition set, which certainly also means that b and a are in the same partition set, or $b \sim a$. Finally, if a and b are in the same partition set, and b and c are as well, then certainly a and c must be as well; that is, if $a \sim b$ and $b \sim c$, then $a \sim c$.

Given the partition into sets P_i, we have created a relation which gives this partition and verified that it is indeed an equivalence relation. This completes the proof. □

9.2 Abstraction and Generalization

As mathematicians sought abstraction and generalization, they began to ask what happens when you peel back layers of structure and complexity. Considering the equivalence of integers modulo 5 gives three important properties: reflexivity, symmetry and transitivity. Equivalence of integers modulo 6 also has these properties. The floor function also has these properties, as does the property that pairwise sums are even, and the rhyming property in the dictionary.

The partition property of Theorem 9.5 then immediately holds for each of these. Modular arithmetic, floor functions and dictionaries are different in so many ways — but they're the same where it matters. This allows us to avoid the nitty gritty details of each. By focusing on a small collection of abstract properties, and seeing what just those imply, one can create an extremely versatile and beautiful theory.[16]

Mathematicians began to apply this type of thinking all over mathematics. What are the most important properties of the real numbers, and what can we prove by assuming only those properties? What about for the rational numbers? For the

[16] It's kind of like the classic Disney princess movies, like Snow White, Cinderella and Sleeping Beauty, that I was forced to watch because I have twin older sisters who would always outvote me if I suggested Space Jam for the 200th time. A young, isolated female with a beautiful voice and animal friends finds herself in distress, only to be rescued by a strapping barrel-chested man. They fall madly in love within 6-12 hours of meeting and with a kiss they live happily ever after. Sure, the details vary from movie to movie, but the main plot line remains constant. This is the big idea with abstraction and generalization. Oftentimes, the details don't matter so much, like whether you have a floor function or a mod-5 function. No matter how you dress it up, when you focus on what really matters, you're left with something worse than Space Jam.

integers? For matrices? Henri Poincaré said that "mathematics is the art of giving the same name to different things." For each of the above, what other mathematical structures are exactly the same where it really matters? If you can identify that essence, then let's give it a name and study it![17]

If you peel back even more from the idea of an equivalence relation, you arrive at the extremely general idea of a relation on a set A. In Definition 9.4, we said that a *relation* is any relationship between ordered pairs of elements of A. It is denoted $\sim$ and can mean anything. For each $a, b \in A$, either $a \sim b$ or $a \not\sim b$—all we require is that each pair of elements is "related" or not. It *could* have the three extra properties to be an equivalence relation, but it does not need to. For instance, you might have $a \sim b$ but also $b \not\sim a$ (like in Example 9.2, where $a \sim b$ if $a \geq b$). Or you might even have $a \not\sim a$ (for example, if $a \sim b$ whenever $a > b$).

A relation was a very general idea, but I would like to peel back just one more layer.[18] So far we have defined a relation on just one set. Let's generalize this to a relation between two sets.

Definition.

Definition 9.12. A *relation* from a set A to a set B is any ordered relationship between each element of A and each element of B, where each pair is either *related* or is *not related.* If $a \in A$ and $b \in B$, then we denote $a \sim b$ if these elements are related, and $a \not\sim b$ if these elements are not related.

Note that if $B = A$, then this matches Definition 9.4, showing that this is a true generalization.

Functions and Relations

Functions and relations were presented much differently, but one goal of abstraction and generalization is to find connections between seemingly disparate objects. And with some thought, you might start noticing a few vague similarities between the two:

1. A relation from A to B and a function $f : A \to B$ both involve a pair of sets.

2. Both allow for some sort of connection between an element in A and an element in B. For a relation, it is $a \sim b$; for a function, it is $f(a) = b$. And for both, some elements might have the connection while others do not.

[17] Indeed, when you take abstract algebra you'll learn names likes "group" and "ring" and "field" to describe some of these essential properties. You will probably begin with groups, which is the topic of the *Introduction to* following this chapter.

[18] Shrek: Relations are like onions. Donkey: They stink? Shrek: No! Layers! Because relations have *layyyerss.* Donkey: Ohhhhhh...Cakes! Cakes have layers! Relations are like cakes? Shrek: No!

3. Each operates on just *one* thing from A and *one* thing from B at a time (one $a \in A$ and one $b \in B$). So two elements in total are considered at a time, never more and never less. We called this a *binary* relationship.

4. Order matters: For relations, we have seen that $a \sim b$ and $b \sim a$ are asserting two different things (unless possibly when the relation is known to be symmetric). Likewise, certainly $f(a) = b$ does not imply that $f(b) = a$.

At this point, there is a way to think about both of these as *ordered pairs*.[19]

Note.

Note 9.13.

- A *relation* $\sim$ from A to B can be thought of as a subset R of $A \times B$ where $(a, b) \in R$ provided $a \sim b$.
- A *function* $f : A \to B$ can be thought of as a subset F of $A \times B$ where $(a, b) \in F$ provided $f(a) = b$.

It may look weird to say that a function is nothing more than a set of ordered pairs from $A \times B$ (domain $\times$ codomain), but when you graph a function from $\mathbb{R} \to \mathbb{R}$, what you are seeing is exactly that! You are seeing a plot of all the ordered pairs!

Despite these similarities, there are two important differences between functions and relations, which brings us to our final recurring theme alert.

Recurring Theme Alert. The definition of a function (Definition 8.1) demanded that each input must have an output: $f(a)$ has to be equal to some $b \in B$. This was the "existence" part of that definition. For a relation, though, there is no such existence requirement. It may very well be that $a \not\sim b$ for every $b \in B$.

Also due to that definition, if $f(a) = b$, then it is impossible for us to also have $f(a) = c$ (for $b \neq c$); this was the "uniqueness" part of that definition. Each input can have only one output. For a relation, though, there is again no such requirement. It may very well be that $a \sim b$ and $a \sim c$ (for $b \neq c$).

In fact, it is the case that *any* subset of $A \times B$ is a relation, while only very special subsets of $A \times B$ would constitute a function. This realization also shows us that a function is a special type of relation.[20] The following definition drives this point home by providing yet another definition of a function — this time in terms of relations. And while the below looks different than Definition 8.1 (our original function definition), they are indeed equivalent.

[19] We are tying together all sorts of ideas at this point, and that now includes the Cartesian product of sets! These were introduced way back in Definition 3.13.

[20] All functions are relations, but not all relations are functions.

Definition.

Definition 9.14. A *function* f from a set A to a set B is a relation $F \subseteq A \times B$ satisfying the property that for every $a \in A$ there exists a unique[21] element $b \in B$ for which $(a, b) \in F$.

And with that — as far as functions are concerned — your undergraduate brain expansion is now complete.

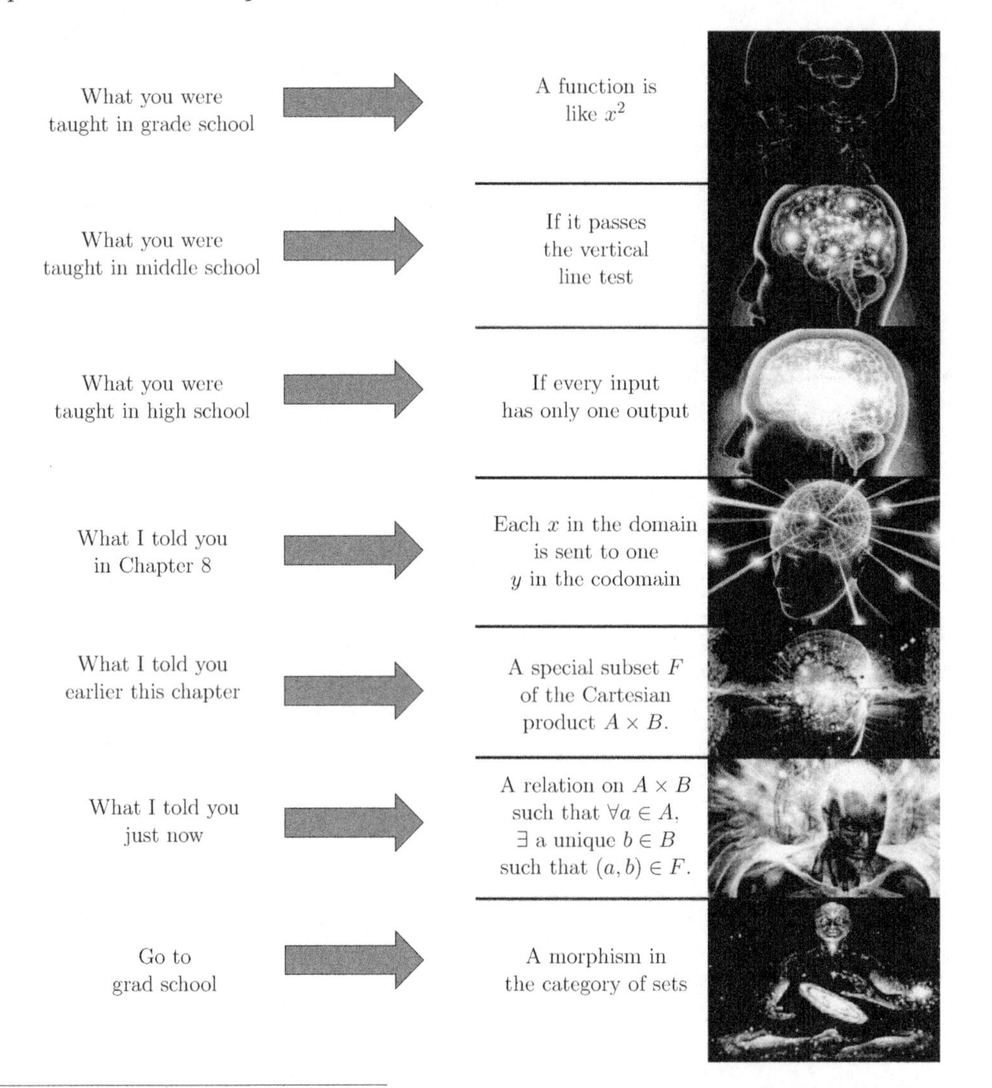

[21] The word "unique" here is saying that for each a there exists one and only one b where $(a, b) \in F$. But this does *not* prevent some b from corresponding to more than one a. It may be the case that $(a_1, b) \in F$ and also $(a_2, b) \in F$. For example, if F is the subset of $\mathbb{R} \times \mathbb{R}$ representing the function $f : \mathbb{R} \to \mathbb{R}$ where $f(x) = x^2$, then notice that $(2, 4) \in F$ and also $(-2, 4) \in F$. However, if this doesn't happen, then we did have a name for such a function: an injection!

It also does not mean that every b has at least one corresponding a. Perhaps there is a b for which $(a, b) \notin F$ for all $a \in A$. But if this doesn't happen, then we again have a special name for such a function: a surjection!

9.3 Bonus Examples

Let's do some examples on small sets.

Example 9.15. Consider the relation $\sim$ on the set $\{x, y, z\}$ such that this is the complete list of related elements:

$$x \sim x \qquad\qquad y \sim y \qquad\qquad z \sim z$$

Is $\sim$ reflexive? Symmetric? Transitive?

Solution. Take another look at Definition 9.3 where we listed the three conditions for a relation to be an equivalence relation, and note the difference between the first condition and the last two conditions. For $\sim$ to be reflexive on A, the condition is straightforward: You need $a \sim a$ for all $a \in A$, otherwise the condition fails. Thus, for the example that we are solving, $\sim$ is reflexive, since $x \sim x$, $y \sim y$ and $z \sim z$.

It turns out that $\sim$ is also symmetric, but explaining why is a little more subtle. Take another look at Definition 9.3. Both symmetry and transitivity are of the "If, then" variety. Symmetry says that *if* you have $a \sim b$, *then* you must also have $b \sim a$. But without the "if," you don't need the "then."[22] So, is $\sim$ symmetric? Sure! The only way it is not symmetric is if you had, say, $y \sim z$ and $z \not\sim y$; that would be a problem. But that doesn't occur here. The only "if $a \sim b$, then $b \sim a$" that occurs is where a and b are both x, or both y, or both z. And in those cases it is trivial that "if $x \sim x$, then $x \sim x$" is satisfied, and likewise for y and z.

Similarly, $\sim$ is transitive. Transitive says that "if $a \sim b$ and $b \sim c$, then $a \sim c$." Again, there are no interesting "if" cases that need to be checked. The only ones are again trivial, like "if $y \sim y$ and $y \sim y$, then $y \sim y$."

Thus, the conclusion is that $\sim$ is reflexive, symmetric and transitive. It is therefore an equivalence relation with equivalence classes $\{x\}$, $\{y\}$ and $\{z\}$. □

This thinking is toughest in these extreme cases, but it is not so different than our examples earlier in this chapter. For instance, the floor function relation was symmetric, even though it was not the case that $2.2 \sim 3.5$ and $3.5 \sim 2.2$. In fact, both $2.2 \not\sim 3.5$ and $3.5 \not\sim 2.2$. It's perfectly fine that all pairs aren't related both ways. It is only when you have the "if" that you need the "then." For example, it *is* the case that $2.2 \sim 2.7$, and so, in order for the floor function relation to be symmetric we would need $2.7 \sim 2.2$ as well—which is true.

As a more interesting finite example, see if you can convince yourself that if $\sim$ is a relation on $\{w, x, y, z\}$ and the below is a complete list of related elements, then $\sim$ is symmetric, but is not reflexive or transitive.

$$x \sim x \qquad y \sim y \qquad z \sim z \qquad z \sim y \qquad y \sim w$$

$$w \sim x \qquad x \sim y \qquad w \sim y \qquad y \sim z \qquad x \sim w \qquad y \sim x$$

[22] If it helps, this is similar to how our truth tables in Chapter 5 counted "False $\Rightarrow$ (True or False)" as true. If that doesn't help, then ignore it.

Similar Matrices

Many of you have probably taken a course in matrix theory, or a course in linear algebra which contained a lot of theory on matrices.[23] You may have learned about *similar* matrices, which is the topic we will discuss now.[24]

Definition.

Definition 9.16. Assume that A and B are $n \times n$ matrices of real numbers. Consider the relation where $A \sim B$ if there exists an invertible $n \times n$ matrix P where

$$P^{-1}AP = B.$$

If this happens, we also say that A is *similar* to B.

For example, if $A = \begin{pmatrix} 1 & 2 \\ 0 & -1 \end{pmatrix}$ and $B = \begin{pmatrix} 1 & 0 \\ -2 & -1 \end{pmatrix}$, then $A \sim B$. Indeed, it turns out that the matrix $P = \begin{pmatrix} 1 & -1 \\ 1 & 1 \end{pmatrix}$ satisfies the definition. You might recall that there is formula for the inverse of a 2×2 matrix. It gives this:

$$P^{-1} = \frac{1}{(1)(1) - (-1)(1)} \begin{pmatrix} 1 & 1 \\ -1 & 1 \end{pmatrix} = \begin{pmatrix} 1/2 & 1/2 \\ -1/2 & 1/2 \end{pmatrix}.$$

And note that

$$\begin{aligned} P^{-1}AP &= \begin{pmatrix} 1/2 & 1/2 \\ -1/2 & 1/2 \end{pmatrix} \begin{pmatrix} 1 & 2 \\ 0 & -1 \end{pmatrix} \begin{pmatrix} 1 & -1 \\ 1 & 1 \end{pmatrix} \\ &= \begin{pmatrix} 1/2 & 1/2 \\ -1/2 & -3/2 \end{pmatrix} \begin{pmatrix} 1 & -1 \\ 1 & 1 \end{pmatrix} \\ &= \begin{pmatrix} 1 & 0 \\ -2 & -1 \end{pmatrix} \\ &= B. \end{aligned}$$

This shows that $A \sim B$. Moreover, by swapping P and P^{-1}, a similar calculation shows that $B \sim A$.

It turns out that if two matrices are similar, then they share a lot of the same properties. If $A \sim B$, then A and B have the same determinant, rank, trace and eigenvalues; they represent the same linear transformation, under different bases; and more. The question whose answer I am sure you are all burning to know: Is $\sim$ an equivalence relation?? It is! Let's prove it.

[23]If you took a course in linear algebra and never saw a matrix, you had a badass prof. (But if you never saw a matrix *or* a linear transformation, then you may have just gotten the room wrong...)

[24]Note: If all of this is foreign to you, just skip over it. This page and the next assumes that you have seen some of this stuff before.

Proposition.

Proposition 9.17. Let $M_n(\mathbb{R})$ denote the set of $n \times n$ matrices of real numbers. Define the relation $\sim$ on $M_n(\mathbb{R})$ for which

$$A \sim B \qquad \text{if} \qquad A \text{ is similar to } B.$$

Then, the relation $\sim$ is an equivalence relation.

Proof. Suppose A, B and C are $n \times n$ matrices.

Reflexive: To see that $A \sim A$, let I be the $n \times n$ identity matrix, and recall that I is invertible and $I^{-1} = I$. Then, $I^{-1}AI = IAI = AI = A$. Thus, we have found an invertible matrix P for which $P^{-1}AP = A$, showing that $A \sim A$, as desired.

Symmetric: Assume that $A \sim B$. Then, there exists an invertible matrix P for which $P^{-1}AP = B$. Observe that by multiplying on the left by P and on the right by P^{-1}, the equation $P^{-1}AP = B$ turns in to

$$\begin{aligned} PP^{-1}APP^{-1} &= PBP^{-1} \\ IAI &= PBP^{-1} \\ A &= PBP^{-1}. \end{aligned} \qquad (🚘)$$

Our goal is to show that $B \sim A$; that is, we want to show that there is an invertible matrix Q such that $Q^{-1}BQ = A$. I claim that $Q = P^{-1}$ works. Recall that since P is invertible, P^{-1} is also invertible and $\left(P^{-1}\right)^{-1} = P$. Now, simply observe that

$$Q^{-1}BQ = \left(P^{-1}\right)^{-1}BP^{-1} = PBP^{-1} = A,$$

the last equality by (🚘).

Transitive: Assume that $A \sim B$ and $B \sim C$. Then, there exist invertible matrices P and Q for which

$$P^{-1}AP = B \qquad \text{and} \qquad Q^{-1}BQ = C. \qquad (✈)$$

Our goal is to show that $A \sim C$; that is, we want to show that there is an invertible matrix R such that $R^{-1}AR = C$. I claim that $R = PQ$ works. Recall that since P and Q are invertible, PQ is also invertible and $(PQ)^{-1} = Q^{-1}P^{-1}$. Now, simply observe that

$$R^{-1}AR = (PQ)^{-1}A(PQ) = Q^{-1}P^{-1}APQ = Q^{-1}(P^{-1}AP)Q = Q^{-1}BQ = C,$$

the last two equalities by the two identities in (✈).

We proved that $\sim$ is reflexive, symmetric and transitive, and hence is an equivalence relation. □

Partial Orders

A relation is a very general idea. An equivalence relation is a special case of a relation and, while still quite general, has many more concrete properties. There is another special case of a relation that is worth mentioning, even though it does not fit into the main storyline of this chapter and hence is relegated to the Bonus Examples section.

In Example 9.2, we saw that the "$\leq$" inequality is *not* an equivalence relation. It is reflexive ($a \leq a$) and transitive (if $a \leq b$ and $b \leq c$, then $a \leq c$), but it is not symmetric (if $a \leq b$, there is no guarantee that $b \leq a$). In fact, one of the main properties of the inequality '$\leq$' is that the only possible way for $a \leq b$ and $b \leq a$, is if $a = b$. This property is called *antisymmetry*.

The equal sign ($=$) is so important, that we identified its three most important properties (reflexivity, symmetry and transitivity) and asked what else has those big three properties, and called anything that does an equivalence relation. In the same way, the inequality sign ($\leq$) is so important that we will identify *its* three most important properties (reflexivity, antisymmetry and transitivity) and ask what else has those big three properties. Anything that does also gets a fancy name: a *partial order*. Finally, to help distinguish it from equivalence relations, instead of using the $\sim$ symbol, we will use this symbol: $\lesssim$.

Definition.

Definition 9.18. Let $\lesssim$ be a relation on a set A. We say that $\lesssim$ is a *partial order* on A if it satisfies the following three properties.

- Reflexive: $a \lesssim a$ for all $a \in A$;
- Antisymmetric: If $a \lesssim b$ and $b \lesssim a$, then $a = b$.
- Transitive: If $a \lesssim b$ and $b \lesssim c$, then $a \lesssim c$ for all $a, b, c \in A$.

Lastly, if A is a set which has a partial order $\lesssim$, then A is called a *poset*.

The most important example of a partial order is what a partial order is designed to mimic: $\leq$ is a partial order on $\mathbb{R}$. Below is the second most important example: $\subseteq$ is a partial order on $\mathcal{P}(\mathbb{N})$.

Example 9.19. Let $\lesssim$ be the relation on $\mathcal{P}(\mathbb{N})$ where

$$a \lesssim b \qquad \text{if} \qquad a \subseteq b.$$

That is, a and b come from the set $\mathcal{P}(\mathbb{N})$, meaning that they are both sets containing only natural numbers. We say that $a \lesssim b$ if the set a is a subset of the set b. For example, $\{1, 4, 6\} \lesssim \{1, 2, 4, 6, 7\}$ since $\{1, 4, 6\} \subseteq \{1, 2, 4, 6, 7\}$.

Let's show that $\lesssim$ is a partial order on $\mathcal{P}(\mathbb{N})$. To do so, we must show that it satisfies the three properties from Definition 9.18.

Reflexive: To see that $a \lesssim a$ for all $a \in \mathcal{P}(\mathbb{N})$, simply note that every set is a subset of itself. Therefore, $a \lesssim a$ for any $a \in \mathcal{P}(\mathbb{N})$, which proves that $\lesssim$ is reflexive.

Antisymmetric: Assume that $a \lesssim b$ and $b \lesssim a$ for some $a, b \in \mathcal{P}(\mathbb{N})$. This means that $a \subseteq b$ and $b \subseteq a$. This is precisely what we need to prove that $a = b$, according to our summary on page 105. This proves that $\lesssim$ is antisymmetric.

Transitive: Assume that $a \lesssim b$ and $b \lesssim c$, for some $a, b, c \in \mathcal{P}(\mathbb{N})$. That is, $a \subseteq b$ and $b \subseteq c$. To see that $a \subseteq c$, pick any element $x \in a$. Since $x \in a$ and $a \subseteq b$, by the definition of a subset (Definition 3.4) we have $x \in b$. Then, since $x \in b$ and $b \subseteq c$, by the definition of a subset (Definition 3.4) we have $x \in c$. We have proven that if $x \in a$, then $x \in c$, hence showing that $a \subseteq c$, again by the definition of a subset (Definition 3.4). This proves that $\lesssim$ is transitive.

Since $\lesssim$ is reflexive, antisymmetric and transitive, $\lesssim$ is a partial order. □

The one crucial difference between a partial order $\lesssim$ and the typical inequality $\leq$ on $\mathbb{R}$ is that in $\mathbb{R}$, every two numbers a and b will either have $a \leq b$ or $b \leq a$. One of the two is bigger, and the inequality detects that.

Meanwhile, we made no such demands of a partial order. It could very well be that $\lesssim$ is a partial order on some set A, and for two of the elements $a, b \in A$, we have $a \not\lesssim b$ and $b \not\lesssim a$. Indeed, we saw this in the last example: note that $\{1,2,3\} \not\subseteq \{2,3,4\}$ and $\{2,3,4\} \not\subseteq \{1,2,3\}$. Thus, $\{1,2,3\}$ and $\{2,3,4\}$ are "incomparable."

This allows for some cool pictures of posets. In the diagram below, we are looking at a partially ordered set on $A = \{1,2,3\}$, where the ordering is again $a \lesssim b$ if $a \subseteq b$. In this diagram, if $a \lesssim b$, then there is a line between a and b, and b appears in a more vertical position. If a and b are incomparable, then they are drawn at the same vertical level.

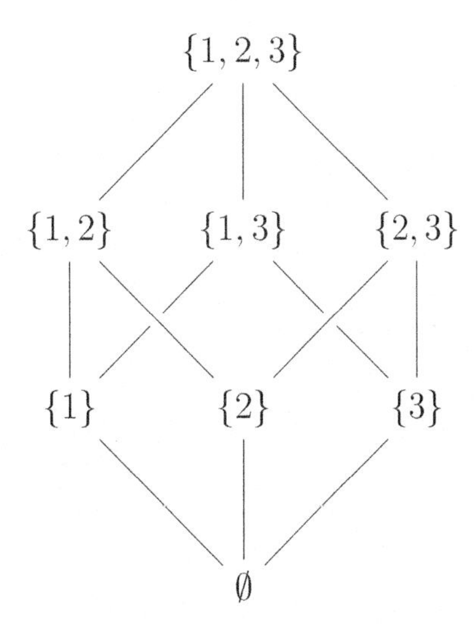

These are called *Hasse diagrams.*

In Exercise 9.36 you will be asked to show that if $A = \mathbb{N}$ and $\lesssim$ is the relation where $a \lesssim b$ whenever $a \mid b$, then $\lesssim$ is a partial order on A. You will also be asked to draw a Hasse diagram for a small case of this.

— Chapter 9 Pro-Tips —

- Equivalence relations are a really important topic that you will encounter in future courses, most notably in abstract algebra. Taking the time now to understand them deeply will pay dividends in your future.

- Many budding mathematicians struggle to appreciate abstraction, in part because an abstract mathematical object can represent so many different things that it is easy to think it represents nothing. When you think of an equivalence relation, keep concrete examples like the equivalence of integers modulo 5, and floor functions, at the front of your mind. When you take abstract algebra, you will learn about many abstract algebraic objects, and I recommend that you constantly refer back to concrete examples. This should improve your ability to learn and appreciate the material.

  I also encourage you to find value and beauty in abstraction. If you knew that an apple was a fruit but did not know of any other fruits, then you might guess that every fruit is spherical, has a stem, is crunchy, and tastes just-ok. But once you know that bananas and peaches and mangoes and watermelon and raspberries are all fruit, then you will have a richer understanding of what "fruit" really is. And with some thought, you can even gain a better understanding of what apples really are.

  This is the value of abstraction. If you thought every animal looked like a squirrel, then imagine your wonder as you strolled through a zoo, and discovered the whole animal kingdom, and began to note the many ways that these animals are different, yet the essential ways that they are the same. This mindset will help you appreciate learning about abstract mathematical objects.

- It's good for mathematicians to know some of our field's history. I've tried to drizzle some in throughout this text, but I will steal an opportunity now to share a bit more. Following this chapter is an introduction to group theory, so here is some interesting history about one of the most important mathematicians in the history of group theory: Évariste Galois.

  Galois was a transcendent mathematician who developed much of group theory as a teenager, yet struggled to get his work noticed; he was a political firebrand with conspiratorial tendencies and a thirst for revolution during a turbulent time in French politics; and he was a romantic who fell in love with a young lady, but whose love she never returned. In fact, I believe out of everyone in the history of mathematics, Galois would have been, hands down, the most exciting to follow on Twitter.[25]

  There are many fascinating aspects to Galois' life, but the one fact about his life that nearly every working mathematician can immediately recite... is how he died.[26] Galois died in a duel at the age of 20, which was likely related to one of these two non-math passions which drove so many of his emotions.

[25]While I can't promise conspiratorial revolutions or Earth-shaking mathematics, here is my Twitter handle: @LongFormMath.

[26]And that's never a good sign. If people can recite many facts about you, but nobody can

A tragic story to lose a life so young, and there is little doubt that this young man, who changed the mathematical landscape with just a few years of work, would have left a colossal legacy if he had decades more.

- Being humble is virtuous, but you should still pat yourself on the back and celebrate your successes. You just read a whole freaking book on advanced mathematics! That's a heck of an achievement, and you should smile and be proud of yourself for accomplishing it.

  I would also like to emphasize that this is more than an isolated thought — it really is a pro-tip. It can be tough to be a pro mathematician, as you are constantly trying to tackle harder and harder math problems. You experience this as a student too, when you take harder and harder classes. Gone are the days when you would solve 30 problems on your homework, and get 30 reassuring moments when you could feel happy about your complete solution. Solutions are harder to come by now — you might work well over an hour for a single proof. Be happy when you get one! Smile, and give your study mates a high five!

  This continues when you begin research. The wins are even less frequent, but they can also be that much more gratifying... assuming you give yourself the freedom to congratulate yourself. Doing so is even more important than when a student, because if you start missing opportunities to feel pleased with your hard work, who knows how much longer it will be until the next chance. So stay on top of it! A positive mindset and a healthy mind are important.

- Your final pro-tip of the book is simply this: Try to find joy in your mathematics. Try to make friends with classmates to form study groups and enjoy the material together. The joy is in feeling a camaraderie among your classmates, rather than competition.[27]

  If you are working on a homework problem or research project, embrace the adventure of not knowing an answer, and searching for that answer on your own or with friends, rather than trying to find a solution online. The joy is in the hunt, and in realizing that the search itself is when the most important learning takes place.

  Math is a huge field, and no one likes all of it. Don't be discouraged if one area seems too dull or difficult. The joy is in finding an area that excites you, and pursuing it further.

  Go out of your way to teach others — both younger math students and those that are not studying math at all. And when you do, be energetic and enthusiastic. More than anything, it is your attitude about math that will resonate with them. A week or two later, they may forget every detail about the math that you shared — but they will remember one thing: your joy.

remember how you died, then you probably lived a good, long life. Hope to be a George Washington over an Abraham Lincoln.

[27]Maybe the real theorems were the friends you made along the way.

— Exercises —

Exercise 9.1. Write a reflection on your experience in this course. In what ways do you now view math differently? What is a proposition/theorem that you liked? What is a proof that you liked? What was a time in which you struggled but found the experience valuable? What areas of math are you excited to learn more about? What else would you like to share?

Exercise 9.2. Give four examples of relations that we did not mention in the chapter. Have two of them be real-world examples and two of them be math examples.

Exercise 9.3. Let $A = \{1, 2, 3, 4, 5\}$. Each part below is a separate definition for the relation "$a \sim b$." For each, write out all pairs that are related.

(a) $a \sim b$ when $a < b$

(b) $a \sim b$ when $a \mid b$

(c) $a \sim b$ when $a \geq b$

(d) $a \sim b$ when $a + b$ is odd

Exercise 9.4.

(a) List all partitions of the set $\{1, 2\}$.

(b) List all partitions of the set $\{a, b, c\}$.

Exercise 9.5. Give four examples of partitions that we did not mention in the chapter. Have two of them be real-world examples and two of them be math examples.

Exercise 9.6. Suppose a group of 10 people are in a room, and some of them shake hands with some of the others (but not everyone shakes hands with everyone). If A is the set of 10 people, consider the relation $\sim$ on A where $a \sim b$ if person a shook hands with person b. Given just this information, is $\sim$ guaranteed to be reflexive? Symmetric? Transitive? Give a brief justification for each.

Exercise 9.7. Consider the relation $\sim$ on the set $\{w, x, y, z\}$ such that this is the complete list of related elements:

$$z \sim z \qquad x \sim y \qquad y \sim x$$

$$w \sim w \qquad x \sim x \qquad y \sim y$$

Is $\sim$ reflexive? Symmetric? Transitive? If a property holds, you do not need to justify it. If it doesn't, say why it fails. If all three hold, then $\sim$ is an equivalence relation; in this case, list the equivalence classes.

Exercise 9.8. Consider the relation $\sim$ on the set $\{w, x, y, z\}$ such that this is the complete list of related elements:

$$y \sim y \qquad y \sim x \qquad w \sim y$$

$$w \sim x \qquad x \sim y \qquad x \sim x$$

Is $\sim$ reflexive? Symmetric? Transitive? If a property holds, you do not need to justify it. If it doesn't, say why it fails. If all three hold, then $\sim$ is an equivalence relation; in this case, list the equivalence classes.

Exercise 9.9. Consider the following equivalence relation $\sim$ on the set $\{1, 2, 3, 4, 5, 6\}$ such that this is the complete list of related elements:

$$1 \sim 1 \qquad 2 \sim 2 \qquad 3 \sim 3 \qquad 4 \sim 4 \qquad 5 \sim 5 \qquad 6 \sim 6 \qquad 1 \sim 2$$

$$2 \sim 1 \qquad 4 \sim 5 \qquad 5 \sim 4 \qquad 5 \sim 6 \qquad 6 \sim 5 \qquad 4 \sim 6 \qquad 6 \sim 4$$

Determine the equivalence classes of $\sim$.

Exercise 9.10. Let $\sim$ be the relation on $\mathbb{N}$ where the complete set of related pairs is

$$\{a \sim a : a \in \mathbb{N}\}.$$

That is, $1 \sim 1$ and $2 \sim 2$ and $3 \sim 3$ and so on. Is $\sim$ an equivalence relation?

Exercise 9.11.

(a) Give an example of a relation on the set $\{1, 2, 3, 4\}$ which is reflexive and symmetric, but not transitive.

(b) Give an example of a relation on the set $\{1, 2, 3, 4\}$ which is reflexive and transitive, but not symmetric.

(c) Give an example of a relation on the set $\{1, 2, 3, 4\}$ which is transitive and symmetric, but not reflexive.

(d) Give an example of a relation on the set $\{1, 2, 3, 4\}$ which is not reflexive, symmetric or transitive.

Exercise 9.12. Let $A = \mathcal{P}(\mathbb{N})$. Let $\sim$ be the relation on A where $a \sim b$ provided $a \subseteq b$. Is $\sim$ reflexive? Symmetric? Transitive? For each property, prove that it holds or find a counterexample. Is $\sim$ an equivalence relation?

Exercise 9.13. Let $\sim$ be a relation on $\mathbb{N}$ where $a \sim b$ when $a \mid b$. Is $\sim$ reflexive? Symmetric? Transitive? For each property, prove that it holds or find a counterexample. Is $\sim$ an equivalence relation? If so, what are its equivalence classes?

Exercise 9.14. Each of the following rules defines a relation on $\mathbb{R}$. Determine which define an equivalence relation. If one does, prove that it is an equivalence relation and find its equivalence classes. If one does not, then show by example which of the reflexive/symmetric/transitive properties does not hold.

(a) $a \sim b$ when $a - b \in \mathbb{N}$

(b) $a \sim b$ when $a - b \in \mathbb{Z}$

(c) $a \sim b$ when $a - b \in \mathbb{Q}$

(d) $a \sim b$ when $a - b \in \mathbb{R}$

Exercise 9.15. Each of the following rules defines a relation on $\mathbb{Z}$. For each part, prove that $\sim$ is an equivalence relation and find its equivalence classes.

(a) $a \sim b$ when $a \equiv b \pmod 6$

(b) $a \sim b$ when $7a - 3b$ is even

(c) $a \sim b$ when $a^2 \equiv b^2 \pmod 4$

(d) $a \sim b$ when $a^2 + b^2$ is even

(e) $a \sim b$ when $2a + b \equiv 0 \pmod 3$

(f) $a \sim b$ when $a + 3b \equiv 0 \pmod 4$

Exercise 9.16. Each of the following rules defines a relation on $\mathbb{Z}$. For each, is $\sim$ reflexive? Symmetric? Transitive? If a property holds, provide a brief justification. If it doesn't, say why it fails. If all three hold, then $\sim$ is an equivalence relation; in this case, list the equivalence classes.

(a) $a \sim b$ when $a^2 = b^2$

(b) $a \sim b$ when $|a - b| \leq 5$

(c) $a \sim b$ when $a \neq b$

(d) $a \sim b$ when $ab \geq 0$

Exercise 9.17. Let $\sim$ be the relation on $\mathbb{R} \times \mathbb{R}$ where $(a, b) \sim (c, d)$ when $|a| + |b| = |c| + |d|$. Prove that $\sim$ is an equivalence relation.

Exercise 9.18. Let A be a nonempty set and let P be a partition of A, written as a collection of sets. For example, if $A = \{1, 2, 3, 4\}$, then perhaps $P = \{\{1, 3\}, \{2\}, \{4\}\}$.

Let $\sim$ be the relation on A where $a \sim b$ if there is some $Q \in P$ such that both $a \in Q$ and $b \in Q$. Prove that $\sim$ is an equivalence relation on A.

Exercise 9.19. Let $d \in \mathbb{N}$ and consider the set P containing an infinite arithmetic progression:

$$P = \{\ldots, -3d, -2d, -d, 0, d, 2d, 3d, \ldots\}.$$

Let $\sim$ be the relation on $\mathbb{N}$ where $a \sim b$ if $a - b \in P$. Is $\sim$ reflexive? Symmetric? Transitive? If a property holds, you do not need to justify it. If it doesn't, say why it fails. If all three hold, then $\sim$ is an equivalence relation; in this case, list the equivalence classes.

Exercise 9.20. If P_{EEPS} is the set of people in the world, and we define a relation as $a \sim b$ if person a has the same birthday as person b, then $\sim$ is an equivalence relation on P_{EEPS}. Give three other real-world examples of an equivalence relation.

Exercise 9.21. Let $\mathcal{D}$ be the set of words in the English dictionary and $\sim$ be the relation for which

$$a \sim b \qquad \text{if} \qquad \text{the word } a \text{ rhymes with the word } b.$$

Use a website like `rhymezone.com` to determine the equivalence class [exponent].

Exercise 9.22. Let $A = \{a, b, c, d, e\}$, and suppose that $\sim$ is an equivalence relation on A. Assume that $\sim$ has two equivalence classes, and that $b \sim e$, $c \sim d$ and $a \sim e$. Determine all related pairs.

Exercise 9.23. In this exercise we will put some rigor behind the practice of thinking of fractions in their "lowest terms," which was a central idea in the proof that $\sqrt{2}$ is irrational. We will represent a fraction $\frac{a}{b}$ as an ordered pair (a, b) where $b \neq 0$, and the equality $\frac{a}{b} = \frac{c}{d}$ will be thought of as $ad = bc$.

Let $A = \{(a, b) : a, b \in \mathbb{Z} \text{ and } b \neq 0\}$. Define the relation $\sim$ on A to be

$$(a, b) \sim (c, d) \qquad \text{if} \qquad ad = bc.$$

Prove that $\sim$ is an equivalence relation.

Exercise 9.24. Let A be an infinite set.

(a) Give an example of an equivalence relation on A which has finitely many equivalence classes.

(b) Give an example of an equivalence relation on A which has infinitely many equivalence classes.

Exercise 9.25.

(a) Let $\sim$ be the relation on $\mathbb{Z}$ where $a \sim b$ when $a \equiv b \pmod 2$ and $a \equiv b \pmod 3$. Is $\sim$ an equivalence relation?

(b) Let $\sim$ be the relation on $\mathbb{Z}$ where $a \sim b$ when $a \equiv b \pmod 2$ or $a \equiv b \pmod 3$. Is $\sim$ an equivalence relation?

Exercise 9.26. Suppose $\sim_1$ and $\sim_2$ are equivalence relations on a set A. Let $\sim$ be the relation on A where $a \sim b$ if both $a \sim_1 b$ and $a \sim_2 b$. Is it true that $\sim$ is an equivalence relation on A? Either prove that it is an equivalence relation, or give a counterexample. (For the counterexample, you would provide a set A and equivalence relations $\sim_1$ and $\sim_2$ on A. Justify that $\sim_1$ and $\sim_2$ are equivalence relations on A, and that $\sim$ is not an equivalence relation on A.)

Exercise 9.27. Suppose $\sim_1$ and $\sim_2$ are equivalence relations on a set A. Let $\sim$ be the relation on A where $a \sim b$ if either $a \sim_1 b$ or $a \sim_2 b$. Is it true that $\sim$ is an equivalence relation on A? Either prove that it is an equivalence relation, or give a counterexample. (For the counterexample, you would provide a set A and equivalence relations $\sim_1$ and $\sim_2$ on A. Justify that $\sim_1$ and $\sim_2$ are equivalence relations on A, and that $\sim$ is not an equivalence relation on A.)

Exercise 9.28. Let A and B be sets. Suppose $\sim_1$ is an equivalence relation on A and $\sim_2$ is an equivalence relation on B. Define a relation $\sim$ on $A \times B$ where $(a, b) \sim (c, d)$ if $a \sim_1 c$ and $b \sim_2 d$.

(a) Prove that $\sim$ an equivalence relation on $A \times B$.

(b) Describe the equivalence classes of $\sim$ in terms of the equivalence classes of $\sim_1$ and $\sim_2$.

Exercise 9.29. Determine a familiar equivalence relation whose equivalence classes are the following:

$$\{\dots, -6, -3, 0, 3, 6, \dots\}\ ,\ \{\dots, -5, -2, 1, 4, 7, \dots\}\ ,\ \{\dots, -4, -1, 2, 5, 8, \dots\}.$$

Exercise 9.30. Determine a familiar equivalence relation whose equivalence classes are the following:

$$\{0\}\ ,\ \{-1, 1\}\ ,\ \{-2, 2\}\ ,\ \{-3, 3\}\ ,\ \{-4, 4\}\ ,\ \dots$$

Exercise 9.31. Explain the error in the following "fake proof" that if $\sim$ is a relation on A that is both symmetric and transitive, then $\sim$ is guaranteed to be reflexive.

Fake Proof. Assume that $\sim$ is symmetric and transitive. By symmetry, if $a \sim b$, then $b \sim a$. By transitivity, since $a \sim b$ and $b \sim a$, also $a \sim a$. We have shown that $a \sim a$, proving that a is reflexive. □

Exercise 9.32. Definition 9.14 provided a connection between functions and relations. Give an example of a function $f_1 : A \to B$ for which

$$a \sim b \qquad \text{if} \qquad f_1(a) = b$$

is an equivalence relation. And then give an example of a function $f_2 : A \to B$ for which

$$a \sim b \qquad \text{if} \qquad f_2(a) = b$$

is not an equivalence relation.

Exercise 9.33. How many relations are there from $\{1, 2, 3\}$ to $\{1, 2, 3\}$? For $n \in \mathbb{N}$, how many functions are there from $\{1, 2, \dots, n\}$ to $\{1, 2, \dots, n\}$? How many relations from $\{1, 2, \dots, n\}$ to $\{1, 2, \dots, n\}$ are not functions?

Exercise 9.34. Definition 9.14 showed us that a function is a special type of relation. In the below, let $\mathbb{R}^+$ be the set of positive real numbers.

(a) Consider the function $f : \mathbb{R} \to \mathbb{R}$, where $f(x) = x$. When viewed as a relation, is f reflexive? Symmetric? Transitive? Is f an equivalence relation?

(b) Consider the function $f : \mathbb{R}^+ \to \mathbb{R}^+$, where $f(x) = \dfrac{1}{x}$. When viewed as a relation, is f reflexive? Symmetric? Transitive? Is f an equivalence relation?

(c) Which functions $f : \mathbb{R} \to \mathbb{R}$ are equivalence relations?

Exercise 9.35. Note 9.13 allows us to think about a relation on $\mathbb{R}$ or $\mathbb{Z}$ as a subset of $\mathbb{R} \times \mathbb{R}$ or $\mathbb{Z} \times \mathbb{Z}$. This in turn allows to graph a relation on the xy-plane, either as a shaded region or points of the integer grid. Each of the following corresponds to a familiar relation on $\mathbb{R}$ or $\mathbb{Z}$. Determine this relation for each.

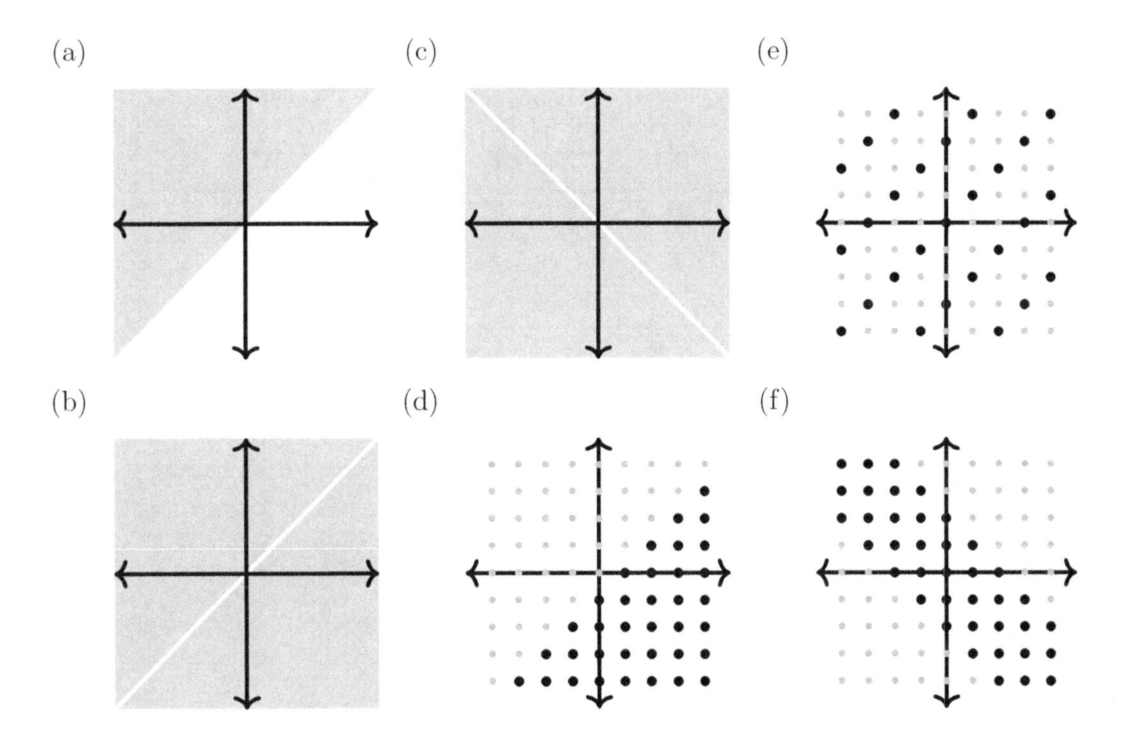

Exercise 9.36. In the Bonus Examples section for this chapter, we discussed partial orders. Read that section before answering the questions below.

(a) Suppose $A = \mathbb{N}$ and $\lesssim$ is the relation on A where $a \lesssim b$ whenever $a \mid b$. Prove that $\lesssim$ is a partial order on A.

(b) Suppose $A = \{1, 2, 3, 5, 6, 10, 15, 30\}$ and $\lesssim$ is the partial order on A where $a \lesssim b$ whenever $a \mid b$. Draw a Hasse diagram of this partial order.

Exercise 9.37. For $n \in \mathbb{N}$, let $\sim_n$ be the relation $a \sim_n b$ when $a \equiv b \pmod{n}$. Using Notation 9.9, the equivalence classes of $\sim_n$ are $[0]$, $[1]$, $[2]$, $\ldots$, $[n-1]$.

(a) Write out the addition and multiplication tables for the equivalence classes for the $n = 4$ case.

(b) Write out the addition and multiplication tables for the equivalence classes for the $n = 5$ case.

(c) Write out the addition and multiplication tables for the equivalence classes for the $n = 6$ case.

(d) Looking at the example below and your tables from parts (a), (b) and (c), name a few things that you find interesting.

As an example, here are the addition and multiplication tables for $n = 3$:

+	[0]	[1]	[2]
[0]	[0]	[1]	[2]
[1]	[1]	[2]	[0]
[2]	[2]	[0]	[1]

·	[0]	[1]	[2]
[0]	[0]	[0]	[0]
[1]	[0]	[1]	[2]
[2]	[0]	[2]	[1]

Exercise 9.38. Read the *Introduction to Group Theory* following this chapter. In this section, D_6 was defined in terms of an equilateral triangle. The group D_8 is the analogous group defined as the group of rotations and reflections of a square to it itself. How many elements are there in D_8? Describe and draw them.

— Open Question —

Suppose I partitioned $\mathbb{N}$ into two sets, A and B. And suppose you were hoping that I partitioned this set in such a way so that either A or B (or both) had numbers x and y for which $x + 3 = y$. Here's a question: Is it guaranteed that you will get your wish? The answer is no. Sure, it's possible. As two examples:

- If I chose $A = \{1, 2, 3, 4, 5\}$ and $B = \{6, 7, 8, 9, 10, 11, \dots\}$ as my partition of $\mathbb{N}$, then both sets have the property! For the first set, if $x = 1$ and $y = 4$, then $x + 3 = y$. And for the second set, if $x = 15$ and $y = 18$, then $x + 3 = y$.

- If I instead chose $A = \{3, 4, 9\}$ and $B = \{1, 2, 5, 6, 7, 8, 10, 11, 12, 13, 14, \dots\}$, then A does not contain any x and y for which $x+3 = y$, but that's ok because B does. For example, if $x = 5$ and $y = 8$, then $x + 3 = y$.

The subtle point: Although it is *possible* that I partitioned $\mathbb{N}$ so that you get your wish (at least one of the sets A or B contains numbers x and y for which $x + 3 = y$), you are not *guaranteed* that I did so. For example, perhaps this was my partition:

- If I chose $A = \{1, 3, 5, 7, 9, \dots\}$ and $B = \{2, 4, 6, 8, 10, \dots\}$ as my partition of $\mathbb{N}$, then neither set has the property. Why? Well, the first set only contains odd numbers, and there is no way that x and y are both odd, and $x + 3 = y$; an odd number plus three is never an odd number. So if x and y are from A, they can't have the property. Likewise, if x and y are from B, then they must be even numbers, and so we can't possibly have $x + 3 = y$, since an even plus three is never an even.

This shows that if we partition $\mathbb{N}$ into two sets, we are not *guaranteed* that one of the two sets contains two elements x and y for which $x + 3 = y$. It's possible, but not guaranteed. (You can think about it as us finding a counterexample.)[28]

However, if we asked whether every partition of $\mathbb{N}$ into two sets has the property that one of the two sets contains two elements x and y for which $x \mid y$, then now the answer is yes. Why? Well, let's again call the two sets A and B. Since it is a partition of $\mathbb{N}$, every element from $\mathbb{N}$ has to go into one of these two sets. It turns out we can prove it simply by asking this: In which sets did 3, 6 and 12 go?

Since we have three numbers being placed into two sets, by the pigeonhole principle one of these sets received at least two of these three elements. If one of the two sets has 3 and 6 in it, then we are done, as we can let $x = 3$ and $y = 6$ and then we have found an x and y from the same set for which $x \mid y$. Likewise, if 3 and 12 are in the same set, then $x = 3$ and $y = 12$ gives $x \mid y$. And if 6 and 12 are in the same set, then $x = 6$ and $y = 12$ gives $x \mid y$. Thus, in all the cases, we are guaranteed that two of these three numbers will end up in the same set, and when they do, the smaller will divide the larger, proving that the property holds. Since our reasoning did not depend on which partition we chose, we are guaranteed that *any* partition of $\mathbb{N}$ into

[28]FYI: That's not the only counterexample. Another one: $A = \{1, 2, 3, 7, 8, 9, 13, 14, 15, 19, \dots\}$ and $B = \{4, 5, 6, 10, 11, 12, 16, 17, 18, 21, \dots\}$.

two sets A and B has the property that one of these two sets (or both!) contains numbers x and y for which $x \mid y$.

The last two examples show that some properties (like $x \mid y$) are guaranteed to occur no matter how you partition your set. But other properties (like $x + 3 = y$) are not guaranteed — maybe some partitions have the property, but not *all* partitions.

Beyond the "$x + 3 = y$" and "$x \mid y$" properties, what other properties should we look at? One that mathematicians are especially interested in are arithmetic progressions. An *arithmetic progression* is a sequence of numbers that are equally spaced. For example, $1, 3, 5, 7$ is a sequence of 4 numbers, with a gap of 2 between consecutive terms. Likewise, $3, 6, 9, 12, 15$ is a sequence of 5 numbers, with a gap of 3 between consecutive terms.[29] The first example is a 4-term arithmetic progression, the second example is a 5-term arithmetic progression.

If we partition $\mathbb{N}$ into two sets, are we guaranteed that one of the two sets contains a 3-term arithmetic progression? Answer: yes! In fact, we can do better. If we partition just the set $\{1, 2, 3, 4, 5, 6, 7, 8, 9\}$ into two sets, we are guaranteed that one of the two sets contains a 3-term arithmetic progression. We are not able to go any smaller, though. It is not true that every partition of $\{1, 2, 3, 4, 5, 6, 7, 8\}$ into two sets has this property; for example, $A = \{1, 2, 5, 6\}$ and $B = \{3, 4, 7, 8\}$ is a partition and neither of these two sets has three numbers which are equally spaced.

As it turns out, if we partition $\{1, 2, 3, 4, \dots, 35\}$ into two sets, we are guaranteed that one of these two sets contains a 4-term arithmetic progression. But not every partition of $\{1, 2, 3, 4, \dots, 34\}$ has this property.

Likewise, if we partition $\{1, 2, 3, 4, \dots, 178\}$ into two sets, we are guaranteed that one of these two sets contains a 5-term arithmetic progression. But not every partition of $\{1, 2, 3, 4, \dots, 177\}$ has this property.

Likewise, if we partition $\{1, 2, 3, 4, \dots, 1132\}$ into two sets, we are guaranteed that one of these two sets contains a 6-term arithmetic progression. But not every partition of $\{1, 2, 3, 4, \dots, 1131\}$ has this property.

What is the tipping point for 7-term arithmetic progressions? For 8-term? For 100-term? We would like to continue this list. Thus, here is a problem: For $k \in \mathbb{N}$, determine the value of N which has the property that, if we partition $\{1, 2, 3, \dots, N\}$ into two sets, we are guaranteed that one of these two sets contains a k-term arithmetic progression, but not every partition of $\{1, 2, 3, \dots, N - 1\}$ into two sets has this property. This N is denoted $W(2, k)$ and called a *van der Waerden number*.

Now, we could generalize this further. What if instead of partitioning into two sets, we partitioned it into three sets? Or four sets? In general, we could try to determine the value of N which has this property: If we partition $\{1, 2, 3, \dots, N\}$ into r sets, we are guaranteed that one of these r sets contains a k-term arithmetic progression, but not every partition of $\{1, 2, 3, \dots, N-1\}$ into r sets has this property. The N value that does this is denoted $W(r, k)$. It turns out that finding exact values for $W(r, k)$ is really hard; $W(r, k)$ is known for just a handful of values of r and k. Can we improve the known bounds on $W(r, k)$? Below is one notable conjecture.

Open Question. Is it true that $W(2, k) < 2^{k^2}$ for all $k \in \mathbb{N}$?

[29] Reader: "Huh, I learned a different definition." Me: "Eh, same difference."

Introduction to Group Theory

In the late 1800s, mathematicians kept noticing that the tools used to solve problems from one area were also being used to solve problems from another. Must they keep reinventing the wheel? Although these problems looked different, they were alike in some fundamental ways. To motivate this, let's look at four mathematical structures which, despite their many differences, share five important characteristics.

The integers, with addition

- Binary operation: Addition is an operation which combines *two* numbers to create another number.
- Closure: If m and n are integers, then $m + n$ is also an integer.
- Identity element: The number 0 is an integer, and it has the property that $n + 0 = n$ and $0 + n = n$ for every integer n.
- Invertibility: If n is an integer, then $-n$ is also an integer, and $n + (-n) = 0$. That is, every integer has an *inverse* and, when combined with its inverse, produces the identity element 0.
- Associativity: If k, m and n are integers, then $(k + m) + n = k + (m + n)$.

Nonzero real numbers, with multiplication

We will use the symbol $\times$ for multiplication of real numbers.

- Binary operation: Multiplication is an operation which combines *two* numbers to create another number.
- Closure: If x and y are nonzero real numbers, then $x \times y$ is also a nonzero real number.
- Identity element: The number 1 is a nonzero real number, and it has the property that $x \times 1 = x$ and $1 \times x = x$ for every nonzero real number x.
- Invertibility: If x is a nonzero real number, then $\frac{1}{x}$ is also a nonzero real number, and $x \times \frac{1}{x} = 1$. That is, every nonzero real number has an *inverse* and, when combined with its inverse, produces the identity element 1.
- Associativity: If x, y and z are nonzero real numbers, then $(x\times y)\times z = x\times(y\times z)$.

The set $\{0, 1, 2, 3, 4, 5\}$, with addition modulo 6

We will use the symbol $+_6$. For example, $4 +_6 5 = 3$ because $4 + 5 \equiv 3 \pmod 6$.

- Binary operation: Addition modulo 6 is an operation which combines *two* numbers to create another number.
- Closure: If m and n are in $\{0, 1, 2, 3, 4, 5\}$, then $m +_6 n$ is also in $\{0, 1, 2, 3, 4, 5\}$.
- Identity element: The number 0 is a number in $\{0, 1, 2, 3, 4, 5\}$, and it has the property that $n +_6 0 = n$ and $0 +_6 n = n$ for every n in $\{0, 1, 2, 3, 4, 5\}$.
- Invertibility: If n is a number in $\{0, 1, 2, 3, 4, 5\}$, then there is a number m from $\{0, 1, 2, 3, 4, 5\}$ for which $n +_6 m = 0$, and $m +_6 n = 0$. That is, every integer has an *inverse* and, when combined with its inverse, produces the identity element 0. Indeed, here is the m for each n:

n	m	$n +_6 m = 0$ and $m +_6 n = 0$?
0	0	✓
1	5	✓
2	4	✓
3	3	✓
4	2	✓
5	1	✓

- Associativity: If k, m and n are numbers in $\{0, 1, 2, 3, 4, 5\}$, then $(k +_6 m) +_6 n = k +_6 (m +_6 n)$.

Real 2×2 matrices with nonzero determinant, with matrix multiplication[30]

- Binary operation: Matrix multiplication is an operation which combines *two* matrices to create another matrix.
- Closure: If A and B are 2×2 matrices with nonzero determinant, then AB is also a 2×2 matrix with nonzero determinant. (cf. Linear algebra.)
- Identity element: The matrix $\begin{pmatrix} 1 & 0 \\ 0 & 1 \end{pmatrix}$ is a 2×2 matrix with nonzero determinant (and commonly called the *identity matrix*), and it has the property that

$$\begin{pmatrix} a & b \\ c & d \end{pmatrix}\begin{pmatrix} 1 & 0 \\ 0 & 1 \end{pmatrix} = \begin{pmatrix} a & b \\ c & d \end{pmatrix} \quad \text{and} \quad \begin{pmatrix} 1 & 0 \\ 0 & 1 \end{pmatrix}\begin{pmatrix} a & b \\ c & d \end{pmatrix} = \begin{pmatrix} a & b \\ c & d \end{pmatrix}$$

  for every 2×2 matrix $\begin{pmatrix} a & b \\ c & d \end{pmatrix}$ with nonzero determinant (or without!).

[30]By "real" 2×2 matrices, we simply mean that the four entries in the matrix are real numbers.

- Invertibility: If $\begin{pmatrix} a & b \\ c & d \end{pmatrix}$ is a 2×2 matrix with nonzero determinant, then $\begin{pmatrix} \frac{d}{ad-bc} & \frac{-b}{ad-bc} \\ \frac{-c}{ad-bc} & \frac{a}{ad-bc} \end{pmatrix}$ is also a 2×2 matrix with nonzero determinant, and

$$\begin{pmatrix} a & b \\ c & d \end{pmatrix}\begin{pmatrix} \frac{d}{ad-bc} & \frac{-b}{ad-bc} \\ \frac{-c}{ad-bc} & \frac{a}{ad-bc} \end{pmatrix} = \begin{pmatrix} 1 & 0 \\ 0 & 1 \end{pmatrix} \quad \text{and} \quad \begin{pmatrix} \frac{d}{ad-bc} & \frac{-b}{ad-bc} \\ \frac{-c}{ad-bc} & \frac{a}{ad-bc} \end{pmatrix}\begin{pmatrix} a & b \\ c & d \end{pmatrix} = \begin{pmatrix} 1 & 0 \\ 0 & 1 \end{pmatrix}.$$

  That is, every 2×2 matrix with nonzero determinant has an *inverse* and, when combined with its inverse, produces the identity element $\begin{pmatrix} 1 & 0 \\ 0 & 1 \end{pmatrix}$.

- Associativity: If A, B and C are 2×2 matrices with nonzero determinant (or without!), then $(AB)C = A(BC)$.

We just considered four sets, each with an operation: First, $\mathbb{Z}$ with $+$. Second, $\mathbb{R} \setminus \{0\}$ with $\times$. Third, $\{0, 1, 2, 3, 4, 5\}$ with $+_6$. Last, the set of real 2×2 matrices with nonzero determinant with matrix multiplication.

There are so many ways in which each of these is different, but in the above five ways they are the same. In fact, these five characteristics are so important that mathematicians have given a special name to a set and operation which satisfy these five characteristics, and we make every undergrad math major spend a month studying them. We call them *groups*.

First, if G is some set, then $*$ is called a *binary operation* on G if it combines two elements from G to create a single element of G. It will look like this: $a * b = c$. Addition, multiplication, addition modulo 6, and matrix multiplication are all binary operations.

Definition.

Definition 9.20. Let G be a set and let $*$ be some binary operation on G. We say G is a *group* under $*$ if it satisfies the following four properties.

1. Closure:[31] For every $a, b \in G$, we have $a * b \in G$.
2. Identity: There exists some $e \in G$ for which $a * e = e * a = a$ for all $a \in G$.
3. Inverses: For every $a \in G$, there exists some $b \in G$ such that $a*b = b*a = e$.
4. Associativity: For every $a, b, c \in G$, we have $(a * b) * c = a * (b * c)$.

These four properties are called the *group axioms*. And once we have this definition, we could, if we wanted, jump right into proving things about groups. For example, the second axiom tells us that there must be an identity element, like 0 for addition,

[31] Technically, the way we defined a binary relation automatically means it is closed, but we keep it here as emphasis.

or 1 for multiplication, or 0 for addition modulo 6, or $\begin{pmatrix} 1 & 0 \\ 0 & 1 \end{pmatrix}$ for 2×2 matrix multiplication. Is it the case that the identity element is always unique? Or could a group have two identity elements? Answer: There can only be one.

Proposition.

Proposition 9.21. Assume G is a group with operation $*$. Then G has only one identity element.

Proof. Assume for a contradiction that e_1 and e_2 are two different identity elements for G. Then, by simply stating what it means to be an identity element:

- $a * e_1 = a$ and $e_1 * a = a$ for all $a \in G$, and
- $a * e_2 = a$ and $e_2 * a = a$ for all $a \in G$.

But since these hold for every $a \in G$, they also hold for e_1 and e_2, since those must also be in G! Substituting $a = e_2$ into the first equality in the first bullet point, this means that $e_2 * e_1 = e_2$. And by substituting $a = e_1$ into the second equality in the second bullet point, this means that $e_2 * e_1 = e_1$. We have shown that

$$e_2 = e_2 * e_1 = e_1,$$

and so $e_2 = e_1$.

We had assumed that e_1 and e_2 were different identity elements, but have proved that they must be the same, giving the contradiction. This proves that a group's identity element must be unique. □

Even for matrix multiplication, it takes some thought to convince yourself that there can't be a second identity matrix. But by Proposition 9.21, there is only one identity matrix. And for *any* other set and operation which satisfy the group axioms, you immediately know that the identity element is unique. And it doesn't stop there... there are whole books filled with theorems about groups! We have already seen four groups — the four sets and operations with which we began this introduction — and every theorem from those books on group theory applies equally to each of those four groups.

There are two special classes of groups that we turn to next, called *dihedral groups* and *permutation groups*. First, the dihedral group D_6, which looks at rigid motions of a triangle.[32]

[32]Note: Some people call this group D_3. I like to really think about these things and decide what I think is best. But this is a really tough one. Compelling arguments on both sides...

The Dihedral Group

Consider an equilateral triangle sitting in the plane.

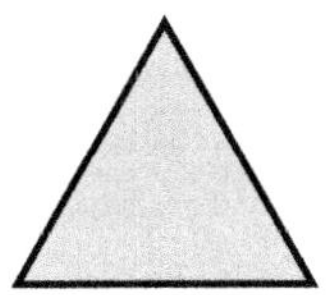

In a moment we are going to be rotating and flipping this triangle, so let's label it in some way so that we can tell how it was moved.[33]

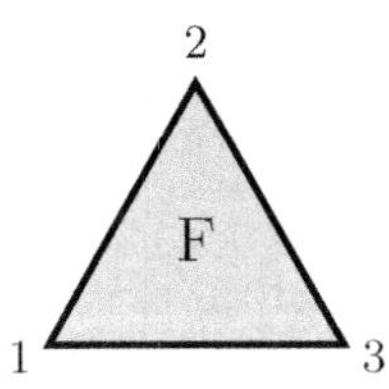

We are interested in describing, in some reasonable fashion, every way in which we can pick up this triangle, move it around however we please, and then place it back onto the same region of the plane it originally occupied.

The first thing we could do is the most boring one: We could pick it up, do nothing to it, and then set it back down. Then it would again look like this:

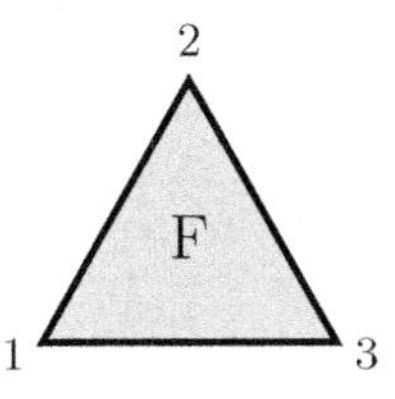

This may seem silly, but if you remember that we are discussing groups, and that every group has an identity element, this suddenly seems important. Next, we could pick up the triangle and rotate it 120° clockwise,[34] and then place it back down.

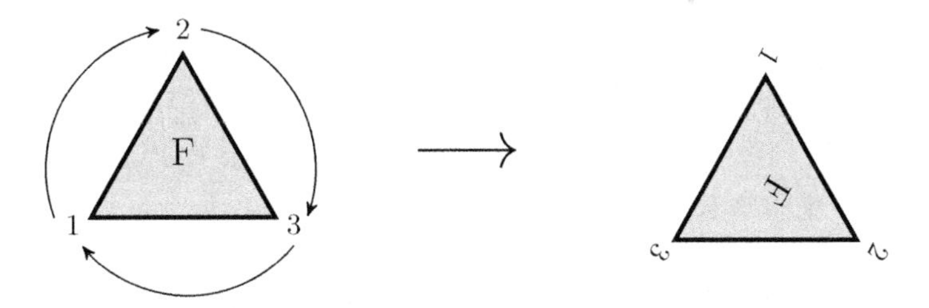

[33]Sorry, haven't figured out how to integrate .gif files into a printed book yet...

[34]More self-dating: It made me feel old when I realized that "clockwise" and "counterclockwise" are becoming antiquated terms for students, since they don't see many analog clocks anymore. So I included arrows in case it helps.

If we picked up the original triangle and rotated it 240° clockwise and placed it back down, we would get this:

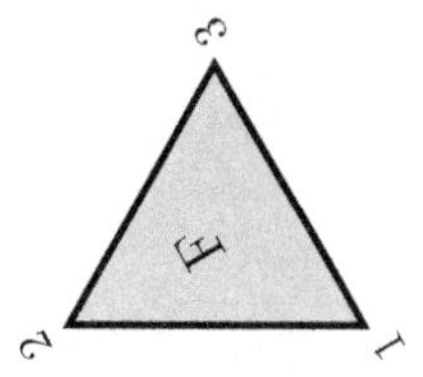

If we rotated the above triangle another third-turn we would get back to where we started, which we have already talked about. So, is that all we can do? Nope! We could also flip the triangle over! And, while we're at it, we could add in some rotations as well. By flipping the triangle over the vertical access, this is what we get:

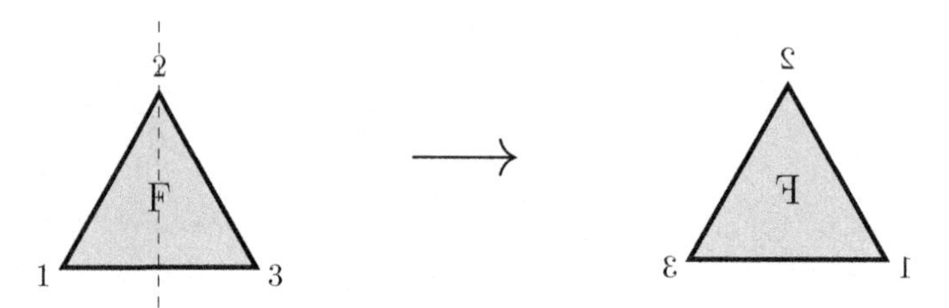

And by taking this and rotating it 120° degrees or 240° degrees, we see two more ways that the triangle could look after we picked it up, moved it around, and placed it back onto the same region it originally started. Here are those two:

And, finally, we are done. There are six possible ways that the triangle could look, and we have now identified them all. Because we are talking about group theory, let's identify each of these with some symbols. For the identity, in which we just picked up the triangle and placed it right back where it started, let's call that 1. For the 120° rotation, let's call that r. Then, since the 240° rotation is just r twice, let's call that $r * r$, or r^2. If we rotate the triangle a third turn we would get back to the identity, which shows that $r^3 = 1$.

Next, we have flips. The triangle resulting from a flip over the vertical axis we will call f. Flipping twice brings the triangle back to the identity, so $f^2 = 1$. The last two triangles above would then be $f * r$ and $f * r^2$.

With this, we have identified what is called the *dihedral group of order 6*, and is denoted D_6. It is:

$$\{1, r, r^2, f, f * r, f * r^2\},$$

and the way we combined elements is by using the logic of triangles. For example, to figure out which of the six elements $r^2 * (f * r)$ is equal to, we could start with the r^2 triangle, and then perform a $f * r$ — that is, perform a flip and then a rotation:

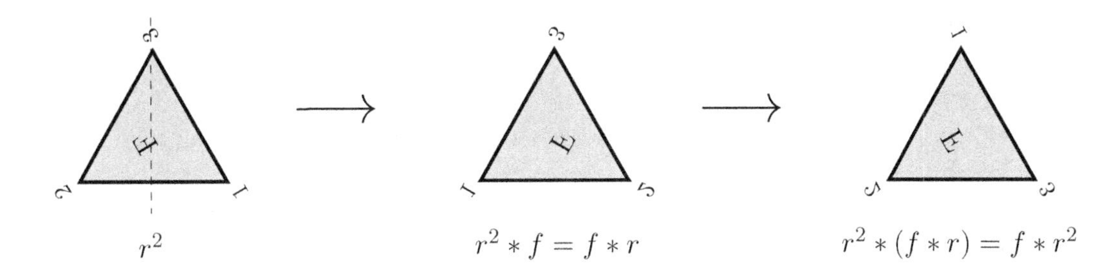

This shows that $r^2 * (f * r)$ is the element $f * r^2$.

The Symmetric Group

The *symmetric groups* are groups where each element is a bijection. For example, let's think about the collection of bijections from $\{1, 2, 3\}$ to $\{1, 2, 3\}$. One such bijection, g, is the one where $g(1) = 2$, $g(2) = 1$ and $g(3) = 3$. To condense our function notation, we will write this bijection like this:

$$\begin{pmatrix} 1 & 2 & 3 \\ 2 & 1 & 3 \end{pmatrix}$$

Each number on top is being mapped to the number right below it. This gives a group called the *symmetric group of order 6*, denoted S_3, and here are its elements:

$$\left\{ \begin{pmatrix} 1 & 2 & 3 \\ 1 & 2 & 3 \end{pmatrix}, \begin{pmatrix} 1 & 2 & 3 \\ 2 & 3 & 1 \end{pmatrix}, \begin{pmatrix} 1 & 2 & 3 \\ 3 & 1 & 2 \end{pmatrix}, \begin{pmatrix} 1 & 2 & 3 \\ 2 & 1 & 3 \end{pmatrix}, \begin{pmatrix} 1 & 2 & 3 \\ 1 & 3 & 2 \end{pmatrix}, \begin{pmatrix} 1 & 2 & 3 \\ 3 & 2 & 1 \end{pmatrix} \right\}$$

Let's call the first function above g_1, the second g_2, and so on. What's the operation for this group? It is function composition; note that the composition of two bijections is a bijection. For example, what is $g_2 \circ g_2$? For starters, g_2 is the function

$$g_2 : \{1, 2, 3\} \to \{1, 2, 3\} \qquad \text{where} \qquad g_2(1) = 2\ ,\ g_2(2) = 3 \text{ and } g_2(3) = 1.$$

So $g_2 \circ g_2$ is the function where $(g_2 \circ g_2)(1) = 3$, $(g_2 \circ g_2)(2) = 1$ and $(g_2 \circ g_2)(3) = 2$. Which is g_3! So $g_2 \circ g_2 = g_3$. Moreover, $g_2 \circ g_2 \circ g_2 = g_1$.

Notice that g_1 is the identity function, and hence will be the identity element in this group. So, g_2 composed with itself three times produced the group's identity element — does this property sound familiar? It kind of looks like the rotation element, r, from the dihedral group. That element had $r^3 = 1$, just like g_2 does here.

The next element, g_4, has the property that $g_4 \circ g_4 = g_1$, our identity element. This is just like how $f^2 = 1$ in the dihedral group. Moreover, g_5 is what you get when you apply g_4 and then g_2 — just like $f * r$. And, likewise, g_6 exactly mirrors

$f * r^2$. In this way, we have a correspondence between elements: g_1 is like 1, g_2 is like r, g_3 is like r^2, g_4 is like f, g_5 is like $f * r$, and g_6 is like $f * r^2$.

This leads to a really big idea in group theory: Not only do D_6 and S_3 both have six elements, but the two operations even match up! That is, if you combine two elements in D_6, and then you combine the corresponding elements in S_3, the two answers will also correspond to each other. Thus, these are not two different groups, they are the *same* group. Yes, the elements look different and were denoted with different symbols, but these amount to nothing more than superficial differences in notation. As far as the operation is concerned, they are the same. When two groups have such a correspondence, we say they are *isomorphic*.

What about the group $\{0, 1, 2, 3, 4, 5\}$ with addition modulo 6, that we talked about at the start? This group is denoted $\mathbb{Z}_6$. So, is $\mathbb{Z}_6$ isomorphic to D_6 and S_3? The answer is no. There is no way to pair up the elements so that the operations behave the same.

One way to see this is to realize that if two groups are the same in every way, then if one group has some property, then the other group must have that property as well. Therefore, to prove that two groups are *not* isomorphic, it suffices to identify a single property from one group that does not exist in the other. Indeed, observe that the group $\mathbb{Z}_6$ has the element 1, which has a special property: it *generates* the entire group. If you take 1 and just keep combining it with itself, you eventually get every element of the group:

$$\begin{aligned}
1 &= 1 \\
1 +_6 1 &= 2 \\
1 +_6 1 +_6 1 &= 3 \\
1 +_6 1 +_6 1 +_6 1 &= 4 \\
1 +_6 1 +_6 1 +_6 1 +_6 1 &= 5 \\
1 +_6 1 +_6 1 +_6 1 +_6 1 +_6 1 &= 0
\end{aligned}$$

Note, however, that D_6 and S_3 do not have any such element. So $\mathbb{Z}_6$ can't possibly be isomorphic to D_6 and S_3.

It took a long time for mathematicians to realize exactly which axioms to include in their definition of a group and which to exclude. If they did not insist that the group operation was associative, then the definition would be far too general, and we would be unable to prove many interesting theorems. However, notice that we did not insist that $a * b = b * a$ for the elements in a group. If this happens, like with $\mathbb{Z}_6$, or $\mathbb{Z}$ under $+$, or $\mathbb{R} \setminus \{0\}$ under $\times$, then the group is called *abelian*. But this is not included as an axiom because doing so would exclude groups like D_6 or S_3 or the set of real 2×2 matrices with nonzero determinant under matrix multiplication (which is denoted $(\mathrm{GL}_2(\mathbb{R}))$).

The time mathematicians spent identifying the best definition of a group has paid off immensely. Not only is group theory versatile and deep, but it is one of the most beautiful theories in all of mathematics. I believe you will all enjoy learning it.

Appendices

Appendix A: Other Proof Methods

Throughout this text we have discussed many approaches to prove a theorem. These include:

- Pigeonhole Principle
- Direct Proof
- Proof by Cases
- Principle of Mathematical Induction
- Proof by Contraposition
- Proof by Contradiction (and, relatedly, proof by minimal counterexample)

These methods were mostly justified through our discussion on logic, and they are sufficient to prepare you for later math courses. There is another class of proof methods, which are justified using the theorems/techniques from some area of math, like probability, linear algebra or combinatorics. If these approaches prove useful enough, they might be given the lofty title of a "method." Indeed, the first three sections of this appendix will be on the probabilistic method, the linear algebra method, and the combinatorial method. The probabilistic and combinatorial methods are really important and used a lot. Connections to linear algebra appear all over mathematics, but the linear algebra method that we will be discussing does not play a central role in math. Nevertheless, it is worth including because most of you have seen linear algebra and it is another good example of applying one area of math to another.

We have studied proofs by cases, and an extreme version of this is a proof with so many cases it will feel like a brute-force argument. In recent decades, computers have proven useful at tackling such problems. As such, we will discuss computer-assisted proofs.

Finally, we will discuss proofs that rely almost entirely on a picture. Calling this a proof *method* is stretching the term to its breaking point, but they are useful and fun and it's my book so I'm doing it.

To keep the focus on the methods rather than on theory-building, we will focus our attention on applications from areas of math that have already been discussed in this book and/or have lower points of entry: Ramsey theory, number theory, set theory, coding theory and combinatorics.

A.1 Probabilistic Method

Suppose a great survey was done of the island of Mene to determine the characteristics of their people, the Mennonites. If the survey found that the Mennonites, on average, have 1.999 hands, what would that tell us? It would tell us that not every Mennonite has two hands — someone was probably in an accident. If the survey reported that the average Mennonite has 1.1 testicles, then you could conclude that there are more men than women on the island.[1] In general, if you find the average of a collection of numbers and you get an answer of m, then some of those numbers must be at least m, and some must be at most m. And if you know that the numbers are all integers (say, counting hands), then an average of 1.999 tells us that at least one of the numbers must be at least 2, and at least one of the numbers must be at most 1.

A slightly different way to think about it: If a randomized procedure has output some numbers whose average value is m, then there is no way that *all* of the numbers are less than m — some must be greater than or equal to m. And, conversely, they cannot all be greater than m — some must be less than or equal to m. Being the average means it is "in the middle" of the values — some above, some below. This is called the *averaging principle*.

Principle.

Principle A.1 (*The averaging principle*). Suppose a random event, denoted X, takes an average value of m; we denote this average by $E[X]$ and refer to it as the "expected value" of the random event. Then, this random event must sometimes take a value at least m, and sometimes take a value at most m.

If I rolled a 6-sided dice a bunch of times and the average roll was a 4.2, then you would know that 5 or 6 had to have come up. Indeed, one reason this is true is simply by the averaging principle: At some point the dice must have taken a value at least 4.2, which can only mean 5 or 6.

If an NBA player makes 31.4% of their shots, then you could think about each shot attempt as a random event which outputs a 0 if they miss and a 1 if they make. This particular player then averages a 0.314; and since all the numbers in the average are 0s and 1s, the player must have at least one make (a one) and at least one miss (a zero).

As simple as the averaging principle seems, it can be amazingly powerful. Our first example comes from Ramsey theory, so please review the *Introduction to Ramsey Theory* beginning on page 41.

[1]Fact: According to surveys, 75% of Americans think that they are above-average drivers. What does this tell you? Something informative? If Jeff Bezos walks into an orphanage, then the average person in the orphanage is a billionaire. What does this tell you? Something uninformative?

Theorem.

Theorem A.2. The (symmetric) Ramsey number $R(t) > 2^{t/2}$.

Proof. Consider a random 2-coloring of the edges of K_n, in which we decide on the color of each edge by flipping a fair coin — each edge has a 50-50 chance of being red, and a 50-50 chance of being blue. Given any set of t vertices, the probability that all of the $\binom{t}{2}$ edges between them are red is

$$\underbrace{\frac{1}{2}\cdot\frac{1}{2}\cdot\frac{1}{2}\cdot\dots\cdot\frac{1}{2}}_{\binom{t}{2}\text{ times}} = \left(\frac{1}{2}\right)^{\binom{t}{2}}.$$

Likewise, the probability that the edges are all blue is $\left(\frac{1}{2}\right)^{\binom{t}{2}}$. Thus, the probability that the this K_t is monochromatic — all red or all blue — is

$$\left(\frac{1}{2}\right)^{\binom{t}{2}} + \left(\frac{1}{2}\right)^{\binom{t}{2}} = 2\left(\frac{1}{2}\right)^{\binom{t}{2}}.$$

So far we have focused only on a single K_t inside K_n. However, there are $\binom{n}{t}$ different copies of K_t, since every collection of t vertices induces a K_t. Thus, the probability that at least one of these K_t is monochromatic is certainly no more than

$$\underbrace{2\left(\frac{1}{2}\right)^{\binom{t}{2}} + 2\left(\frac{1}{2}\right)^{\binom{t}{2}} + 2\left(\frac{1}{2}\right)^{\binom{t}{2}} + \dots + 2\left(\frac{1}{2}\right)^{\binom{t}{2}}}_{\binom{n}{t}\text{ times}} = \binom{n}{t}\cdot 2\left(\frac{1}{2}\right)^{\binom{t}{2}}.$$

Therefore, if n is chosen so that

$$\binom{n}{t}\cdot 2\left(\frac{1}{2}\right)^{\binom{t}{2}} < 1,$$

then there's less than a 100% chance that our coin flips produced at least one monochromatic K_t. Meaning, there is some chance that *none* of these K_t are monochromatic. Which n allows this inequality to hold? If $n = \lfloor 2^{t/2}\rfloor$, then

$$\begin{aligned}
\binom{n}{t}\cdot 2\left(\frac{1}{2}\right)^{\binom{t}{2}} &< \frac{n^t}{t!}\cdot 2\left(\frac{1}{2}\right)^{\binom{t}{2}} \\
&= \frac{\lfloor 2^{t/2}\rfloor^t}{t!}\cdot 2\cdot\frac{1}{2^{t(t-1)/2}} \\
&\le \frac{2^{t^2/2}}{t!}\cdot\frac{2}{2^{t^2/2-t/2}} \\
&= \frac{2^{1+t/2}}{t!} \\
&< 1,
\end{aligned}$$

for $t \ge 3$. Thus, for all $t \ge 3$, we have $R(t) > 2^{t/2}$. □

The end requires some careful algebra, which is common for problems like this. But the main idea is straightforward: Pick a random color for each edge. It is then a simple product to determine the chance that any particular K_t is monochromatic, and with just a little more work you can determine an upper bound on the chance that at least one K_t is monochromatic. And, sure enough, for $n = \lfloor 2^{t/2} \rfloor$ the chance that every K_t is monochromatic is less than 100%. Therefore, by randomly choosing colors, there is some small chance that we avoided any monochromatic K_t. And since this small chance has to come from somewhere, there must be some way to coloring K_n to avoid any monochromatic K_t.

This is pretty magical, because it tells us nothing about how to actually find such a coloring, just that by coloring in the stupidest way possible we *could possibly* get lucky and stumble upon one. Which is enough for us, because the problem asks whether it is true that *every* coloring of K_n has a monochromatic K_t or not, and this indirectly shows that (for $n = \lfloor 2^{t/2} \rfloor$) there must indeed be some coloring which produces no monochromatic K_t, proving the result.[2]

Sum-Free Sets

A second example deals with so-called *sum-free sets*. A set S is said to be sum-free provided there does not exist $a, b, c \in S$ for which $a+b = c$. For example, $\{1, 3, 5, 6, 8\}$ is not sum-free, because $3 + 5 = 8$, while all three of these numbers are in the set. Also, note that a, b and c do not need to be unique, and so $3 + 3 = 6$ is another reason why this set is not sum-free.

Meanwhile, $\{1, 3, 5, 7, 9\}$ is sum-free, as you can check by writing down all possible sums and noticing that none of them are in $\{1, 3, 5, 7, 9\}$, or by observing that the sum of any two odds is even which immediately guarantees the sum-free property.

Given a set of natural numbers, one might wonder how large of a subset we could find which is sum-free. For $\{1, 3, 5, 7, 9\}$, the entire set is a sum-free subset. For $\{1, 3, 5, 6, 8\}$, the entire set is not sum-free and every subset of it which contains four elements is also not sum-free. However, there are five subsets of size three which are sum free: $\{1, 3, 5\}$, $\{1, 3, 8\}$, $\{1, 5, 8\}$, $\{1, 6, 8\}$ and $\{5, 6, 8\}$. For $\{1, 2, 3, 4\}$, the largest sum-free subset has size two, with four subsets realizing this: $\{1, 3\}$, $\{1, 4\}$ $\{2, 3\}$ and $\{3, 4\}$.

In this last example, we were only able to keep 50% of the elements to form a sum-free subset. Is it always possible to find a sum-free subset containing half the elements? If not, is it always possible to find a sum-free subset containing, say, 25% of the elements? How high can we go? And how would one prove such a thing? If you start thinking about what an arbitrary set of natural numbers looks like, and then start trying to get a grasp on all sums of pairs of elements from that set... things start getting really hairy really fast.

Nevertheless, the probabilistic method can prove that it is always possible to find more than a third of the elements which form a sum-free subset.

[2]We are like the classic Batman villain, Two-Face (AKA Harvey Dent), who flipped his lucky coin to make important decisions, while also claiming "I don't leave anything up to chance, I make my own luck." Indeed, proof used chance to make our own luck!

Theorem.

Theorem A.3. If $A \subseteq \mathbb{Z}$ is a set of N natural numbers, then there exists a subset B of A for which $|B| > N/3$ and B is sum-free.

Proof. Let $a_{\max}$ be the largest element of A. It is a straightforward theorem from number theory that there are infinitely many primes congruent to 2 (mod 3), and so there exists a prime p such that $p = 3k + 2$ for some $k \in \mathbb{N}$ and $p > 2a_{\max}$.

We are looking for a subset B with more than a third of the elements which is sum-free. In fact, we will make our job harder: We will prove that there exists a set B of this size which is sum-free when the sum is performed modulo p. Why does this make our job harder? Note that if $a + b = c$, then also $a + b \equiv c \pmod{p}$, so we are not losing any sums. However, while it is not true that $3 + (p - 1) = 2$, we do have $3 + (p - 1) \equiv 2 \pmod{p}$. Thus, our set B will also avoid having all of 2, 3 and $(p - 1)$ within it, even though this is allowed by the official sum-free rules.

Nevertheless, making our job harder does have some advantages. Consider the set

$$S = \{k + 1, k + 2, k + 3, \dots, 2k + 1\}.$$

Observe that S is sum-free modulo p, as the smallest value any sum of two elements could equal is

$$(k + 1) + (k + 1) = 2k + 2,$$

which is just barely not in S. And using the definition $p = 3k + 2$, the "largest" value this sum could have is

$$(2k + 1) + (2k + 1) = 4k + 2 \equiv k \pmod{p},$$

which wrapped all the way around the integers modulo p, but is again just barely not in S.

Notice that if $x \in \{1, 2, 3, \dots, p - 1\}$, then when you reduce every number in $\{x, 2x, 3x, \dots, (p-1)x\}$ modulo p, you will get precisely the set $\{1, 2, 3, \dots, p\}$ — each element once, perhaps in a different order. (See Theorem 2.19's proof for details).[3]

Next, choose a number t from $\{1, 2, 3, \dots, p - 1\}$ at random, so that each number has a $\frac{1}{p-1}$ chance of being chosen. Instead of $\{x, 2x, 3x, \dots, (p - 1)x\}$ like in the last paragraph, consider the set $\{a_1 t, a_2 t, \dots, a_N t\}$, where the a_i are the elements of A. How many elements from this set should we expect be elements of $S = \{k + 1, k + 2, k + 3, \dots, 2k + 1\}$? Fix any $s \in S$. According to the last paragraph, since each $a_i \in \{1, 2, 3, \dots, p - 1\}$, there is precisely one $x \in \{1, 2, 3, \dots, p - 1\}$ for which $a_i x = s$. Thus, since t is chosen at random, each $a_i \in A$ has a $\frac{1}{p-1}$ chance that $a_i t = s$. Therefore, the chance that a_i is sent to one of the $|S|$ elements in S, must be $\frac{|S|}{p-1}$. And with N such a_i terms, the number of elements in A which we expect[4] to be sent to elements of S is equal to $N \cdot \dfrac{|S|}{p - 1}$.

[3]For example, if $p = 5$ and $x = 3$, then $\{x, 2x, 3x, 4x\} = \{3, 6, 9, 12\}$, and when reduced modulo 5 this gives $\{3, 1, 4, 2\}$, which equals the set $\{1, 2, 3, 4\}$. Or if we instead chose $x = 2$, then $\{x, 2x, 3x, 4x\} = \{2, 4, 6, 8\} \equiv \{2, 4, 1, 3\} = \{1, 2, 3, 4\}$.

[4]Technically we are using something called *linearity of expectation* here.

And the important observation is that because $|S| = k + 1$ and $p = 3k + 1$, the number of elements from A which we should expect to land inside S is

$$N \cdot \frac{|S|}{p-1} = \frac{N(k+1)}{3k} > \frac{N}{3}.$$

That is, more than a third of the elements! And so, by the averaging principle, there must be some t for which more than a third actually do land within S. For this value of t, let

$$B = \{a \in A : at \text{ is in } S, \text{ when taken modulo } p\}.$$

Then, not only is $|B| > N/3$, but we can also see that B is sum-free. Indeed, we can prove this final claim by a short proof by contradiction. Assume for a contradiction that we did have $a + b = c$ where $a, b, c \in B$, then you would also have $at + bt = ct$, but by the definition of B, these are all elements inside S which we know to be sum-free, giving the contradiction.[5] □

I love this proof. Trying to construct a set which is sum-free is hard. Trying to find a big set which is not sum-free, and where all of its large subsets are also not sum-free, is really, really hard. Just try it on your own. Try to come up with a set S of 10 natural numbers which is not sum-free, and moreover none of its 5-element subsets is sum-free. Yeah, pretty dang hard.

There are just so many pairs of elements, and the summing property induces all sorts of overlapping restrictions. There is structure and rules everywhere! But yet this proof says, "I don't care. I am going to ignore allllll that supposed structure. I am going to just throw all the elements into a box, shake them up at random using this t scalar, and see what happens." A third of these shaken-up elements have a good chance of landing on the sum-free set S. And poof! That's it!

According to a famous adage, any sufficiently advanced technology is indistinguishable from magic. If the same can be said for mathematical ideas, then the probabilistic method is a piece of mathematical technology of the first order.

Generalizations

In closing, I will note that both of the theorems we have discussed in this section have more general forms. An edge in a graph can be thought of as a set of two vertices, and Theorem A.2 can be generalized to the case where each "edge" contains three vertices, or four vertices, or any other fixed number of vertices. As for the second theorem, Theorem A.3, this result can be generalized to include sets of nonzero integers, rather than just natural numbers. (But it is standard to not include 0 in our sets, since $0 + 0 = 0$ immediately prohibits it from being sum-free.)

5

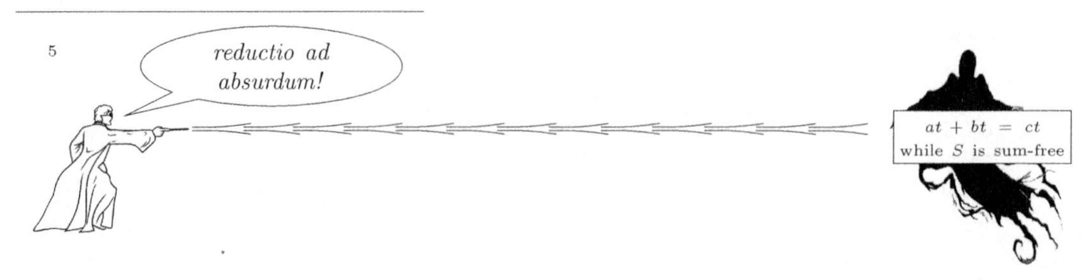

A.2 Linear Algebra Method

Mathematics is not as neatly partitioned as the standard undergraduate curriculum suggests. While there are perfectly good reasons to have separate courses for abstract algebra and real analysis and complex analysis and combinatorics and statistics and probability and the rest, the mathematics castle is not a collection of distantly-separated towers. The areas of math connect, overlap and interact. When the theory of one area overlaps well with the theory of another, then a new area of math is born. When the theory of one area is useful as a tool to solve the problems of another, then a new method of proof is born.

The probabilistic method uses probability to solve problems outside of probability. We did not prove a theorem which tied together probability and Ramsey theory, but rather used probability in the proof of a theorem on Ramsey theory. In this way, probability was a tool. There are many more examples of this throughout mathematics, and in this section we will discuss one more. In this case, the tool is linear algebra, which is a course that I trust most of you have taken.

After you study real analysis, you may be amazed that we could have an entire class devoted to the ridiculously narrow topic of linear functions. But while linear functions may seem remarkably simple and specific, the field is worth studying in-part because some really amazing things happen when you think about collections (or "systems") of linear functions, and in-part because linear functions present themselves in unexpected places throughout mathematics.

k-Intersecting Sets

Recall that a family of sets is a set for which each element is itself a set. A common "extremal" question is, how big can such a family of sets be if we demand that the sets it contains satisfy some property?

For our first example, begin by fixing some natural number k. Next, suppose our family of sets is $\{A_1, A_2, A_3, \dots, A_m\}$, where each of the m elements is a set. Finally, here are the two properties we demand:

1. $A_i \subseteq \{1, 2, 3, \dots, n\}$ for all i, and

2. $|A_i \cap A_j| = k$ for any pair of distinct sets A_i and A_j.

If $\mathcal{F}$ is a family of sets with these properties, then $\mathcal{F}$ is called a *k-intersecting family*. For example, if $n = 5$, then here is a 2-intersecting family:

$$\mathcal{F}_1 = \big\{\{1, 2\}, \{1, 2, 3\}, \{1, 2, 4, 5\}\big\}.$$

In this case, any pair of sets has an intersection of size 2. Here is another 2-intersecting family where $n = 5$:

$$\mathcal{F}_2 = \big\{\{1, 2, 3\}, \{2, 3, 4\}, \{1, 3, 4\}\big\}.$$

Again, the intersection of any pair of sets has size 2.

Here is the main question we are interested in: How big can $\mathcal{F}$ be? That is, what is the maximum number of sets that you can include in $\mathcal{F}$ before it ceases to be k-intersecting?

To solve this, we are going to use linear algebra. At this point this should be quite surprising, because linear algebra deals with linear functions and vector spaces... but where are the linear functions in this problem?? Where is the vector space?? We weren't given any linear functions like $y = 2x$, or $x + 2y + 3z = 0$. We have nothing but a family of sets, and this k-intersecting property on those sets.

To locate the linearity, first recall that a vector represents a linear function, and if the vectors contain n real numbers, then they are vectors in the vector space $\mathbb{R}^n$. For example,

$$\begin{bmatrix} 1 \\ 2 \\ 3 \end{bmatrix}$$

is a vector in $\mathbb{R}^3$. But all this means is that it is acceptable to search for vectors in our problem rather than linear functions. Do you see the vectors? You might guess that the family

$$\mathcal{F}_1 = \big\{\{1, 2\}, \{1, 2, 3\}, \{1, 2, 4, 5\}\big\}$$

could be turned into the vectors

$$\begin{bmatrix} 1 \\ 2 \end{bmatrix}, \begin{bmatrix} 1 \\ 2 \\ 3 \end{bmatrix}, \begin{bmatrix} 1 \\ 2 \\ 4 \\ 5 \end{bmatrix}.$$

However, recall that in linear algebra you always worked with vectors of the same dimension. You would never mix vectors from $\mathbb{R}^2$ with $\mathbb{R}^3$ and $\mathbb{R}^4$. Likewise, it will be worthwhile to ensure all of our vectors here are the same length.

The big idea is to recall that the sets are all subsets of $\{1, 2, 3, \ldots, n\}$. For example, in the family $\mathcal{F}_1$ above, we had said $n = 5$; consequently, we will be turning each set into a vector with 5 rows. Next, note that the only possible elements in these sets are 1, 2, 3, 4 and 5. So rather than including the number in our vectors, we could instead just include whether or not each number appeared, by using a 1 if it was included and a 0 if it was not.

For example, we will be turning the set $\{1, 2, 3\}$ into the vector

$$\begin{bmatrix} 1 \\ 1 \\ 1 \\ 0 \\ 0 \end{bmatrix}.$$

Since 1, 2 and 3 all appeared in the set, the first, second and third rows each received a 1. Since 4 and 5 did not appear, those rows received a 0. Likewise, the set $\{1, 2, 4, 5\}$

will be represented by the vector

$$\begin{bmatrix} 1 \\ 1 \\ 0 \\ 1 \\ 1 \end{bmatrix}.$$

These are called *incidence vectors*, and they are part of the vector space $\{0, 1\}^n$, which is an n-dimensional vector space just like $\mathbb{R}^n$, except that the vectors can only include the numbers 0 and 1, the only allowed scalars are 0 and 1, and all arithmetic is done modulo 2. This is a clever way to do it, because now every set corresponds uniquely with a vector, and every vector in this vector space corresponds uniquely to a set. Moreover, just like with all n-dimensional vector spaces, this vector space satisfies the following fundamental property:

> *If $v_1, v_2, v_3, \ldots, v_m$ are linearly independent vectors in an n-dimensional vector space, then $m \leq n$. That is, it is impossible to have more than n linearly independent vectors in an n-dimensional vector space.*

We began by asking how many sets can there be in a k-intersecting family of sets, provided each is a subset of $\{1, 2, 3, \ldots, n\}$. We can now answer that question.

Theorem.

Theorem A.4. If $\mathcal{F}$ is a k-intersecting family of sets for which each set is a subset of $\{1, 2, 3, \ldots, n\}$, then $|\mathcal{F}| \leq n$.

Proof. Let $\mathcal{F} = \{A_1, A_2, \ldots, A_m\}$. Thus, our goal is to show $m \leq n$. For each set A_i, let $v_i \in \{0, 1\}^n$ be the $n \times 1$ incidence vector of A_i. If we are able to show that the vectors $\{v_1, v_2, \ldots, v_m\}$ are linearly independent, we would have m linearly independent vectors in an n-dimensional vector space, which would imply $m \leq n$, as desired. Therefore, in order to conclude this proof, all we must show is that the vectors $\{v_1, v_2, \ldots, v_m\}$ are linearly independent. We will show that now.

Assume for a contradiction that this collection of sets is linearly dependent. By definition, this means that there exist constants λ_i, not all 0, for which

$$\sum_{i=1}^{m} \lambda_i v_i = \vec{0},$$

where $\vec{0}$ is the $n \times 1$ vector containing all zeros.

Now, if you recall some basic properties of the dot product of two vectors, you will notice that the k-intersecting property of the A_i produces these two properties of the incidence vectors: $v_i \cdot v_j = k$ for all $i \neq j$, and $v_i \cdot v_i = |A_i|$ for all i. This allows us to do the following, where we begin with the simple fact that dot product of two zero vectors equals 0, followed by some fancy (linear) algebra.

$$\begin{aligned}
0 &= \vec{0} \cdot \vec{0} \\
&= \left(\sum_{i=1}^{m} \lambda_i v_i\right) \cdot \left(\sum_{j=1}^{m} \lambda_j v_j\right) && \text{(See previous page)} \\
&= \sum_{i=1}^{m} \lambda_i^2 (v_i \cdot v_i) \;+\; \sum_{i \neq j} \lambda_i \lambda_j (v_i \cdot v_j) && \text{(FOIL)} \\
&= \sum_{i=1}^{m} \lambda_i^2 |A_i| \;+\; \sum_{i \neq j} \lambda_i \lambda_j k && \text{(See previous page)} \\
&= \sum_{i=1}^{m} \lambda_i^2 |A_i| \;+\; \left[k\left(\sum_{i=1}^{m} \lambda_i\right)^2 - \sum_{i=1}^{m} \lambda_i^2 k\right] && \text{(Rewrite)} \\
&= \sum_{i=1}^{m} \lambda_i^2 (|A_i| - k) \;+\; k\left(\sum_{i=1}^{m} \lambda_i\right)^2 && \text{(Rewrite)} \\
&> 0. && \text{(See below)}
\end{aligned}$$

As for the six equalities, don't worry if one or two of them were confusing.[6] For the final inequality, note that both terms are nonnegative, so all we have to show is that one or the other is positive. First, notice that $|A_i| \geq k$ for all i, and at most one A_i has size exactly equal to k, since otherwise $|A_i \cap A_j| \neq k$ for any other such A_j. Without loss of generality, suppose $A_2, A_3, \ldots, A_m$ all have more than k elements.

Recall that the λ_i are not all zero. Consider two cases: Only λ_1 is nonzero, or $\lambda_j \neq 0$ for some $j > 1$. In the first case, note $k\left(\sum_{i=1}^{m} \lambda_i\right)^2 = k(\lambda_1)^2 > 0$. In the second case, note $\sum_{i=1}^{m} \lambda_i^2 (|A_i| - k) \geq \lambda_j^2(|A_j| - k) > 0$. In both cases, we have confirmed the final inequality. Thus, by looking above at our big computation, we showed that $0 > 0$, which gives us our contradiction[7] and completes the proof. □

This proof is due to R. C. Bose in 1949, and his two-page paper was the very first to use a linear argument to solve a problem in combinatorics. This type of problem in which one attempts to maximize (or minimize) the size of a set of objects, given that the objects satisfy some restrictions, is called an *extremal* problem.

The linear algebra method uses the linear independence properties of vector spaces, dimensionality and rank properties, properties of special classes of matrices, orthogonality properties, and more. Coding theory contains many examples, and while many are complicated, the following is a brief introduction.

[6] And if you understood even five out of the six of them, then your linear algebra prof deserves major props.

[7]

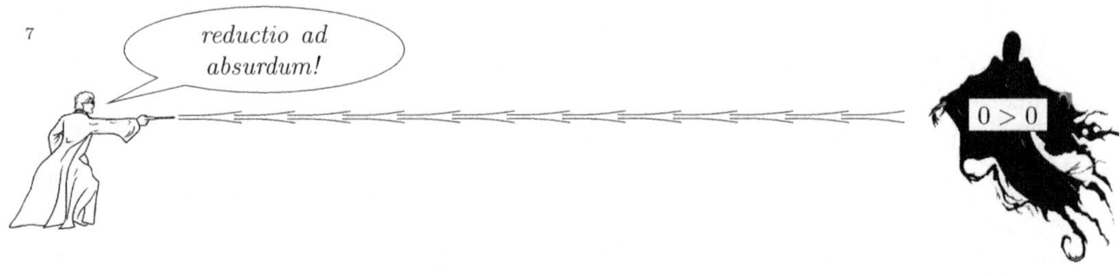

Linear Algebra in Coding Theory

There are many applications of the linear algebra method to coding theory, in which 0-1 vectors are again used. This is a quick introduction. A *binary linear code of length* n is one in which the codewords — each of which is a vector of length n containing only 0s and 1s as entries — form a linear subspace of the vector space $\{0, 1\}^n$. Said differently, the codewords have the property that any linear combination of codewords is itself a codeword. This is particularly useful for detecting errors in transmission.

Given this setup, it feels more natural that linear algebra could play an important role, as a vector space is present. To give an example of a theorem whose proof makes use of the linear algebra method, we need to state a few more definitions.

- The *Hamming distance* between two codewords is equal to the number of coordinates in which these codewords differ. For example,

$$\begin{bmatrix} 1 \\ 0 \\ 0 \end{bmatrix} \quad \text{and} \quad \begin{bmatrix} 0 \\ 0 \\ 1 \end{bmatrix}$$

  have Hamming distance 2, since they differ in two coordinates — the first and third.

- Given a binary linear code of length n, we could compute the Hamming distance for every pair of codewords. The minimum value over all pairs is called the *minimal distance* for the linear code.

- Given C, a binary linear code of length n, its *dual* is the binary linear code of length n which is defined using a dot product like so:

$$C^{\perp} = \{x \in \{0, 1\}^n : x \cdot c = 0 \text{ for all } c \in C\}.$$

- A binary linear code of length n, denoted C, is called (n, k)-*universal* if, for any subset of k coordinates $S = \{i_1, i_2, \ldots, i_k\}$, if one takes each codeword of C and removes every coordinate except for the k coordinates listed in S, then the length-k binary strings they get include all 2^k possible binary strings.

Codes which are (n, k)-universal play an important role in coding theory, and their applications include testing logical circuits, construction of k-wise independent random variables, and more. Given this long-winded setup, below is the theorem.

Theorem.

Theorem A.5. If C is a linear code of length n, and its dual $C^{\perp}$ has minimal distance $k + 1$, then C is (n, k)-universal.

While we won't go through the proof, this setup suggests how useful linear algebra can be in a field like coding theory — and *many* other areas of math.[8]

[8] Advice for the healthy mathematician: Exercise regularly, eat healthy, learn linear algebra.

A.3 Combinatorial Method

Suppose you wish to prove that $1+2+3+\cdots+n = \dfrac{n(n+1)}{2}$ for every $n \in \mathbb{N}$. One option is to prove it by induction, as we did in Proposition 4.2. But there is another way, which I find more enlightening. First, observe that each side of this equal sign is always an integer. And, if you have a set with finitely many elements, then the cardinality of that set is also an integer. These two minor observations turn out to be crucial.

If you can find a set S for which $1+2+3+\cdots+n$ counts the elements in S, and you can show that $\frac{n(n+1)}{2}$ also counts the elements in S, then they would have to be equal to each other! A set can only have one cardinality, so if

$$1+2+3+\cdots+n = |S| = \frac{n(n+1)}{2},$$

then of course

$$1+2+3+\cdots+n = \frac{n(n+1)}{2}.$$

This is an approach to proving an equality holds. Let's continue the above equality as our first example. Suppose a group of $n+1$ people are in a room, and every one goes around shaking hands until everybody has shaken the hand of everyone else exactly once. I claim that both sides of $1+2+3+\cdots+n = \dfrac{n(n+1)}{2}$ count how many handshakes occurred.

On the one hand,[9] you could imagine one person at a time going around shaking hands. The first person would shake hands with all n other people, and then go stand in the corner. The second person would shake hands with the remaining $n-1$ people, and then join the first person in the corner. The third person would shake hands with the remaining $n-2$ people, and then join the first two people in the corner. And so on. Eventually, the nth person would shake hands with the last person, head to the cramped corner, and the final person will have already shaken everybody's hand and so will go straight to the corner. Adding up all these handshakes, we get

$$n+(n-1)+(n-2)+\cdots+2+1+0,$$

which does indeed equal the left side of the equality: $1+2+3+\cdots+n$.

Next we will show that $\frac{n(n+1)}{2}$ also counts this same thing, although in a slightly different way. Indeed, we can, on the other hand,[10] count this set by listing all pairs of people, such as

$$\begin{gathered}\text{Chanratha} \leftrightarrow \text{Anna}\\ \text{Shawheen} \leftrightarrow \text{Duong}\\ \text{Sayonita} \leftrightarrow \text{Corey}\\ \vdots\end{gathered}$$

[9] Terrible pun intended.

[10] Sorry not sorry.

We can think about this as first choosing the first person, and then choosing the second person:

Choose first person	Choose second person
↑	↑
$n+1$ options	n options

The total number of ways is thus the product of these: $n(n+1)$ ways to pick both. However, this overlooks one thing: choosing Sayonita and then Corey is the same as choosing Corey and then Sayonita, because all that matters is that that pair shook hands. We don't want to count Sayonita $\leftrightarrow$ Corey and Corey $\leftrightarrow$ Sayonita, we only want to count one of these. In fact, the above will count each pair twice, and hence is overcounting by a factor of 2. Thus, the true number of handshakes is

$$\frac{n(n+1)}{2}.$$

We first showed that there were $1+2+3+\cdots+n$ handshakes, and showed that there were $\frac{n(n+1)}{2}$ handshakes. Since these both count the same collection, they must be equal to each other:

$$1+2+3+\cdots+n = \frac{n(n+1)}{2}.$$

This strategy of counting the same set in two different ways, and then concluding that the two counts must be equal, is either called the *double counting method* or the *combinatorial method.*

Binomial Coefficients

Suppose you have a set of n elements (like $\{1,2,3,\ldots,n\}$), and you want to pick k elements from this set. The number of different ways you can do this is denoted $\binom{n}{k}$. In words, $\binom{n}{k}$ is called "n choose k." These are called *binomial coefficients.* Here are a couple examples:

- In the set $\{1,2,3,4,5\}$, there are 5 ways to choose one element from this set—the options are $\{1\}$, $\{2\}$, $\{3\}$, $\{4\}$, $\{5\}$. Therefore $\binom{5}{1} = 5$.
- There are 10 ways to pick 2 elements from this set—the options are $\{1,2\}$, $\{1,3\}$, $\{1,4\}$, $\{1,5\}$, $\{2,3\}$, $\{2,4\}$, $\{2,5\}$, $\{3,4\}$, $\{3,5\}$, $\{4,5\}$. Therefore $\binom{5}{2} = 10$.

What follows is a fundamental property of binomial coefficients.

Proposition.

Proposition A.6. Suppose $n, k \in \mathbb{N}_0$. Then,

$$\binom{n}{k} = \binom{n}{n-k}.$$

Proof. We will use double counting by showing that both sides of the proposition count the number of subsets of $\{1, 2, \ldots, n\}$ which have k elements. The left-hand side counts this by the very definition of $\displaystyle\binom{n}{k}$.

As for $\binom{n}{n-k}$, it counts the number of subsets of $\{1, 2, \ldots, n\}$ which have $(n-k)$ elements. However, observe that the complement turns a set with $(n-k)$ elements into a set with k elements. For instance, as a subset of $\{1, 2, 3, 4, 5, 6\}$, the set $\{1, 3, 4, 6\}$ has complement $\{2, 5\}$ (notation: $\{1, 3, 4, 6\}^c = \{2, 5\}$), and the complement of $\{1, 2, 3\}$ is $\{4, 5, 6\}$. Notice that if you take the complement of a set and then take the complement of the result, you get the original set back. In this way, each subset of size $(n-k)$ is exactly paired with one of size k. Thus, the number with size $(n-k)$ is equal to the number with size k. And so, $\displaystyle\binom{n}{n-k}$ also counts the number of sets of size k. □

Pascal's Rule

There are many more nice combinatorial identities involving binomial coefficients. Here's another fundamental one:

Proposition.

Proposition A.7 (*Pascal's Rule*). Suppose $n, k \in \mathbb{N}_0$. Then,

$$\binom{n}{k} = \binom{n-1}{k-1} + \binom{n-1}{k}.$$

Proof Idea. Again, we prove this with double counting. The left-hand side is the number of ways to choose a subset of $\{1, 2, 3, \ldots, n\}$ which contains k elements. To give this a name, let's call $\mathcal{S}$ the set of all k-element subsets of $\{1, 2, 3, \ldots, n\}$. Let's show that the right-hand side also counts the sets in $\mathcal{S}$. We will do this by breaking up the sets in $\mathcal{S}$ into two groups: one of group of size $\binom{n-1}{k-1}$ and one of size $\binom{n-1}{k}$.

The proof is clever, so let's first think about how one might come up with it. The $\binom{n-1}{k-1}$ counts the number of ways to choose $k-1$ elements from a set of size

$n-1$ (but every set in $\mathcal{S}$ has k elements, so having $k-1$ elements is one less than we need). The $\binom{n-1}{k}$ counts the number of ways to choose k elements (the correct number) from a set of size $n-1$.

Both parts dealt with a set of $n-1$ elements, so let's pick one. Consider the set $\{2, 3, \ldots, n\}$, which has size $n-1$. If we choose k-element subsets from this set, then we get $\binom{n-1}{k}$ sets, all of which are in $\mathcal{S}$; in particular, this gives us all the sets in $\mathcal{S}$ which does not contain a 1. Meanwhile, if we choose a $(k-1)$-element subset, then we get $\binom{n-1}{k-1}$ sets, but none of these are in $\mathcal{S}$ since they have just $k-1$ elements. But... if we throw 1 into each of these, then they are! Ok, now here's the proof:

***Proof*.** The left-hand side counts all k-element subsets from $\{1, 2, 3, \ldots, n\}$. Partition all these k-element subsets by whether the subset contains the element 1. If it contains 1, then there are $k-1$ other elements to select from the $n-1$ elements in $\{2, 3, \ldots, n\}$, giving $\binom{n-1}{k-1}$ sets. If it doesn't contain 1, then there are k elements to select from the $n-1$ elements in $\{2, 3, \ldots, n\}$, giving $\binom{n-1}{k}$ sets.

Since every set either contains 1 or doesn't contain 1, this is a complete partition of the collection of all k-element subsets from $\{1, 2, 3, \ldots, n\}$. Thus,

$$\binom{n}{k} = \binom{n-1}{k-1} + \binom{n-1}{k},$$

completing the proof. $\square$

Once again, we proved an identity not by using a bunch of algebra but by interpreting one of the two sides as the number of objects in a particular set, and then working to show that the other side also counts the objects in that set.

Pascal's Triangle

Note that $\binom{n}{0} = 1$, since there is one way to choose zero elements (i.e., do nothing, producing the empty set). Also, $\binom{n}{n} = 1$, since there is one subset containing all the n elements (the whole set itself). Using these two properties and the proposition we just proved $\left(\text{that } \binom{n}{r} = \binom{n-1}{r-1} + \binom{n-1}{r}\right)$, we can generate something called *Pascal's triangle*, which we first met in the Chapter 4 Bonus Examples. Below are the first five rows of the triangle. As binomial coefficients, it looks like the triangle on the left; as integers, it looks like the triangle on the right.

$$\begin{array}{ccccccccc} & & & & \binom{0}{0} & & & & \\ & & & \binom{1}{0} & & \binom{1}{1} & & & \\ & & \binom{2}{0} & & \binom{2}{1} & & \binom{2}{2} & & \\ & \binom{3}{0} & & \binom{3}{1} & & \binom{3}{2} & & \binom{3}{3} & \\ \binom{4}{0} & & \binom{4}{1} & & \binom{4}{2} & & \binom{4}{3} & & \binom{4}{4} \end{array} \quad = \quad \begin{array}{ccccccccc} & & & & 1 & & & & \\ & & & 1 & & 1 & & & \\ & & 1 & & 2 & & 1 & & \\ & 1 & & 3 & & 3 & & 1 & \\ 1 & & 4 & & 6 & & 4 & & 1 \end{array}$$

Both $\binom{n}{k} = \binom{n-1}{k-1} + \binom{n-1}{k}$ and $\binom{n}{k} = \binom{n}{n-k}$ are visible in these triangles. Can you spot any other identities? Below is another one, which we will prove via the combinatorial method (i.e., the double counting method).

Proposition.

Proposition A.8. The sum of the nth row in Pascal's triangle equals 2^n. That is,

$$\sum_{k=0}^{n} \binom{n}{k} = 2^n.$$

In the triangle, the very top 1 is considered the 0th row, then the row with just two 1s is the 1st row, and so on.

Proof. We will prove that the sum of the numbers in the nth row counts the number of subsets of $\{1, 2, \ldots, n\}$. We begin with the left-hand side. This side counts the number of subsets by partitioning them by their size; their size can range from 0 to n. The number of size k is equal to $\binom{n}{k}$, by the definition of the binomial coefficient. Therefore, adding up the number of sets for each possible k gives

$$\sum_{k=0}^{n} \binom{n}{k},$$

which is the left-hand side, as desired.

The right-hand side, 2^n, also counts the number of subsets of $\{1, 2, \ldots, n\}$. Indeed, every such subset can be realized by going through the elements of $\{1, 2, \ldots, n\}$ one-by-one and asking whether or not that element is in the subset. For each element, the answer is yes or no.

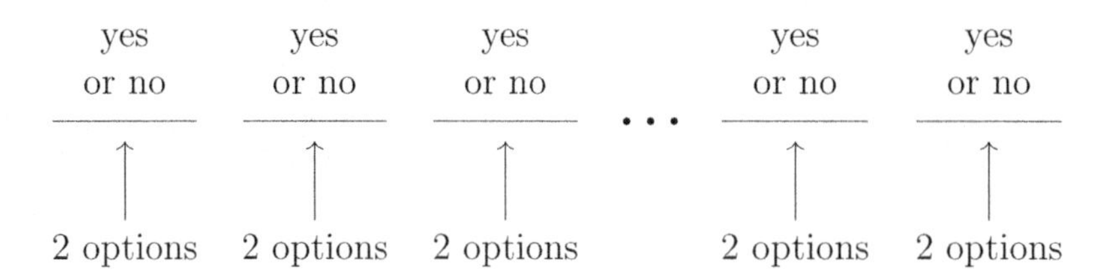

For example, if $n = 5$, the subset $\{1, 2, 5\}$ is obtained by answering "yes, yes, no, no, yes." In general, there are two options for each of the n elements, and each subset corresponds to a unique selection of yeses and noes, and vice versa. Thus, the total number of subsets is

$$\underbrace{2 \cdot 2 \cdot 2 \cdot \ldots \cdot 2}_{n \text{ times}} = 2^n.$$

Since we counted the same set in two different ways, by the combinatorial method

$$\sum_{k=0}^{n} \binom{n}{k} = 2^n. \qquad \square$$

There's another summation property that can be seen in Pascal's Triangle.

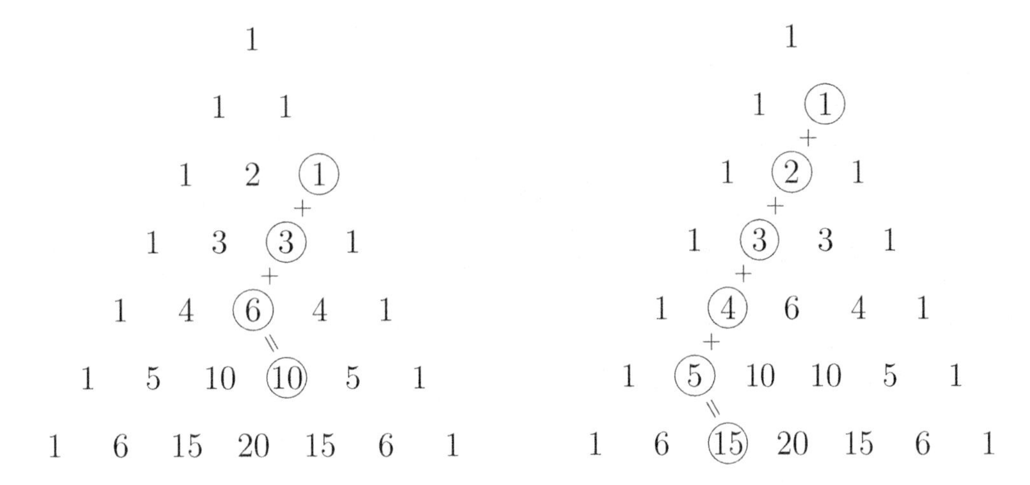

In the first triangle, if you add up the first three circled numbers, you get the fourth circled number. In the second triangle, the first five circled numbers add up to the sixth circled number. Moreover, you can do this for any diagonal: Start on the right of the triangle, moving down-left and add up all the numbers as you go, and the answer will be the number down and to the right. This is called *the hockey stick property*, which can be stated in terms of binomial coefficients like this:

Proposition.

Proposition A.9 (*The Hockey Stick Property*). Pascal's triangle satisfies the "hockey stick property" above. In particular,

$$\sum_{i=k}^{n} \binom{i}{k} = \binom{n+1}{k+1}.$$

What is the combinatorial proof of this? Notice that the right-hand side counts the number of $(k+1)$-sized subsets of $\{1, 2, \ldots, n+1\}$. I'll end this section by challenging you to come up with a way to partition these subsets into groups of size $\binom{k}{k}, \binom{k+1}{k}, \binom{k+2}{k}, \ldots, \binom{n}{k}$.

A.4 Computer-Assisted Proofs

How many colors do you need to color a map? To keep this book inexpensive, I won't use actual colors — just different shades of gray — but here is a map of the contiguous United States (meaning, we left off the states of Hawaii and Alaska[11]) where each of these 48 states was filled in with a color/shading.

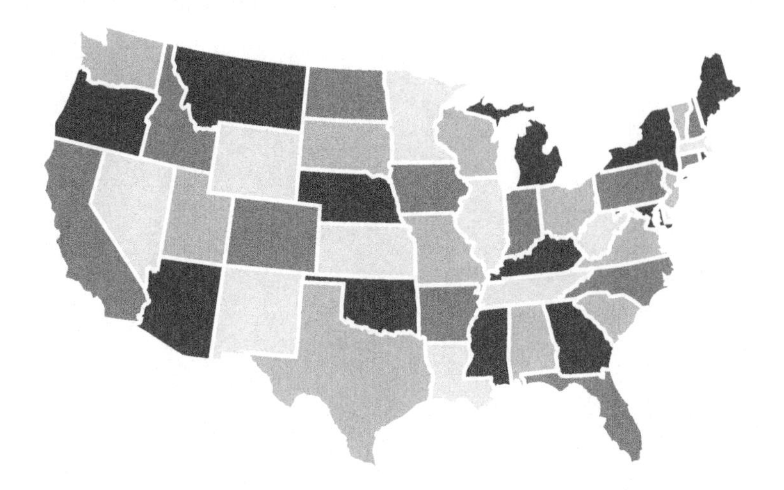

Notice that only four colors/shadings were used. Is it possible to color the map of the United States with just three or fewer colors? Well, we could certainly give every state the same color, but we don't want that. The standard mapmakers' rule is that if two states share a border, they should get different colors (if their corners meet at a single point, then it's ok if they have the same color).

Can the USA be colored with three colors under this condition? The answer is no. In fact, looking at just the southwest is enough to see that. Try to convince yourself that this region alone can not be colored with three colors:

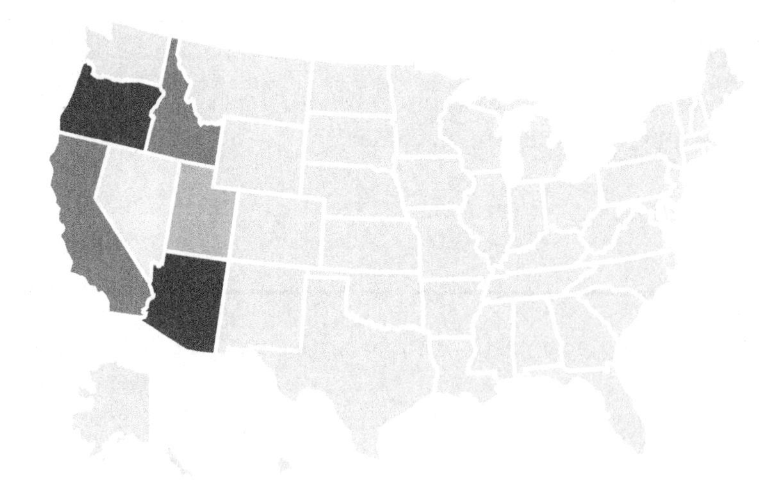

[11]And who knows, perhaps by the time you're reading this we will have left off the states of Washington D.C. and/or Puerto Rico as well.

Since it is possible to color the USA with four colors, but not possible with three colors, this means that four is the minimum number of colors that a mapmaker could use for this particular map. If you try coloring a map of African countries, you will find that four is again the minimum number of colors that can properly color the map. A map of Canadian provinces can also be colored with four colors (although their minimum number is actually three). Interestingly, you never seem to need more than four. There is no map that would require a mapmaker to use five colors. Even crazy, made-up maps like this one only need four colors.

In 1852, a man named Francis Guthrie was coloring a map of the counties of England, and even though it was a complicated map, four colors were enough. This intrigued Francis, who mentioned it to his brother Frederick, but neither could prove it to be true, or find a map that needed more than four colors. Frederick was a student at University College, London, and he decided to tell the problem to his math professor... who just happened to be Augustus De Morgan (of *De Morgan's Laws*, which we discussed in the sets and logic chapters)! De Morgan was also stumped, so he wrote to William Hamilton (who you'll meet in later courses) and said:

> A student of mine [Guthrie] asked me today to give him a reason for a fact which I did not know was a fact — and do not yet. He says that if a figure be any how divided and the compartments differently colored so that figures with any portion of common boundary line are differently colored — four colors may be wanted but not more — the following is his case in which four colors are wanted.

In fact, this letter has even been preserved for history. Here is a scan of the actual letter De Morgan wrote. His writing is not always the clearest, but you can hopefully read some of it, and you can see some maps that he drew which require four colors; he labeled the colors A, B, C and D.

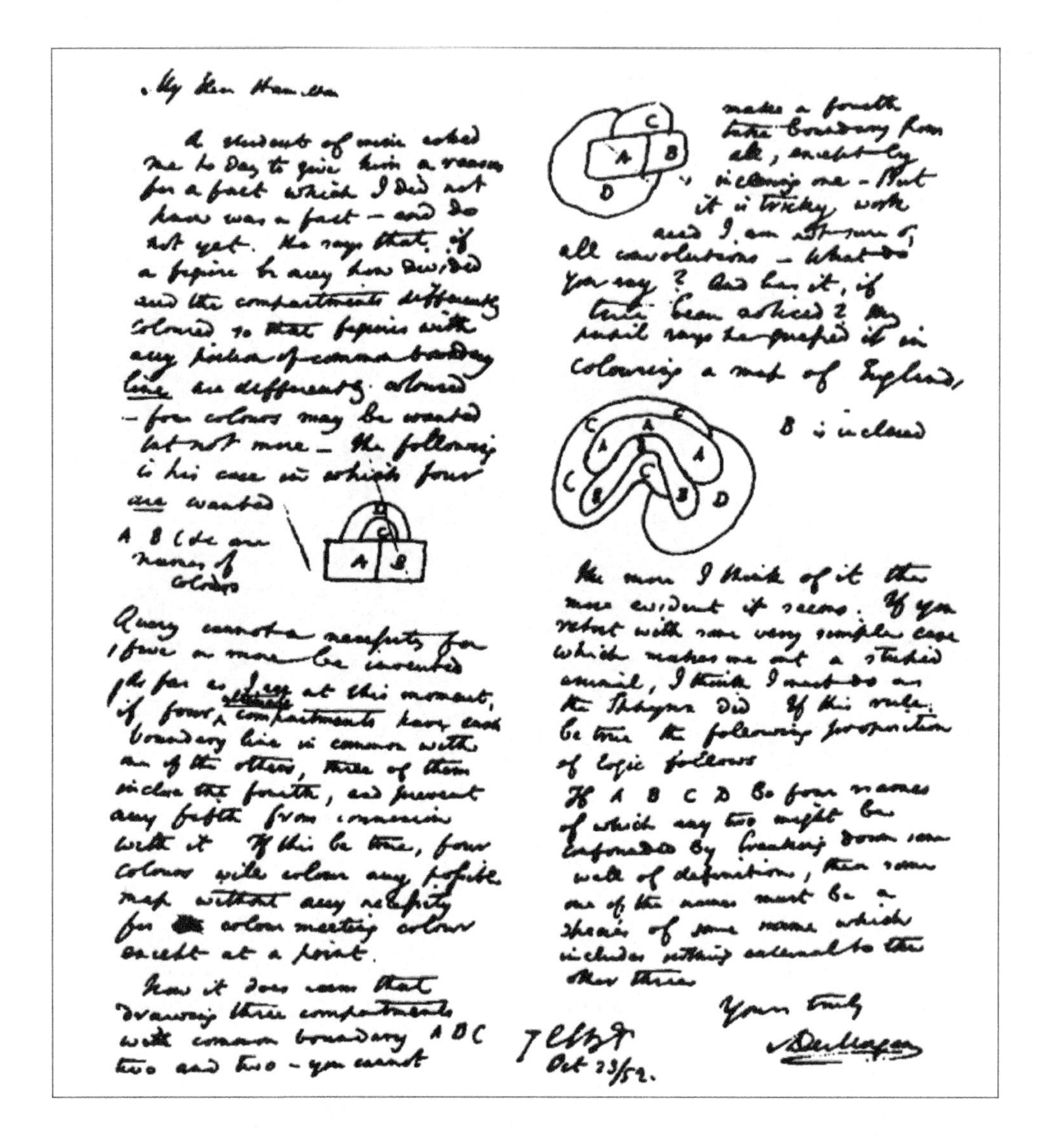

My dear Hamilton

A student of mine asked me to day to give him a reason for a fact which I did not know was a fact — and do not yet. He says that if a figure be any how divided and the compartments differently coloured so that figures with any portion of common boundary line are differently coloured — four colours may be wanted but not more — the following is his case in which four are wanted

A B C &c are names of colours

Query cannot a necessity for five or more be invented. As far as I see at this moment, if four ultimate compartments have each boundary line in common with one of the others, three of them inclose the fourth, and prevent any fifth from connexion with it. If this be true, four colours will colour any possible map without any necessity for colour meeting colour except at a point.

Now it does seem that drawing three compartments with common boundary A B C two and two — you cannot make a fourth take boundary from all, except by inclosing one — But it is tricky work and I am not sure of all convolutions — What do you say? And has it, if true been noticed? My pupil says he guessed it in colouring a map of England.

B is inclosed

The more I think of it the more evident it seems. If you retort with some very simple case which makes me out a stupid animal, I think I must do as the Sphynx did. If this rule be true the following proposition of logic follows

If A B C D be four names of which any two might be confounded by breaking down some wall of definition, then some one of the names must be a species of some name which includes nothing external to the other three

Yours truly

A De Morgan

Oct 23/52

And yet, these two world-class minds also failed to prove or disprove the conjecture. Thirty eight years after this initial action, in 1890, Percy Heawood proved that no map requires more than five colors. This is progress — it shows that no map needs six or seven or more colors — but his proof did not rule out the possibility that there is a map that requires exactly five colors.

In the 1960s and 70s — over a century after the problem was initially posed — some new ideas were found useful. Heinrich Heesch had the idea to use the technique of *discharging*,[12] and several thought to use a *reducibility* argument. The general idea is to use a *proof by minimal counterexample.* Assume that there are maps that require more than four colors. Then, among all these counterexample maps, one of them will

[12]One of my favorite math techniques!

contain the fewest number of countries (let's assume it is mapping countries, rather than states or another region).

Next, work was done to identify *reducible configurations*, which are special arrangements of countries for which, if any map contains such a configuration, it can be transformed to a map with fewer countries, and the minimum number of colors needed for the smaller map is the same as with the bigger map. Thus, no minimal counterexample can contain any reducible configuration — if it did, then that counterexample could be reduced to an even smaller counterexample, meaning the original was not the *minimal* counterexample.

Nevertheless, they discovered that there was a large set of *unavoidable configurations* that had to be considered. They identified 1,936 configurations which none of their theoretical arguments could handle. On the one hand, this is great! They took infinitely many possible maps and reduced them to fewer than two thousand maps! On the other hand, this is terrible! Each of these unavoidable configurations was a map that would need to be shown to be four-colorable, and many of these were big, complicated maps. This was far too much work for any human — or even for any reasonably-sized team of humans.

Fortunately for them, computers were available and were already powerful enough to check all 1,936 cases. It was still a big computation (it took their computers 40 days to complete), but it could be done. Once the computer completed its massive calculation, the conjecture was proved. Then, because it was proved, it moved from being a *conjecture* to a *theorem*, and became known as *the four color theorem.*

This proof bothered some mathematicians. This was the first major theorem whose proof relied on a computer. Up until then, proofs could (at least in theory) be verified by other mathematicians. But for this proof, you would have to also verify that the computer program was coded correctly, and that the computer ran the code properly. The computation was far too complicated to check by hand, but it would also be impractical to verify that there were no bugs in the code that the computer runs on. Today, many mathematicians use programs like Maple to do mathematics. Could you comb through all of Maple's code to confirm that there is no strange bug that has garbled a massive computation?

Now, to be fair, there are some ways to at least partially-check computer-ran proofs for correctness, and there are of course many ways that a non-computer proof can be flawed. The objections to the proof of the four color theorem went beyond correctness. Math prides itself on being a product of the mind. It is pure thought and reason. It is *human.* Do computer-based proofs take us away from that? Should such a thing be allowed in math?

In 1979, a paper was published entitled *The Four Color Theorem and its Philosophical Significance.* In it, the author makes an analogy. Suppose the Martians had a style of mathematics very similar to our own, but with one big difference. Suppose that the greatest Martian mathematician is named Simon, and while most of his proofs look just like our own, occasionally Simon tells his fellow Martians that he has proved a theorem but its proof is too complicated to explain, so they should just take his word on it. And they do, because of his great reputation. Thus, in Martian math textbooks, many proofs look like ours, but some simply say "Simon says."

This article's author argues that Simon is like our computers. And sure, for the four color theorem we are able to understand the broad idea of what the computer is doing, and if we really wanted to dig in and check any one of the 1,936 cases, it would be possible to brew 3 big pots of coffee, find a comfortable armchair, and then email the work to a group of grad students to do for you. Sure, it's *possible.* But is this a slippery slope?

What if in ten years our computers prove theorems that would take a human decades of work to understand? What if in twenty years our even more powerful computers prove the Riemann hypothesis (the most sought-after proof in mathematics), but it is so complicated that humans would never be able to unwind the computer's argument to understand even the broadest strokes of what the computer did? Such a proof would be no more instructive to us than "Simon says." Are we ok with that?

Theorem-Proving Computers

I am far from an expert, but I did spend a couple weeks as a grad student working with my friend Andy Soffer to try to build a rudimentary theorem-proving program. We wondered if we could create a big database of theorems from graph theory, all of the form

$$\langle\!\langle \text{these graph characteristics} \rangle\!\rangle \text{ imply } \langle\!\langle \text{these other graph characteristics} \rangle\!\rangle.$$

We wondered, if we properly entered all of these theorems, whether our computer could find us more theorems. Maybe our program could be taught to realize that if one assumes conditions C_3, C_{12} and C_{43}, then by combining theorems T_{23}, T_{48} and T_{102}, these conditions can be shown to imply condition C_{54}. Maybe there are all sorts of theorems waiting out there, that don't require any sophisticated ideas to prove — they just need to be pieced together properly from things that are already known. And maybe eventually we could integrate in some semi-sophisticated techniques.

It kind of worked, but would have needed a lot more work than the couple of weeks we put into it. Nowadays, many people smarter than me (but not as smart as Andy) are working on far more sophisticated forms of our long-abandoned project.

Will these programs eventually be able to have novel ideas to prove things? Piecing together ideas (i.e., "diagram chasing") is something they can do well. But what about real, genuine, novel ideas? The best computers can beat the best humans at chess, but they are not being *smart* per se. They can analyze a huge amount of possible positions and they have a system to determine which positions are best, but is that cleverness or just brute force? On the other hand, even if it is brute force, can tools from machine learning help a computer mimic cleverness well enough that, to humans, it *looks* clever?

When you enter something into Google Search, their software is really good at figuring out what you might have meant, and suggesting an alternative search for you. I can imagine something like this for research, where a program suggests techniques for you to try, theorems for you to look at, and papers for you to read. But again, that's far from letting a program loose and sitting back as it wins you the Fields Medal.

Could programs become good at checking for errors? Is it possible that math journals will some day require you to submit your work to a computer program to check for correctness before they consider its quality for publication? Could such a program allow me to spend half as much time grading each week? We can all dream, but is it actually possible?

Earlier we asked this: Should computers have a role in our proofs? The four color theorem was the first major theorem whose proof relied on a computer, but it was certainly not the last. The Kepler conjecture,[13] the double bubble conjecture[14], and the boolean Pythagorean triples problem[15] were all proven with the help of a computer, plus many more. The proof of the Pythagorean triples problem was *200 terabytes* in size. These "successes," have not unified the math community. Some resist the change, some embrace it, and some ignore it altogether. Which brings me to our next character in this story.

Shalosh B. Ekhad

Shalosh B. Ekhad is an author on dozens of published research articles. For example, Ekhad and Doron Zeilberger wrote a paper entitled "A WZ Proof of Ramanujan's formula for π." As of December 2020, this article has been cited by 46 other research articles. Another paper, with coauthors George Andrews and Doron Zeilberger, entitled "A short proof of Jacobi's formula for the number of representations of an integer as a sum of four squares" has been cited 35 times. Ekhad seems like quite the accomplished researcher!

What is interesting is, Ekhad is not a person. Ekhad... is a computer. Or, perhaps more precisely, some computer software and code that is run through various (super)computers, as the need demands. To understand Ekhad, you must first understand Doron Zeilberger. Zeilberger is an eccentric math professor at Rutgers University who believes that computers should be much more involved in our mathematics. He is part of a growing number of mathematicians who believe that we are not going to dig deep enough to find the best mathematical gold unless we use our most powerful tool: computers.

Zeilberger believes this so passionately that he has gone so far as to name his computer. The name Shalosh B. Ekhad is Hebrew for "three B one," which is a reference to AT&T's 3B1 desktop computer, which was the earliest incarnation of Ekhad. Zeilberger believes his computer should get credit for the help it provided, and so he has listed Ekhad as a coauthor on dozens of his papers, dating back to the late 1980s. In fact, there are even papers where Ekhad is the *sole* author! Zeilberger did not believe his own contributions even warranted coauthorship! Computers have the potential to change many of math's millennia-long traditions.

[13]Which says that the way oranges are stacked in the grocery store (each orange stacked on top of three oranges below it, in the gap they form) is the most efficient way to stack spheres.

[14]Which deals with soap bubbles and minimal surface areas.

[15]Which says that every time you color each of the natural numbers either red or blue, you are guaranteed that there are three of them which form a Pythagorean triple and which all received the same color.

Computing Conjectures

After some clever theoretical work, the proof of the four color theorem was essentially a huge proof by cases. Humans told a computer which cases to check, and it computationally verified each one. This is not the only use of computers in mathematics, though.

On homework and exams, you are usually told what to prove. But in research, you often don't know ahead of time what the theorem is going to be. There are certainly many conjectures in the world which await a proof, so yes, you could simply pick one, hope it's true, and try to prove it. But research often involves coming up with your own conjectures, or improving someone else's. Often, you must discover your own theorems, not just your own proofs.

Sometimes, these theorems come about naturally: You start with some conditions, you have a general idea of where you want to go, and you see what you can prove in that direction. If you keep deducing new things, then at the end you will have proven *something*, and if that something is interesting, then that's your theorem! Do you see? The proof was developed along the way, and what you get at the end is the theorem. But the proof came first! At times, this is how a theorem is discovered.

Other times, you have to make a conjecture without knowing if it's true, and then set out to prove it. As it happens, computers can be very useful at coming up with these conjectures. I first learned this in graduate school. My Ph.D. advisor and I would be working on something (and since we researched in combinatorics, it was often something in discrete math, and hence well set up for a computer), and we would want to know some specific values for the first 20 cases. Rather than spend two weeks slaving over the computation (and probably making several mistakes along the way), we simply let a computer do the work for us![16]

Our goal was often to prove that something was true for every $n \in \mathbb{N}$ (or perhaps every $m, n \in \mathbb{N}$, or perhaps with even more variables). Once you have the first 20 cases, you can stare at them, looking for patterns. Perhaps there's a power-of-2 or a factorial you can pull out. Perhaps the binomial coefficients are appearing. If you factor the numbers and all the primes are small, that's a good sign. If some are big, perhaps the answer is a sum of two (or more) nice things. There is some detective work involved,[17] but eventually you might notice a pattern, a general formula, and eventually a conjecture. Once you have a conjecture, all you need is a proof. But if the formula involved factorials, binomial coefficients, powers-of-2, or anything else with many combinatorial meanings, then you have a lot to work with.

In this way, even papers which do not mention a computer might have relied on one. Computers work naturally in discrete math, but they are beginning to be used

[16]Of course, you can't just *tell* a computer to do it for you, you have to write a program. Fortunately for us, one of my advisor's collaborators really enjoyed coding. So we would typically just send our coding requests to him, and he would do them for us. As someone who enjoys proving more than coding, this was a *delightful* way to do research.

[17]There is also some software to help with this step. But my biggest shout-out goes to the website `OEIS.org`, which lets you enter the start of a numerical sequence, and it tells you what the sequence could mean. If people gave OEIS coauthorship like Zeilberger does with his computer, OEIS might be the greatest combinatorist in history.

in more and more fields as data-generators and hence conjecture-finders, so learning some basics of coding could be quite useful if you pursue math research.

There are communities of mathematicians who are particularly interested in these computer-based investigations. Indeed, journals like *Experimental Mathematics* publish papers on "formal results inspired by experimentation; conjectures suggested by experiment; descriptions of algorithms and software for mathematical exploration; surveys of areas in mathematics from the experimental point of view; and general articles of interest to the community." Math has thus joined physics, chemistry and the other sciences in having an experimental component. Our laboratories are quite different (both in type, size and price tag), and the experiment's certainly feel different, although perhaps many readers of a proofs textbook will find those differences appealing.

Now, while I proudly sing the praises of computers (my dissertation would have been worse without them), a word of caution is in order. Pablo Picasso once said, "Computers are useless; they can only give you answers." As far as math is concerned, it might be better to say, "Computers are *limited.* They can only give you *partial* answers." Oftentimes, the code you run never tells you why something was true — you have to figure that out on our own. Moreover, it can usually only tell you a few specific cases — the general case must be attacked in a more theoretical way.

The problem is that the data might mislead you. Our first example of this was way back on pages 10 and 11, when we gave the example of a circle with chords, and how a conjecture that looked good... was false. Just for fun, let's close out this section with eight integrals. Each of these integrals is more difficult for a computer to determine than the previous ones. If you computed the first five, six or seven of these integrals, it would be reasonable to stop there, confident that you had discovered the answer for the rest. And yet, the eighth integral breaks the pattern and shows us again why one must be careful when given just a partial answer.

$$\int_0^\infty \frac{\sin(x)}{x}\,dx = \frac{\pi}{2}$$
$$\int_0^\infty \frac{\sin(x)}{x}\frac{\sin(x/3)}{x/3}\,dx = \frac{\pi}{2}$$
$$\int_0^\infty \frac{\sin(x)}{x}\frac{\sin(x/3)}{x/3}\frac{\sin(x/5)}{x/5}\,dx = \frac{\pi}{2}$$
$$\int_0^\infty \frac{\sin(x)}{x}\frac{\sin(x/3)}{x/3}\frac{\sin(x/5)}{x/5}\frac{\sin(x/7)}{x/7}\,dx = \frac{\pi}{2}$$
$$\int_0^\infty \frac{\sin(x)}{x}\frac{\sin(x/3)}{x/3}\frac{\sin(x/5)}{x/5}\frac{\sin(x/7)}{x/7}\frac{\sin(x/9)}{x/9}\,dx = \frac{\pi}{2}$$
$$\int_0^\infty \frac{\sin(x)}{x}\frac{\sin(x/3)}{x/3}\frac{\sin(x/5)}{x/5}\frac{\sin(x/7)}{x/7}\frac{\sin(x/9)}{x/9}\frac{\sin(x/11)}{x/11}\,dx = \frac{\pi}{2}$$
$$\int_0^\infty \frac{\sin(x)}{x}\frac{\sin(x/3)}{x/3}\frac{\sin(x/5)}{x/5}\frac{\sin(x/7)}{x/7}\frac{\sin(x/9)}{x/9}\frac{\sin(x/11)}{x/11}\frac{\sin(x/13)}{x/13}\,dx = \frac{\pi}{2}$$
$$\int_0^\infty \frac{\sin(x)}{x}\frac{\sin(x/3)}{x/3}\frac{\sin(x/5)}{x/5}\frac{\sin(x/7)}{x/7}\frac{\sin(x/9)}{x/9}\frac{\sin(x/11)}{x/11}\frac{\sin(x/13)}{x/13}\frac{\sin(x/15)}{x/15}\,dx \approx \frac{0.99999999998\pi}{2}$$

A.5 Proofs by Picture

At times, the most convincing and remarkable proofs are ones that you can *see*. Many people are visual learners or at least appreciate pictorial explanations. A visual proof may really click. For example, you have been told for years that if you FOIL the expression $(x + y)^2$, you would get $x^2 + 2xy + y^2$. When x and y are positive, there is a slick way to prove this by dividing up the area of an $(x + y) \times (x + y)$ square:

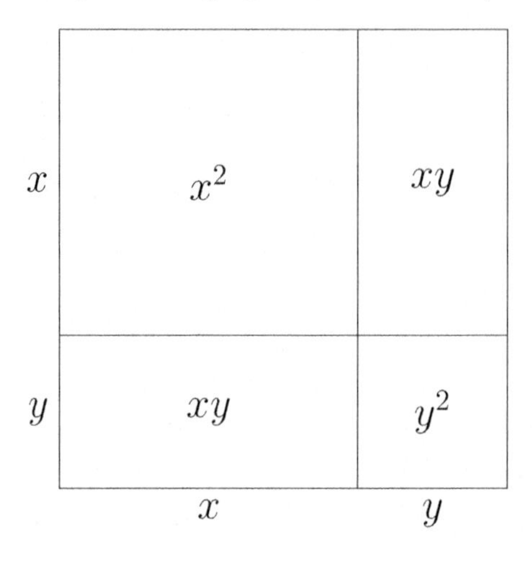

In the Chapter 2 Bonus Examples we proved what was called the AM-GM inequality, which was good practice using a direct proof to prove an inequality. The theorem says that if x and y are positive real numbers, then $\sqrt{xy} \leq \frac{x+y}{2}$. There are also a couple proofs by picture of this theorem. Here is the first one:

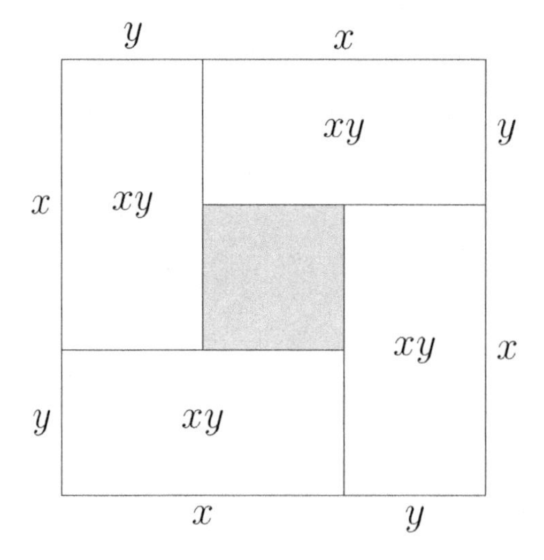

This proves that $4xy \leq (x+y)^2$, and by dividing over the 4 and taking the square root of both sides, this gives the result. As you will notice, some of these proofs require you to do a little more algebra at the end. Also, it should be noted that in some settings that picture alone would suffice as a proof, while in other cases you would be expected to explain more.

Below is a second proof of the AM-GM inequality. In the picture you can see that $\sqrt{xy} \leq \frac{x+y}{2}$, since the vertical segment of length $\sqrt{xy}$ is shorter than the vertical segment of length $\frac{x+y}{2}$. But in order to believe this proof, the reader must understand why those segments have those lengths. The length $\frac{x+y}{2}$ is because that vertical segment is a radius, and so its length is half the length of the diameter, which is $x + y$. As for the length $\sqrt{xy}$, this is because of similar triangles. Notice that $\triangle ADC \sim \triangle CDB$. So, if we call h the length of $\overline{CD}$, then $\dfrac{x}{h} = \dfrac{h}{y}$, which implies $h^2 = xy$, which implies $h = \sqrt{xy}$.

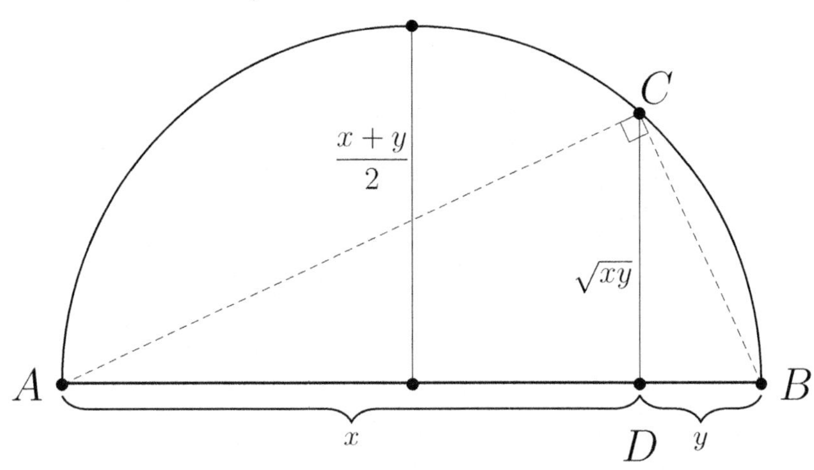

Is this a proof by picture or just a proof that uses a picture? The distinction might depend on the level of your audience.

Proofs by picture also work well when dealing with sums of numbers following a simple pattern. See if you can convince yourself that the below is a proof by picture that $1 + 2 + 3 + \cdots + n = \dfrac{n^2}{2} + \dfrac{n}{2}$.

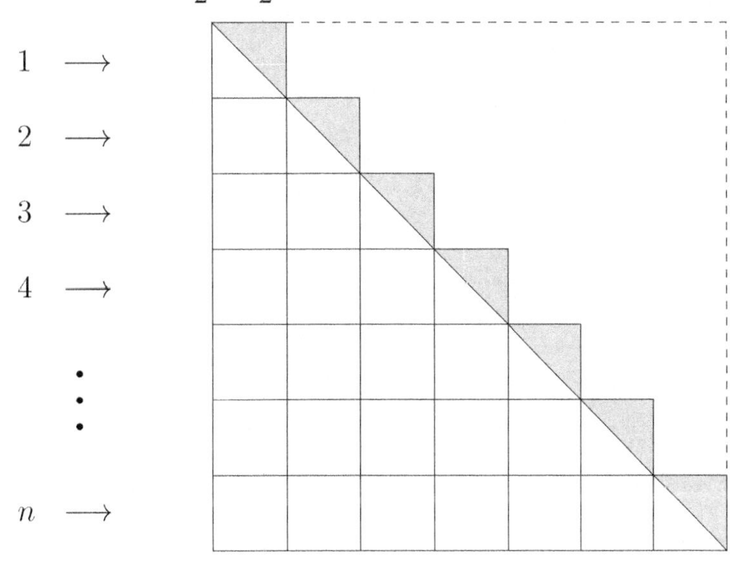

Proofs by picture which rely on dissecting a shape into components are especially common and beautiful. The last example showed that $1+2+3+\cdots+n = \frac{n^2}{2}+\frac{n}{2}$. Let's let $T_n = 1+2+3+\cdots+n$, and prove by picture that

$$(2n+1)^2 = 8T_n + 1.$$

Here's that:

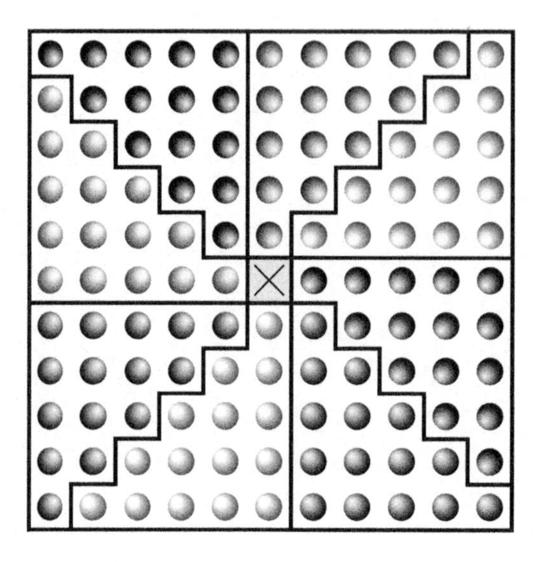

Many geometric series can be seen through a picture. For example, the fact that

$$\frac{1}{2}+\frac{1}{4}+\frac{1}{8}+\frac{1}{16}+\cdots = 1$$

can be seen by taking a square of area 1, and chopping up that area into pieces of size $\frac{1}{2}$, $\frac{1}{4}$, $\frac{1}{8}$, $\frac{1}{16}$, and so on.

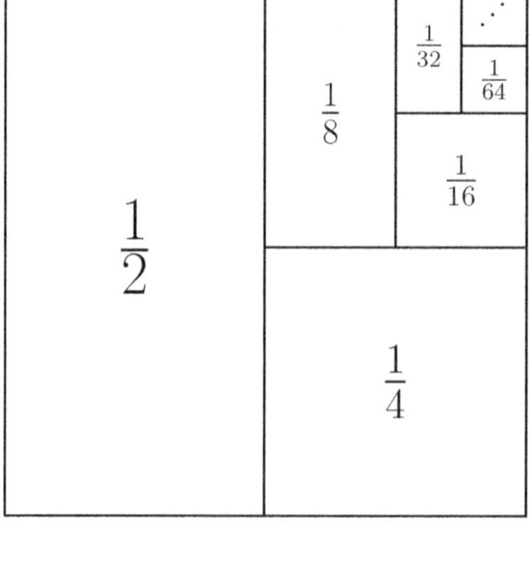

$\frac{1}{32}$ $\frac{1}{64}$ $\frac{1}{8}$ $\frac{1}{16}$ $\frac{1}{2}$ $\frac{1}{4}$

Our final example is similar to the last one, except in this case we will divide up a square of area 4. The result is this:

$$1 + 2\left(\frac{1}{2}\right) + 3\left(\frac{1}{4}\right) + 4\left(\frac{1}{8}\right) + 5\left(\frac{1}{16}\right) + \cdots = 4.$$

This is not a geometric series, but can be viewed as a *differentiated* geometric series.

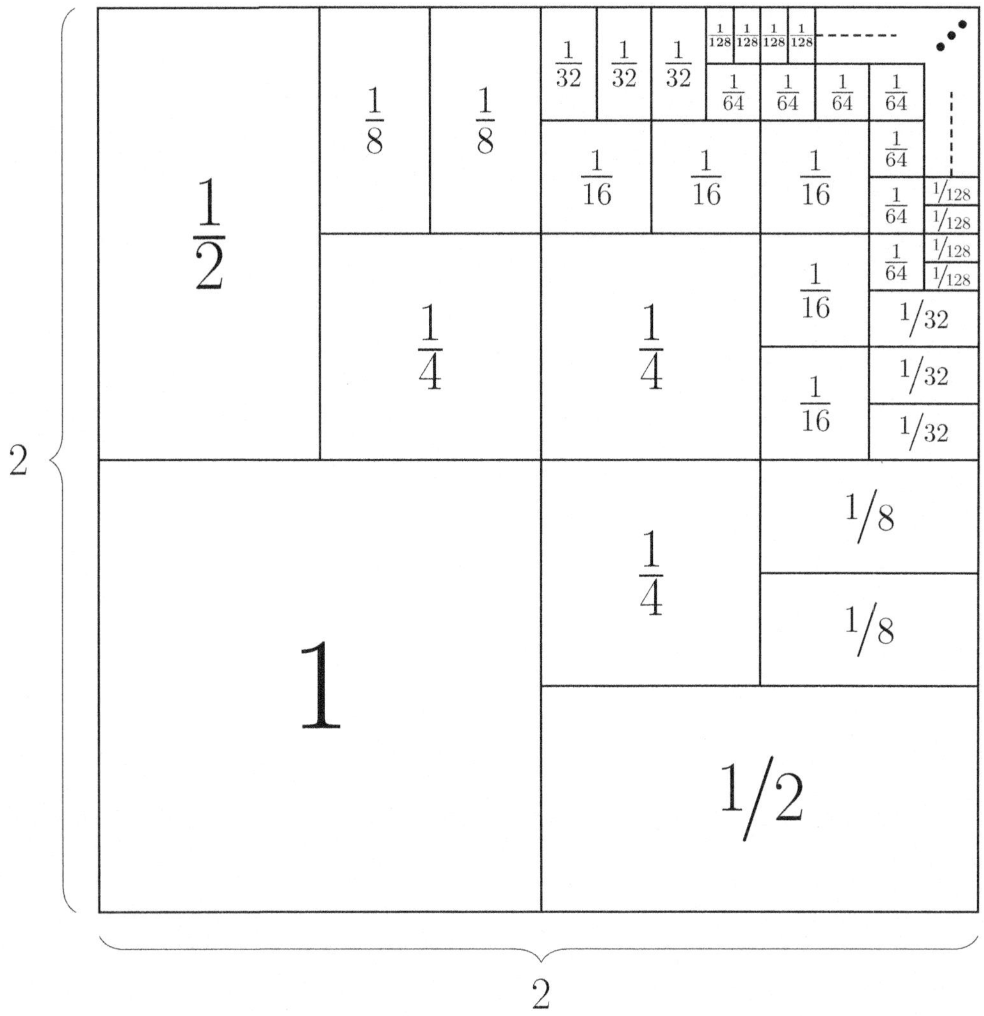

$\frac{1}{2}$ $\frac{1}{8}$ $\frac{1}{8}$ $\frac{1}{32}$ $\frac{1}{32}$ $\frac{1}{32}$ $\frac{1}{128}$ $\frac{1}{128}$ $\frac{1}{128}$ $\frac{1}{128}$

$\frac{1}{64}$ $\frac{1}{64}$ $\frac{1}{64}$ $\frac{1}{64}$ $\frac{1}{64}$

$\frac{1}{16}$ $\frac{1}{16}$ $\frac{1}{16}$ $\frac{1}{64}$ $1/128$ $\frac{1}{64}$ $1/128$ $\frac{1}{64}$ $1/128$ $1/128$

$\frac{1}{4}$ $\frac{1}{4}$ $\frac{1}{16}$ $1/32$ $1/32$ $\frac{1}{16}$ $1/32$

2

$\frac{1}{4}$ $1/8$ $1/8$

1

$1/2$

2

Appendix B: Proofs From The Book

Paul Erdős (who passed away in 1996) imagined a book in which God wrote down every theorem, and following each theorem He wrote down the best, most beautiful, most elegant proof of that theorem. Ironically, Erdős was an agnostic atheist, yet his idea of "The Book" caught on, and soon it entered the standard vocabulary of research mathematicians. It also highlights a common belief among mathematicians that proofs should be beautiful. As G.H. Hardy said, "there is no permanent place in this world for ugly mathematics."

Indeed, I once attended a math conference and was chatting with the photographer who was hired to document the event. She said that academic conferences are her specialty and had taken pictures at conferences in dozens of different areas, from the sciences to the humanities. I asked her what stood out about the math conferences, expecting us to spend the next couple minutes joking around about mathematicians' lack of fashion, or something. But she remarked that the thing which most surprised her was how often mathematicians talked about beauty in their field. She said that only art conferences talked about beauty more than us.

If Erdős read a proof and found it correct yet ugly, he might comment that it is good that we know the theorem is true, but we should still search for "the proof from The Book!" This refrain has echoed through the decades.

The title of this appendix was stolen[1] from Martin Aigner and Günter M. Ziegler, who had the wonderful idea to write a book with the same name and idea. It's the type of idea which, if I had had it first, I probably would have dropped all other projects to focus exclusively on it. A couple proofs in this appendix are also included in their delightful book, although the collection in this appendix is much more geared to mathematicians at your stage of learning (and I again take a long-form approach to the task). But I encourage you to check out their book; it is broken up by subject, so after you finish each of your upper division math courses you can read through the Book Proofs they amassed from that field. Enjoy!

[1]Well, let's call it a tribute.

B.1 Merry Madness from March

The first Book Proof is one that I have shared many times. My research is in combinatorics, which I find to be the most fun and puzzle-like area of math, but saying that to a non-math person doesn't really do my field justice. Therefore, if someone at a party asks what math I study, and they seem sufficiently interested and sober, I often tell them the following problem, which I claim conveys what it *feels* like to do combinatorics.

Example B.1. In the NCAA's March Madness Tournament there are 64 teams that play. The tournament is single-elimination (once you lose, you're out). How many games are played throughout the tournament?

(Sometimes the person will be able to realize that in the first round 32 games are played (since 64 teams pair up and play a round). I then note that this is indeed a way to solve the problem: If they apply the same reasoning they could find the number of games in the second round, the third round, and so forth, and then just add up the number of games from each round $(32 + 16 + 8 + \dots)$. But this, I tell them, is boring, it can't be easily generalized to other tournaments, and it doesn't give any insight into *why* the answer is what it is.

Furthermore, I and many other combinatorists do not like computational solutions. Computations are boring. Plus, computers can do them better than I can, while also committing fewer errors. And computations often hide what is really going on. I want to know how many games are played not only because the answer might be of interest, but because I want to *understand* how many games are played.

Finally, because people in the wild tend to like when math has applications, I sometimes also mention how the answer has practical value — the number of games played translates directly into the number of times the NCAA has to reserve TV time slots, how much revenue they should expect, the number of times they need to reserve a court, etc.

But usually within a minute, and sometimes within 7 seconds, I give them the following neat solution. Here it is.)

Solution. Note that in each game, precisely 1 team loses, and of course each loss happens in a game. So, the number of games equals the number of losses. Now, the champion never loses, but every other team loses one game and then they're out. So with 64 teams, 63 of them lose once, so there must be 63 losses, so there must be 63 games.[2] □

That's it! Don't count games, count losses. There is a bijection between the games and the losses (every game has 1 loss, and every loss occurs in 1 game), so counting losses is the same as counting games. And since 63 teams lose one time, it's easy to count those losses!

[2] **drops mic**

B.2 Significant Sets of Shifting Shapes

The Pythagorean theorem says if a and b are the lengths of the legs of a right triangle, and c is the length of the hypotenuse, then $a^2 + b^2 = c^2$.

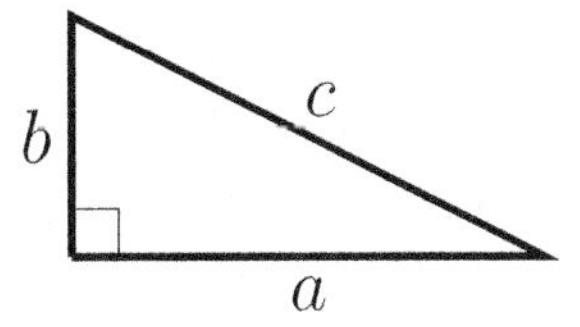

In Chapter 7 we saw Pythagoras' proof of the Pythagorean theorem. This was one of the first theorems in mathematics, and it was independently discovered all around the world. The Pythagoreans did indeed discover this result very early, and you might recall that their argument came down to this picture:

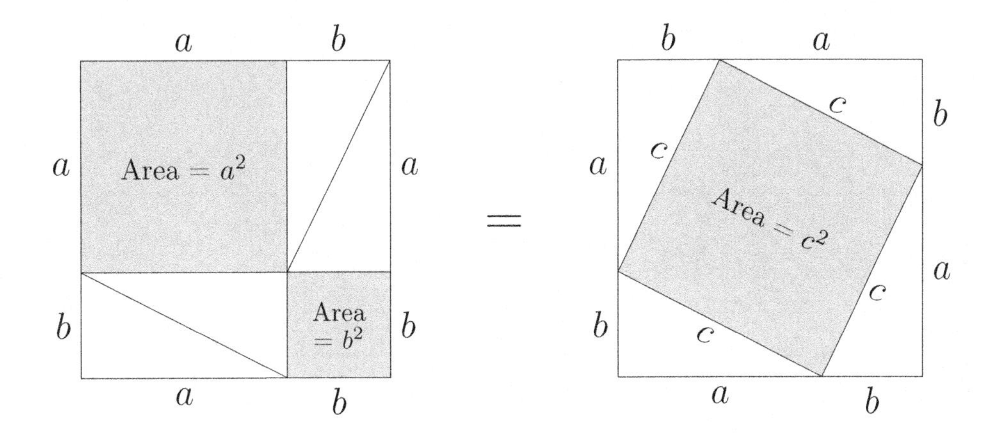

It is curious that nearly every culture that did mathematics discovered some form of this theorem. It seems remarkably fundamental or human or both. Although the Pythagoreans provided the first known general proof, the Babylonians knew of it as early as 2000 B.C.—a full 1500 years before Pythagoras was born. A few hundred years after Pythagoras, a geometric explanation of the 3-4-5 case of the theorem was independently discovered, and was recorded in one of the oldest Chinese mathematical texts, the *Zhoubi Suanjing*. Here is its author's drawing:

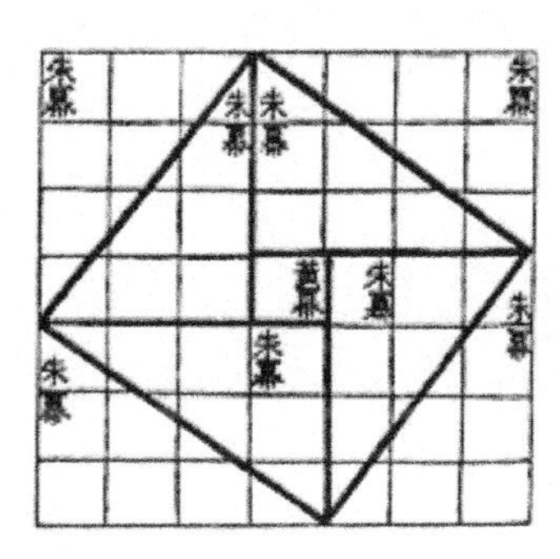

Around 300 B.C., Euclid gave his own proof — the first known *axiomatic* proof. As with the Pythagoreans, Euclid did not think about this algebraically in the "a squared plus b squared equals c squared" way that any 9th grader can rattle off today. He imagined there were actual squares on the sides of a right triangle, and the theorem asserts that the big square is equal in area to the sum of the other two. Indeed, his proof relied on the following diagram:

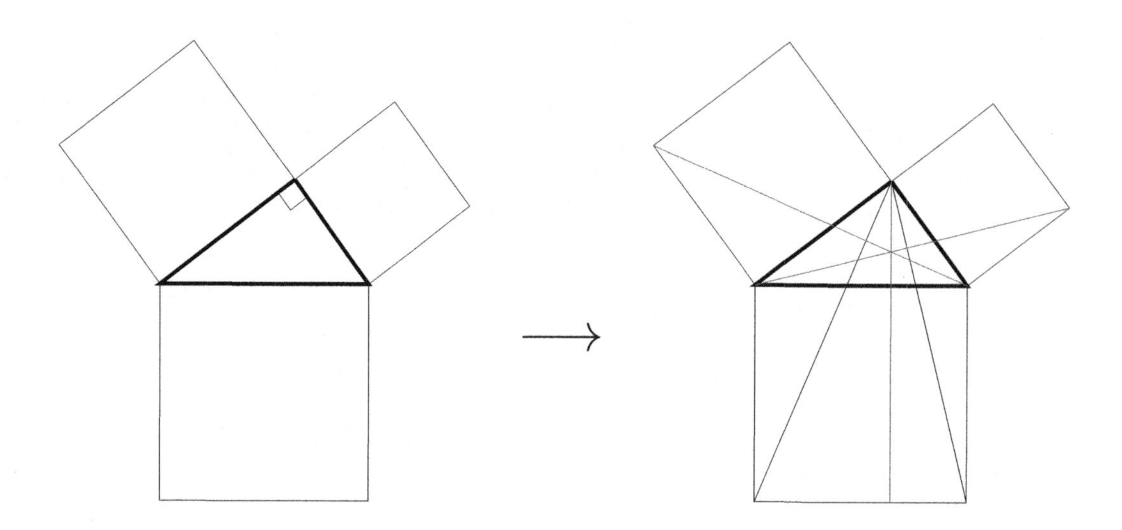

Today, there are over 370 proofs of the Pythagorean theorem! U.S. President James Garfield discovered a proof, as did Albert Einstein. A more modern proof of the result is found by considering the matrices

$$\begin{pmatrix} a & b \\ -b & a \end{pmatrix} \quad \text{and} \quad \begin{pmatrix} c & 0 \\ 0 & c \end{pmatrix}.$$

These matrices have equal determinants when $a^2 + b^2 = c^2$, and once you consider these linear transformations geometrically, they can be seen to be rotations of each other.

I would now like to outline two more proofs. The first uses similar ideas to the 3-4-5 proof from *Zhoubi Suanjing*, but comes from Indian mathematician Bhāskara II from the 11th century AD. His argument also resembled the Pythagoreans, but in his case he cut up a single $c \times c$ square to form an $a \times a$ square and a $b \times b$ square.

Proof by Bhāskara II

First, here is the triangle:

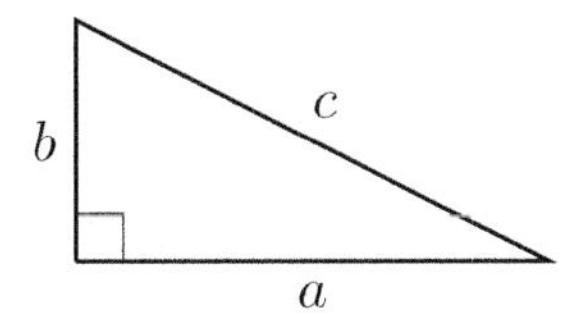

Placing four copies of this triangle into a $c \times c$ square, by placing each hypotenuse along one of the sides, leaves one square in the middle that is uncovered:

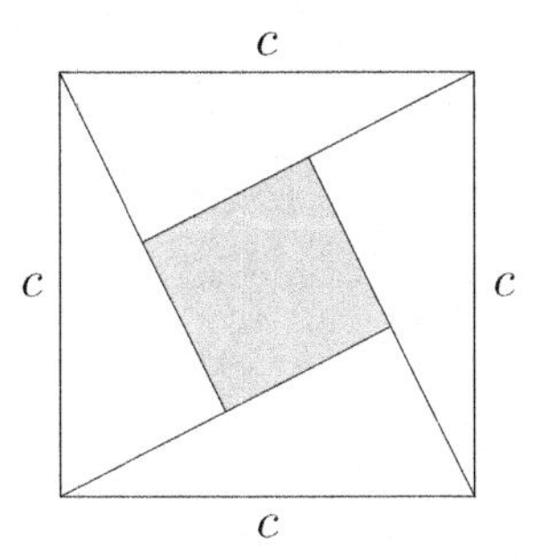

The big square has area c^2. And notice that the square in the middle has dimensions $(a - b) \times (a - b)$. Next, we move around these five pieces to get this:

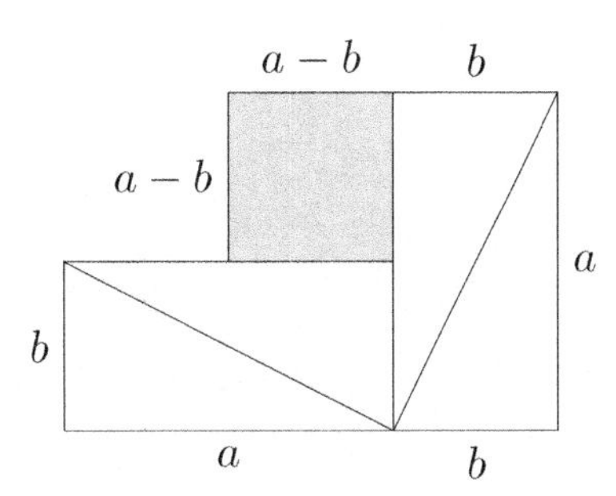

And if you squint just right, this is indeed an $a \times a$ square and a $b \times b$ square:

Area $= a^2$

Area $= b^2$

Thus, the c^2 area is equal to an a^2 area plus a b^2 area, completing the proof.

Proof by Geoffrey Margrave

We have seen several The Book-eligible proofs, but if I had to choose one, it would be this one. It comes together so beautifully. Take any right triangle:

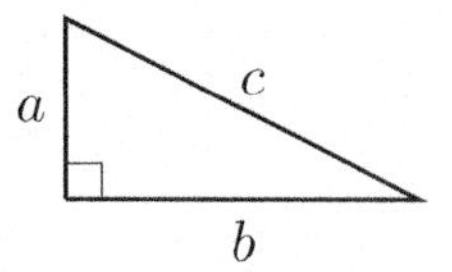

Now, scale up this triangle three times, first by a factor of b, next by a factor of c, and last by a factor of a. This produces these three similar triangles:

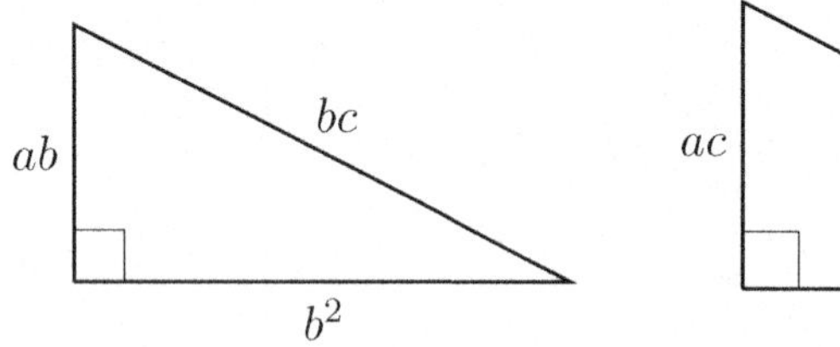

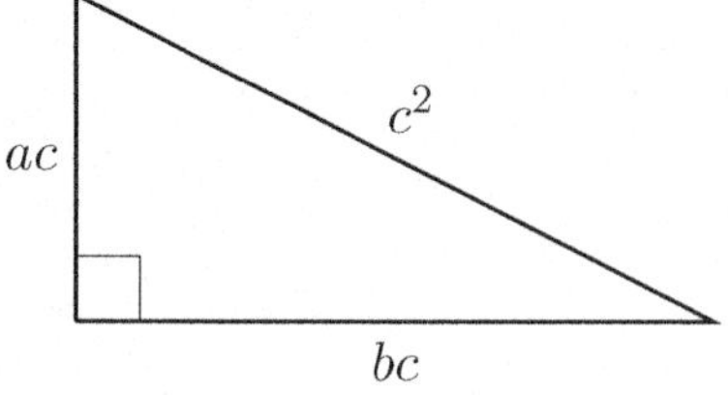

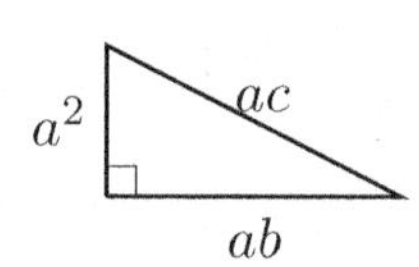

Next, by simply rotating or mirroring them, we arrive here:

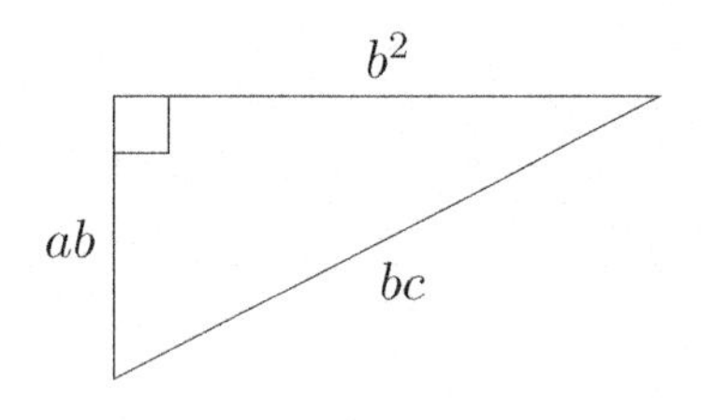

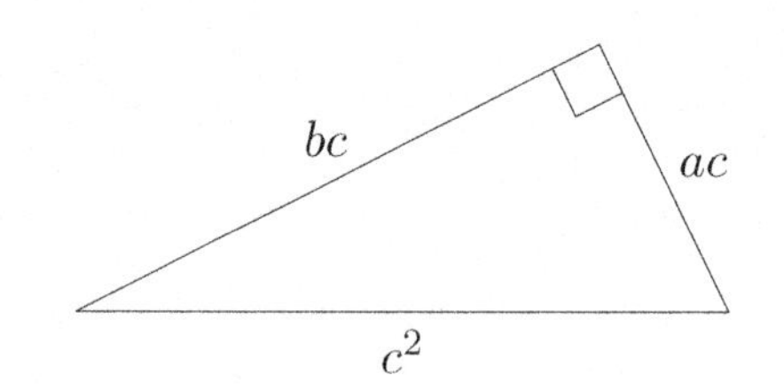

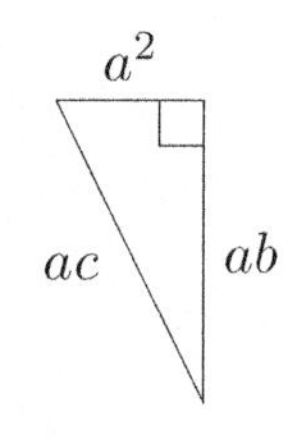

And, finally, piece them together:

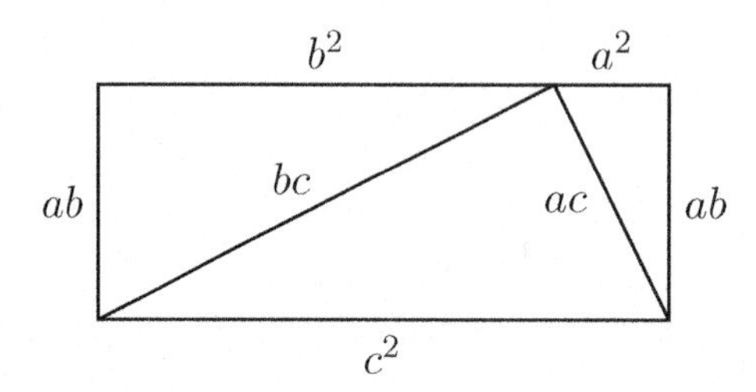

Because the angles of a triangle add up to 180°, and the two non-right angles add up to 90°, this shape must be rectangle. Now simply note that the top and bottom edges of a rectangle are the same length, which in this case means that

$$a^2 + b^2 = c^2,$$

completing the proof.[3]

[3] **drops mic**

A Final Fun Fact

Do you remember how, a few pages back, we said that Euclid thought of the Pythagorean theorem as attaching literal squares to the edges of a right triangle? Well, it turns out that the theorem generalizes, allowing you to attach *any* shape to the edges, provided the three are simply scaled and re-angled versions of each other; these are called *similar* shapes. For example, all semicircles are similar, so in the below, the semicircle on the hypotenuse has precisely as much area as the other two combined.

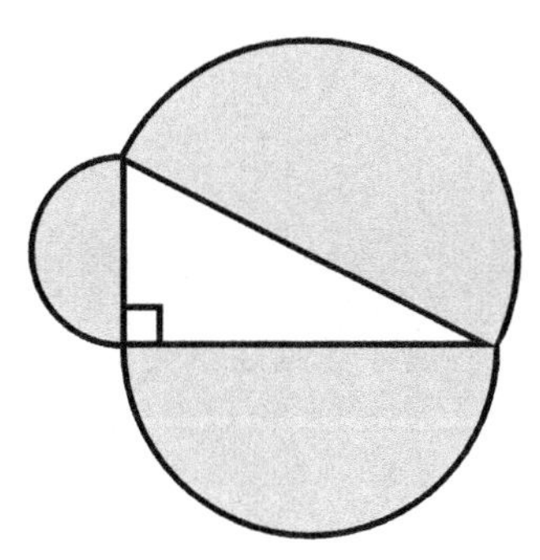

Or, even better, the area inside the hypotenuse Homer Simpson below is equal to the sum of the areas inside the other two Homer Simpsons.

This fun fact was known by Hippocrates of Chios in the 5th century BC, and Euclid included a proof of it in *Elements*.

B.3 A Flow of Factors From Fermat

There is a sequence of numbers called *Fermat numbers* which are denoted $F_0, F_1, F_2, F_3, \dots$ and which have this form:

$$F_n = 2^{2^n} + 1.$$

Here are the first six Fermat numbers:

- $F_0 = 2^{2^0} + 1 = 2^1 + 1 = 3$
- $F_1 = 2^{2^1} + 1 = 2^2 + 1 = 5$
- $F_2 = 2^{2^2} + 1 = 2^4 + 1 = 17$
- $F_3 = 2^{2^3} + 1 = 2^8 + 1 = 257$
- $F_4 = 2^{2^4} + 1 = 2^{16} + 1 = 65537$
- $F_5 = 2^{2^5} + 1 = 2^{32} + 1 = 4294967297$

This sequence has some simple properties, like how each number is odd, and how the numbers are growing reeeeaallllyy fast. But although this may seem like a pretty unremarkable sequence, it has some really interesting properties. For example, look at the numbers above and notice that

- $F_1 = F_0 + 2.$
- $F_2 = F_0 \cdot F_1 + 2.$
- $F_3 = F_0 \cdot F_1 \cdot F_2 + 2.$
- $F_4 = F_0 \cdot F_1 \cdot F_2 \cdot F_3 + 2.$

This pattern continues. In general,

$$F_n = F_0 \cdot F_1 \cdot F_2 \cdot F_3 \cdot \ldots \cdot F_{n-1} + 2.$$

This can be proven by induction. The base case ($F_1 = F_0 + 2$) checks out, and the inductive hypothesis is the assumption that

$$F_k = F_0 \cdot F_1 \cdot F_2 \cdot F_3 \cdot \ldots \cdot F_{k-1} + 2.$$

In the induction step we aim to prove that this must also hold when k is replaced with $k+1$. But we will prove it in reverse: We will start with $F_0 \cdot F_1 \cdot F_2 \cdot F_3 \cdot \ldots \cdot F_k + 2$ and work to show that equals F_{k+1}. One last preliminary note: When we apply the induction hypothesis, we will be using a slightly rewritten version of the above:

$$F_0 \cdot F_1 \cdot F_2 \cdot F_3 \cdot \ldots F_{k-1} = F_k - 2.$$

Ok, here is that induction step:

$$\begin{aligned} F_0 \cdot F_1 \cdot F_2 \cdot F_3 \cdot \ldots \cdot F_k + 2 &= (F_0 \cdot F_1 \cdot F_2 \cdot F_3 \cdot \ldots F_{k-1}) \cdot F_k + 2 \\ &= (F_k - 2) \cdot F_k + 2 && \text{(Induction hyp.)} \\ &= \left(2^{2^k} - 1\right)\left(2^{2^k} + 1\right) + 2 && \text{(Def: } F_k = 2^{2^k} + 1) \\ &= 2^{2^{k+1}} - 1 + 2 \\ &= F_{k+1}, \end{aligned}$$

completing the proof by induction. (This is just a warm-up. A Book Proof is soon!)

This recurrence implies something else interesting about Fermat numbers: Every pair of Fermat numbers are relatively prime! That is, if F_m and F_n are two distinct Fermat numbers, then any prime that divides F_m does not divide F_n, and vice versa. Why is this true? Suppose that between these two Fermat numbers, F_n is the larger one. And suppose that p is some prime that divides both F_m and F_n, meaning that $F_n = pk$ and $F_m = p\ell$, for some $k, \ell \in \mathbb{N}$. Well, by the above recurrence we know that

$$F_n = F_0 \cdots F_{m-1} \cdot F_m \cdot F_{m+1} \cdots F_{n-1} + 2.$$

That is, there is an F_m in the midst of that big product. Therefore the above is the same as

$$pk = F_0 \cdots F_{m-1} \cdot p\ell \cdot F_{m+1} \cdots F_{n-1} + 2.$$

By moving everything but the 2 from the right to the left, and factoring out a p, we get

$$p(k - F_0 \cdots F_{m-1} \cdot \ell \cdot F_{m+1} \cdots F_{n-1}) = 2.$$

Or, since that gigantic parenthetical is some integer, we can write this more simply as

$$pt = 2 \qquad \text{for some } t \in \mathbb{Z}.$$

The only prime that could do this is $p = 2$. This gives us a contradiction because each Fermat number is odd! So there is no way that 2 divides both. So they must indeed be relatively prime, completing the proof. (Was that the Book Proof? Nope! Still warming up!)

Let's now end our background discussion on Fermat numbers and move on to a Book Proof of the famous result that there exist infinitely many prime numbers. This is Theorem 7.5, and the proof we gave in Chapter 7 is certainly worthy of being included in The Book. Nevertheless, for a theorem of this stature, I'd like to think it deserves some "honorable mention" proofs in The Book. If so, below is such a proof.

Theorem.

Theorem 7.5. There are infinitely many prime numbers.

Proof. Since any two Fermat numbers are relatively prime, no two Fermat numbers share any prime factors. Moreover, simply because they are integers larger than 2, each Fermat number is prime or a product of primes. Finally, notice that there are infinitely many Fermat numbers:

$$F_0, F_1, F_2, F_3, F_4, \ldots.$$

We have an infinite list of numbers, each of which contains at least one prime, and no primes appear in more than one number in the list. This implies that there must exist infinitely many prime numbers.[4] □

[4] *drops mic*

B.4 A Pinpointed Proof Pausing Prussian Parades

The city of Königsberg in Prussia (modern-day Russia) has a river that flows through it. The river twice forks and rejoins creating two large islands. The different parts of this city are linked by seven bridges.

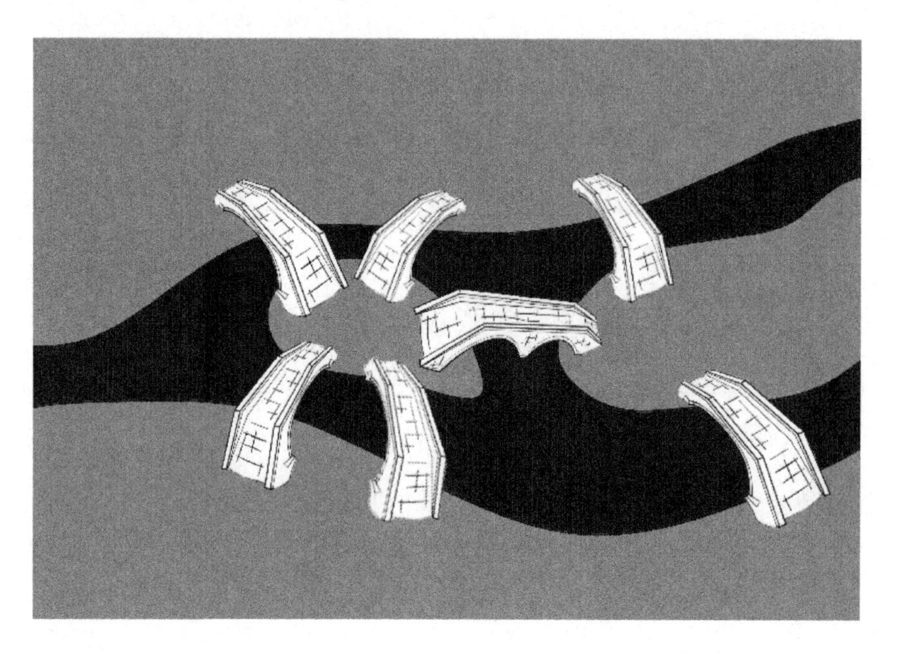

An 18th century question was this:

> Is it possible to have a parade through this city in such a way that the parade route crosses each bridge exactly once?

Leonhard Euler solved this problem, and in doing so the field of graph theory was born. His first insight was that the land masses can be reduced to a single point and the bridges can be drawn as arcs or lines, giving what we now call a *graph*.

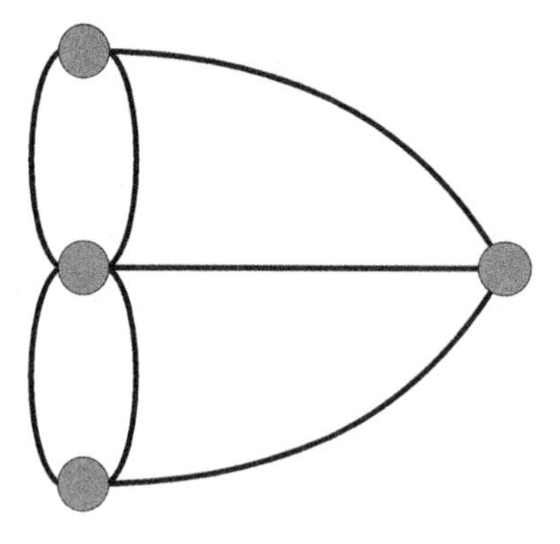

In today's graph theory lingo, the dots are called *vertices* and the lines are called *edges*.

Euler's next insight was to think about how many edges are touching each vertex.

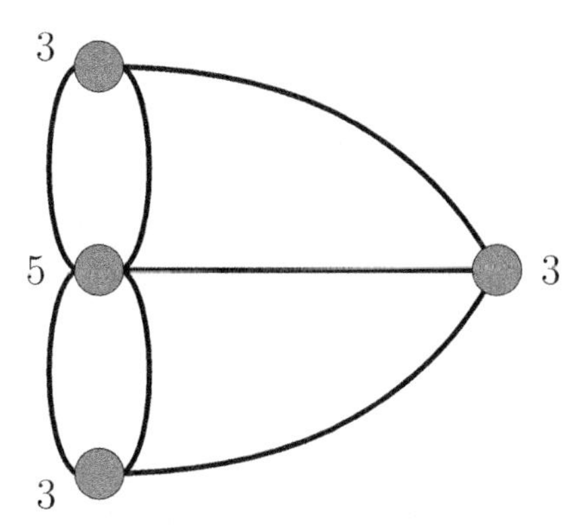

With one final insight, Euler arrived at a Book Proof that no such parade route exists. What follows was his reasoning.

Since a parade entering one vertex will then leave the same vertex, except for possibly the vertices on which the parade begins or ends, the number of times a parade enters a node is equal to the number of times a parade leaves that node. So if we assume there there is a parade route, and v is a vertex for which the parade does not begin or end at v, then all of the edges touching v must come in pairs—if the parade comes to v along edge e_1, and leaves along edge e_2, then we can pair up e_1 and e_2. And note that if all the edges touching v can be paired up, then there must be an *even* number of edges touching v.

In summary, we have shown that except for possibly the starting and ending vertices, the degree of each vertex must be *even* if there is any hope of such a parade route existing. But if you look at the graph modeling Königsberg, all four vertices are touching an *odd* number of edges! So there cannot possibly be such a parade route.[5]

It is a really clever argument. Instead of immediately getting your hands dirty trying to map out all the possible routes and trying to characterize what actions would produce a dead end, Euler found a criterion that any successful parade route must satisfy. The vertex on which the parade begins and ends could possibly touch an odd number of edges, but every other vertex has to touch an even number. With four odd-numbered vertices, a parade must be impossible.

Leonhard Euler was born in Basel, Switzerland in 1707. He first made a name for himself when at the age of 24 he proved that $\sum_{k=1}^{\infty} \frac{1}{k^2} = \frac{\pi^2}{6}$. Later, he famously proved that $e^{i\pi} + 1 = 0$, and his solution to the Königsberg bridge problem is considered the earliest contribution to the field of graph theory. But these are just drops in the ocean of his accomplishments. Euler is the most productive mathematician in history, publishing about 800 pages of mathematical work per year. His collected works comprise 92 volumes! Now *that's* long-form mathematics!

[5] **drops mic**

B.5 Cleverly Cutting the Cruising Coins

Imagine you have a chessboard that has a lower-left square, but extends infinitely up and to the right. Also, imagine that the three lower-left squares are shaded, and that a coin is placed on these three squares.

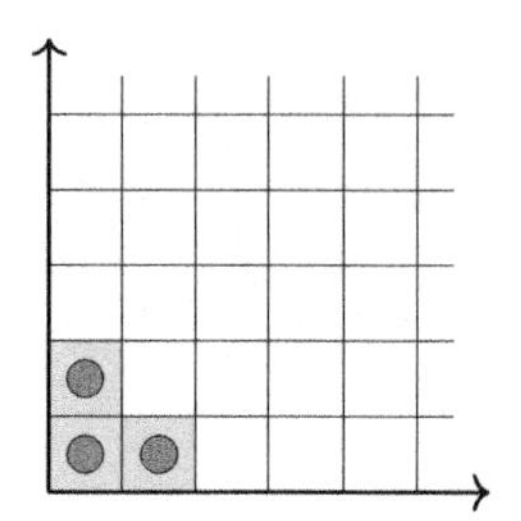

Now, let's play a game. Your goal is to make a series of moves so that the three shaded squares are empty of coins. How? Well, there is one type of move you are allowed to make. If a coin has no coin above it or to its right, then you may remove that coin and place a new coin in each of those empty squares. For example, here is a sequence of five moves:

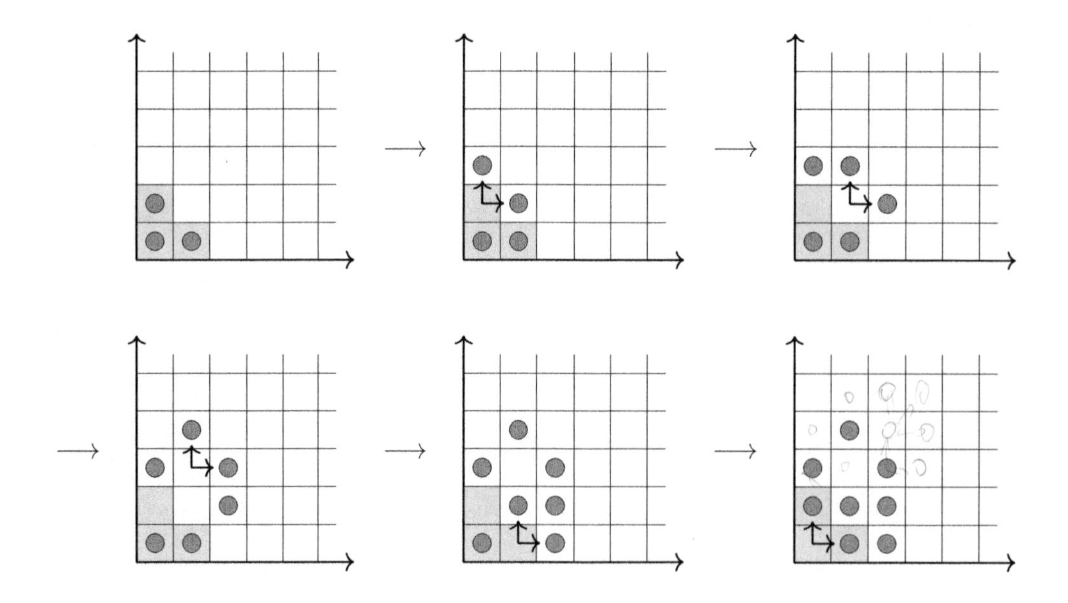

So far, we have not succeeded. By continuing this move sequence or starting over on your own, can you successfully vacate the three shaded squares? Give it a real try before looking at the answer on the next page!

As you might have figured out, it is impossible.

Theorem.

Theorem. There is no way to vacate the three shaded squares using legal moves.

How can we prove that no sequence of moves achieves this goal? Proving a negative is much harder than proving a positive. The idea, though, is beautiful. We consider an equivalent problem.

Imagine that you begin with one coin in the corner square, and half-coins on the other two shaded squares.

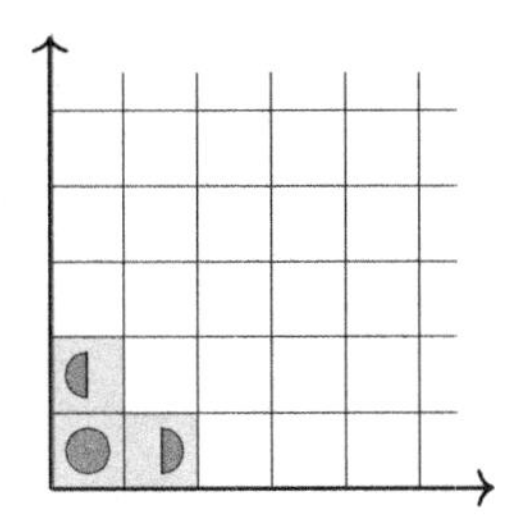

And suppose that when you make a move, you split your (partial) coin in two, and place half on above square and half on the square to the right. For example, here are four such moves:

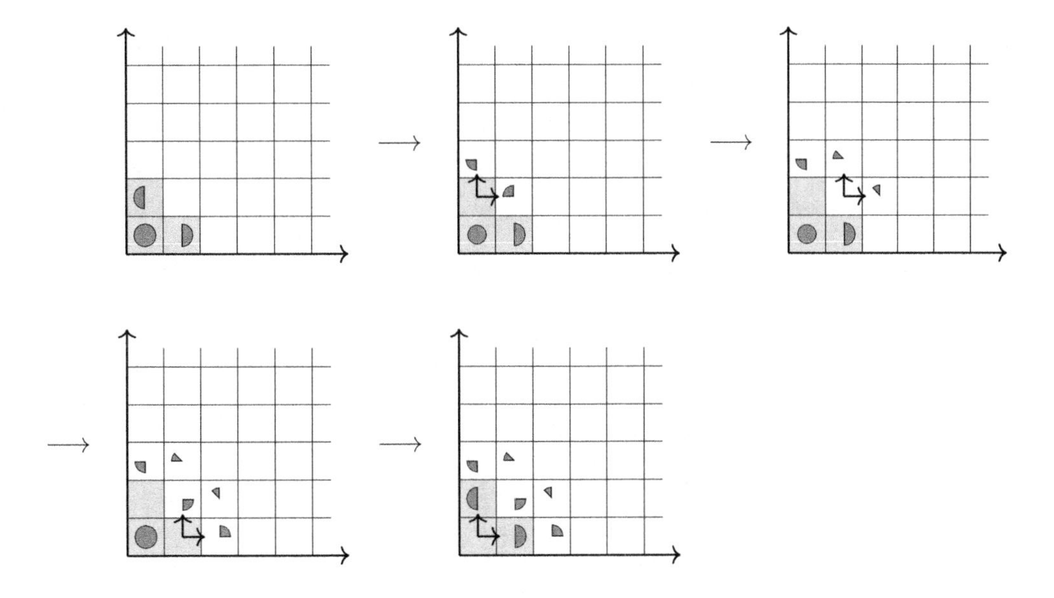

In the original setup of the problem, the above sequence of moves would produce the exact same board, except with full coins instead of partial coins. But whether or not a square has a (partial) coin on it does not change. And therefore, whether or not the three shaded squares are vacated does not change.

For this reason, this new formulation of the problem is equivalent to the original, provided we tweak the question to this: Is there a sequence of moves for which the three shaded squares have no coins *or partial coins* on them?

Why does this help? One helpful technique in proofs is to find what is called an *invariant*—something that does not change no matter what you do. In the original problem, the coins kept multiplying; but in our new formulation, the total number of coins is always two, if you add up all the pieces. The reason is simple: there were two coins at the start (a full coin and two halves), and each move neither creates nor destroys any coins—it only splits a (partial) coin into two pieces.

And because the pieces of a divided coin are placed above and to the right, each square can only hold a predictable coin-fraction:

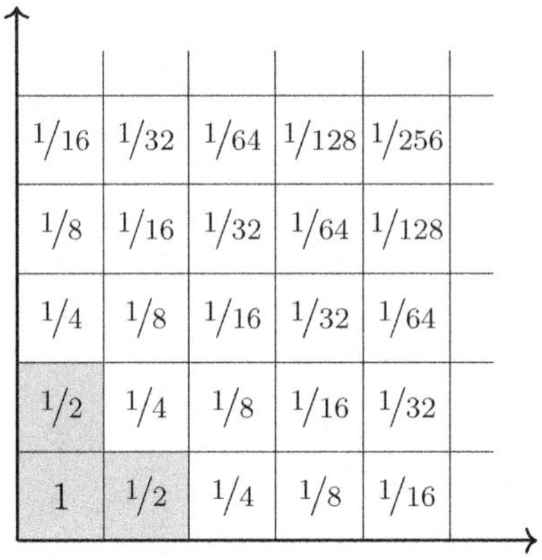

1/16	1/32	1/64	1/128	1/256
1/8	1/16	1/32	1/64	1/128
1/4	1/8	1/16	1/32	1/64
1/2	1/4	1/8	1/16	1/32
1	1/2	1/4	1/8	1/16

As we already said, since we started with two coins in total, at every stage we still have two coins in total. Let's now go to the other extreme. Given that each square can only ever have one (partial) coin on it, and the size of a partial coin is determined by the above chart, what is the total number of coins that the infinite board can hold if every square on the board contained its (partial) coin? In the bottom row, there would be

$$1 + \frac{1}{2} + \frac{1}{4} + \frac{1}{8} + \frac{1}{16} + \dots = 2 \text{ coins.}$$

In the second row, there would be

$$\frac{1}{2} + \frac{1}{4} + \frac{1}{8} + \frac{1}{16} + \frac{1}{32} + \dots = 1 \text{ coin.}$$

In the third row, there would be

$$\frac{1}{4} + \frac{1}{8} + \frac{1}{16} + \frac{1}{32} + \frac{1}{64} + \dots = \frac{1}{2} \text{ coin.}$$

In the fourth row, there would be

$$\frac{1}{8} + \frac{1}{16} + \frac{1}{32} + \frac{1}{64} + \frac{1}{128} + \dots = \frac{1}{4} \text{ coin.}$$

And so on. Therefore, by adding the number in the first row, the second row, the third row, and so on, the total number of coins on the entire board would be

$$2 + 1 + \frac{1}{2} + \frac{1}{4} + \frac{1}{8} + \dots = 4 \text{ coins.}$$

This shows something interesting. While the entire board can hold 4 coins, notice that the three shaded squares can themselves hold $1 + \frac{1}{2} + \frac{1}{2} = 2$ coins. Half of the total value of the board lives on just those three shaded squares![6] And so all of the non-shaded squares combined have a total coin-value of 2 coins. In fact, this is the key contribution of this new formulation of the problem.

To prove the theorem, we must show that it is impossible for our (partial) coins to vacate the shaded squares. Assume for a contradiction that there *is* a sequence of moves that vacates these three shaded squares; that is, at the end of this sequence of moves, all of the coins are on non-shaded squares. Call this sequence of moves S. Then, as we discussed above, both of these would have to be true:

- The partial coins must always add up to exactly two full coins. Thus, after the completion of S, the coin pieces *must add up to two full coins.*
- Notice that at every stage of a sequence of moves, only a finite number of squares have coins on them. Thus, at the completion of S, only finitely many squares will be occupied. Therefore, S is assumed to successfully escape to only non-shaded squares, which have a combined coin-value of 2 coins; however, S will only occupy a finite number of the infinitely many non-shaded squares, and so the squares occupied at the conclusion of S *must have a coin-value of less than two full coins.*

This is our contradiction.[7] Said differently, if we imagine the process as dividing our coins in two, we showed that if our procedure *did* vacate the shaded squares, then in order for them to total two full coins, they would have to cover *all* of the non-shaded squares. But this is impossible, because the procedure can only ever cover finitely many squares.[8]

[6]How's that for income inequality!

[7]
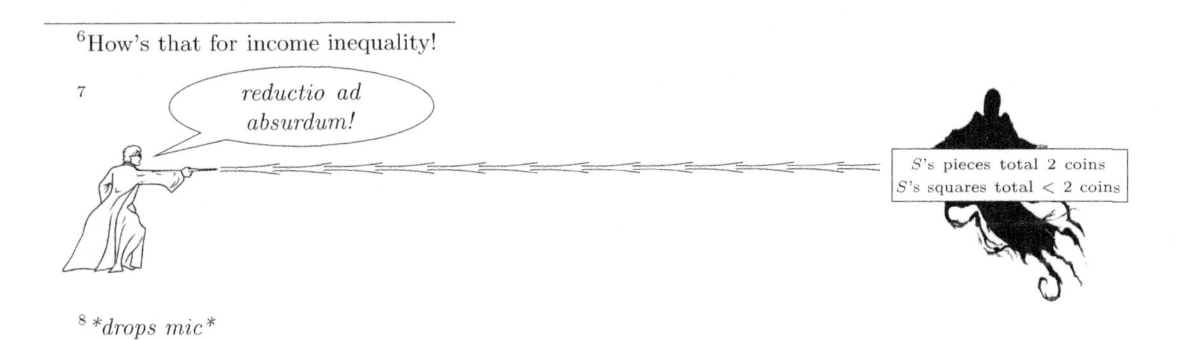

[8] **drops mic**

B.6 An Antisocial Ant Avalanche

Imagine you had some ants on a 1 meter stick, and assume that each ant is either walking toward the left end or toward the right end of the stick.

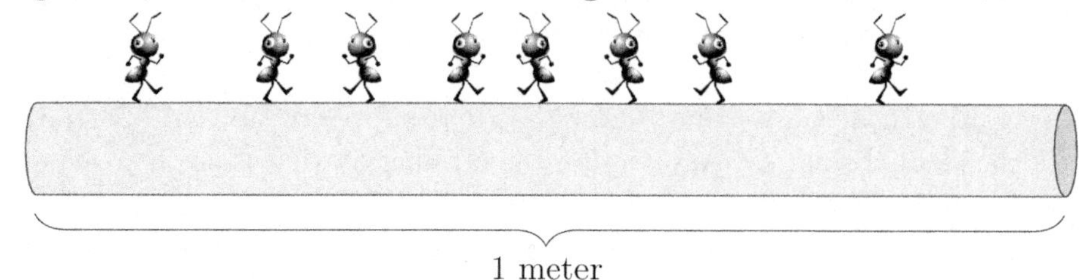

Also, let's assume that the stick is not very thick, so that the ants cannot pass by each other when they meet. Instead, whenever two ants bump into each other, they simply turn around and walk in the opposite direction. And if an ant reaches either end of the stick, assume they walk right off the edge.

Here's the question we want to answer: Assuming each ant walks at 1 meter per minute, how long will it take until all of the ants have walked off the ends?

Let's do some examples. Suppose there was just one ant, and it was right on the edge facing toward the great unknown.

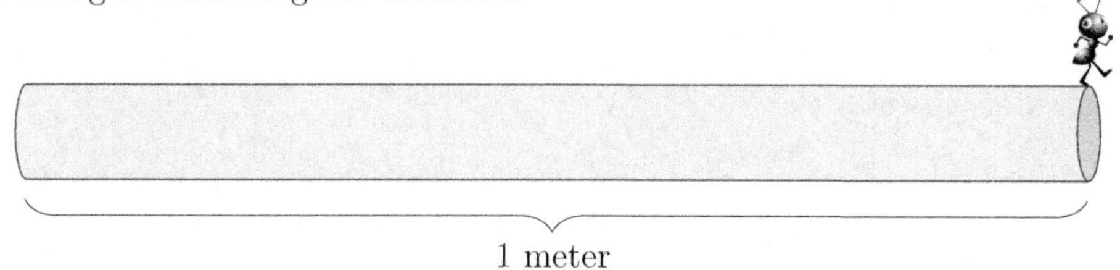

In this case, that one ant will fall almost immediately off the end. Furthermore, if you had several ants all on one side and facing that same direction, it could be over nearly as quickly.[9]

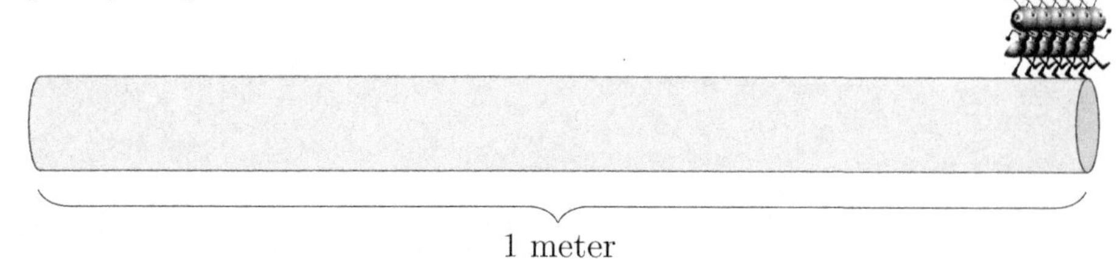

On the other hand, if you had one ant on one end, facing back toward the far end, then the ant has the entire 1 meter stick to traverse, and since the ants move at 1 meter per minute, it will take precisely 1 minute until it falls off the end.

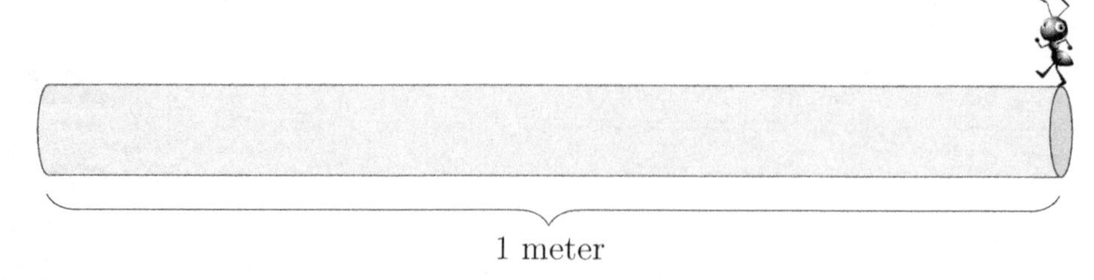

[9] ♫ The ants go marching one by one, hurrah, hurrah. ♫

What if there are two ants on the far ends, each facing the other?

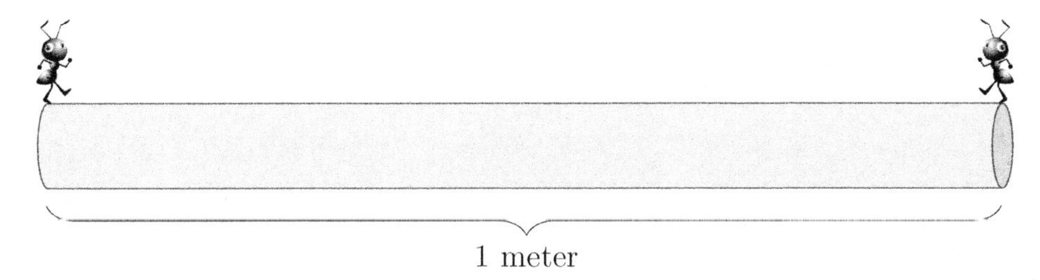

Since they walk at 1 meter per minute, after 30 seconds they will both reach the middle of the stick (we won't worry about each ant's thickness). Thus, right before the mid-stick collision, this is the picture:

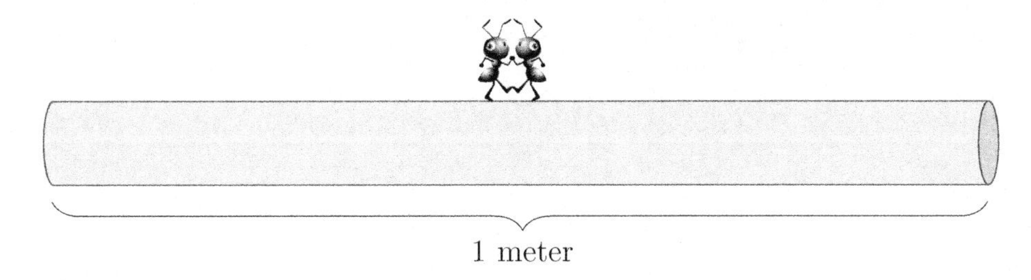

After they collide, they immediately turn around and head back in the opposite direction.[10]

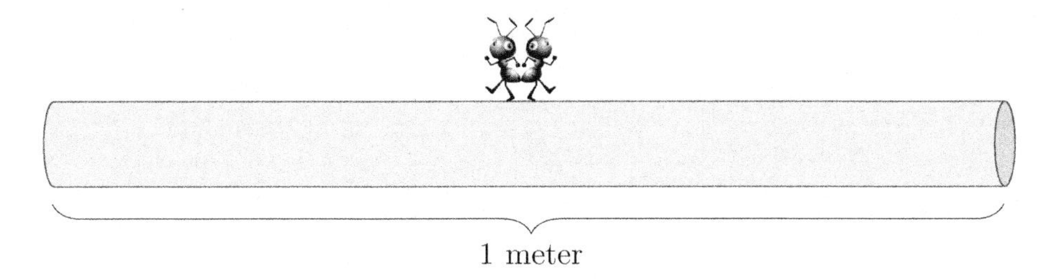

It will take another 30 seconds for them to walk from the middle of the stick to the ends, at which point they will fall off the ends.

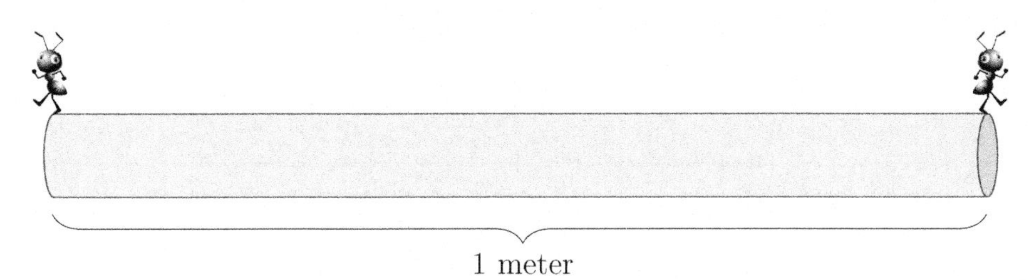

Thus, the entire amount of time until these two ants fall off is $30 + 30$ seconds, which is 1 minute.

[10]If this were a physics book, I would call this an *elastic* collision; the ants immediately bounce off each other and head back where they came at the same speed. Also, if this were a physics book I might call the ants "point masses," since we are essentially assuming they have no thickness.

What about if I added another ant in the middle? Then how long?

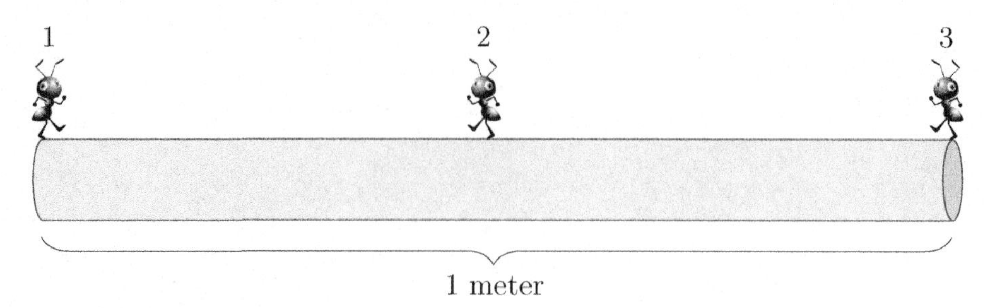

In 15 seconds the picture will be this:

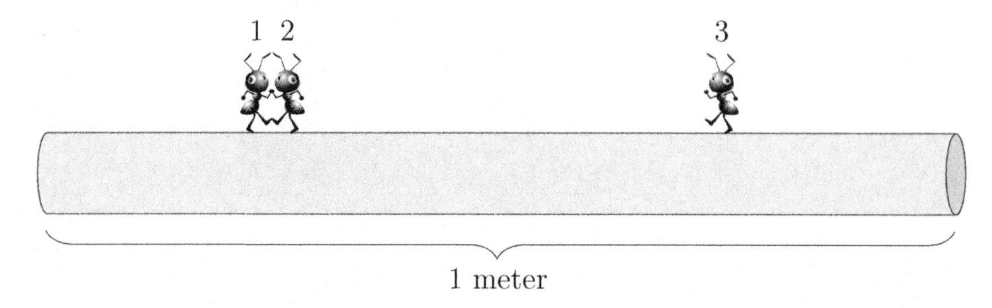

After another 15 seconds, this would be the picture:

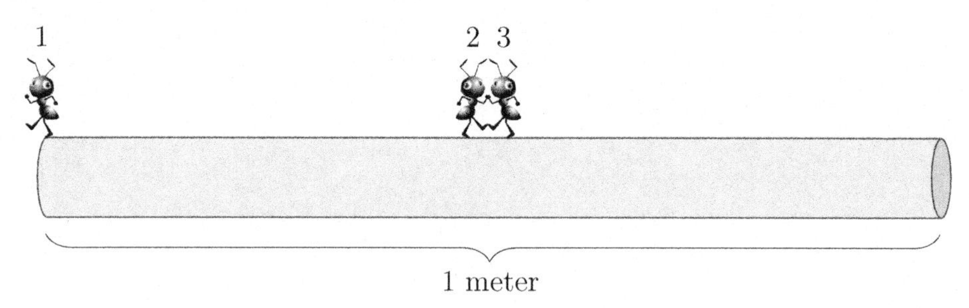

And after another 30 seconds, this is the picture:

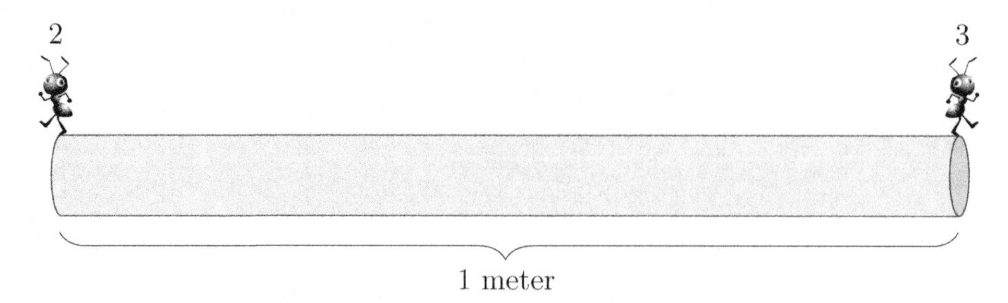

Add it all up, and this arrangement of ants requires $15 + 15 + 30 = 60$ seconds. That is, it again requires 1 minute until they have all fallen off.

What if I had 7 ants? Or 100 ants? The middle ants in these cases could have a huge number of collisions with the other ants before they finally fall off the end. Just imagine all the collisions in the below arrangement, which has only 14 ants!

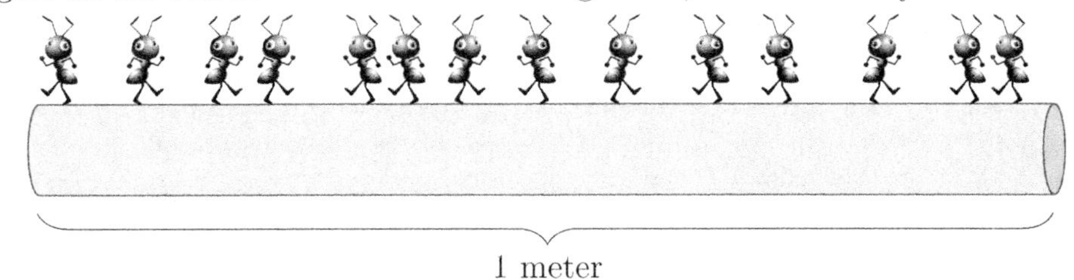

We have seen that different arrangements require different amounts of time, so let's amend our previous question:

Question. Among *all* possible ant arrangements, what is the *maximum* amount of time until all the ants have fallen off the stick?

So far, we have seen two arrangements that take 1 minute. Is there an arrangement that takes 2 minutes? 10 minutes? With a billion ants, could there be ants pinging around for hours or days until the last one falls off? The answer might surprise you:

Answer. The maximum amount of time until all of the ants have fallen off the ends is 1 minute. That is, for every single arrangement, with any number of ants, after 1 minute the stick is guaranteed to be cleared of ants.

The Book Proof of this is simply delightful. There seems to be all sorts of things going on, and trying to track a central ant, with all of its collisions, seems enormously difficult. The Book Proof, however, finds a way around all of that. Here it is:

When two ants collide, they turn around and walk back where they had come, always moving at exactly 1 meter per minute.

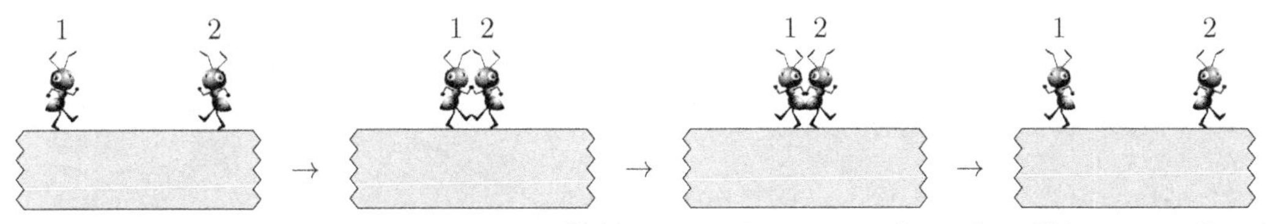

However, notice that two ants colliding, turning around, and walking away has the exact same effect as the two ants being transparent and simply *passing through each other*.

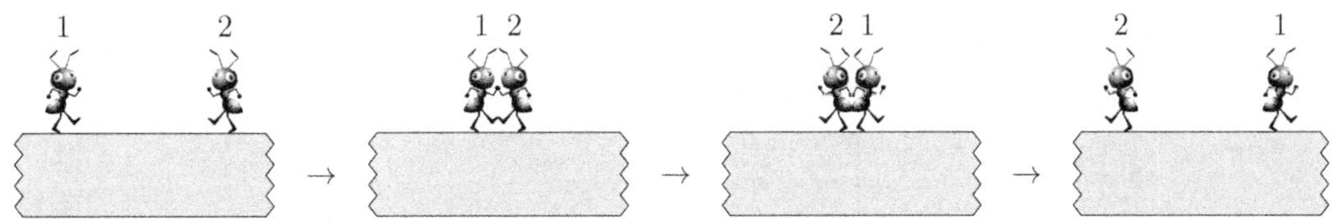

With this new perspective, there are zero collisions! Every single ant passes right through every other ant. And if there are no collisions, then every ant can stay on the stick for at most 1 minute (and only if it walks the entire length of the stick). This proves that 1 minute is the maximum amount of time for any configuration.[11]

[11] **drops mic**

B.7 A Pack of Pretty (Book) Proofs by Picture

In the last appendix, on proof methods, I included a discussion on proofs by picture. Not only can you view proofs by picture as a strategy to prove things, but there are many Book Proofs by Picture. Our first example is this remarkable fact:

$$\arctan(1) + \arctan(2) + \arctan(3) = \pi.$$

Before the proof, let's say a few words so that you appreciate what is going on. First, let's do a refresher on the arctangent function. You might recall that[12]

$$\tan(\pi/4) = \frac{\sin(\pi/4)}{\cos(\pi/4)} = \frac{\sqrt{2}/2}{\sqrt{2}/2} = 1.$$

Since $\tan(\pi/4) = 1$, we have $\arctan(1) = \pi/4$. What is $\arctan(2)$? This is not a standard unit-circle answer. Now, we could represent it with a triangle, as shown on the right.

In that triangle, $\tan(\theta) = \frac{\text{opposite}}{\text{adjacent}} = \frac{2}{1} = 2$, and so $\arctan(2) = \theta$. So $\arctan(2)$ is the angle θ in the picture (it happens to be a little more than $\pi/3$ radians, but it's nothing nice). Likewise, $\arctan(3)$ is nothing nice. But yet, magically,

$$\arctan(1) + \arctan(2) + \arctan(3) = \pi.$$

Somehow, the two not-nice terms balance each other out perfectly to give a really nice answer. Now, π should be thought of as an angle; recall that π radians is the angle of a straight line, it is 180°. This is the key — the proof by picture shows that the three terms add up to a straight line.

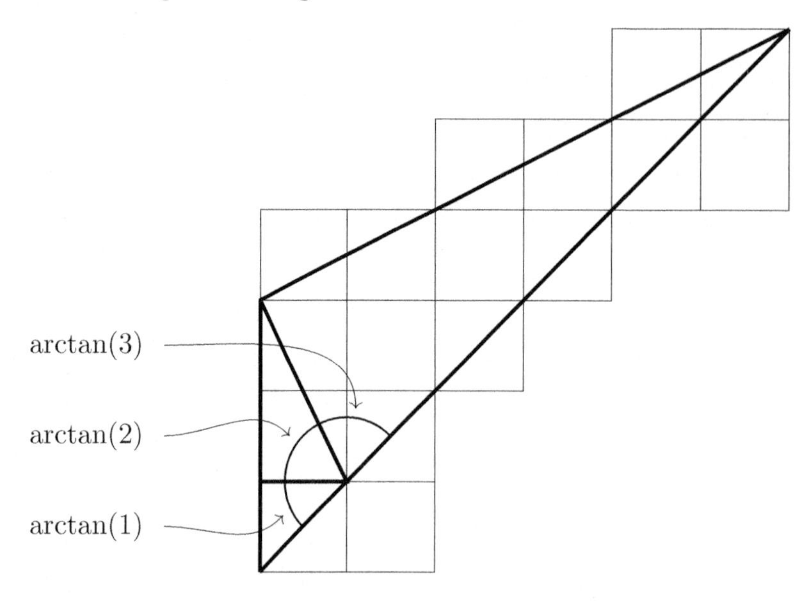

[12]Note: Standard practice in advanced math is to use radians instead of degrees. For example, we refer to $\pi/4$ radians instead of 45 degrees.

A problem from calculus

A classic exercise from calculus is to show that if $x > 0$, then $x + \frac{1}{x} \geq 2$. This can be done by finding the derivative of $f(x) = x + x^{-1}$, which is $f'(x) = 1 - x^{-2}$. This function has a critical point when $0 = 1 - x^{-2}$, which can be solved:

$$x^{-2} = 1 \qquad \text{implies} \qquad x^2 = 1 \qquad \text{implies} \qquad x = 1,$$

the final implication using the assumption that $x > 0$, hence ruling out the $x = -1$ possibility. One can then argue (using the first or second derivative test) that this critical point is an absolute minimum of f, which means $f(x) \geq f(1) = 1 + \frac{1}{1} = 2$ for all $x > 0$.

That's a perfectly fine proof, but it requires that the reader know calculus, and it basically just comes down to a calculation. To me, it's not inspiring. If I were God, I wouldn't put it in My book.

The picture proof below, on the other hand, is quite slick and beautiful. The only thing it relies on is the observation that if a square's area is at least 4 (as the one below is), then it must be at least a 2×2 square. Then, since each dimension must be at least 2, the bottom edge (of length $x + \frac{1}{x}$, in the square below) must be at least 2, which gives the result. The below picture captures all this at once.

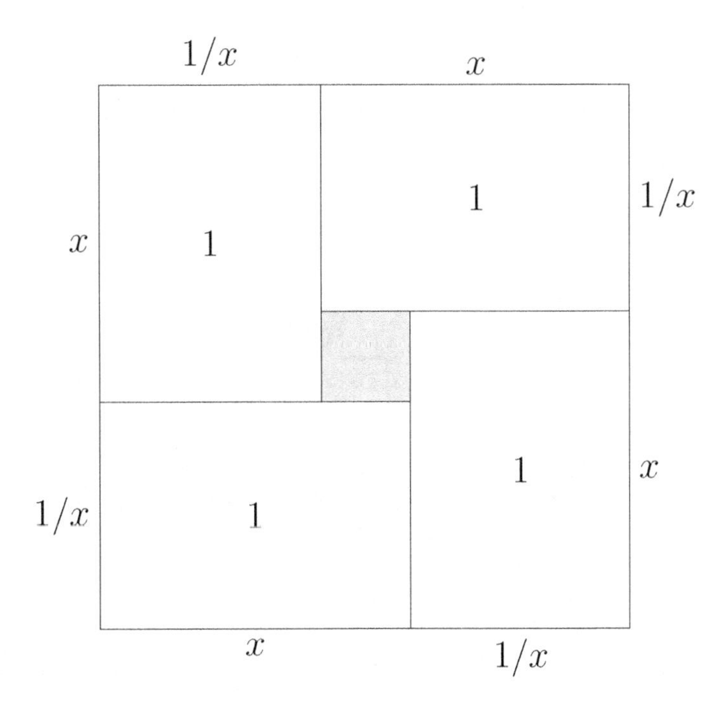

Sums of odd natural numbers

There are also some beautiful proofs by picture involving sums of numbers. In the footnote on page 148, we saw a proof by picture that $1+3+5+\cdots+(2n-1)=n^2$. For a similar problem, consider a $2n \times 2n$ grid of balls. These balls can be divided up to show that $1+3+5+\cdots+(2n-1)=\frac{1}{4}(2n)^2$, which is the same as saying

$$1+3+5+\ldots(2n-1)=n^2.$$

Here is that proof picture:

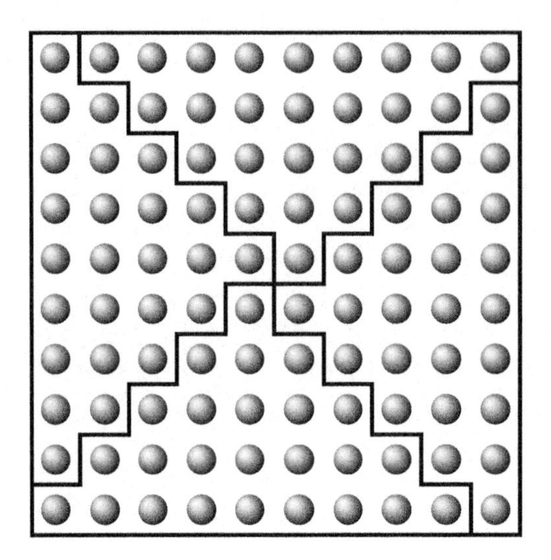

A pyramidal proportion

In 1615, Galileo proved something similar: $\dfrac{1+3}{5+7}=\dfrac{1}{3}$, and $\dfrac{1+3+5}{7+9+11}=\dfrac{1}{3}$, and $\dfrac{1+3+5+7}{9+11+13+15}=\dfrac{1}{3}$, and, in general, $\dfrac{1+3+\cdots+(2n-1)}{(2n+1)+(2n+3)+\cdots+(4n-1)}=\dfrac{1}{3}$. His argument:

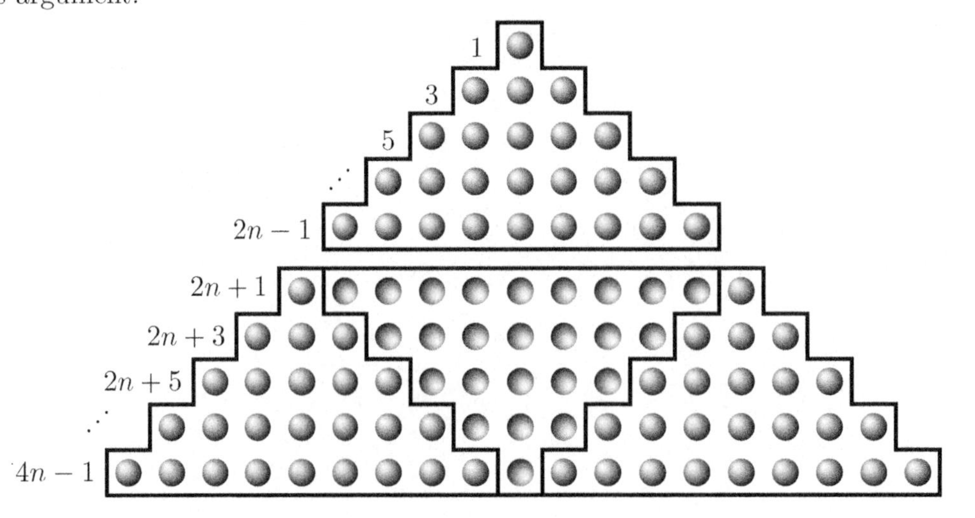

Sums of natural numbers

The ancient Greeks found another similar proof. They showed that $1+2+1=2^2$, and $1+2+3+2+1=3^2$, and $1+2+3+4+3+2+1=4^2$, and, in general,

$$1+2+3+\cdots+(n-1)+n+(n-1)+\cdots+3+2+1=n^2.$$

Here was their visual argument:

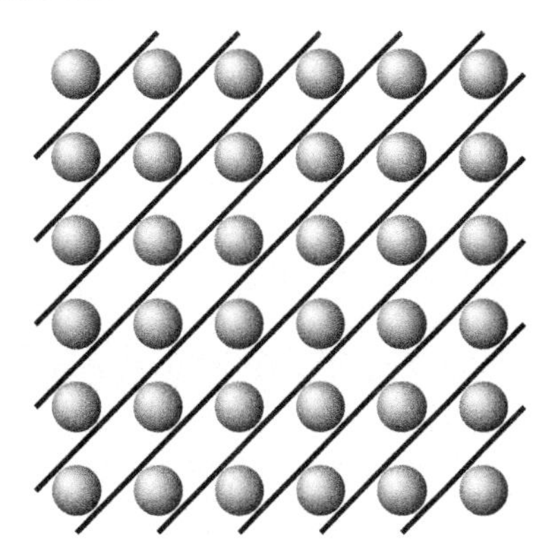

Up-and-down sums of odd natural numbers

A similar method can show that $1+3+1=1^2+2^2$:

And $1+3+5+3+1=2^2+3^2$:

And $1+3+5+7+5+3+1=3^2+4^2$:

And, in general, $1+3+5+\cdots+(2n-1)+(2n+1)+(2n-1)+\cdots+5+3+1=n^2+(n+1)^2$.

B.8 An Image's Insightful Illusion

We are going to discuss another tiling problem. This time, the tiles are equilateral diamonds, which can be thought of as two equilateral triangles glued together: . We are going to use these shapes to tile a regular hexagon whose side lengths are n times larger that the side length of the tiles. (The pictures below are the $n = 5$ case.)

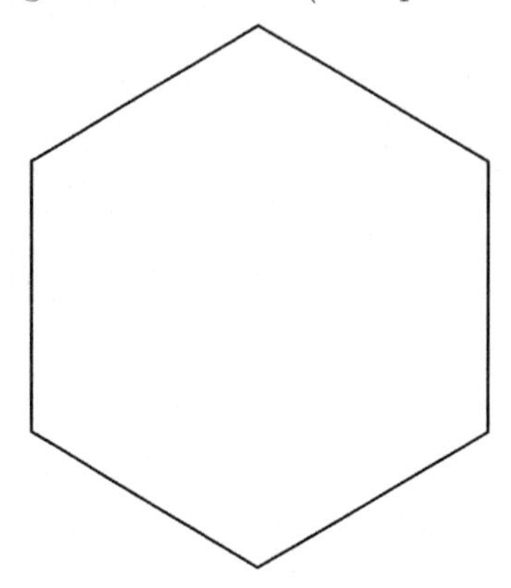

There are three different orientations for these tiles which could be used in a perfect covering: , and . Here is one way a perfect cover might begin:

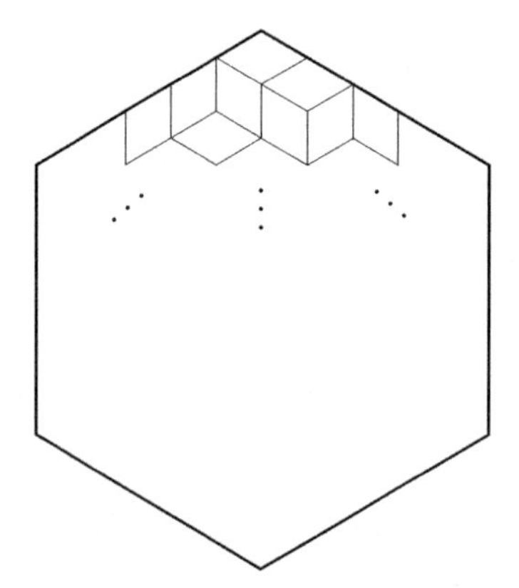

You might notice that it is impossible to perfectly cover this hexagon with tiles of only one orientation. For example, if you tried for the horizontal orientation, then starting from the top corner, here is where you get into trouble.

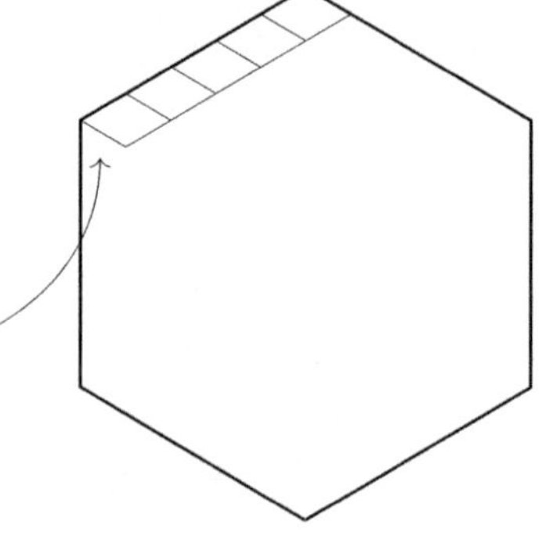

In fact, you need to use some tiles of each of these three orientations. Moreover, it turns out that it is necessary that each orientation be used the exact same number of times! This is the theorem we wish to prove.

Theorem.

Theorem. In any perfect covering of the above hexagon with the above diamonds, exactly one third of the diamonds are of each of the three orientations.

Why is this true? What is the Book Proof? Consider any such covering:

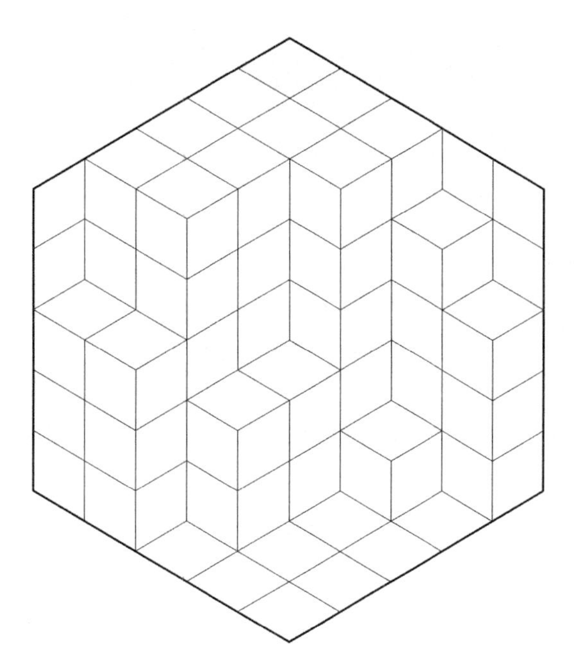

There are sections of this perfect covering where many horizontally-oriented pieces are grouped together, and sections where another orientation are grouped together. There doesn't seem to be any rhyme or reason to it. And unless you can spot a pattern, how can you prove it?

The trick is to color/shade the shapes based on their orientation. That is, all the shapes of the same orientation get the same shading. This is the magical idea, and because I didn't want you jump ahead and see it, I enlarged the pictures so that you have to turn the page to see this next step...

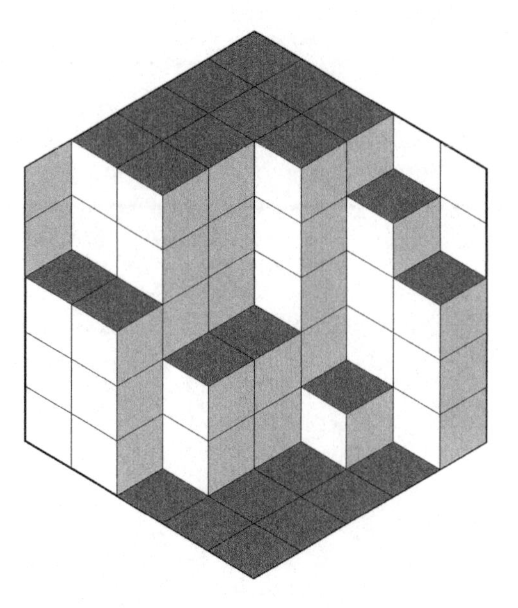

Because our brains evolved in a three-dimensional world, an optical illusion pops right out! This looks like boxes stacked into the corner of a room, with a bright light shining from the left, a dimmer light shining from the right, and no light shining from the top. But with a square floor and walls, a viewer from the left, right or top would see the same amount of stuff — exactly one square's worth. Thus, there must be the same number of tiles of each color, completing the proof.[13] It's a proof by optical illusion!

This is where the proof would typically end, although if you want a touch more justification, you can also use your three-dimensional mind to project each of these box faces against the wall it is parallel to (move the dark shapes down to the floor, the lightest shapes to the right right wall, and the gray shapes to the left wall). Doing so immediately gives the one-third relationship that the theorem asserts:

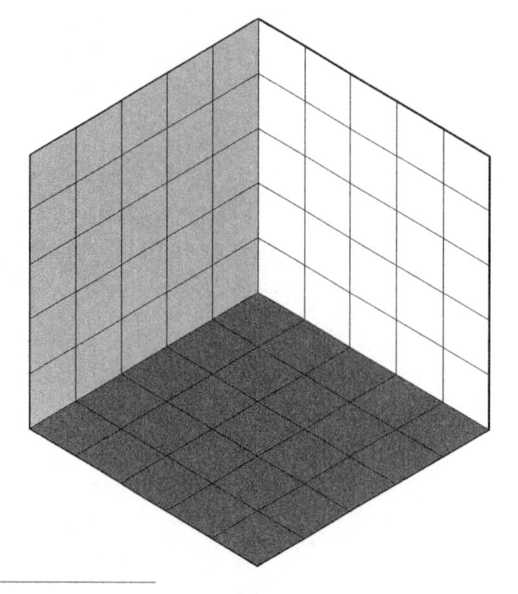

[13] **drops mic**

B.9 Monotone Marches through Muddled Marks

An *infinite sequence* is an infinite list of numbers, like

$$2, 4, 6, 8, 10, 12, 14, \ldots.$$

Given an infinite sequence like the one above, a *subsequence* is a list like

$$4, 6, 10, 20, 22, 146, \ldots.$$

That is, it is an infinite sequence where each of the terms comes from the original sequence, and they are in the same order as in the original sequence. Thus, $4, 2, 8, 6, 10, \ldots$ is not a subsequence, because the terms are out of order.

As another example, the infinite sequence

$$1, -2, 3, -4, 5, -6, 7, -8, \ldots$$

has the subsequence

$$1, 3, 5, -8, 11, -20, \ldots.$$

We are going to be interested in *monotone* subsequences. A subsequence is monotone if the terms are always increasing, or are always decreasing. For example, the infinite sequence

$$1, -2, 3, -4, 5, -6, 7, -8, \ldots$$

has many monotone subsequences. For example, the subsequence

$$1, 3, 5, 7, 9, \ldots$$

is always increasing ($1 < 3 < 5 < 7 < \ldots$), so it is a monotone subsequence. Alternatively, the subsequence

$$-4, -8, -12, -16, -20, \ldots$$

is always decreasing ($-4 > -8 > -12 > -16 > -20 > \ldots$), so it is a monotone subsequence. And here's one final example:

$$-8, 10, 12, 204, 848, \ldots$$

is a monotone subsequence. Notice that the terms in this last subsequence did not follow any discernible pattern, but that's ok. As long as they come from the original sequence, and are in the same order as they were in the original sequence, then that counts as a subsequence.

Let's now prove something!

Theorem.

Theorem B.2. Every sequence has a monotone subsequence.

Scratch Work. The proof of this is just straight-up snappy. Once you know precisely what to look for, it falls through perfectly. First, here's the idea behind it:

Usually we draw a sequence in the xy-plane by just plotting the points (e.g., if $a_3 = 6$, then we put a point at $(3, 6)$). This gives a picture like this:

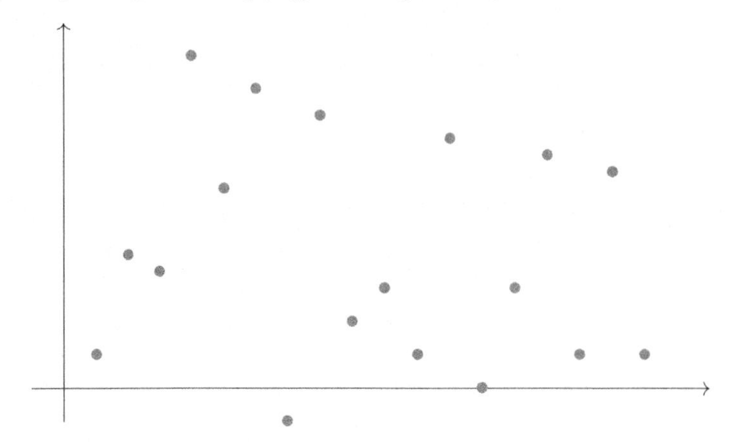

This time, though, we will connect the dots to make a zig-zagged line:

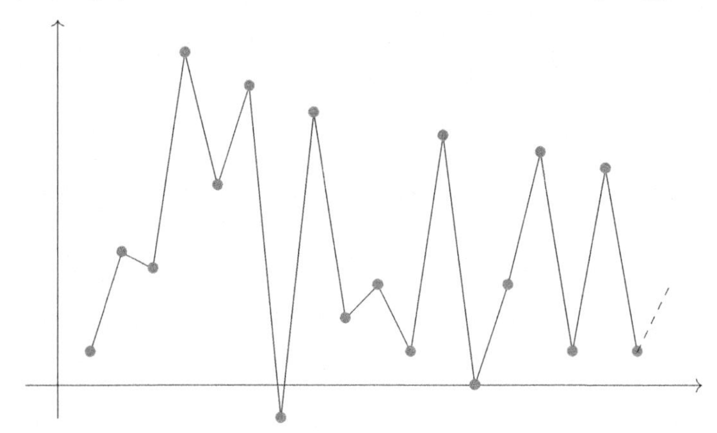

From this picture you can maybe spot a nice *decreasing* subsequence (hence, a monotone subsequence):

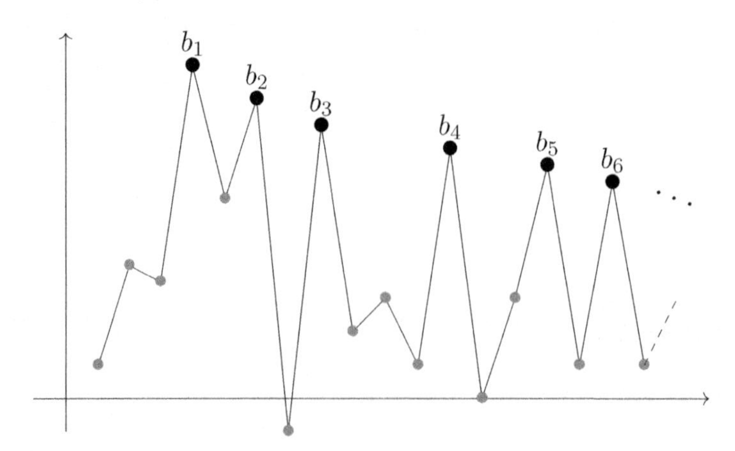

The magical definition that will solve everything is that of a *peak*; we want those labeled points to form *peaks*, and we want to be able to say that if we have an infinite sequence of peaks, then we do indeed have a decreasing sequence. The definition that does it is this: define a *peak* to be a point a_n which is larger than every later point; that is, a_n is a peak if $a_n \geq a_m$ for all $m > n$.

So if we have infinitely many peaks, then we obtain a sequence like the one above, which will be decreasing. So what if we don't? Then we only have finitely many peaks. In this case, we can find an *increasing* sequence, which is again monotone. To see how, just note that if you're past the last peak, then any point you pick is not a peak, which means there is some point after it which is larger. So one at a time you can pick larger and larger points, giving an increasing sequence.

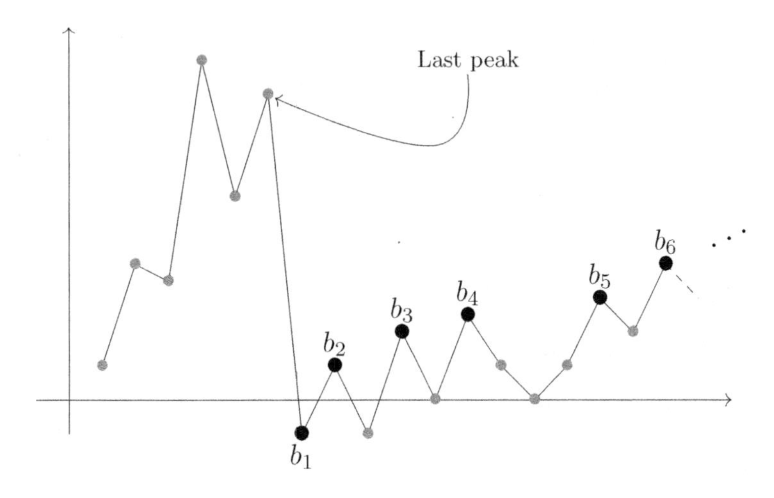

At last, here is the Book Proof of Theorem B.2.

Proof. Call a_n a *peak* if a_n is larger than every later point. That is, if $a_n \geq a_m$ for all $m > n$.

Either there are infinitely many peaks or finitely many peaks. Assume first that the sequence $a_1, a_2, a_3, \dots$ has infinitely many peaks. Then, let b_k be the kth peak. Then, by the definition of a peak, $b_k \geq b_{k+1}$, implying that $b_1, b_2, b_3, \dots$ is a decreasing subsequence.

Now assume that $a_1, a_2, a_3, \dots$ has finitely many peaks, and let a_N be the last one. Then let $b_1 = a_{N+1}$. Since a_N was the last peak, b_1 is *not* a peak, implying that there is some later point b_2 that is larger than it. Likewise, since b_2 is after the last peak, it is also not a peak, and so there must be some later point b_3 that is larger than it. Continuing in this way we construct an increasing subsequence $b_1, b_2, b_3, \dots$.

In either case we found a monotone subsequence, so we are done.[14] □

[14] **drops mic**

Finite sequences

We can also ask similar questions about finite sequences and their subsequences. If you have a finite sequence with, say, 9 terms, $a_1, a_2, a_3, \ldots, a_9$, must there exist a monotone subsequence of length 3? Length 4? Length 5? What can we guarantee?

For example,

$$1, 4, 2, 7, 6, 9, 3, 8, 5$$

is a sequence of 9 numbers. This sequence's longest increasing subsequence has length 4. In fact, there are several of this length, but here's one: $1, 2, 6, 8$. This sequence's longest decreasing subsequence has length 3. Again, there are a few examples, but here's one: $7, 6, 3$.

If our sequence was $9, 8, 7, 6, 5, 4, 3, 2, 1$, then its longest increasing subsequence has length 1, since it is just a single number. But its longest decreasing subsequence has length 9, since the entire sequence counts as a subsequence and is decreasing.

You see, there is often a trade-off between having a long increasing subsequence (a subsequence $b_1, b_2, \ldots, b_m$ where $b_i \leq b_{i+1}$ for all i), and a long decreasing subsequence (a subsequence $b_1, b_2, \ldots, b_m$ where $b_i \geq b_{i+1}$ for all i).

Our next theorem deals with these sequences of finite length, and tells us just how long of an increasing/decreasing subsequence we can guarantee.

Theorem.

Theorem B.3. If $a_1, a_2, a_3, \ldots, a_{kn+1}$ is a finite sequence with $kn + 1$ terms, then there exists an increasing subsequence of length $n + 1$, or a decreasing subsequence of length $k + 1$.

According to this theorem, if we let $k = 3$ and $n = 3$, then we will be referring to a sequence of length 10 (since $kn + 1 = 10$). And we will be guaranteed either an increasing subsequence of length 4 (since $n + 1 = 3 + 1 = 4$), or a decreasing subsequence of length 4 (since $k + 1 = 3 + 1 = 4$). As it turns out, 10 really is the lowest you can go to guarantee this. With 9 numbers, it is possible that the longest increasing subsequence and the longest decreasing subsequence are both of length 3. Here's an example of that:

$$3, 2, 1, 6, 5, 4, 9, 8, 7.$$

How do we prove the theorem? I chose n and k for the variables so that, perhaps, "$kn + 1$" will ring a distant bell in your head... $kn + 1$ looks a lot like the generalized pigeonhole principle! But how could we possibly use the pigeonhole principle in a problem like this? It is kind of like the "peaks" idea from our last example, but with a clever twist.

Proof. Consider a finite sequence $a_1, a_2, a_3, \ldots, a_{kn+1}$ with $kn+1$ terms. For each term a_i of the sequence, let $L(a_i)$ be the length of the longest increasing subsequence that begins at a_i. Even though we can't say for certain what any of the values $L(a_i)$ are, giving them a name is quite valuable.

Case 1: If $L(a_i) \geq n+1$ for any i, then we immediately know that there is an increasing subsequence of length $n+1$—that's what the L-function means! Thus, in this case, we have completed the proof.

Case 2: The other possibility is that $L(a_i) \leq n$ for all i. Said differently, this is the case for which $L(a_i) \in \{1, 2, 3, \ldots, n\}$ for every i. This allows us to use the general form of the pigeonhole principle. We are placing $kn+1$ objects (the terms of the sequence) into n boxes (the values in $\{1, 2, 3, \ldots, n\}$), where a_i is placed into box m if $L(a_i) = m$. By the general form of the pigeonhole principle, there must be at least one box with at least $k+1$ objects in it. Let's suppose it is Box s.

Each a_i in Box s is from the sequence $a_1, a_2, \ldots, a_{kn+1}$, and so we can take $k+1$ things out of this box and form a subsequence with them! Let's call this subsequence $b_1, b_2, \ldots, b_{k+1}$. And don't forget what it means for them to have been in Box s—each b_i is a term of the sequence $a_1, a_2, \ldots, a_{kn+1}$, and the longest increasing subsequence starting at b_i has length s.

The proof now comes down to a single observation: It must be that $b_i \geq b_{i+1}$ for all i. Why? Assume for a contradiction that $b_i < b_{i+1}$. Let's suppose that

$$b_{i+1}, c_2, c_3, \ldots, c_s$$

is the longest increasing sequence starting at b_{i+1} (which, once again, is assumed to be of length s). Then, since we are assuming that $b_i < b_{i+1}$, this must be an increasing sequence starting at b_i:

$$b_i, b_{i+1}, c_2, c_3, \ldots, c_s.$$

But that sequence has length $s+1$, while $L(b_i) = s$, which is a contradiction.

In Case 2 we have shown that if $L(a_i) \leq n$ for all i, then by the general pigeonhole principle there must be $k+1$ terms of the sequence which all have the same L-value. By turning these $k+1$ terms into a subsequence, $b_1, b_2, \ldots, b_{k+1}$, we showed that this must be a *decreasing* subsequence. That is, we showed that $b_i \geq b_{i+1}$, since otherwise it is impossible for all the b_i to have $L(b_i) = s$. Thus, we are done. Case 1 found an increasing subsequence of length $n+1$, and Case 2 found a decreasing subsequence of length $k+1$. This proves the theorem.[15] □

[15] **drops mic**

B.10 Zigging Zeniths and Zagging Zones

The following is a theorem that I learned on Twitter, proving that, at least once, my procrastination paid off. Thanks to Joel Hamkins for tweeting it!

Imagine you have a rectangle, and you draw a line which zig-zags between the bottom edge to the top edge, beginning at the bottom-left and eventually ending up at the bottom-right. Doing this divides up the rectangle into two regions.

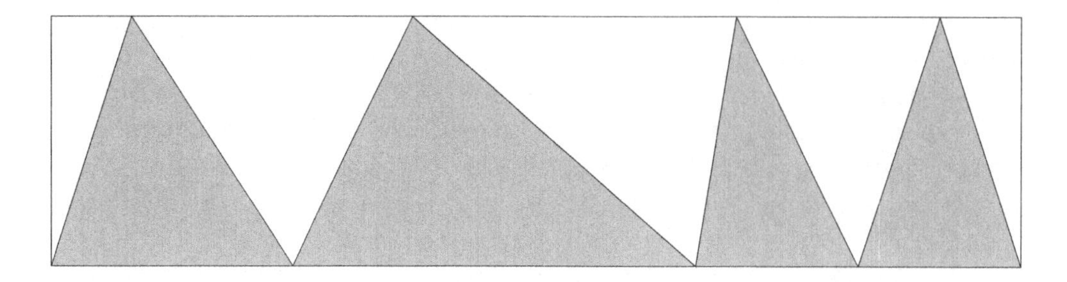

A really cool theorem says that the area of the shaded region is always equal to the area of the non-shaded region. Hamkins calls this the *zig-zag theorem.*

What is the Book Proof of the zig-zag theorem? If the picture looks like the above, then you can divide up the big rectangle into a collection of smaller rectangles, each of which is half-shaded and half-not-shaded.

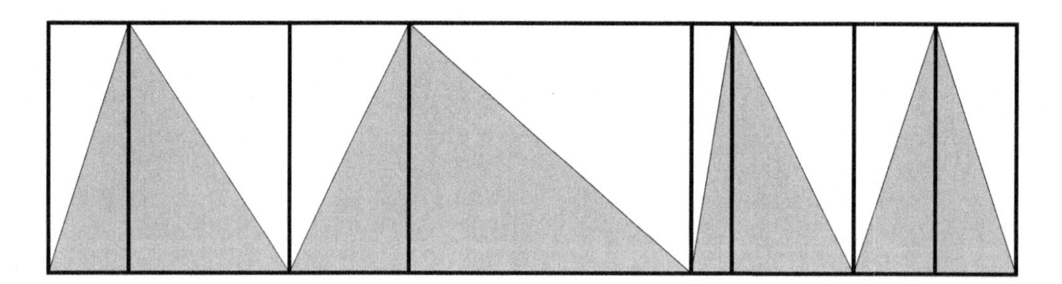

That's it! Each of these eight rectangles has half-and-half, and they add up to the whole, so the whole must be half-and-half. This argument can be immediately generalized to any number of zigs and zags. Here's another example:

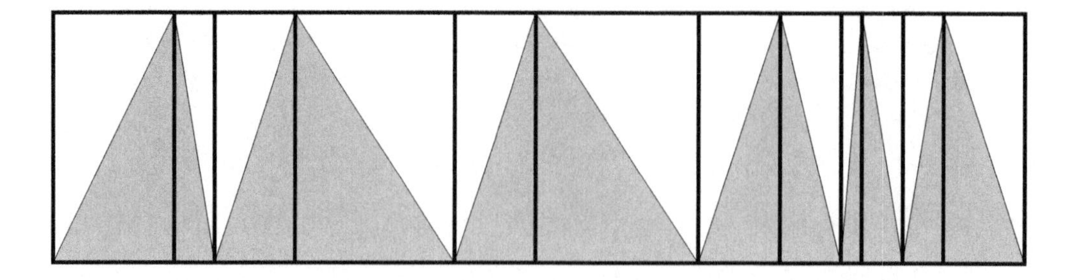

Notice, though, that in stating this problem we did not assume that each zig-zag moved us to the right. What if the picture "doubles back"? For instance, what if the below is what our picture looks like?

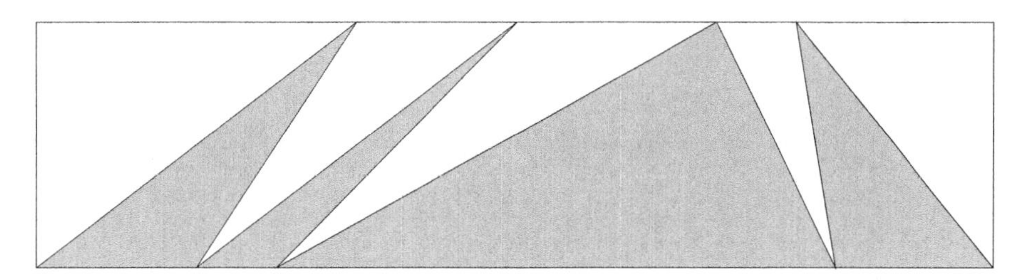

Our previous argument no longer directly applies, but there is another slick argument that does. Let's say the big rectangle has a width of w and a height of h. Note that the area of each shaded triangle is $\frac{1}{2}bh$, where b is the width of the triangle's base and h is the height of the rectangle (notice that all triangles have this same height). Let's suppose there are n triangles, with bases of length b_1, b_2, ..., b_n.

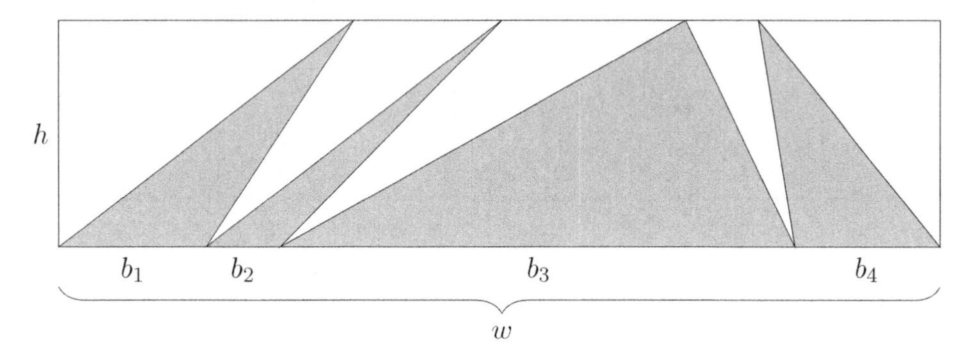

Notice that the shaded triangles' bases collectively stretch the entire base of the rectangle. Thus:

$$b_1 + b_2 + \cdots + b_n = w.$$

Then,

$$\begin{aligned}
\text{The total shaded area} &= \frac{1}{2}b_1h + \frac{1}{2}b_2h + \frac{1}{2}b_3h + \cdots + \frac{1}{2}b_nh \\
&= \frac{h}{2}(b_1 + b_2 + b_3 + \cdots + b_n) \\
&= \frac{h}{2}w \\
&= \frac{1}{2}wh \\
&= \text{half of the area of the rectangle.}
\end{aligned}$$

The rectangle is divided into two parts: the shaded and the non-shaded. We just proved that the shaded region is half of the rectangle's area, which means the non-shaded must also be half of the area, which proves the zig-zag theorem.[16]

[16] **drops mic**

Appendix C: Writing Advice

Ever since you were young, you have been taught how to write. You have told stories, summarized readings and composed speeches, and your success at each of these was in-part dependent on your ability to communicate well. Some day you may write essays to get a scholarship, go to graduate school, or earn a job. It can be challenging to summarize a book you read or tell a story from your life. It can be even more challenging to communicate a complicated mathematical argument in a way that is clear and mistake-free. In this appendix I have collected some advice to improve your proof writing.

C.1 Writing Proofs

For now, be as rigorous as possible

As someone taking their first proof-based math course, you should strive to be as rigorous and thorough as possible. You should begin every proof by stating your assumptions. You should end every proof with a summarizing sentence noting that you reached the conclusion. Clearly cite every used definition and work out all of your algebraic steps. If your proof uses the fact that "an even plus an even is an even," and that has not been proven yet in this course, then you should prove it. In future courses, it will be ok to say "an even plus an even is an even" with no further justification. In this course, though, while you are learning proof methods, facts like that have to be proven or cited. This is good — when learning a new proof method, it can be helpful to prove things which are mathematically simple, to allow you to put all your focus on the method and on your proof writing.

As I said in the Chapter 3 Pro-Tips, this is how I think about it: When my dad taught me how to drive, he insisted that I do everything *perfectly*. Hand placement, mirrors, speed limit, spacing, signs, blinkers, lights, focus, radio, ... every last thing should be done perfectly. It's not that being soooo meticulous is crucial; less so would still be plenty safe. It's because everyone relaxes this alertness eventually — and if you start by driving perfectly, then once you relax you will still end up in a great place. This was my dad's reasoning.

The same reasoning holds with proof writing. If your proofs begin with surgical precision, then once you inevitably relax a bit, you won't do so to a point that mistakes are introduced or your readers are confused.[1] Moreover, even if certain aspects of your proof-writing relax, it is important that your thinking stays just as sharp. This ultra-rigorous approach will not only improve your proof-writing, but will help train your brain to be logically meticulous.

While some of the advice below is also tailored to where you are now, most of it is solid advice for your entire math career. That certainly includes this piece of timeless advice: know your audience.

Know your audience

As with all communication, what counts as "effective" proof-writing depends on who will likely be reading your work. What terminology is common knowledge, and what should be redefined for the reader? What level of rigor is expected or required?

When you are writing up a homework assignment, you can generally use all of the definitions and theorems that have been covered up to that point in the course (although you should cite them when you use them). If you are writing a research article to submit to the *AKCE International Journal of Graphs and Combinatorics*, then your audience certainly knows what a graph's vertices and edges are, and so you may omit that detail. When writing for a class, you are expected to prove more of the details than you would for a research article. Conversely, in a research article it would be more common to accompany your solutions with your own examples, which illustrate some of your proof's abstract ideas in a concrete setting.[2]

For your homework, a good rule-of-thumb is to write as if you were trying to convince a classmate.

When taking a writing class, you may have learned about writing with "voice." What tone do you want to set? If you are writing a text message, your tone is probably much more informal than if you are applying for a job. You should feel welcome to develop your own style and voice to your proof-writing, but make sure not to lose the rigor when doing so.[3]

Grammatical rules

It is common to think that rules of English (or whatever language you are writing in) are not as important when you are writing about math. This is false. You should

[1] When my grandma taught my dad to drive, her advice was simpler: "Assume every other driver is an idiot." While I did consider making this the lesson for your proof writing... I ultimately chose to go with my dad's more wholesome take.

[2] Feel free to do this in your homework, too! Just make sure it doesn't take away from the proof itself. It should be thorough without the example.

[3] If you read some other math textbooks and then read one of mine, you'll probably discover that in my view math should be much more conversational than it is in most published textbooks — especially when your audience is college students. So I hesitate to say too much right now, since the main point of this appendix is to convey the broadly-held views of mathematicians on writing, not push my own outlier views, and of course you should check with your professor when it comes to your coursework. That said, in a footnote I am comfortable saying that I think almost all math is too formal. And I hope the next generation of mathematicians moves the norm towards a lighter tone.

write in complete sentences. You should have natural paragraph breaks. You should capitalize and punctuate just as you would in a writing class.[4] If your sentence ends with a mathematical expression, it still needs a period. Example: If $|x| = 2$, then $x = 2$ or $x = -2$.

There are also some math-specific rules which should be observed. Most importantly, you should not start a sentence with a variable or symbols; having it appear after a period can be confusing to the reader. So instead of starting a sentence saying "f satisfies the intermediate value property" you could say write something like "The function f satisfies the intermediate value property." Usually the way around it is to add a couple of words stating what the variable/symbol means. Two more examples: instead of starting a sentence with "$\mathbb{Q}$ is a subset of $\mathbb{R}$" or "$x^2 - 2x + 1 = 0$ has one solution," you could write "The set $\mathbb{Q}$ is a subset of $\mathbb{R}$" or "The equation $x^2 - 2x + 1 = 0$ has one solution."

When writing a long equation with many parts or indices, it often wise to place it in its own line. So instead of saying "notice that $\sum_{n=1}^{\infty} \int_0^n \cos^n(x) \frac{x^n}{n}\, dx = f_2(x, y)$," you would write it like this: "notice that

$$\sum_{n-1}^{\infty} \int_0^n \cos^n(x) \frac{x^n}{n}\, dx = f_2(x, y)."$$

Giving it its own line allows the details to be larger and situated better. (Also, notice that the expression ended a sentence and hence ended with a period.)

Finally, when you define a term, you should *italicize* it, underline it, or make it **bold**. In this book, I italicized my terms when I defined them, but bold is also common. In handwritten work, underlining is common.

Using symbols

On a chalkboard or in your scratch work, it is common to use symbols like $\forall$ for "for all," $\exists$ for "there exists," $\Rightarrow$ for "implies," and $\Leftrightarrow$ for "if and only if." You may use abbreviations like iff for "if and only if," wlog for "without loss of generality," and s.t. for "such that." These allow you to write things quickly, keeping your attention on the math.[5] But in formal writing, they should not be used.

Incidentally, mathematicians have agreed that *some* symbols are always ok. For example, you may use numbers (-2, 3, 43.4534, $\frac{2}{3}$, etc.), you may use numbers which have been given their own symbol (π, e, φ, etc.), you may use the standard arithmetic symbols ($+$, $-$, $\cdot$) and variable names (a, n, x, etc.), you may use symbols for named sets ($\mathbb{N}$, $\mathbb{Z}$, $\mathbb{Q}$, $\mathbb{R}$, $\mathbb{C}$) and the standard set symbols ($\cup$, $\cap$, $\setminus$, $\in$). Therefore, you may write "there exists $n \in \mathbb{N}$," but you should not write "$\exists n$ in the set $\mathbb{N}$."

[4] One exception that I have adopted: If I want to refer to many versions of, say, the variable k, it can confuse the reader to write "the ks all have the property...," since this looks like k is multiplied by s. Thus, at times I break the apostrophe/pluralization rule and write "the k's all have the property...." This is not done widely, but if you must decide between clarity and grammar, and there is no good way to satisfy both, then at times it may be reasonable to side with clarity. (As you'll see, the main times I break from convention is when I am concerned about clarity.)

[5] And writing things like "$\forall x \in \mathbb{N}, \exists y \in \mathbb{R}$, s.t. $x^2 = y$" lets you feel like you know how to write in a second language. And that second language is hieroglyphics.

We also allow (and embrace) the use of two particular abbreviations: i.e. (which means "that is") and e.g. (which means "for example").[6]

Understandably, these rules may seem arbitrary and hard to remember. But with a little experience reading books/articles/websites on proof-based math, you will quickly pick up these customs.[7] In closing this section, here are a few more small tips when working with symbols:

- Make sure you do not introduce a new symbol without defining it in some way. In some cases, a symbol x could be defined as simply as this: "Note that there is some $x \in \mathbb{N}$ such that $x > 100$." This has given x a meaning. In other cases, you may formally define what a set S is. Every symbol without a known definition (like $\mathbb{N}$) must be introduced in some way.

- While using a symbol like $\subseteq$ is acceptable when writing, say, "Note that $\{1, 2\} \subseteq \{1, 2, 3\}$, which implies...," it would be improper to write "Since our first set is $\subseteq$ our second set...." You should also not write things like "Note that the two sets are $=$." Those symbols can be use as parts of a mathematical expression, but not as simple word-replacements.

- This is a perfectly clear English sentence: "Because it snowed last night, I have to shovel the driveway." Here is a very similar math sentence which is less clear: "Because $x^2 = 4$, $x = 2$ or $x = -2$." The comma makes it look like it is part of a list. Plus, we use commas for other purposes in math,[8] so scrunching one between two math symbols could confuse your reader on their first read—is that a grammatical comma or a mathematical comma? Thus, instead of "Because $x^2 = 4$, $x = 2$ or $x = -2$," you could write "Because $x^2 = 4$, we see that $x = 2$ or $x = -2$." Or, "Because $x^2 = 4$, we have $x = 2$ or $x = -2$." Or, "Because $x^2 = 4$, it follows that $x = 2$ or $x = -2$." There are many options, but it is advised to insert a couple words to avoid confusion.

Don't say too little

A common difficulty is deciding how much you should include in your proof. This is especially difficult in an Intro to Proofs class—if you have to say why the sum of two evens is even, do you also have to say why the sum of two integers is an integer? The sum of two real numbers is a real number? Is there anything you *don't* have to justify? In later classes, when you are allowed to assume the basic properties of arithmetic, things are a little easier, but still it can be a challenge. The best general advice that I can give is to check with your professor when you are unsure. Instead, let's talk about general math writing tips.

---

[6]Grammar Pro-Tip: Always add a comma after you use "e.g." or "i.e." in a sentence. E.g., the sentence "An even times an even is an even; e.g., $3 \cdot 4 = 12$."

[7]Learning customs is important, too. This book is teaching you what it is like to be a mathematician, and our customs are part of it.

[8]Main examples: in ordered pairs, like $(2, 3)$; in a function which has multiple arguments, like $f(x, y)$; in sets appearing as a list, like $\{1, 2, 3, \ldots\}$; or intervals, like the closed interval $[a, b]$.

In most writing classes, your teachers tell you that your writing should be short, punchy, to-the-point. If you can use one word instead of two, do it. If you can use one sentence instead of two, *definitely* do it. In math, the advice is not so straightforward. To begin, you should make sure to say enough that your solution is complete. You should cite every definition and theorem you use — and you should say why they apply. You should work out any algebra that is needed, and you should ensure that it is clear how the logic takes the reader from one step to the next. On this point, students usually err on the side of justifying too little than justifying too much.

That said, it is certainly possible for a proof to be too wordy.[9] If you can directly say what you mean in a clear and concise way, then do it. But oftentimes in math, clarity is the cost of concision. Every mathematician is familiar with the experience of having to read a single sentence three or four[10] times to understand it. Instead of one confusing sentence, if the author could have explained the same idea in two clear sentences which the reader would have understood without having to reread them, then using one sentence is *less* efficient for your readers than using two. You should optimize your writing for your readers, not for the word-counter.

In general, my advice is to first explain your reasoning so that it is clear and complete, regardless of how much writing it takes. Then, look back over your proof and see if you can find ways to make it concise without sacrificing its clarity or correctness. Usually you can! Often, you can reduce its length significantly. Having several drafts of your work before your final write-up is quite common; try to embrace the fact that writing is a sequence of small steps to the final product, and it never comes out perfectly at the start.[11]

This is an art, not a science, and practice will go a long way. Try to put yourself in your readers' shoes, and try to spot the moments where confusion could arise. Or, better yet, let your peers read over your work to see if they understand it. Thoughtful feedback is invaluable. Also, read over your peers' work to see how they proved the same result. You can pick up tips from each other, and then everyone improves.

Avoid extraneous statements

Related to being too wordy, you should avoid saying things that have no bearing on your solution. Suppose you are trying to prove that if c is a rational number, then c^2 is also rational. Suppose you start your proof like this:

> Assume that c is rational; then, by the definition of a rational number, $c = \frac{m}{n}$ for some integers m and n where $n \neq 0$. Furthermore, we may assume that m and n are chosen so that $\frac{m}{n}$ is written in lowest terms.

Everything you said is correct, but if you worked out the rest of the proof (basically, $c = \frac{m}{n}$ implies $c^2 = \frac{m^2}{n^2}$ which is again a ratio of integers and $n^2 \neq 0$), at no point did you use the fact that $\frac{m}{n}$ was written in lowest terms. Since that was never used, you should remove it. It was true, but not needed.

[9] And that's coming from a guy who is vigorously promoting "long-form" math.

[10] or five or six or ...

[11] I wish you could see how many small edits it took to arrive at the final version of this paragraph!

When one reads a confusing proof with extraneous statements, it can feel like trying to solve a messy puzzle. You want your proofs to be clear, not a jumbled mess to decipher. Can you imagine trying to solve a 500-piece jigsaw puzzle which includes 700 pieces in it, and you have to identify and remove the 200 extraneous pieces in order to complete the puzzle? You don't want your proofs to feel like that. Include no red herrings in your story.

Break up complicated proofs into parts

If a proof is longer than a couple paragraphs, it is easy for a reader to get lost in your argument. Thus, it often helps to highlight the big structure of the problem.

For example, in our induction proofs we indicated clearly when we were doing our base case, inductive hypothesis, induction step and conclusion (e.g., the proof of Proposition 4.4); in our proofs by cases, we indicated clearly when we were doing Case 1, Case 2, etc. (e.g., the proof of Proposition 2.7); in some of our proofs of if-and-only-if statements we indicated clearly when we were doing the forward direction and the backward direction (e.g., the proofs of Proposition 6.3 and Theorem 9.5); and in some of our more complicated proofs, we broke up the argument into parts, and explained how those parts fit together (e.g., the proof of Theorem 2.13).

There were also many times where the first or second paragraph of the proof was simply an explanation of what was to come, containing the big picture without any of the details (e.g., the proof of Lemma 9.10). This foreshadowing can really help the reader understand what you are doing and why.

Don't stop with your first draft

I've said it before, but it is worth repeating: Writing is a process, and your first draft should not be your last draft. Look back over your work and search for ways to improve it. Was your notation optimal? Is each part clear enough? Is there a sentence that was never used and can be removed? Could it be reorganized or restructured in a way to make it clearer?

Miscellaneous Advice

- When using a word which has a mathematical definition, take extra care that you are using it correctly. You must say what you mean, and mean what you say.[12]

- In math, it is common to use "we" when writing proofs, rather than "I." You will see passages like, "by plugging in $x = 2$, we see that f is discontinuous." Research articles with a single author will even use "we," even though it is clear they did the work by themselves. This can help create an invitational tone, where you are bringing the reader along with you.[13] Even if you solved a homework problem by yourself, you may use "we."

[12] Say it if and only if you mean it.
[13] Note: Most STEM fields do not do this.

- When possible, avoid using "it" in your writing. When you have used several nouns up to that point, it can be hard to know what "it" is referring to. For example, suppose I wrote "Since $f'(x) = g(x)$, it must be continuous." What does "it" refer to here? The function f or the function g? There are times when the reader could work harder to deduce what "it" must be referring to, but don't make them do so. Make it easy on your readers. In our example, it would be better to write "Since $f'(x) = g(x)$, we see that f is continuous."

- Mathematicians love short words to begin a new sentences. Examples: Therefore, thus, hence, consequently, however. It is good to mix up which of these you use. Doing so doesn't change the math, but it makes for a more engaging read for your readers. We even use "and" and "but" to start a sentence more than in most writing, because ending a sentence can help the reader segment off an idea in their head, separating it more clearly from the idea that you are sticking in the next sentence.

- When you took geometry in high school, you may have written proofs as two columns, where on the left you wrote a single sentence or statement, and on the right you justified it. You should not do this. All proof-based math is written in paragraph form.

- Did I mention that clarity is the most important thing? These rules are here to help you write clearly. If they get in the way of clarity, you may break them.

- Communication of all kinds is cultural, and this includes math communication. I have shared some generally-adopted practices of math writing, but these practices may depend on where you work or study, and who is reading your proofs. Be aware and responsive to the culture in which your work lives.

- It is good to thank those who contributed to your work. Books have editorial boards, and research articles have acknowledgments sections.[14] Indeed, the idea for this very bullet point came from the article *Some Guidelines for Good Mathematical Writing*, authored by Francis Su and published in the newsmagazine *MAA Focus*.

C.2 Writing in LaTeX

All formal math writing is done in a program[15] called LaTeX. This book was written in LaTeX, every other math book you have read was likely written in LaTeX, and the exams your math professors have given you were very likely written in LaTeX. It

[14]Deciding who to acknowledge can be difficult. In an age where there are so many resources that you can pull ideas from, it is impossible to thank every article, YouTube video, talk, tweet or personal conversation which may have inspired a sentence here or a paragraph there, but you should at least cite those whose work you directly use or those who reviewed your work. If you are wondering what level of assistance on your homework deserves an acknowledgment... ask your professor.

[15]To be pedantic, it is technically a "language" or a "software system."

is used for every math journal in the world, both for the submissions and the final journal publication. I once read a blog post from the editor of a math research journal, who wrote about how many crank math papers they receive each year. He said that if they read every paper from someone claiming to have solved some centuries-old conjecture, they would spend all their time on such nonsense. How do they quickly spot the submissions that can be immediately discarded? Their number 1 rule: If it is not written in LaTeX, discard it.[16]

In fact, LaTeX is nearly universal throughout the STEM fields. I once even received an email from a professor of deaf studies at my university who was writing a paper for a deaf studies journal which only accepted papers written in LaTeX. She was pretty knowledgeable already, but she had some technical questions for me about the software.

It is well worth your time to learn LaTeX. I first learned it as an undergrad. One semester I chose one of my math classes and decided to type up all of my homework for that class in LaTeX. This was a lot of extra work, as it takes time to learn. Yet with just one semester of this, I was already typing up most of my homework about as fast as it would have taken to handwrite it. From that point on, I typed up all my homework.[17] This time investment paid off big in graduate school, when LaTeX proficiency is necessary. I encourage you to try it out.

For instructions on how to write in LaTeX, and templates that you can use for your homework, check out this book's page on the `LongFormMath.com` website.

[16]No matter how good Microsoft Equation Editor gets, it will forever send shudders down the backs of mathematicians of a certain age.

[17]Which is still stored on my computer to this day! Mistakes and all!

Index

Notation:

$a \mid b$, a divides b, 55
$a \nmid b$, a does not divide b, 55
$a \equiv b \pmod{m}$, a is congruent to b mod m, 62
$\gcd(a, b)$, the greatest common divisor of a and b, 59
$\text{LCM}(a, b)$, the least common multiple, 85
$\binom{n}{k}$, the binomial coefficient, n choose k, 178, 435
$n!$, the factorial of n, 155
$f : A \to B$, a function f from A to B, 332
$|x|$, the absolute value, 83
$\phi(N)$, the Euler totient function, 93
$\pi(N)$, the prime counting function, 90
$R(t)$, the Ramsey number, 46, 425
$f \circ g$, function composition, f of g, 347
$\lfloor x \rfloor$, the floor function of x, 334
$i_A : A \to A$, identity function on A, 352
$f(X)$, the image of X, 366
$f^{-1}(Y)$, inverse image of Y, 366
f^{-1}, inverse function, 352
$W(r, k)$, Van der Waerden's function, 411
$E[X]$, expected value, 424
$P \vee Q$, P or Q, 209
$\sim P$, not P, 209
$P \wedge Q$, P and Q, 209
$\Leftrightarrow$,if and only if, 210
$\Rightarrow$, implies, 52, 210
$\exists$, there exists, 220
$\nexists$, there does not exist, 220
$\forall$, for all, 220
ϕ, the golden ratio, 205
$\sim$, relation, 383
$[a]$, equivalence class of a, 387
$\lesssim$, partial order, 398
$[a, b]$, the closed interval from a to b, 101
(a, b), the open interval from a to b, 101
$(a, b]$ and $[a, b)$, the half open intervals from a to b, 101
$\emptyset$, the empty set, 97
$\mathbb{N}$, the set of natural numbers, 97
$\mathbb{N}_0$, the set of nonnegative integers, 99
$\mathbb{Q}$, the set of rational numbers, 100
$\mathbb{R}$, the set of real numbers, 100
$\mathbb{R}^2$, the set of ordered pairs of real numbers, 100
$\mathbb{Z}$, the set of natural numbers, 97
S^1, the unit circle, 100
$|A|$, the cardinality of A, 109
$A \times B$, the Cartesian product of A and B, 109
A^c, set complement, 108
$\mathcal{P}(A)$, the power set of A, 109
$A \triangle B$, the symmetric difference of A and B, 132
$A = B$, set equality, 105
$x \in S$, x is an element of S, 97
$A \cap B$ and $\bigcap_{i=1}^{n} A_i$, union of sets, 106
$A \setminus B$, set subtraction, 108
$A \subseteq B$, A is a subset of B, 101

$A \cup B$ and $\bigcup_{i=1}^{n} A_i$, union of sets, 106
K_n, the complete graph, 41
D_{2n}, the dihedral group, 416
$\mathrm{GL}_2(\mathbb{R})$, general linear group, 420
S_n, symmetric group, 419
□, ■, ▯, ▮, , Q.E.D. symbols, 30

Topics:

Absolute value, 83
Algebraic number, 308
AM-GM inequality, 75
And, logical connective, 209
Antimagic square, 36, 193
Antisymmetric, 398
Aperiodic perfect covering, 38
Archimedean principle, 193
Arctangent function, 354
Associativity, 415
Averaging principle, 424

Bézout's identity, 59
Biconditional statement, 212
Big data, 285
Bijection principle, 372
Bijective function, 337
Binary, 176
Binary code, 433
Binary operation, 413
Binary representation, 177
Binomial coefficient, 435
Binomial symmetry, 436
Binomial theorem, 178
Bridges of Königsberg, 462

Cantor's diagonalization argument, 376
Cardinality, 109, 371
Cartesian product of sets, 110
Center of mass, 256
Chessboard problems, 2
Chomp game, 329
Class, 383
Closed interval, $[a, b]$, 100
Closure, 415
Closure in the integers, 48
Coding theory, 433
Codomain, 332
Collatz conjecture, 282
Combinatorial method, 434
Combinatorics, 371
Common divisor, 59
Common multiple, 85
Complete graph, 41
Composite number, 67
Computer science, 311
Conditional statement, 212
Congruence modulo m, 62
Conjecture, 10
Continuous function, 253
Contrapositive, 225, 262
Converges, 234
Converse, 212
Coprime, 33
Corollary, 10
Counterexample, 11, 269
Cryptography, 91

Data science, 285
De Morgan's law, general form, 190
De Morgan's laws, 112
De Morgan's logic laws, 216
Definitions, 48
Dense, 134
Dihedral group, 417
Divisibility, 55
Division algorithm, 58, 315
Domain, 332
Double counting method, 435

Edge, 24
Eigenvalue, 288
Eigenvector, 288
Element of a set, 97
Empty set, $\emptyset$, 97
Encryption, 92
End of proof symbols, 30
Equivalence class, 383
Equivalence relation, 383
Erdős, Paul, 20, 46, 453
Escape game, 464

Euclid, 302
Euler brick, 323
Euler's theorem, 94
Galois, Évariste, 400
Even integer, 48
Existence and uniqueness, 332, 360
Existential quantifier, 220
Exponential function, 354
Exponential growth, 205
Exponentiation, 271
Extremal problem, 432

Factorial, 155
Family of sets, 118
Fermat number, 189, 460
Fermat's little theorem, 71, 180
Feynman technique, 80
Fibonacci sequence, 200
Floor function, 334
Func-y pigeonhole principle, 345
Function, 331
Function as a subset, 393
Function composition, 347
Function definition, 332
Function inverse, 352
Function, relation definition, 394
Fundamental theorem of arithmetic, 165, 314

Game, 326
Game theory, 325
Gauss, Carl Friedrich, 153
Generating functions, 203
Goldbach's conjecture, 90
Golden ratio, 205
Google PageRank, 288
Graph, 24, 334, 462
Greatest common divisor, 59
Group, 415

Halmos tombstone, 4
Halting problem, 311
Hamming distance, 433
Harmonic series, 174, 258
Hasse diagram, 399
Hilbert's hotel, 372
Hilbert, David, 377
Hockey stick property, 439
Horizontal line test, 340

Identity element, 415
Identity function, 352
If and only if, logical connective, 210
If, then statements, 51
Image, 366
Implies, logical connective, 210
Incidence vectors, 431
Indexed families of sets, 118
Induction, 147
Induction, multiple base cases, 171
Infinitude of primes, 300, 461
Injective function, 335
Integers, $\mathbb{Z}$, 97
Intersection of sets, 106
Inverse image, 366
Invertibility, 415
Invertibility of functions, 352
Irrational number, 305
Isomorphic groups, 420

k-intersecting sets, 429
Klein bottle, 145

LaTeX, 493
Least common multiple, 85
Lemma, 10
Limit, 234
Linear algebra method, 429
Logic, 207
Logical 'and', $\wedge$, 209
Logical 'if and only if', $\Leftrightarrow$, 210
Logical 'implies', $\Rightarrow$, 210
Logical 'not', $\sim$, 209
Logical 'or', $\vee$, 209
Logical equivalence, $\Leftrightarrow$, 215
Lone vertex, 24

Magic square, 36, 193, 322
Manifolds, 144
Mantel's theorem, 183
Markov chains, 292
Mathematical arguments, 1

Mersenne prime, 87
Minimax theorem, 328
Mixed strategy, 328
Möbius strip, 144
Mod, 62
Modular arithmetic, 62
Modular arithmetic properties, 65
Modular cancellation law, 70
Monochromatic graph, 43
Murphy's law, 80
Mutually prime, 33

Nash equilibrium, 327
Nash, John, 327
Natural logarithm, 354
Natural numbers, $\mathbb{N}$, 97
Negation, 222
(Non)sandwich, 47
Not, logical connective, 209

Odd integer, 48
One-to-one function, 335
Onto function, 336
Open interval, (a, b), 101
Open question, 38
Open sentence, 209
Open set, 140
Or, logical connective, 209
Other proof methods, 423

Paradox, 228
Partial order, 398
Partition, 380
Pascal's rule, 436
Pascal's triangle, 437
Penrose tiling, 39
Perfect cover, 3
Perfect cover of the plane, 38
Perfect number, 87
Periodic perfect covering, 38
Permutation, 365
Pigeonhole principle, 14
Pigeonhole principle, function version, 345
Power set, 109
Prime number, 67
Prime number theorem, 90
Probabilistic method, 424
Proof by cases, 53
Proof by contradiction, 293
Proof by contrapositive, 261
Proof by induction, 148
Proof by minimal counterexample, 314
Proof by optical illusion, 476
Proof structure: contradiction, 297
Proof structure: contrapositive, 264
Proof structure: direct, 52
Proof structure: induction, 149
Proof structure: injection, 339
Proof structure: set equality, 105
Proof structure: strong induction, 165
Proof structure: subset, 102
Proof structure: surjection, 339
Proofs from The Book, 453
Proper divisor, 87
Proposition, 10
Pythagorean theorem, 455
Pythagorean triple, 323

Quantifiers, 219
Quotient, 58

Ramsey number, 46, 425
Ramsey theory, 41, 425
Ramsey, Frank, 41
Range, 332
Rational numbers, $\mathbb{Q}$, 100
Real analysis, 253
Real numbers, $\mathbb{R}$, 100
Recursion, 500
Reduced form, 85
Reductio ad absurdum, 293
Reflexive, 383
Relation as a subset, 393
Relation from A to B, 392
Relation on A, 383
Relatively prime, 33
Remainder, 58
RSA algorithm, 92
Russell, Bertrand, 207

Sequence, 199

Sequence convergence, 234
Sequence, recursively defined, 191
Series, 258
Set, 97
Set complement, 108
Set subtraction, 108
Set-builder notation, 99
Similar matrices, 396
Similar triangles, 386
Space-filling curve, 104
Statement in logic, 208
Strategy-stealing argument, 329
Strong induction, 164
Subset, 101
Sum-free set, 426, 427
Surjective function, 336
Symmetric, 383
Symmetric difference, 132
Symmetric group, 419

Tangent function, 354
Tautology, 210
Tetris, 31
The Book, 453
The Math Castle, 29
Theorem, 10
Tiling, 159
Tofu, Porco and Dragon, 125
Topology, 137, 142
Transcendental number, 308
Transitive, 383
Tree, 194
Triangular numbers, 149
Truth table, 214

Undecidable problem, 313
Union of sets, 106
Universal lossless compression algorithm, 356
Universal quantifier, 220
Universal set, 108

Vacuously true, 217
Van der Waerden number, 411
Venn diagram, 107
Vertex, 24
Vertical line test, 334

Writing advice, 487

Zeno's paradox, 229

Printed in Great Britain
by Amazon